T0234378

A Modern Approach to Classical Mechanics

Second Edition

A Modern Approach to Classical Mechanics

Second Edition

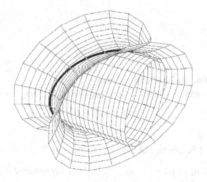

Harald Iro

retired from

Institute for Theoretical Physics, Johannes Kepler University Linz, Austria

World Scientific

NEW JERSEY · LONDON · SINGAPORE · BEIJING · SHANGHAI · HONG KONG · TAIPEI · CHENNAI

Published by

World Scientific Publishing Co. Pte. Ltd.

5 Toh Tuck Link, Singapore 596224

USA office: 27 Warren Street, Suite 401-402, Hackensack, NJ 07601

UK office: 57 Shelton Street, Covent Garden, London WC2H 9HE

Library of Congress Cataloging-in-Publication Data
Iro, Harald, author.
    [Mechanik. English]
    A modern approach to classical mechanics / Harald Iro, Johannes Kepler
University Linz, Austria. -- Second edition.
        pages cm
    Includes bibliographical references and index.
    ISBN 978-9814696289 (hardcover : alk. paper) -- ISBN 978-9814704113 (pbk : alk. paper)
    1. Mechanics. 2. Statistical mechanics. I. Title.
    QA805.I7613 2015
    531--dc23
                          2015011461

British Library Cataloguing-in-Publication Data
A catalogue record for this book is available from the British Library.

Copyright © 2016 by World Scientific Publishing Co. Pte. Ltd.

All rights reserved. This book, or parts thereof, may not be reproduced in any form or by any means, electronic or mechanical, including photocopying, recording or any information storage and retrieval system now known or to be invented, without written permission from the publisher.

For photocopying of material in this volume, please pay a copying fee through the Copyright Clearance Center, Inc., 222 Rosewood Drive, Danvers, MA 01923, USA. In this case permission to photocopy is not required from the publisher.

In-house Editor: Christopher Teo

Printed in Singapore

From the Preface of the first edition

The discovery of chaotic behavior in nonlinear dynamical systems is the third great revolution in physics in the twentieth century – after the theory of relativity and quantum mechanics. Even though, as early as the turn of the 20th century, H. Poincaré had written papers on the predictability of natural phenomena – and even of the universe at large – it wasn't until the 1960s, with the onset of easily available computing resources, that clear evidence of nonpredictability appeared. After that, the theory of chaotic behavior leaped forward, became a part of every branch of science where dynamical systems are studied mathematically. The evidence of nonpredictability also stimulated a great change in the field of classical mechanics itself. ...

Chaotic motion should not be banished as a mere curiosity to some small part of the book. The reader or student ought to be acquainted with chaotic behavior at an early stage of their study of classical mechanics. She or he should be able to understand when and why a physical system may behave chaotically. ...

Of course, the ideas and methods of classical mechanics are developed in this book also largely within conventional, integrable systems; but the concepts of phase space, first integral and conserved quantity are stressed right from the beginning. Linear stability analysis is introduced as a first tool to investigate the stability of orbits. With these concepts at hand, chaotic behavior in nonlinear systems is discussed early on in the treatment. Moreover, topics that are relatively uncommon in presentations of classical dynamical systems – such as a particle in a homogeneous magnetic field; various cases of the spinning top; the problem of two centers of force; and the restricted three-body problem – are considered. The general conditions for integrability of a dynamical system are presented in the framework of Hamilton-Jacobi theory, and we touch on the stability of planetary motion. Canonical perturbation theory leads finally to the KAM theorem on motion in a system that is nearly integrable....

Linz, June 2002

Preface to the second edition

When asked by World Scientific Publishers to prepare a second edition I checked my book for the first time after several years. And I found it worthwhile to remove errors, make changes and polish the text and the structure. Since, in the meantime I became more and more interested in the history of mechanics, I also took the opportunity to improve and amplify the presentation of the historical background. But the main goal remains the same: To stress the possibility of chaotic behavior in nonlinear systems by investigating the integrability or non-integrability of the systems discussed. Concerning the last point in the book, the stability of the solar system, the question still is: How long will the solar system maintain its familiar configuration?

Problems and examples are included at the end of each chapter; in the text, the symbol (E) denotes that a related exercise appears at the end of the chapter. Often references to books listed in the Bibliography are also denoted [author/s]. When proper names of people appear in the index, the page number given refers to the biographical data. The mathematical appendices are not intended to give an introduction to the respective topics, but serve merely as compilations and references.

I thank Christopher Teo from World Scientific for his assistance in preparing this second edition. And still I remember gratefully the support and help of Prof. Bernhard Schnizer, Yurij Holovatch, and John Wojdylo in editing the first edition and thus also contributing to this one.

Linz, Summer 2015 Harald IRO

Contents

1 Basic considerations and concepts   **1**

1.1   Why classical mechanics is still challenging   1

1.2   The birth of classical mechanics   3

1.3   Observations and the resulting pictures   5

1.4   Time, space and motion   9

     1.4.1   Newton's concepts   9

     1.4.2   The mathematical pictures of space and time . .   11

     1.4.3   Kinematics   12

2 Foundations of classical mechanics   **15**

2.1   Mass, quantity of motion, and force   15

2.2   Newton's laws   17

2.3   Analytical mechanics   19

     2.3.1   The basic equations of mechanics   20

     2.3.2   Point masses and forces   22

2.4   Constants of motion   23

     2.4.1   Constants of motion and conserved quantities . .   23

     2.4.2   Conservation of energy   27

     2.4.3   Angular momentum and its conservation   31

3 One-dimensional motion of a particle   **35**

3.1   Examples of one-dimensional motion   35

3.2   General features   37

3.3   Back to the examples   42

     3.3.1   The inclined track   42

     3.3.2   The plane pendulum   43

     3.3.3   The harmonic oscillator   46

3.4 The driven, damped oscillator 49
    3.4.1 The driven oscillator with linear damping 50
    3.4.2 The periodically driven, damped oscillator 52
3.5 Stability of motion 55
    3.5.1 Two examples 55
    3.5.2 Linear stability analysis 57
3.6 Anharmonic one-dimensional motion 63

4 Peculiar motion in two dimensions   **71**
4.1 The two-dimensional harmonic oscillator 71
4.2 The Hénon-Heiles system 79
4.3 A 'useless' conserved quantity 86
4.4 Chaotic behavior 89
4.5 Laplace's clock mechanism does not exist 97

5 Motion in a central force   **101**
5.1 General features of the motion 102
    5.1.1 Conserved quantities 102
    5.1.2 The effective potential 105
    5.1.3 Properties of the orbits 108
5.2 Motion in a $1/r$ potential 110
    5.2.1 The case $L \neq 0$ 110
    5.2.2 Bounded motion for $L = 0$ 117
5.3 Motion in the potential $V(r) \propto 1/r^\alpha$ 118
    5.3.1 Mechanical similarity 120
5.4 The Runge-Lenz vector 121
5.5 Integrability vanishes 125
    5.5.1 The homogeneous magnetic field as the sole force 126
    5.5.2 Addition of a central force 133
    5.5.3 Motion in the symmetry plane 134

6 Gravitational force between two bodies   **139**
6.1 Two-body systems 139
    6.1.1 Center of mass and relative coordinates 143
    6.1.2 Conserved quantities 145
6.2 The gravitational interaction 147
6.3 Kepler's laws 151
    6.3.1 Beyond Kepler's laws 159

6.4   Gravitational potential of large bodies 160
       6.4.1   The potential of a homogeneous sphere 161
       6.4.2   The potential of an inhomogeneous body 164
6.5   On the validity of the gravitational law 166

7 Collisions of particles. Scattering   171
7.1   Unbounded motion in a central force 171
7.2   Kinematics of two-particle-collisions 176
       7.2.1   Elastic collisions of two particles 177
       7.2.2   Kinematics of elastic collisions 179
7.3   Potential scattering 185
       7.3.1   The scattering cross section 186
       7.3.2   Scattering in the $1/r$ potential 189

8 Changing the frame of reference   195
8.1   Inertial frames 197
8.2   Changing the inertial frame 198
8.3   Linear transformations of the coordinates 200
       8.3.1   Translation of the coordinate system 200
       8.3.2   Rotation of the coordinate system 201
8.4   The Galilean group 205
       8.4.1   Transformation of forces 207
8.5   Transformations to non-inertial frames 209
       8.5.1   Accelerated frames of reference 209
       8.5.2   Rotating frames of reference 211
       8.5.3   Motion in a rotating frame 216

9 Lagrangian mechanics   223
9.1   Constrained motion 223
9.2   Calculus of variations 231
       9.2.1   The Euler-Lagrange equation 233
       9.2.2   Transforming the variables 238
       9.2.3   Constraints 240
9.3   The Lagrangian 243
       9.3.1   Inverse problem in the calculus of variations . . . 243
       9.3.2   Inverse problem for Newton's equation of motion 244
       9.3.3   The Lagrangian for a single particle 247
       9.3.4   Hamilton's principle 249

9.3.5 The Lagrangian in generalized coordinates 250
9.3.6 Further applications of the Lagrangian 255
9.3.7 Nonuniformly moving frames of reference 256

10 Conservation laws and symmetries   **261**
10.1 Equations of motion for N point masses 261
10.2 The conservation laws 265
    10.2.1 The motion of the center of mass 265
    10.2.2 Conservation of angular momentum 267
    10.2.3 Conservation of energy 271
10.3 The Lagrangian of a system of N particles 275
10.4 Infinitesimal transformations 278
    10.4.1 Infinitesimal translations of time 278
    10.4.2 Infinitesimal coordinate transformations 279
    10.4.3 Galilean transformations and constants of motion 282

11 The rigid body   **287**
11.1 Degrees of freedom of a rigid body 288
11.2 Some basics of statics 289
    11.2.1 Historical survey 289
    11.2.2 The basic physical principles 290
    11.2.3 Simple machines 292
11.3 Dynamics of the rigid body 296
    11.3.1 Historical landmarks 296
    11.3.2 The motion of a rigid body 298
    11.3.3 The inertia tensor 303
    11.3.4 Euler's equations of motion 309
    11.3.5 The motion of a spinning top 317
    11.3.6 The symmetric spinning top 324

12 Small oscillations   **337**
12.1 The double pendulum 337
12.2 The harmonic approximation 341
    12.2.1 The general theory 341
    12.2.2 The double pendulum (again) 346
    12.2.3 Vibrations of a triatomic molecule 349
12.3 From linear chain to vibrating string 357
    12.3.1 The vibrating string 361

13 Hamiltonian mechanics   **365**
  13.1 Hamilton's equations of motion 365
      13.1.1 A particle in a central force field 368
      13.1.2 The rigid body 369
      13.1.3 Central force and homogeneous magnetic field . . 371
  13.2 Poisson brackets 377
  13.3 Canonical transformations 380
      13.3.1 The generating function of a transformation . . . 381
      13.3.2 Canonical invariants 386
      13.3.3 Infinitesimal canonical transformations 387
  13.4 Symmetries and conservation laws 390
  13.5 The flow in phase space 392

14 Hamilton-Jacobi theory   **397**
  14.1 Integrability 397
      14.1.1 Liouville's theorem on integrability 398
      14.1.2 Sketched proof of the theorem 400
  14.2 Time-independent Hamilton-Jacobi theory 403
      14.2.1 The Hamilton-Jacobi equation 403
      14.2.2 Separation of variables 405
  14.3 The problem of two centers of gravity 411

15 Three-body systems   **419**
  15.1 The restricted three-body problem 420
  15.2 Solutions of the problem 426
  15.3 Is our planetary system also chaotic? 434

16 Approximating non-integrable systems   **439**
  16.1 Action-angle variables 439
      16.1.1 Definition and general properties 439
      16.1.2 Transforming to action and angle variables . . . 443
  16.2 Dynamics on the torus 452
  16.3 Canonical perturbation theory 455
      16.3.1 The one-dimensional anharmonic oscillator . . . 459
      16.3.2 First order corrections 462
  16.4 The KAM theorem 463
  16.5 Is the solar system stable? 467
      16.5.1 A few historical landmarks 467

16.5.2 On the stability of planetary orbits 469

In retrospect   **475**

Appendix   **477**

A Coordinates; vector analysis   **477**
  A.1 The Euclidean space $\mathbb{E}^3$ 477
  A.2 Cartesian coordinates 479
  A.3 Orthogonal, curvilinear coordinates 481
     A.3.1 General relations 481
     A.3.2 Spherical coordinates 483
     A.3.3 Cylindrical coordinates 485

B Rotations and tensors   **487**
  B.1 Rotations 487
  B.2 Tensors 490

C Green's functions   **493**
  C.1 The Dirac δ-function 493
     C.1.1 Distributions 493
     C.1.2 The δ-function 494
  C.2 Fourier transforms 495
  C.3 Linear differential equations 497
     C.3.1 The Green's function 497
     C.3.2 The equation of the damped oscillator 498

Bibliography   **503**

Index   **509**

1

Basic considerations and concepts

1.1 Why classical mechanics is still challenging

In the latter half of the 19th century, physics was widely considered to be complete. The basic pillar of the physics was Newtonian Mechanics, and this was augmented by the theories of Lagrange, Hamilton and Jacobi. So at this time, increased focus was directed at the epistemology of physics, particularly mechanics. E. MACH[1], to mention just one prominent thinker, thoroughly expounded epistemological aspects of mechanics, in his book, "The Science of Mechanics – a Critical Account of its Development" (see Bibliography).

At the beginning of the 20th century two epistemological issues proved to be very fruitful for the future of physics: The existence of the ether and the existence of atoms. The former was settled by Einstein's theory of relativity, partly motivated by Mach's criticism of Newton's absolute space, and the second found its resolution in quantum mechanics.

But there was also a further issue at the turn of the century, one that was to have a major impact on physics only much later. The question of the stability of planetary motion was studied in particular by H. POINCARÉ[2]. In 1889 he submitted to the Swedish Academy a prize-

---

[1]Ernst Mach (1838-1916), Austrian physicist and epistemologist.

[2]Henri Poincaré (1854-1912), French mathematician, physicist, and philosopher.

winning paper on the stability of the motion of three gravitationally interacting bodies. In later work, he concluded that the stability question is of a fundamental nature. In "Science and Method" he anticipated an essential characteristic of chaotic behavior: '*small differences in the initial conditions*' may '*produce very great ones in the phenomena*'.

Half a century passed before the conception of classical mechanics changed fundamentally in the eyes of the majority of physicists[3]. This occurred with the appearance of chaotic behavior in deterministic equations such as those of Newton. It all started with E. Lorenz's observation in 1963 that a simplified atmospheric model – derived from hydrodynamic equations and consisting of three coupled nonlinear first-order ordinary differential equations – turned out to show quite different numerical solutions for extremely tiny changes in the initial conditions. This discovery precipitated a tremendous amount of research into dynamical systems worldwide. We mention only two research fields directly related to classical mechanics: chaotic behavior in a particular nonlinear two-dimensional Hamiltonian system and chaos in the restricted three-body systems with gravitational interactions. We discuss both these systems in detail later. At the same time the celebrated Kolmogorov-Arnold-Moser (KAM) theorem (1954-63) solved Poincaré's convergence problem in the power series treatment of dynamical systems that are not exactly solvable.

As a result of this discovery, the generally accepted presumption of **predictability of mechanical systems** was overturned. At least since P. S. LAPLACE[4] expressed that predictability[5] in 1814, it had been universally believed that given all the initial conditions, and given sufficiently powerful calculational tools, one could predict the future state of any classical system. This is still *true in principle*, but the fact remains that due to extreme sensitivity to changes in the initial conditions, it is *practically not possible* to predict the future for many systems, since the initial values are always only known to a certain accuracy.

The new picture also has implications for other fields of physics. For example, in statistical physics, it sheds new light on ergodic theory,

---

[3]Starting with Poincaré mainly mathematicians were aware that the solutions of the fundamental equations may show a new behavior.

[4]Pierre Simon Laplace (1749-1827), French mathematician and physicist.

[5]For Laplace's statement see Chapter4, Section 4.5.

and allows a new understanding of the arrow of time – that is, the apparent irreversible direction of time in the macroscopic world, despite the microscopic time reversibility of the fundamental laws.

Chaotic behavior is ubiquitous even in rather simple mechanical systems. Most textbooks on classical mechanics only consider the small minority of so-called integrable systems. In the light of the 'chaos revolution', this cannot be considered adequate in a modern approach. Therefore, we introduce concepts and tools necessary to understand integrability and chaotic behavior quite early in the treatment, along with examples of chaotic systems. Presenting this modern view of classical mechanics is our chief goal.

1.2 The birth of classical mechanics

It was an outstanding genius who gave rise not only to mechanics but to physics as the science where starting from basic laws the observed phenomena are derived: I. NEWTON[6]. Newton's "Philosophiae naturalis principia mathematica" (Mathematical principles of natural philosophy, London 1687; "Principia" for short) terminated a period of about two millennia[7]. Newton's 'Magnum Opus'[8] replaced the dicta of antiquity and the subsequent suppositions about the nature of motion, its properties and causes, by a few axioms, exploited their consequences and compared these successfully with physical reality. Classical mechanics is based on the central axioms (or laws) of the "Principia".

---

[6]Isaac Newton (1642-1727), English physicist and mathematician. Newton is one of the most important and influential scientists of all time.

An often-quoted biography of Newton is R. Westfall's "Never at Rest". Also recommendable is the perspective on his life and work presented in "Let Newton be!" (see the Bibliography for both).

[7]Of course there is a series of preceding achievements in the field we call today classical mechanics (see e.g. the comprehensive work of R. Dugas "A history of mechanics"). The situation is properly described by Newton's dictum in a letter to R. Hooke: 'If I have seen further it is by standing on the shoulders of Giants.

[8]Newton's main works are the "Principia" and the "Opticks". Also famous is his mathematical contribution to the field of analysis. Actually he, together with Leibniz, laid its foundation stone (see the end of this chapter). But during his life time he published only small accounts of his theories of fluxions and of the quadrature of curves.

Little known, but nevertheless a part of his personality, are his activities in alchemy and chronology.

There were three editions of the "Principia" (1687, 1713, 1726) followed by translations into English (1727 by Andrew Motte), French (1759 by the Marquise Émilie du Châtelet), German (1871 by Jakob Ph. Wolfers), and Russian (1915/16 by Aleksei N. Krylov). The latter three include also instructive commentaries and additions. Another remarkable Latin edition is the so-called 'Jesuit' edition by François Jaquier and Thomas Le Seur (Geneva 1739-42) which provides a huge amount of valuable annotations doubling the volume of the "Principia".

The "Principia" starts with two sections followed by three books. The first section entitled "Definitions" deals with the concepts of mass, motion, force, time, and space. In the second section "Axioms, or Laws of Motion" the famous, three basic laws are presented.

Then follows the first book "Of the Motion of Bodies" dedicated to the motion of a body subject to central forces. In its first section Newton introduces eleven important Lemmata needed for his proofs[9]. Afterwards he starts discussing various aspects of the motion of a body subject to a 'centripetal force' (i.e. central force, see Eq. 2.31). Subsequently he considers bodies mutually attracting each other with central forces, switching then to forces between solid bodies[10].

The second book, also entitled "Of the Motion of Bodies", treats the motion in resisting media. Fluids and the motion of bodies in a fluid are discussed. In particular the circular motion of a fluid is examined, thereby disproving R. DESCARTES'[11] vortex theory.

---

[9]These Lemmata serve to introduce, in disguise, the calculus in Newton's deductions. They present the notions of vanishing small quantities and their convergence to a limit. Curves are approximated by sequences of polygons tending in the limit of increasingly small steps to the curves (like Archimedes's calculation of the area of a parabolic section nearly two thousand years ago). Then the main part of the Lemmata treats the relationship between an arc a, its chord c, and the tangent t (cf. the Figure below). In the limit of vanishing size these three become equal.

[10]Apparently so far Newton has point-like bodies in mind without stating it explicitly.

[11]René Descartes (1596-1650) advocated in his book "Principia Philosophiae" that the planets are carried around the sun by the vortex motion of a subtle celestial

The third book "On the System of the World", i.e. the planetary system, contains the law of the gravitational force and is devoted to celestial mechanics. Topics are: The shape of the earth and the weight of bodies in different regions of the earth, moon's motion and its inequalities, the tides, and comets[12]. The motion of the moon is considered as a three body system (sun, earth, and moon) and observed time dependencies of the parameters of the lunar orbit are calculated.

Before we proceed, following and extending Newton's ideas, we consider some general premises.

1.3 Observations and the resulting pictures

The laws of physics inevitably have a mathematical form, since physics aims to be quantitative, and even precise. But behind the mathematical form stands concepts and reasoning. A (theoretical) physicist's everyday routine is mainly concerned with mathematics. But particularly when new theories are formed or old ones are scrutinized, it is the concepts and their understanding that is important. What is the relation between physical reality and its mathematical image?

Everyone who observes, conceptualizes or even changes the surrounding nature develops an imagination that goes beyond simple cognition of the environment. They construct pictures of nature – for example, suppositions are made about connections between processes; conclusions are drawn about influences between processes. A picture of the world emerges that often also includes metaphysical currents, perhaps even religious ones. Physics is restricted to the rational part of the world view. We quote below views held by outstanding physicists on the relation between the exterior world and the individual's knowledge of the world. Presumably, most physicists share these views. The selection is far from complete: it can only touch on the epistemology of mechanics

---

matter. In Descartes' view there is no void; space **is** the presence of this subtle matter.

At least in this respect Newton's "Principia" are a reaction to Descartes' work. In the full title of Newton's "Principia" the words *Principia* and *Philosophiae* were printed in red.

[12]Newton published this part in a separate version not depending on the first two books. Its English translation "A Treatise on the System of the World" appeared 1728 in London.

(and physics). Our selection attempts to illustrate the philosophy basic to this book.

In the following, we refer to introductions of two important books on the principles of mechanics – written by H. HERTZ[13] and L. BOLTZMANN[14] at around the turn of the 20th century (see Bibliography). Both these physicists are certainly among the most prominent physicists of that era.

In his book on the principles of mechanics, Hertz starts with a model of the epistemological process in physics[15]:

> *As a basis ... we make use of ... preceding experiences, obtained by an accidental observation or by an intentional experiment. The method, however, that we constantly use to deduce the future from the past, thereby obtaining the desired foresight, is this: we attribute inner images or symbols to the exterior objects; and this we do in such a manner that the consequences of the images necessary by thought are just the images of the consequences necessary by nature of the objects represented. In order that this demand can be fulfilled, certain reconciliations (accords) between nature and our mind have to exist. Experience teaches us that this demand can be fulfilled and that hence such accords exist.*

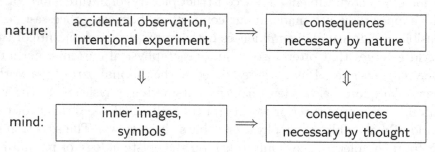

Hertz's scheme of the process of gaining knowledge

In his lectures on the principles of mechanics, Boltzmann refers to Hertz's view, and presents his own:

[13]Heinrich Hertz (1857-1894), German physicist.

[14]Ludwig Boltzmann (1844-1906), Austrian physicist.

[15]In the following, all citations that do not refer to an English edition were translated by the present author.

It has hardly ever been doubted ..., that our thoughts are mere images of the objects (rather symbols for these) having at most a certain kinship to them, but can never coincide with them and can be related only to them like the letters to the sound or the note to the tone. Because of the limitedness of our intellect, they can only reflect a small part of the objects.

We may now proceed in two ways: 1. We may leave the images more general. Then there is less risk that they later turn out to be false, since they are more adaptable to the newly discovered facts; yet because of their generality the images become indeterminate and pale and their further development will be connected with a certain uncertainty and ambiguity. 2. We particularize the images and finish them in detail to a certain degree. Then we have to add much more arbitrariness (hypothetical), thus perhaps not fitting to new experiences; however, we have the advantage, that the images are as clear and distinct as possible and we can deduce from them all consequences with complete certainty and uniqueness.

Furthermore, Boltzmann points out that the requirement *'to register only the directly given phenomena and not to add something arbitrary'*, to omit all hypothetical, does not lead beyond *'simply marking every phenomenon'*.

So, when constructing images, symbols, and models, a certain freedom in the choice of hypotheses exists. If the results of distinct hypotheses do not differ, then criteria from outside physical reasoning are used for the selection. Something like aesthetics of explanations, models, and theories also exists. A very important criterion, according to Mach, is the 'economy of thought' in theories, in particular, that the number of suppositions should be minimal. Already Newton advocated simplicity of explanation in two rules of reasoning in the beginning of Book III of the "Principia"[16]:

---

[16]In the present book, all the citations from Newton's "Principia" are taken from the English translation "The Principia" by A. Motte (cf. Bibliography). When stressing that we refer explicitly to that translation we simply write: The Principia. Occasionally we use more modern words here.

Rule I:

*We are to admit no more causes of natural things than such
as are both true and sufficient to explain their appearances.*

Rule II:

*Therefore to the same natural effects we must, as far as
possible, assign the same causes.*

The hypotheses and postulates resulting from creation of the images
of nature are refined and made more precise in a dialectic way. Progress
in physics results from an interplay between gaining experience with
the process considered (increasingly detailed experience in the sense of
Hertz) and improving the model (image) of the process accordingly by
comparing the consequences with the increasing experience of physical
reality. Thus, although originally imprecise, ordinary concepts are refined, their meaning becomes more exact. For example, the concepts
momentum, force and energy of a body are easily distinguished by a
physicist – but in common usage, these three concepts are almost always confounded.

With regards to observations and experiments, it was G. GALILEI[17]
who analyzed the data of his measurements mathematically according
to his believe that *'the book of nature is written in the language of
mathematics'*[18]. In Galileo's time, systematic observation was already
established in astronomy, as was the use of mathematics to represent and
summarize the results; but observation of the motion of bodies under
controlled and controllable conditions, and the mathematical representation of physical process such as those in Galileo's experiments (cf.
Section 3.3.1), was completely new.

The mathematical symbols and images of classical mechanics were
formulated by Isaac Newton. In this book, we will start from his laws
and dwell mainly on the elaboration of the (mathematical) consequences
that, according to the philosophical sketch just presented, should fit the
experimental observation - the consequences in nature. But still we have
to clarify the stage on which Newton's laws act.

---

[17]Galileo Galilei (1564-1642), Italian physicist and astronomer.

[18] "Il Saggiatore", Rome 1623. The quotation is taken from Stilman Drake's translation "The Assayer".

1.4 Time, space and motion

The meaning of the basic concepts time, space and motion is hard to outline exactly. Philosophers, as well as physicists, have devoted studies to these concepts[19]. In physics, these concepts have a more precise usage – and are also more restricted in content (see Boltzmann's view above) – than in everyday life. So the physical concepts of *time* and *space* have different and fewer characteristics. There are no 'good' or 'bad' times. And physical space is not attributed with 'wideness', as in common speech[20]. The central properties of time and space in physics are to be measurable and mappable onto mathematical entities. Although these quantities are prerequisites to describe physical processes, their conceptualization may change as physics progresses. At the end of the 19th century, concepts of both time and space seemed to be well established – and were revised by Einstein's theories of relativity.

1.4.1 Newton's concepts

Below, we present Newton's concepts – essentially, the underlying concepts of classical mechanics – expounded in his "Principia", and confront them with Mach's critique in "The Science of Mechanics".

Newton on **time** (The Principia, Definitions, Scholium):

> *Absolute, true, and mathematical time, of itself and from its own nature, flows equably without relation to anything external, and by another name is called duration; relative, apparent, and common time is some sensible and external (whether accurate or unequable) measure of duration by the means of motion, which is commonly used instead of true time, such as an hour, a day, a month, a year.*

These sentences may make sense intuitively; however, they seem to be inconsequential if one takes into account Newton's intention *'to investigate only actual facts'*. Presumably, the statements reflect Newton's

---

[19] An extensive treatment of the history of theories of space is given in M. Jammer's recommendable book "Concepts of Space" (cf. Bibliography).

[20] Unfortunately, physicists, too, keep talking about 'empty' space.

contemporary philosophical and ideological background. Mach[21] argues, that Newton's concept of time is an abstract (useless) concept:

> *This absolute time can be measured by comparison with no motion; it has therefore neither a practical nor a scientific value; and no one is justified in saying that he knows aught about it. It is an idle metaphysical conception.*

Newton on **space** (ibid.):

> *Absolute space in its own nature, without relation to anything external, remains always similar and immovable. Relative space is some movable dimension or measure of the absolute spaces; which our senses determine by its position to bodies; and which is vulgarly taken for immovable space;*
> *...*

And on **motion**, which connects space and time, Newton wrote (ibid.):

> *Absolute motion is the translation of a body from one absolute place into another; and relative motion, the translation from one relative place into another.*

The objections raised by Mach to absolute space:

> *No one can say anything about absolute space and absolute motion, they are just objects in thoughts that cannot be demonstrated in practice. As has been pointed out in detail, all our principles of mechanics are experiences about relative positions and motions of bodies.*

And as to uniform motion:

> *Motion can be uniform with respect to other motion. The question of whether a motion as such is uniform has no meaning at all...*

---

[21]Mach took the view that science should not ask 'why?' but 'how?' (R. Carnap, An Introduction to the Philosophy of Science, Dover Publications, New York 1995).

Also, Mach did not consider **Newton's pail experiment** as compelling proof of absolute space: a bucket filled with water is spun, then suddenly stopped; before being spun, the water surface is flat, but as being spun, the surface forms a concave shape that remains for some time after stopping. According to Mach, the curvature of the surface of the rotating water only indicates motion relative to the fixed stars, and not relative to a fictitious absolute space[22]. The *a priori* existence of space and time is a feature of Newtonian mechanics; in the general theory of relativity, space and time are 'produced' by the masses. Although an absolute conception of space and time is certainly the background for Newton's laws, these laws can be applied to physical situations without referring to that background[23].

In classical physics, one describes the behavior of (the images of the) objects in a 'space', and the change of their positions in that space with 'time': **Space** and **time** or, to be more precise, their images are the very basic concepts of physics. **Motion** connects space and time. The epistemological background of the concepts of space and time is certainly interesting; nevertheless, it is the mapping of space and time onto mathematical entities that is essential for theoretical mechanics.

1.4.2 The mathematical pictures of space and time

To describe the position as a mathematical object we proceed in the common way: We choose some point of reference in space, the origin **O**, and a system of three orthogonal axes intersecting each other in the origin. This is our **frame of reference**[24]. In our experience, we know, that then every point **r** in space is uniquely labelled by three real numbers, the projections of **r** onto the chosen axes. All these points in space form now a vector space $\mathbb{R}^3 = \{\mathbf{r} = (x, y, z) \mid x, y, z \text{ real}\}$, where usually one sets $\mathbf{O} = (0, 0, 0)$. In similar reasoning the instances in time are simply represented by the real axis ($\mathbb{R}^1 = \{t \mid t \text{ real}\}$) with $t_0 = 0$ as the reference instance of time

Two more characteristics of physical space that are to be reflected in the mathematical picture are: the distance between two points, and

---

[22]This controversy was resolved in favor of Mach by the general theory of relativity.

[23]The situation is somewhat similar to quantum mechanics, where the interpretations seem to be not important in practice.

[24]On the influence of the choice of the frame of reference see chapter 8.

the angle between two straight lines[25]. A mathematical object suitable for representing physical space in classical mechanics is

$\mathbb{E}^3$, the **Euclidean vector space in three dimensions**.

The mathematical properties of $\mathbb{E}^3$ are summarized in Appendix A. A point $\mathbf{r} \in \mathbb{R}^3$ is represented in the Euclidean space $\mathbb{E}^3$ by the *radius vector* $\mathbf{r}$, whose components are the *Cartesian coordinates* of that point. This Euclidean space is the mathematical image of the physical *configuration space*. If the motion of a body is restricted, a one or two dimensional space (motion along a line or on a surface) may suffice. On the other hand, it will turn out to be advantageous to extend the set of coordinates (to obtain, say, a 'phase space'), and hence to use a higher dimensional space.

1.4.3 Kinematics

In this book mechanics is presented as the science of the motion of bodies under the influence of forces. Bodies are taken to be characterized by their masses, their appearances, and their constitution, e.g. whether they are rigid or soft. For a complete representation of the motion of a body in classical mechanics, one needs to know its position for any time. The consideration of motion itself (without referring to a cause) is called **kinematics**. If we also take into account forces as the origin of the motion we deal with **dynamics**[26]. This book is devoted mainly to the dynamics of (rigid) bodies. Only a short part is concerned with their **statics**, the theory of equilibrium of bodies and forces (see chapter11).

To measure the time dependence, one compares the motion at various stages with a time standard, which is usually – by definition – a *periodic motion* (e.g. the motion of a pendulum or the oscillations of an atomic clock). The mathematical picture of particle or body motion is the function $\mathbf{r}\,(t)$ for object's position with respect to the origin $\mathbf{O}$. For an extended object, in many cases it is sufficient to consider a representative (mathematical) point of the body.

---

[25]We do not want to discuss the mathematical meaning of 'Straight'. In physical reality, in the context of classical mechanics, light rays between the objects may serve to define straight lines in space ('in vacuo').

[26]The distinction between dynamics and kinematics is historical.

Of particular importance in classical mechanics are the first and second time derivatives of the radius vector: the rate of change of the radius vector $\mathbf{r}(t)$ with time,

$$\text{the } velocity \ \mathbf{v} = \frac{d\mathbf{r}}{dt} = \dot{\mathbf{r}},$$

and the rate of change of the velocity vector with time,

$$\text{the } acceleration \ \dot{\mathbf{v}} = \frac{d^2\mathbf{r}}{dt^2} = \ddot{\mathbf{r}}.$$

Here we used the calculus of infinitesimals invented independently by G. W. LEIBNIZ[27] and Isaac Newton[28].The notation of derivatives is due to Leibniz; Newton denoted the time derivative - *fluxion* as he called it - by a dot on top of the fluctuating quantity. For short we will use very often Newton's dot-notation, but without having his different background in mind. The fundamental laws of classical mechanics are formulated in terms of the variables $\mathbf{r}(t)$, $\dot{\mathbf{r}}(t)$, and $\ddot{\mathbf{r}}(t)$.

The stage is now set with space and time; let us introduce the actors in the next chapter.

---

[27]Gottfried Wilhelm Leibniz (1646 -1716), German polymath and philosopher.

[28]The invention priority of the Calculus - as it is now is called - was the issue of a famous controversy between Leibniz and Newton.

2

Foundations of classical mechanics

Before establishing the laws, Newton introduces the definitions of *mass, force,* and *quantity of motion.* We shall take these as our starting point. Again we also mention Mach's criticisms of Newton's formulation, from his book, "The Science of Mechanics" (see Bibliography). Then we present the Newton's laws in the form used usually in classical mechanics and derive rather general consequences.

2.1 Mass, quantity of motion, and force

Newton's definitions (The Principia, Definitions)

Definition I:

> *The quantity of matter is the measure of the same arising from its density and bulk conjointly.*

> ... It is this quantity that I mean hereafter everywhere under the name of body or mass. And the same is known by the weight of each body, for it is proportional to the weight, as I have found by experiments on pendulums, very accurately made, which shall be shown hereafter.

Definition II:

The quantity of motion is the measure of the same, arising from the velocity and quantity of matter conjointly.

The motion of the whole is the sum of motions of all parts; ...

Definition III:

The **vis insita**, *or innate force of matter, is a power of resisting, by which every body, as much as in it lies, continues in its present state, whether it be of rest, or of moving uniformly forward in a right line.*

Definition IV:

An impressed force is an action exerted upon a body, in order to change its state, either of rest, or of moving uniformly forward in a right line.

Mach criticized Definition I as being a sham, and Definition III as being superfluous in view of Definition IV. Mach was not alone in disliking Newton's concept of mass: Hertz, too, in his "Principles of Mechanics", is of the opinion that Newton must have been embarrassed about his enforced definition of mass as a product of volume times density. Newton did not distinguish between the mass appearing in Definition I and the mass referred to in Definition II. The equivalence of this two masses is not evident; we will discuss this later (Section 6.2). The modern terms for these masses are **gravitational mass** and **inertial mass** respectively.

Nowadays, the 'quantity of motion' is called **momentum**. Also, all three concepts – 'mass', 'momentum', and 'force' – are more precisely defined in physics than in popular usage, not least because they are related by mathematical equations. As to the history of these concepts, we refer to the books by M. Jammer, "Concepts of mass" and "Concepts of force" (see Bibliography).

2.2 Newton's laws

The laws per se are not Newton's main achievement. Some of them were stated (in a different form) already previously. For example in Christiaan HUYGENS' "Horologium oscillatorium"[1] one can find three 'hypotheses', not so different from Newton's laws, as a basis for Huygens' analysis of motion caused by gravity[2]. However it is Newton's merit at least having derived from his three laws a universal theory of motion.

The laws (The Principia, Axioms, or Laws of Motion)

Law I:

Every body continues in its state of rest, or of uniform motion in a right line, unless it is compelled to change that state by forces impressed upon it.

Law II:

The change of motion is ever proportional to the motive force impressed; and is made in the direction of the right line in which that force is impressed.

Law III:

To every action there is always opposed an equal reaction: or the mutual actions of two bodies upon each other are always equal, and directed to contrary parts.

These laws are followed by several 'Corollaries'. The first Corollary is the

Superposition Principle of Forces[3] ('parallelogram law of forces'):

---

[1]Christiaan Huygens (1629-1695), Dutch mathematician and scientist. He invented the pendulum clock about 1656 and published later his "Horologium oscillatorium" (The Pendulum Clock), Paris 1673. We mention also his "Traité de la Lumière", Leiden 1690, on the wave theory of light. This book contains also hie earlier "Discours de la cause de la pesanteur" (Discourse about gravity).

[2]D. Speiser, "Die Grundlegung der Mechanik in Huygens' Horologium Oscillatorium und in Newtons Principia", in "Die Anfänge der Mechanik" ed. K. Hutter, Springer Berlin 1989.

[3]We have formulated the principle in modern language.

*If forces are applied to the same point of a body, the effect
will be same as a force which is equal to their vectorial sum.*

The first law encapsulates what was known about the motion of
a single mass at the time. Galileo had already concluded empirically
that a particle moving without the influence of a force tries to preserve
its velocity[4]. Descartes expanded Gaileo's conclusion by stating that a
body (in the absence of a force) only moves in a straight line and never
along a curve[5].

Regarding the first law, ARISTOTLE[6] believed that in absence of
a motive power all bodies (on earth) would come to rest. As to the second law, Newton's great achievement was to recognize that the 'change
in (the quantity of) motion', the acceleration, is proportional to the
force. It is true however, that already in 1604, Galileo – after at first
incorrectly relating velocity to the distance – recognized the relevance
of acceleration, the increase of velocity with time, in the free fall[7]. *Newton's second law is the basis of dynamics.*

In view of the definitions preceding the laws in Newton's book, Mach
criticized the first law and the second law, in particular, as an 'unnec-
essary tautology'. *He subsequently replaced 'Newton's list of statements
by a much simpler, more systematic, and more satisfactory one'.* But
Mach's laws, in contrast to the 'unsystematic' ones of Newton, are not
easily comprehensible to those unfamiliar with the discussion preceding
them in his book. One could get the (false) impression that Mach was
not exactly an admirer of Newton's accomplishments. Mach, it seems,
was aware of the impression he would give, for after discussing the laws,
he states that the repetitions and tautologies sprang from opposition by
Newton's contemporaries, as well as from Newton's own unclarity with
regards to the significance of the new – nevertheless, that these short-
comings *'cannot cast the faintest shadow on his intellectual greatness'.*

Remarks

---

[4]1612 in a letter on sunspots to the merchant of Augsburg Marcus Welser.

[5]Descartes, Principia Philosophiae, second part, § 39.

[6]Aristotle (384-322 BC), Greek philosopher.
  Here we refer to his "Physics" ($\Phi\upsilon\sigma\iota\kappa\eta\sigma$ $A\kappa\rho\sigma\alpha\sigma\epsilon\omega\sigma$). The title is – for us –
somewhat misleading. The book is philosophical, i.e. insight is obtained by thinking
(only) with a little bit of experience, but not based on experiments!

[7]Stillman Drake, "Galileo at work", Chapter 6 (see Bibliography).

i) In the classical antiquity, as well as much later, it was believed that motion in the absence of forces is circular (Aristotle[8], Ptolemy[9]).

ii) It may seem that force is *defined* by the second law. But a force can be determined by comparison with another force that serves as a reference (in a static measurement); then the second law is more than a mere definition of force.

2.3 Analytical mechanics

Newton's book is hard to read even in its English translation. Its mathematics is hidden in the text, there are nearly no formulas displayed, and it is of a kind not familiar to us anymore. Newton derived the consequences of his verbally presented laws using the so-called *geometrical method*, i.e. he relied heavily on geometric arguments, mainly exploiting proportions between the quantities considered and often referring to figures. Since EUCLID's[10] "Elements" this was still the main method of mathematical reasoning since antiquity. But somehow infinitesimals and their properties had to enter. Newton achieved this by the eleven lemmas on vanishing small quantities he proposed right at the beginning of his first book. Only then he started to investigate particular

---

[8]Aristotle, "Physics", Chapter 8: 'Only circular motion can be infinite and continuous.' A prejudice not really scrutinized until Renaissance times.

[9]Claudius PTOLEMY (about 90-168), Greco-Egyptian astronomer and mathematician, famous for his "Μαθηματικη Συνταξις". Better known is the Arabic name "Almagest". It is a comprehensive astronomic treatise and handbook. The observed motion of the sun, the planets, and the moon is modeled to take place on cycles around the earth. On these circles move epicycles to account for the observed orbit. (Epicycles are smaller cycles whose center move on bigger ones.) Also the "Almagest" was esteemed for its description of the planetary system until Renaissance times.

[10]Euclid of Alexandria (about 323-328 BC), Greek mathematician, 'father of geometry'. His "Elements" are presumably the most influential book in the fields of Mathematics and Physics. Its style was exemplarily for many scientific books, in particular for the "Principia". The Elements consist of 13 books. There are different versions due to the fact that it survived through the Byzantine and Arabic tradition, like so many other antique treasures. Many slightly differing copies have been made since antiquity. There exist numerous translations, amongst them several in English. A rather recent English one is that by R. Catesby Taliaferro from 1939.

systems[11].

The application of analysis began rather shortly after the appearance of the "Principia". The French clergyman P. VARIGNON[12] was the first one writing 1700 Newton's law for rectilinear motion in the form $ddx/dt^2 = y$, where $ddx = d^2x$ and y represents the force. Thus he established analytical mechanics. Using the new mathematical language Varignon published numerous articles on the motion of a body in the Memoires de l'Academie Royale (Paris) from 1700 until 1711.

The breakthrough of the analytic presentation of Newton's mechanics came with L. EULER[13]. In his two volume book "Mechanica sive motus scientia analytice exposita" (Mechanics or the science of motion analytically presented; St. Petersburg 1736) he systematically discussed the motion of a point mass (see below).

The rather final analytical form was given to mechanics by J.-L. LAGRANGE[14] in his famous "Mécanique Analytique" (Analytical Mechanics; Paris 1788). There is not a single figure in this book!

It is this language which was used henceforward in classical mechanics.

2.3.1　The basic equations of mechanics

Interpreting the laws and definitions in modern terms, and casting them in mathematical language, we can write down the basic equations of mechanics. The position, velocity, acceleration, and force appearing in these equations are vectors in Euclidean space[15] (Section 1.4.3); they can be represented in terms of Cartesian components (cf. Appendix A). For a single particle (or body) at position $\mathbf{r} = (x, y, z)$, with velocity $\mathbf{v} = \dot{\mathbf{r}} = (\dot{x}, \dot{y}, \dot{z})$ and acceleration $\ddot{\mathbf{r}} = (\ddot{x}, \ddot{y}, \ddot{z})$, the basic equations are as follows:

---

[11]A work that facilitates the access to the "Principia" is S. Chandrasekhar's "Newton's Principia for the Common Reader" (see Bibliography).

[12]Pierre Varignon (1654-1722), mathematician. Varignon learned about the infinitesimal calculus from private instructions given by Johann Bernoulli to l'Hôpital.

[13]Leonhard Euler (1707-1783), Swiss mathematician and physicist.

[14]Joseph-Louis Lagrange (1736-1813), French mathematician and physicist.

[15]The distinguished role of Euclidean space in Newton's mechanics can be traced back to the first law.

- The **momentum p** $= (p_1, p_2, p_3)$ of a body is the product of its mass m and its velocity **v**:

$$\mathbf{p} = m\mathbf{v}. \tag{2.1}$$

- A **force F** is represented by a vector in configuration space, $\mathbf{F} = (F_1, F_2, F_3)$. If two forces $\mathbf{F}_1$ and $\mathbf{F}_2$ are applied to a point on a body, then the total force **F** is the vectorial sum of both forces:

$$\mathbf{F} = \mathbf{F}_1 + \mathbf{F}_2 \tag{2.2}$$

(*superposition principle of forces*).

- The **first two laws** of Newton relate the change of momentum ('quantity of motion') of a body with the force applied to it; these are summarized in an equation that we call **Newton's equation of motion**[16]:

$$\dot{\mathbf{p}} = \mathbf{F} \tag{2.3}$$

($\mathbf{p} = const$, if $\mathbf{F} = \mathbf{0}$ (1st law)). If the mass m is time independent, then the change in velocity, i.e. the **acceleration** $d^2r/dt^2 = \ddot{\mathbf{r}}$, is a measure of the force:

$$m\ddot{\mathbf{r}} = \mathbf{F}. \tag{2.4}$$

Written in components, we have

$$\begin{aligned} m\ddot{x} &= F_1 \\ m\ddot{y} &= F_2 \\ m\ddot{z} &= F_3. \end{aligned} \tag{2.5}$$

---

[16]One may suppose that there is an unjustified arbitrariness in translating 'the change of motion is **proportional** to the force' of Newton's second law into an equal sign between the change of momentum and the force. From Newton's reference to Galileo's discovery 'that the descent of bodies observed the duplicate ratio of time' as being in accord with the first and the second law (The Principia, Scholium at the end of "Axioms, or Laws of Motion") one can infer the equality. Further, the continuing application of the second law in the "Principia" to the motion in various central forces leads to same results as the analytical treatment. Moreover, in Varignon's communications to the academy and in Euler's "Mechanica..." the equality (apart from a constant) between the time derivative of the momentum and the force appears.

- The **third law** relates the forces that *two bodies* exert on each other:

$$\mathbf{F}_{21} = -\mathbf{F}_{12}, \tag{2.6}$$

where $\mathbf{F}_{ik}$ is the force applied by body k to body i (action = reaction).

The superposition principle for forces represents the starting point for *statics*, a subject of great importance in engineering. We present a short introduction to statics in Section 11.2.

2.3.2 Point masses and forces

In the following, we apply Newton's definitions and laws to so-called **point masses**. By definition, point masses are mathematical points having mass; they have no structure, like a body. Therefore, the only possibility for a point mass to change, is to move. Since motion can occur in three independent directions, the number of **degrees of freedom** of a point mass is three. At first, the concept of point mass is assumed to apply to real situations, when the bodies or particles involved are of sizes negligible with respect to the typical length scales (e.g. the distances between the bodies) of the systems considered – the particles are taken to be mathematical points. Thereby the internal structure and the internal degrees of freedom[17] of the particles are ignored. Then, starting from the concept of point mass, by a natural extension, the motion of extended bodies lies within the scope of the *mechanics of point masses* (see, for instance, the treatment of rigid bodies in Chapter 11).

If a force is present at every point $\mathbf{r}$ in (some part of) configuration space, then the particle is said to be in a **force field** $\mathbf{F} = \mathbf{F}(\mathbf{r})$. Here, it is explicitly assumed that the force only depends on the position[18] $\mathbf{r}$.

---

[17]Roughly speaking, the internal degrees of freedom are the independent possibilities of a body to change its appearance; for example, the three independent rotations of a non-spherically symmetric body (see Chapter 11). We focus on the internal degrees of freedom later.

[18]If the field of force has been determined by static methods, then one must keep in mind that by inserting this force into the equation of motion (2.3) or (2.4), it has been tacitly assumed, that the force does still not depend on t and/or $\mathbf{v}$, when time varies or velocities are involved. In particular, this restriction may be problematic in relativistic mechanics when dealing with large velocities of the particle.

In general, the force may also depend on the time t and velocity of the particle $\mathbf{v}$. A simple example is a particle experiencing the force $\mathbf{F}$ as well as an additional frictional force proportional to the velocity (E)

$$m\ddot{\mathbf{r}} = -\lambda\dot{\mathbf{r}} + \mathbf{F}.$$

In the general case, when $\mathbf{F} = \mathbf{F}(\mathbf{r}, \mathbf{v}, t)$, the equation of motion (2.3) or (2.4) (in coordinate representation a system of three coupled second order ordinary differential equations) can only be solved numerically. If, however, $\mathbf{F}$ has particular simplifying properties, then the equations of motion possess general features that make them more amenable to analytical solution. In the following, we make two restrictive assumptions:

i) We consider *only a point mass with constant mass*; then, since $d(m\dot{\mathbf{r}})/dt = m\ddot{\mathbf{r}}$, the equation of motion reduces[19] to (2.4).

ii) The *force field* $\mathbf{F}$ *depends only on position*[20] $\mathbf{r}$: $\mathbf{F} = \mathbf{F}(\mathbf{r})$.

2.4 Constants of motion

Above we have introduced the basic concepts and relations of classical mechanics. But before we proceed to consider particular mechanical systems in the next chapter, we will point out some general features of the relations between the fundamental quantities.

2.4.1 Constants of motion and conserved quantities

The solution of the second order ordinary differential equations (2.5), in vector form $m\ddot{\mathbf{r}} = \mathbf{F}(\mathbf{r})$, for given initial values of position and velocity,

$$\mathbf{r}_0 := \mathbf{r}(t = 0) \quad \text{and} \quad \dot{\mathbf{r}}_0 := \dot{\mathbf{r}}(t = 0), \tag{2.7}$$

is the **orbit** $\mathbf{r} = \mathbf{r}(t)$ of the particle in 'ordinary' three-dimensional space, i.e. **configuration space**. Introducing the velocity $\mathbf{v} = \dot{\mathbf{r}}$ as an

---

[19]In the case of variable mass, one obtains from (2.3) the equation

$$\dot{\mathbf{r}}\frac{dm}{dt} + m\ddot{\mathbf{r}} = \mathbf{F}.$$

To solve this equation, dm/dt has to be known (E). An easy case is in one dimension, when $dm/dt = const$ and F does only depend on v.

[20]In Electrodynamics the Lorentz force depends on $\dot{\mathbf{r}}$ (compare Section 5.5).

additional variable, the second order differential equation (2.4) can be transformed to an **autonomous system**[21] of two first order equations:

$$\dot{\mathbf{r}} = \mathbf{v}$$
$$\dot{\mathbf{v}} = \mathbf{F}/m. \tag{2.8}$$

It often turns out to be convenient to write the **dynamical variables** $\mathbf{r}$ and $\mathbf{v}$ together as a vector $(\mathbf{r}, \mathbf{v})$ in the six-dimensional **phase space** and to consider $(\mathbf{v}, \mathbf{F}(\mathbf{r})/m)$ as the components of a six-dimensional vector function dependent on $(\mathbf{r}, \mathbf{v})$.

Given the initial values $\mathbf{r}_0$ and $\mathbf{v}_0$, the system of differential equations (2.8) possesses a *unique solution* if $\mathbf{F}$ fulfills certain conditions. We assume that $\mathbf{F}$ is well-behaved. Consequently, every **trajectory**[22] $(\mathbf{r}(t), \mathbf{v}(t))$ in phase space depends in a unique manner on the initial values:

$$\mathbf{r} = \mathbf{R}(\mathbf{r}_0, \mathbf{v}_0; t)$$
$$\mathbf{v} = \mathbf{V}(\mathbf{r}_0, \mathbf{v}_0; t). \tag{2.9}$$

(This is not true for the orbit in configuration space! See, for instance, Figs. 15.9 and 15.10.) The explicit forms of the functions $\mathbf{R}$ and $\mathbf{V}$ are to be calculated by solving the equations of motion. The following statements address the existence and general properties of the solutions.

Since $\mathbf{F}$ is assumed to be independent of time t and velocity $\dot{\mathbf{r}}$, the equations of motion (2.4) are ordinary differential equations involving only second derivatives with respect to time. Therefore, they are *invariant under the* **time reversal** *transformation*

$$t \to -t,$$

because replacing d/dt by $-d/dt$ leaves the second time derivative d^2/dt^2 in (2.4) unchanged. This means that the motion in configuration space is reversible: for the solutions of (2.4), the orbits are the same[23] irre-

---

[21]*Autonomous* means that the equations do not explicitly depend on time t.

[22]Paths in configuration space are termed 'orbits', and paths in phase space are termed 'trajectories'.

[23]A simple example of an equation that is not invariant under time reversal is

$$\frac{d^2x}{dx^2} + \gamma\frac{dx}{dt} = 0,$$

spective of the direction of time. In (2.8) time reversal leads to

$$\dot{\mathbf{r}} = -\mathbf{v},$$
$$\dot{\mathbf{v}} = -\mathbf{F}/m;$$

only the *direction* of the velocity and the acceleration *is changed*. Under time reversal, in phase space, the particle moves along the same trajectory but in the opposite direction.

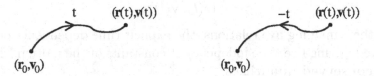

Therefore, under time reversal, if the particle is at $(\mathbf{r}(t), \mathbf{v}(t))$ in phase space at time t, then after evolving for time $-t$ it arrives at the point $(\mathbf{r}_0, \mathbf{v}_0)$, so that

$$\mathbf{r}_0 = \mathbf{R}\left(\mathbf{r}(t), \mathbf{v}(t); -t\right)$$
$$\mathbf{v}_0 = \mathbf{V}\left(\mathbf{r}(t), \mathbf{v}(t); -t\right). \qquad (2.10)$$

This is just the inverse of equations (2.9).

When written in Cartesian components, Eqs. (2.10) represent six equations, where the components of $\mathbf{r}$ and $\mathbf{v}$ together with t appear in the functions $\mathbf{R}$ and $\mathbf{V}$ on the right hand side. These equations show that *the functions* $\mathbf{R}$ *and* $\mathbf{V}$ *are constant in time along a trajectory* $(\mathbf{r}(t), \mathbf{v}(t))$ in phase space, with the constants being the initial values $\mathbf{r}_0$ and $\mathbf{v}_0$. These fix a trajectory for all time. The initial values $\mathbf{r}_0$ and $\mathbf{v}_0$ are a set of six independent[24] **constants of the motion**, which can be written as[25]

$$K_j\left(\mathbf{r}(t), \mathbf{v}(t), t\right) = k_j = const, \qquad j = 1, \ldots, 6; \qquad (2.11)$$

whose solutions are
$$x \propto e^{-\gamma t}.$$
(See also the damped oscillator, Subsection 3.4.1.) Looking into the past, the solution grows exponentially, whereas in the future the solution becomes vanishingly small.

[24]Since $\mathbf{r}_0$ as well as $\mathbf{v}_0$ can be chosen independently!

[25]The initial values $\mathbf{r}_0$ and $\mathbf{v}_0$ are particular examples of the constants k_j in these relations.

all the total time derivatives vanish:

$$\frac{d}{dt} K_j \left(\mathbf{r}(t), \mathbf{v}(t) \right) = \frac{\partial K_j}{\partial \mathbf{r}} \frac{d\mathbf{r}}{dt} + \frac{\partial K_j}{\partial \mathbf{v}} \frac{d\mathbf{v}}{dt} + \frac{\partial K_j}{\partial t} = 0. \qquad (2.12)$$

In general, not all of the functions K_j can be calculated explicitly; in-deed, there are few cases when this is possible. Now, if any one of the relations (2.11), e.g. K_1, is used to express time t in terms of $\mathbf{r}(t)$ and $\mathbf{v}(t)$, i.e.

$$t = f \left(\mathbf{r}(t), \mathbf{v}(t) \right),$$

then in the remaining five relations, the explicit time dependence can be eliminated, giving five time independent constants of the motion. These are the **conserved quantities**

$$I_j \left(\mathbf{r}(t), \mathbf{v}(t) \right) = i_j = const, \qquad j = 1, \ldots, 5. \qquad (2.13)$$

They depend on time only through the dynamical variables $\mathbf{r}(t)$ and $\mathbf{v}(t)$, i.e. $\partial I_j / \partial t = 0$, such that

$$\frac{d}{dt} I_j \left(\mathbf{r}(t), \mathbf{v}(t) \right) = \frac{\partial I_j}{\partial \mathbf{r}} \frac{d\mathbf{r}}{dt} + \frac{\partial I_j}{\partial \mathbf{v}} \frac{d\mathbf{v}}{dt} = 0. \qquad (2.14)$$

A constant of the motion, or a conserved quantity, is also called a (time dependent or time independent) **first integral** or **integral of the motion**.

Each relation (2.13) represents a five-dimensional hypersurface in six-dimensional phase space. The trajectory of the particle lies in each of these hypersurfaces; it is therefore the curve of intersection of all five hypersurfaces[26]. The orbit is the projection of the trajectory in phase space onto configuration space (Note: the uniqueness of the phase space motion is lost in the projection). Thus, knowledge of all the conserved quantities implies possession of the unique solution, even though solutions of the ordinary differential equations were not calculated.

Unfortunately, one cannot, on the basis of just these statements about the existence of a solution, deduce how the conserved quantities

---

[26]The situation is a generalization of the one in three-dimensional space, where the intersection of two (independent) surfaces (i.e. two-dimensional manifolds) is a curve (a one-dimensional manifold). A concrete example of such an intersection is shown in Fig. 4.3.

are to be determined – unless one knows the solution of the equations. This problem is insurmountable for most systems. However, for each known constant of the motion, the number of equations to be solved is decreased by one. It is advantageous to find as many constants of the motion as possible before attempting to solve the ordinary differential equations. For a conservative system (see below), there is always an explicitly known conserved quantity: namely, the *total energy*. In this case, four conserved quantities or five constants of the motion remain to be determined[27]. If the force in a conservative system is a central force (see below), then there are further explicit conserved quantities. A system is said to be **integrable**, if there is a sufficient number of first integrals for the solution to be known without solving the equations of motion (2.8) directly (again, see Section 14.1).

2.4.2 Conservation of energy

The first of the favorable cases mentioned in the previous paragraph is the following. Suppose we multiply equations (2.4) by $\dot{\mathbf{r}}$,

$$m\dot{\mathbf{r}}\ddot{\mathbf{r}} = \frac{d}{dt}(m\dot{\mathbf{r}}^2/2) = \mathbf{F}\dot{\mathbf{r}},$$

and integrate both sides over time from $t_0 = 0$ to t_1:

$$\frac{1}{2}m\mathbf{v}_1^2 - \frac{1}{2}m\mathbf{v}_0^2 = \int_0^{t_1} \mathbf{F}\frac{d\mathbf{r}}{dt}dt = \int_{\mathbf{r}_0,\mathcal{C}}^{\mathbf{r}_1} \mathbf{F}\,d\mathbf{r}, \tag{2.15}$$

where $\mathbf{r}_0 = \mathbf{r}(0)$, $\mathbf{r}_1 = \mathbf{r}(t_1)$, $\mathbf{v}_0 = \mathbf{v}(0)$, $\mathbf{v}_1 = \mathbf{v}(t_1)$, and $\mathcal{C}$ is the path of the particle between t_0 and t_1. The quantity

$$T = m\mathbf{v}^2/2 = m\dot{\mathbf{r}}^2/2 \tag{2.16}$$

---

[27]Later we see that because of the structure of Eqs. (2.8), three conserved quantities with certain properties are sufficient to give a complete solution (see Section 14.1).

is the **kinetic energy**[28] of the particle with velocity $\mathbf{v}$, and the integral

$$\int_{\mathbf{r}_0,\mathcal{C}}^{\mathbf{r}_1} \mathbf{F}(\mathbf{r})d\mathbf{r}$$

is the **work** done by the force $\mathbf{F}$ in moving the particle along path $\mathcal{C}$ from $\mathbf{r}_0$ to $\mathbf{r}_1$. The difference in the kinetic energies at $t = 0$ and $t = t_1$ is therefore equal to the work done by the force in that period.

Let us consider this work more closely. In general, the amount of work done in moving a particle from $\mathbf{r}_0$ to $\mathbf{r}_1$ depends on the path $\mathcal{C}$ taken. If the integral does not depend on the path, i.e.

$$\int_{\mathbf{r}_0,\mathcal{C}}^{\mathbf{r}_1} \mathbf{F}d\mathbf{r} =: \int_{\mathbf{r}_0}^{\mathbf{r}_1} \mathbf{F}d\mathbf{r}, \qquad \forall \mathcal{C}, \qquad (2.17)$$

then the force field $\mathbf{F}$ is **conservative**. In particular, for two arbitrary paths $\mathcal{C}_1$ and $\mathcal{C}_2$ with common starting and end points $\mathbf{r}_1$ and $\mathbf{r}_2$ (see sketch), this implies that

$$0 = \int_{\mathbf{r}_0,\mathcal{C}_1}^{\mathbf{r}_1} \mathbf{F}d\mathbf{r} - \int_{\mathbf{r}_0,\mathcal{C}_2}^{\mathbf{r}_1} \mathbf{F}d\mathbf{r} = \int_{\mathbf{r}_0,\mathcal{C}_1}^{\mathbf{r}_1} \mathbf{F}d\mathbf{r} + \int_{\mathbf{r}_1,\mathcal{C}_2}^{\mathbf{r}_0} \mathbf{F}d\mathbf{r} = \oint_{\mathcal{C}_1+\mathcal{C}_2} \mathbf{F}d\mathbf{r}.$$

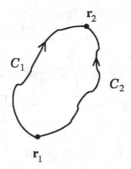

---

[28]Leibniz observed that the quantity $m\dot{\mathbf{r}}^2$ is conserved in many mechanical processes. He called this quantity **vis viva** (living force) and proposed the principle of its conservation in several publications (see e.g. *Specimen Dynamicum*, Part I, Acta Eruditorum 1695). Until the beginning of the 20th century one can find books using this term.

Generally there was a long lasting confusion between the concepts which we call today energy, force, and momentum.

Since the paths are arbitrary, one obtains the condition:

The force field **F** *is conservative, if and only if for every closed path* C

$$\oint_C \mathbf{F} d\mathbf{r} = 0. \tag{2.18}$$

Using **Stokes' theorem**

$$\oint_C \mathbf{F} d\mathbf{r} = \int_O (\boldsymbol{\nabla} \times \mathbf{F}) df$$

(with $\boldsymbol{\nabla} = (\partial/\partial x, \partial/\partial y, \partial/\partial z)$), where O is an arbitrary surface bounded by the closed curve C and df is the vector normal to a surface element, we see that the integral condition (2.18) is equivalent to the differential condition,

$$\boldsymbol{\nabla} \times \mathbf{F} = 0, \tag{2.19}$$

for the force to be conservative.

A mathematical consequence of (2.19) is that there exists a scalar function $V(\mathbf{r})$, such that[29]

$$\mathbf{F}(\mathbf{r}) = -\boldsymbol{\nabla} V(\mathbf{r}); \tag{2.20}$$

$V(\mathbf{r})$ is the **potential** of the force field $\mathbf{F}(\mathbf{r})$. Integrating **F** along some path between $\mathbf{r}_0$ and $\mathbf{r}_1$

$$\int_{\mathbf{r}_0}^{\mathbf{r}_1} \mathbf{F} d\mathbf{r} = -\int_{\mathbf{r}_0}^{\mathbf{r}_1} \boldsymbol{\nabla} V d\mathbf{r} = -\int_{\mathbf{r}_0}^{\mathbf{r}_1} dV = V(\mathbf{r}_0) - V(\mathbf{r}_1), \tag{2.21}$$

results in the difference of the values of V at those points; in this context, the potential $V(\mathbf{r})$ is referred to as the **potential energy** at position **r**. Combining with Eq. (2.15), we now have

$$\frac{1}{2}m\mathbf{v}_1^2 + V(\mathbf{r}_1) = \frac{1}{2}m\mathbf{v}_0^2 + V(\mathbf{r}_0) = const =: E; \tag{2.22}$$

i.e. the sum of the kinetic and the potential energy is the **total energy** E. Since $\mathbf{r}_0$ and $\mathbf{r}_1$ are two arbitrary points, **conservation of energy** follows:

---

[29]The minus sign is chosen by convention.

*For a conservative force, the sum of kinetic and potential
energy, the total energy E, is constant in time:*

$$E = T(\mathbf{v}) + V(\mathbf{r}) = \frac{m}{2}\dot{\mathbf{r}}^2 + V(\mathbf{r}) = const. \qquad (2.23)$$

The total energy E is conserved. The system is said to be
conservative.

Finding this first integral constitutes an important step towards obtaining the solution: the number of equations remaining to be solved is
reduced. This the calculational advantage. Perhaps even more importantly, the conserved quantity has a physical meaning that is recogniz-
able in many different systems – it is always energy, or a quantity quite
analogous to energy. Conservation laws immediately offer a qualitative
understanding of features of the particle motion. For instance, in many
different physical situations, an increase in particle velocity can be interpreted as an increase in kinetic energy at the expense of potential
energy.

Remarks

i) For one-dimensional motion, the force field is always conservative
   since $F(x)$ always has a potential

$$V(x) = -\int F(x')dx' \qquad (2.24)$$

(since $F(x) = -dV(x)/dx$). From the conservation law (2.22), one
can directly deduce (see next chapter) the solution of the equation
of motion in the following form:

$$t - t_0 = \pm \int\limits_0^x dx' \frac{1}{\sqrt{2\left(E - V(x')\right)/m}}.$$

For a one-dimensional system, there are two constants of motion
(x_0 and v_0) and thus only one conserved quantity; this we already
have. Therefore, every one-dimensional system is integrable (solvable). We discuss this further in Section 3.2.

ii) Another special case of energy conservation occurs in three dimensions, when a special property of the force ensures that the three
components (2.5) of the equation of motion are not coupled:

$$
\begin{aligned}
m\ddot{x} &= F_1(x) \\
m\ddot{y} &= F_2(y) \\
m\ddot{z} &= F_3(z).
\end{aligned}
\tag{2.25}
$$

The three-dimensional system separates into three one-dimensional
systems. In the light of remark (i), the problem can always be
solved: for each coordinate, there is a conserved energy E_i, $i =
1, 2, 3$. The total energy is

$$
E = E_1 + E_2 + E_3,
\tag{2.26}
$$

and the potential $V(\mathbf{r})$ is simply the sum of three parts:

$$
\begin{aligned}
V(\mathbf{r}) &= -\left(\int^x F_1(x')dx' + \int^y F_2(y')dy' + \int^z F_3(z')dz' \right) \\
&= V_1(x) + V_2(y) + V_3(z).
\end{aligned}
$$

2.4.3 Angular momentum and its conservation

There is a further class of systems where a useful conserved quantity
appears. The vector product of the equation of motion (2.4) with the
radius vector $\mathbf{r}$,

$$
\begin{aligned}
\mathbf{r} \times m\ddot{\mathbf{r}} &= \frac{d}{dt}(\mathbf{r} \times m\dot{\mathbf{r}}) \\
&= \mathbf{r} \times \mathbf{F},
\end{aligned}
$$

gives

$$
\frac{d}{dt}\mathbf{L} = \mathbf{N},
\tag{2.27}
$$

where we have defined the **angular momentum L**, to be

$$
\mathbf{L} = \mathbf{r} \times m\dot{\mathbf{r}} = \mathbf{r} \times m\mathbf{v} = \mathbf{r} \times \mathbf{p}
\tag{2.28}
$$

and the **torque** or the **moment of the force N** about the origin, to
be

$$
\mathbf{N} = \mathbf{r} \times \mathbf{F}.
\tag{2.29}
$$

The angular momentum **L** is given in Cartesian components (cf. Eq. (A.7)) by

$$\begin{aligned} L_1 &= m\,(y\dot{z} - z\dot{y}) \\ L_2 &= m\,(z\dot{x} - x\dot{z}) \\ L_3 &= m\,(x\dot{y} - y\dot{x})\,. \end{aligned} \tag{2.30}$$

Similar relations hold for the components of the torque **N**. The equation for the angular momentum (2.27) is analogous to the equation for the linear momentum $d\mathbf{p}/dt = \mathbf{F}$. However, one should observe that both quantities[30] in (2.27) are referred to the origin[31] (E). Since $|\mathbf{N}| = rF\sin\alpha$, where α is the angle between the radius vector **r** and the force **F**, the torque produced by a force, which is parallel to **r**, vanishes, and is maximal when the force is perpendicular to **r**.

Central forces

An important special case occurs when the force produces a torque that vanishes at every point. That means that at every point, the force is directed towards (or outwards from) the center about which the torque is defined; it is parallel (or antiparallel) to the radius vector **r**. Such a **central force** has the form

$$\mathbf{F}(\mathbf{r}) = f(r)\frac{\mathbf{r}}{r}. \tag{2.31}$$

Since **F** is parallel to **r**, the torque vanishes,

$$\mathbf{N} = \mathbf{r} \times \mathbf{F} = 0, \tag{2.32}$$

and from (2.27), the **law of conservation of angular momentum** follows:

In a central force field, the angular momentum of a particle is conserved,

$$\mathbf{L} = \mathbf{L}(\mathbf{r}, \mathbf{v}) = const. \tag{2.33}$$

---

[30]That the quantities defined in Eqs. (2.28) and (2.29) are vectors depends strongly on the three-dimensionality of space.

[31]Here, the origin is taken as an example of a point of reference. In practice, the point of reference is some special point of the system – for example, a point of symmetry of the potential, which for convenience, is taken as the origin.

In fact, this provides three conserved quantities, so the number of equations remaining to be solved is considerably reduced.

From its definition (2.28), we recognize that the angular momentum at any instant is perpendicular to the orbit $\mathbf{r}(t)$:

$$\mathbf{r}\mathbf{L} = 0, \quad \forall t. \tag{2.34}$$

Therefore, if the angular momentum is constant in time, the orbit remains in a plane perpendicular to the fixed direction of $\mathbf{L}$; the motion is therefore effectively two-dimensional.

Every central force field is conservative. It follows from (2.31) that $\nabla \times \mathbf{F} = 0$. For a central force field, there exists therefore a potential V such that $\mathbf{F}$ is the gradient of that potential (cf. Eq. (2.20)). The potential can be calculated readily. Multiplying (2.31) by $\mathbf{r}$, one obtains

$$f(r)r = -\mathbf{r}\nabla V(\mathbf{r}) = -r\frac{\partial}{\partial r}V(\mathbf{r}) = -r\frac{d}{dr}V(r), \tag{2.35}$$

where the last step used the fact that since $f = f(r)$, the potential V can depend only on $r = |\mathbf{r}|$. *The potential of a central force is spherically symmetric.* Thus the relation between f and V is

$$f(r) = -\frac{d}{dr}V(r) \tag{2.36}$$

or

$$V(r) = -\int^r f\left(r'\right) dr'. \tag{2.37}$$

In the case of a particle subjected to a central force, we immediately have four conserved quantities: energy and angular momentum (three components). These are sufficient to solve the problem at hand: Eqs. (2.4) or (2.8) are integrable, i.e. reducible to simple integrals (see Section 14.1).

Problems and examples

1. **Variable mass: Rocket motion** ([Sommerfeld], [Corben/ Stehle]).

   Consider the one-dimensional motion of a rocket, perpendicular to the surface of the earth, acted upon by the force of gravity $F = -mg$. The rate of loss of mass due to loss of fuel,

   $$dm/dt = -\mu,$$

as well as the velocity $-w$ of the gases ejected ($-w$ is the velocity with respect to the rocket!) are constant. The total change of momentum with respect to the earth consists of the change in momentum of the rocket,

$$d\,(mv)\,/dt = m\dot{v} - \mu v,$$

and the change in momentum of the ejected gases, $\mu(v - w)$. The only force present is the gravitational force, therefore we have the equation of motion:

$$m\dot{v} - \mu v + \mu(v - w) = m\dot{v} - \mu w = -mg.$$

Discuss this equation and solve it ($m = m_0 - \mu t$).

2. Discuss and solve the equation for one-dimensional motion of a particle in the presence of a frictional force, assuming that the resistance due to friction is proportional to

a) v; b) v^2.

3. Show that a shift in the origin

$$\mathbf{r}' = \mathbf{r} + \mathbf{s} \quad \text{with} \quad d^2\mathbf{s}/dt^2 = \mathbf{0},$$

does not change the form of the equation $d\mathbf{L}/dt = \mathbf{N}$.

4. For a central force $\mathbf{F}(\mathbf{r}) = f(r)\mathbf{r}/r$, calculate the expressions $\nabla\mathbf{F}$, $\nabla \times \mathbf{F}$ and $V(r)$.

5. Determine the angular momentum of a particle moving with constant velocity v

a) Along a straight line;
b) In a circular orbit.

3

One-dimensional motion of a particle

The simplest system consists of a single particle whose motion only has one degree of freedom. In the following, we investigate the constants of the motion and the conserved quantities for Newton's equation in one dimension. Since the representation in phase space here is two-dimensional, the particle's motion in phase space can be visualized graphically.

3.1 Examples of one-dimensional motion

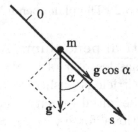

Figure 3.1: The inclined track.

i) The inclined track: A particle of mass m slides without friction on a track inclined at angle α to the direction of the gravitational force $\mathbf{F} = m\mathbf{g}$. Since forces obey vector addition (see Page 17), the force of

35

gravity can be split into components parallel and perpendicular to the track[1]:

$$\mathbf{F} = \mathbf{F}_{\parallel} + \mathbf{F}_{\perp},$$

where (see Fig. 3.1)

$$|\mathbf{F}_{\parallel}| = mg\cos\alpha. \tag{3.1}$$

Motion along the track is influenced only by $\mathbf{F}_{\parallel}$; the perpendicular component $\mathbf{F}_{\perp}$ is exactly counterbalanced by the track[2] (see Fig. 3.1). Let s be the distance along the track (with respect to some initial point $s_0 = 0$). The momentum of the particle is $p = m\dot{s}$. Hence, the equation of motion is

$$m\ddot{s} = mg\cos\alpha. \tag{3.2}$$

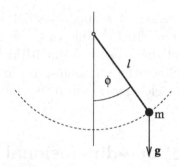

Figure 3.2: The plane pendulum.

ii) The plane mathematical pendulum*:* A particle of mass m is attached to the end of a massless bar of length l that can swing freely about a fixed point. The particle is pushed such that its motion always remains in the plane (i.e. the initial velocity vector lies in the plane containing the bar and gravitational force vector). Since $l\phi$ is the arclength,

---

[1]To stress that the system is one-dimensional, we choose the word 'track'. This setup was ingeniously used by Galileo in his experiments on 'falling' particles to allow him to measure the time intervals (see Section 3.3.1).

[2]The component of the gravitational force perpendicular to the track only serves to keep the particle on the track; it does not influence the dynamics along the track. We will discuss this later (Section 9.1) in more detail.

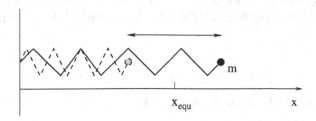

Figure 3.3: An oscillating mass.

where ϕ is the angle between the bar and the vertical, the equation of motion reads (see also Section 3.3.2 below):

$$ml\ddot{\phi} = -mg\sin\phi. \qquad (3.3)$$

iii) The harmonic oscillator: A particle of mass m is confined to move along the x-axis. It is attached to a spring, with equilibrium position x_{equ}. Assume that the spring obeys Hooke's Law, $|F| \propto |x - x_{equ}|$, meaning that the force is **harmonic**, i.e. it is given by $F = -k(x-x_{equ})$. If x_{equ} is chosen as the origin, i.e. $x_{equ} = 0$, then the particle moves according to the law

$$m\ddot{x} = -kx. \qquad (3.4)$$

In examples (i) and (iii), the equation of motion is linear, while in example (ii), the pendulum obeys nonlinear equation of motion. Example (ii) also shows that the dynamical variable does not have to be a Cartesian coordinate. Nevertheless, the equation has the form

$$m\ddot{q} = F(q),$$

where q is some generalized coordinate.

3.2 General features

Consider Newton's equation for the motion of a particle with mass m in a field of force $F(x)$

$$m\ddot{x} = F(x), \qquad (3.5)$$

where the variable (not necessarily a Cartesian coordinate) $x(t)$ denotes the position of the particle at time t. A potential[3] $V(x)$ exists, since (see Eq. (2.24))

$$V(x) = -\int^x F(x')\,dx'. \tag{3.6}$$

The force $F(x)$ is obtained from $V(x)$ by

$$F(x) = -\frac{dV}{dx};$$

hence the force is conservative. Consequently, equation (3.5) is integrable: multiplying (3.5) by $\dot{x}$ (as in the derivation of (2.22)) yields the conservation law of energy,

$$\frac{m}{2}\dot{x}^2 + V(x) = E = \frac{m}{2}v_0^2 + V(x_0), \tag{3.7}$$

with initial values $x_0 = x(t = t_0)$, $v_0 = v(t = t_0)$. The energy E can immediately be used to find the velocity,

$$\dot{x} = \pm\sqrt{\frac{2}{m}(E - V)}, \tag{3.8}$$

and then the solution $x(t)$ by integration,

$$t = t_0 \pm \int_{x_0}^{x} \frac{dx'}{\sqrt{\dfrac{2}{m}(E - V(x'))}}. \tag{3.9}$$

The sign is chosen in accordance with the desired direction of the velocity. The integral can always be computed, either analytically or numerically, when the potential is 'reasonably' well-behaved. (We assume that it is.) Thus, a *one-dimensional problem is always solvable.*

Consider the particular potential shown in Fig. 3.4. Typical trajectories in the 2-dimensional phase space $(x, \dot{x})$, at various energies, are shown in Fig. 3.5. Dependence of the motion on the energy can be inferred from Fig. 3.4. It is important to recognize that at every point x,

---

[3]In the above examples one has:
i) $V(s) \propto -s$;
ii)$V(\phi) \propto -\cos\phi$; and,
iii)$V(x) \propto x^2$.

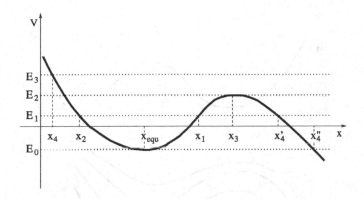

Figure 3.4: A generic potential $V(x)$.

the difference between the total energy E and the potential energy $V(x)$ is the kinetic energy available, $T(\dot{x}) = E - V$. This gives the absolute value of the velocity $|\dot{x}|$ (from (3.8)), as seen in the trajectories of Fig. 3.5 corresponding to different energy values.

Bounded motion

At energies E that are lower than $V(x_3)$ (see Fig. 3.4), and for initial values $x_0 < x_3$, the motion is **bounded**; i.e. motion occurs between two **turning points**. The turning points are the zeroes of the kinetic energy $T = m\dot{x}^2/2 = 0$. Hence, at a turning point, $\dot{x} = 0$, the velocity vanishes. In Fig. 3.4, for the total energy E_1, the turning points are x_1 and x_2. Indeed, the velocity $\dot{x}$ changes direction at a turning point. Accordingly, the sign in (3.8) changes when the particle stops at a turning point. Clearly, when kinetic energy T is maximal, then $|\dot{x}|$ too. If the particle is initially at turning point x_2 with the velocity $\dot{x} = 0$, then it arrives at turning point x_1 after time $t_1 = t(x_1)$ has elapsed (cf. Fig. 3.4). By conservation of energy (3.7),

$$\frac{m}{2}\dot{x}^2 + V(x) = E,$$

(where $E = V(x_1) = V(x_2)$), and hence the velocity at time t, $0 \le t \le t_1$, is

$$\dot{x}(t) = -\sqrt{\frac{2}{m}\left(E - V\left(x(t)\right)\right)}.$$

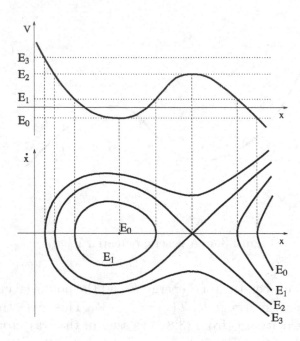

Figure 3.5: The trajectories in phase space for different types of motion in the potential $V(x)$.

The motion of the particle from x_1 to x_2 is the exact reverse of this, which is a consequence of the invariance of the equation of motion (3.5) under time reversal. The time required to go from x_2 to x_1 is therefore half the total time τ required to get back to where it began, namely x_2:

$$\tau/2 = \int_{x_2}^{x_1} \frac{dx}{\sqrt{\frac{2}{m}\left(E - V(x)\right)}}. \tag{3.10}$$

After time τ, the motion repeats itself, and one has

$$x(t + \tau) = x(t). \tag{3.11}$$

τ is called the **period**. *All one-dimensional, bounded motion is* **periodic**.

In general, the period τ, and consequently the **oscillation frequency** $\omega := 2\pi/\tau$, depend on the energy chosen: $\tau = \tau(E)$. Since the motion is periodic, the *trajectories in phase space are closed* (cf. Fig. 3.5). The periodicity of the motion suggests that a Fourier series expansion of $x(t)$ is close at hand. Indeed, this constitutes the first step in the method of *harmonic analysis*.

The point x_{equ} for which $V(x)$ has a minimum (Fig. 3.4) is called the **equilibrium position**,

$$\left.\frac{dV}{dx}\right|_{x=x_{equ}} = 0.$$

No net force acts at x_{equ}, and $\ddot{x}$ vanishes. Expanding $V(x)$ about the equilibrium position x_{equ}, we have

$$V(x) = V(x_{equ}) + \frac{1}{2}m\omega_0^2 (x - x_{equ})^2 + \ldots, \tag{3.12}$$

where the quantity

$$m\omega_0^2 := \left.\frac{d^2V}{dx^2}\right|_{x=x_{equ}} \tag{3.13}$$

is positive (and therefore ω_0 real), as seen in Fig. (3.4). For small energies (i.e. energies close to $V(x_{equ})$), one can use the **harmonic approximation**, in which all terms after the quadratic term in the expansion (3.12) of $V(x)$ are omitted. Shifting the coordinate system such that x_{equ} is the new origin – i.e. setting $x_{equ} = 0$ – one obtains the equation of motion:

$$\ddot{x} = -\omega_0^2 x. \tag{3.14}$$

The point mass oscillates (see Section 3.3.3 below) about the equilibrium with frequency ω_0. The frequency ω_0 is *independent of the energy E* and is determined only by the curvature of the potential in the vicinity of x_{equ}.

Unbounded motion

Motion along the positive x-axis is **unbounded** if for all $x > x_{min}$, $E > V(x)$. (Similarly for unbounded motion along the negative x-axis.) In Fig. 3.4 this occurs at E_3; here, $x_{min} = x_4$. For negative initial velocity at $x > x_4$, the particle reaches the turning point x_4 and subsequently escapes to infinity ($x \to \infty$). If the energy of the particle is $E_2 = V(x_3)$, the motion in the vicinity of x_3 depends on the details of the function $V(x)$. (The shape of the potential then determines, for example, whether x_3 is reached after a finite time interval, or never reached.) A selection of corresponding phase space trajectories is shown in Fig. 3.5. If $E > V(x), \forall x$, the motion is always unbounded.

3.3　Back to the examples

3.3.1　The inclined track

Galileo's studies of particle motion on an inclined plane (in fact, Galileo used a ball rolling along a groove) are probably the first experiments in the history of science that were informed by a modern understanding of how to put questions to nature (cf. Section 1.3). He reported on his experiments at first in his treatise *De motu*, that he began to write in 1586 but never finished it[4].

Consider a body of mass m moving under the influence of gravity, $\mathbf{F} = m\mathbf{g}$, along a straight track inclined at angle α to the gravity vector $\mathbf{g}$ (cf. Fig. 3.1). In the absence of friction, the equation of motion reads (cf. (3.2))

$$m\ddot{s} = mg\cos\alpha. \tag{3.15}$$

The corresponding potential

$$V(s) = -\int_{s_0}^{s} ds'\, mg\cos\alpha = (s_0 - s)\, mg\cos\alpha$$

is monotonically (linearly) decreasing. Consequently, only unbounded motion is possible as $s \to \infty$. The solution

$$s = s_0 + v_0 t + \frac{g\cos\alpha}{2}t^2, \qquad s_0 = s\,(t = 0)\,,\ v_0 = \dot{s}(t = 0) \tag{3.16}$$

is obtained either by direct integration of the equation of motion (3.15) (substituting $\dot{s} = v$; E) or by using conservation of energy (3.7), which here is given by (E)

$$\frac{m}{2}\dot{s}^2 + (s_0 - s)\, mg\cos\alpha = E. \tag{3.17}$$

For initial velocities $v_0 < 0$, the particle moves upward along the track ($s < s_0$) until it reaches a turning point ($\dot{s} = 0$) at time $t_u =$

---

[4]S. Drake, Galileo at work, Chapter 1 (see Bibliography).

$-\dfrac{v_0}{g\cos\alpha}$. After reversing direction, the particle escapes to $s = +\infty$ with constant acceleration. For $\alpha = 0$, the track is vertical and the motion reduces to **free fall** (E).

The trajectories in phase space $(s, \dot{s})$ can be plotted readily (E) using Eq. (3.16) and the velocity

$$\dot{s} = v_0 + gt\cos\alpha. \tag{3.18}$$

3.3.2 The plane pendulum

Galileo brought another object within the scope of scientific investigation: the pendulum[5]. A simple pendulum consists of a body attached to a string or rigid rod, fixed at a pivot point, and allowed to swing freely. In 1602 Galileo found that the time to complete one oscillation depends only on the length[6] (and not on the weight of the swinging body).

The mathematical model of a simple **plane pendulum** consists of a point mass m attached to a string moving under the influence of gravity in an arc of radius l. The motion always remains in this plane, and gravity acts along the plane of motion. Our model pendulum is depicted in Fig. 3.6. At a distance $s = l\phi$ along the arc (let $\phi = 0$ be the equilibrium position, i.e. pendulum's rest position), the only nonzero net force component is F_{tang}, the component of the gravitational force $m\mathbf{g}$ that acts tangentially to the arc:

$$F_{\text{tang}} = -mg\sin\phi.$$

The force component in the direction of the string (perpendicular to the arc) is balanced by the suspension force of the pendulum. Expressing the velocity $\dot{s}$ along the circular orbit in terms of $\dot{\phi}$,

$$\dot{s} = l\dot{\phi},$$

---

[5]It is said that already at the age of seventeen, he had the idea to create a pendulum clock. This occurred to him after the observation a huge, swinging chandelier in the Cathedral of Pisa. He noticed that the time to complete a full swing was independent from the length of the arc traced out by the tip of the chandelier.

But according to S. Drake (*loc. cit.*, Chapter 4) it was not until 1602 that Galileo investigated the oscillations of long pendulums.

[6]S. Drake, *ibid.*, Chapter 4.

the equation of motion[7] is

$$ml\ddot{\phi} + mg\sin\phi = 0.$$

Defining

$$\omega^2 = g/l, \tag{3.19}$$

we have

$$\ddot{\phi} + \omega^2\sin\phi = 0. \tag{3.20}$$

This differential equation is nonlinear in the variable ϕ.

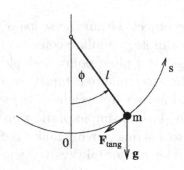

Figure 3.6: The plane pendulum.

Define the potential $\tilde{V}$ to be V measured in energy units mgl:

$$\tilde{V} = V/mgl.$$

Integrating the force F_{tang} over $ld\phi$ (cf. (3.6)), we obtain for $\tilde{V}$,

$$\tilde{V}(\phi) = \int_0^\phi d\phi'\sin\phi' = 1 - \cos\phi, \tag{3.21}$$

where the zero of the energy is chosen such that $\tilde{V}(0) = 0$ (i.e. $E = T$ at $\phi = 0$). The equation for conservation of energy (obtained by multiplying Eq. (3.20) by $l\dot{\phi}$) is

$$\frac{1}{2}\dot{\phi}^2 + \omega^2(1 - \cos\phi) = \varepsilon\omega^2, \qquad \varepsilon := E/ml^2\omega^2,$$

---

[7]This may not appear to be self-evident. However, noting that since $s = l\phi$, the velocity in the direction of F_{tang} is $\dot{s} = l\dot{\phi}$, the law (3.5) can be applied locally. A systematic derivation is given in Section 9.1.

from which the angular velocity follows,

$$\dot{\phi} = \pm\omega\sqrt{2(\varepsilon + \cos\phi - 1)}. \qquad (3.22)$$

After separating the variables ϕ and t, we have the integral

$$t = \pm\frac{1}{\omega}\int_0^{\phi} d\phi' \left/ \sqrt{2\left(\varepsilon - 1 + \cos\phi'\right)}\right.$$

This is an *elliptic integral of first kind* ([Sommerfeld], [Whittaker]). Properties of elliptic integrals are beyond the scope of this book and we do not pursue them here. We mention, however, that in general, the pendulum's period depends on the initial position: only for small values of ϕ may we replace $\sin\phi$ by ϕ, giving the period derived originally by Galileo,

$$T \simeq 2\pi\sqrt{\frac{l}{g}}. \qquad (3.23)$$

(See the treatment below, in particular Eq. (3.27).)

Instead, we discuss qualitatively the trajectories in the phase plane $(\phi, \dot{\phi})$ for several values of the energy ε. The trajectories have the following features, depending on the total energy ε, and considering only the range $|\phi| \leq \pi$:

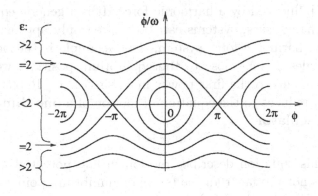

Figure 3.7: The trajectories of the plane pendulum.

a) $\underline{\varepsilon < 2}$: There are turning points ($\dot\phi = 0$) at $\phi = \pm \arccos(1 - \varepsilon) =:$ $\pm\phi_{max}$, with $|\phi_{max}| < \pi$. The motion in configuration space is an *oscillation* between $-\phi_{max}$ and ϕ_{max}. At these extreme angles $d\dot\phi/d\phi = \infty$, i.e. the trajectories in the phase plane are closed curves and intersect the ϕ-axis perpendicularly (cf. Fig. 3.7).

b) $\underline{\varepsilon > 2}$: $\dot\phi$ is always either < 0 or > 0. The pendulum *revolves* in complete circles clockwise or counterclockwise. $\dot\phi$ is minimal at $\phi = \pi$.

c) $\underline{\varepsilon = 2}$: $\dot\phi = 0$ at $\phi = \pm\pi$. The slope of the trajectory at the intersection with the ϕ-axis is finite. Proof: For $\varepsilon = 2$, we have $d\dot\phi/d\phi = \mp\omega \sin(\phi/2)$ and, in particular, at $\phi = \pi$,

$$d\dot\phi/d\phi = \mp\omega.$$

The pendulum *almost completes a revolution*. The time it takes to arrive at $\phi = \pi$ is infinite (E). $\phi = \pi$ is an unstable equilibrium point. A small displacement is sufficient to make the point mass leave (the neighborhood of) $\phi = \pi$.

These features are illustrated in Fig. 3.7.

3.3.3   The harmonic oscillator

We have already derived Eq. (3.4) for a particle acted upon by a force obeying Hooke's Law. However, this equation of motion is valid not just for particles influenced by a harmonic force. It is a generic equation of physics: in many cases, systems can be – often only approximately – thought of as harmonic force system. One example is the .pendulum. For small angles ϕ, i.e. close to the equilibrium position, we expand the force, $\sin\phi$, up to the first power in ϕ, $\sin\phi \simeq \phi$. Replacing ϕ by x, Eq. (3.20) reduces to the equation of motion of a **one-dimensional harmonic oscillator**:

$$\ddot{x} + \omega^2 x = 0. \tag{3.24}$$

Therefore, this equation describes motion in any potential as long as the motion is not too far from the (stable) equilibrium point (recall the derivation of Eq. (3.14) above).

To solve the equation of motion, we make the Euler ansatz,

$$x = Ce^{\lambda t},$$

then from (3.24), we obtain $\lambda^2 + \omega^2 = 0$, or $\lambda = \pm i\omega$. Thus, the general solution is

$$x = Ae^{i\omega t} + Be^{-i\omega t} = C_1 \sin \omega t + C_2 \cos \omega t.$$

Since x must be real, the complex quantities A and B are related by $B = A^*$, with A arbitrary. A and the real constants C_1 and C_2, respectively, are determined by the initial conditions

$$x(t = 0) = x_0 = 2\,\mathrm{Re}\,A = C_2$$

$$\dot{x}(t = 0) = v_0 = 2i\omega\,\mathrm{Im}\,A = \omega C_1.$$

Hence, expressed in terms of the initial values, the solution is

$$x(t) = x_0 \cos \omega t + \frac{v_0}{\omega} \sin \omega t \qquad (3.25)$$

$$v(t) = \dot{x}(t) = v_0 \cos \omega t - \omega x_0 \sin \omega t. \qquad (3.26)$$

This is a parametric representation of an ellipse in the phase plane (x, v), whose center is at the origin.

The period of the motion is independent from the initial position. No matter how large x_0 is, the time it takes for one cycle is

$$T = 2\pi/\omega. \qquad (3.27)$$

Setting $\omega = \sqrt{g/l}$, we recover Galileo's result (3.23) for the period of a pendulum in the linear approximation.

We now give a concrete example of the method discussed in Subsection 2.4.1; namely, we solve Eq. (3.24) with the help of the constants of the motion. It is easy to see that the conservation of energy, given by

$$E = \frac{m}{2}\left(\dot{x}^2 + \omega^2 x^2\right),$$

is an equation for an ellipse in the phase plane. The velocity is $\dot{x} = \sqrt{2E/m - \omega^2 x^2}$. Separating the variables and integrating both sides,

$$\int_{t_0}^{t} dt' = \int_{x_0}^{x} \frac{dx'}{\sqrt{2E/m - \omega^2 x'^2}},$$

we obtain

$$t + I = \frac{1}{\omega} \arcsin\left(\frac{\omega x}{\sqrt{2E/m}}\right), \tag{3.28}$$

where the constant I contains both constants of integration (t_0 and the value of the function at x_0). This can be cast in the form of Eq. (2.11) from Subsection 2.4.1:

$$I = \frac{1}{\omega} \arcsin\left(\frac{\omega x(t)}{\sqrt{2E/m}}\right) - t. \tag{3.29}$$

It is easy to prove that $dI/dt = 0$ **(E)**. We therefore have a second constant of the motion, I, in addition to the energy.

From (3.28), we find for $x(t)$,

$$x(t) = \frac{\sqrt{2E/m}}{\omega} \sin(\omega t + \omega I). \tag{3.30}$$

The relation between (E, I) and the initial values (x_0, v_0) is obtained by comparing this with Eq. (3.25):

$$x_0 = \frac{\sqrt{2E/m}}{\omega} \sin(\omega I), \qquad v_0 = \sqrt{2E/m}\cos(\omega I).$$

Inverting these functions, we have

$$I = \frac{1}{\omega} \arctan\left(\frac{\omega x_0}{v_0}\right) \tag{3.31}$$

and

$$E = \frac{m}{2}\left(v_0^2 + \omega^2 x_0^2\right). \tag{3.32}$$

In (3.25) and (3.26), suppose the initial values are now $(x(t), v(t))$. After time $-t$ has elapsed, the trajectory must arrive at (x_0, v_0), where[8]

$$\begin{aligned}
x_0 &= x(t)\cos(-\omega t) + \frac{v(t)}{\omega}\sin(-\omega t) \\
v_0 &= v(t)\cos(-\omega t) - \omega x(t)\sin(-\omega t).
\end{aligned} \tag{3.33}$$

---

[8]These equations are also obtained by expressing in (3.25) and (3.26) the initial values in terms of $x(t)$ and $v(t)$.

In this representation, the (time independent) initial values x_0 and v_0 are constants of the motion (cf. Eqs. (2.10)). Inserting both values into Eq. (3.32), we see explicitly that the energy is time independent:

$$E = \frac{m}{2} \left(v_0^2 + \omega^2 x_0^2 \right) = \frac{m}{2} \left(v^2 \left(t \right) + \omega^2 x^2 \left(t \right) \right).$$

The energy is the sole conserved quantity (time independent constant of the motion). In agreement with the general considerations of Subsection 2.4.1, the ellipse (3.32) is the trajectory in phase space, and the constant of the motion (3.29) describes the motion of the system on the trajectory.

3.4 The driven, damped oscillator

In the following, we consider the effect on particle motion of a velocity-dependent frictional force that acts together with a time-dependent driving force. In such a system the energy is not conserved.

Frictional forces

Frictional forces **R** are ubiquitous in nature; they dissipate energy. **Static friction** on a *solid surface* is an example of friction: when a body is placed onto an inclined plane, if the angle of inclination is not too great, the body will not move. The static friction force is proportional to the force with which the body is pressed against the surface, and is independent of the velocity. Here, however, we consider a velocity-dependent frictional force, $\mathbf{R} = \mathbf{R}(\mathbf{v})$. The velocity dependence differentiates this force from those considered thus far. The nature of the dependence may differ from system to system. When a particle is placed in a fluid, a frictional forces appears only for $\mathbf{v} \neq 0$. Two cases may be distinguished. (Assume one-dimensional motion.)

- In a *liquid*, at low velocity, in the laminar region, **Stokesian friction**

$$R(v) = -2\gamma mv, \qquad \gamma > 0 \qquad (3.34)$$

occurs. The constant γ is proportional to the viscosity η.

- In *gases*, at high velocity (though less than the speed of sound), **Newtonian friction**

$$R(v) = -\beta v \left| v \right|, \qquad \beta > 0, \qquad (3.35)$$

occurs. The constant β is proportional to the density ρ of the gas.

3.4.1   The driven oscillator with linear damping

We consider a one-dimensional oscillator whose motion is damped by the Stokesian frictional force (3.34), while being subjected to a time-dependent external force $F = mk(t)$. In accordance with the superposition principle, Newton's equation of motion for the particle contains the sum of the three forces (harmonic, Stokesian frictional, and time-dependent forces)

$$\ddot{x} + 2\gamma\dot{x} + \omega^2 x = k(t). \tag{3.36}$$

The *general solution* of this differential equation is

$$x(t) = x_h(t) + x_p(t), \tag{3.37}$$

which is the sum of the *general solution $x_h(t)$ of the homogeneous equation*

$$\ddot{x} + 2\gamma\dot{x} + \omega^2 x = 0 \tag{3.38}$$

and a *particular solution $x_p(t)$ of the inhomogeneous equation* (3.36). Employing the Euler ansatz $x = Ce^{\lambda t}$, the solution of the homogeneous equation is found to be

$$x_h = e^{-\gamma t}(Ae^{-i\bar{\omega}t} + A^* e^{i\bar{\omega}t}), \tag{3.39}$$

with natural frequency

$$\bar{\omega} := \sqrt{\omega^2 - \gamma^2}. \tag{3.40}$$

The damping term shifts the natural frequency of the pure undriven oscillator down (as long as $\gamma < \omega$). The constants $\operatorname{Re} A$ and $\operatorname{Im} A$ in the homogeneous solution are determined from the initial values. The particular solution, however, satisfies $(x_p(0), \dot{x}_p(0)) = (0, 0)$ and is therefore independent of the initial values.

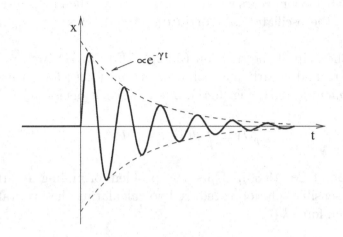

Figure 3.8: The kicking of a damped oscillator.

Kicking the oscillator

Here, we present a mathematical method for obtaining a particular solution of the linear differential equation (3.36). First, we determine the solution for the specific force $k(t) = \delta(t)$, where $\delta(t)$ is the Dirac δ-function (see the Appendix, Section C.1). Newton's equation is then

$$\ddot{x} + 2\gamma\dot{x} + \omega^2 x = \delta(t). \tag{3.41}$$

A particular solution of this equation is called a **Green's function** $G(t)$ (cf. Appendix, Section C.3). The physical interpretation of the equation (3.41) is the following. A damped harmonic oscillator is kicked with a very strong, instantaneous (δ-function-shaped) force. What we want to know is the behavior of the oscillator after the kick, the **response** to the force, if the particle is initially at rest. The particular solution $x_p(t)$ $(= G(t))$,

$$x_p(t) = \frac{1}{\bar{\omega}}\, e^{-\gamma t} \sin(\bar{\omega}t)\, \Theta(t), \tag{3.42}$$

where $\bar{\omega}$ given by (3.40), is calculated in Appendix C (cf. Eq. (C.13)) and graphed in Fig. 3.8. The step function (Heaviside function) $\Theta(t)$ – defined by Eq. (C.1) – is the mathematical expression of causality. Before the kick is delivered, there is no motion (cf. $t < 0$ in Fig. 3.8). For $t > 0$, the particle performs a **damped oscillation** (cf. Fig. 3.8).

Oscillatory behavior occurs when $\gamma < \omega$, since $\bar{\omega}$ is then real. Otherwise, for $\gamma \geq \omega$, the oscillator is **overdamped** and relaxes to equilibrium exponentially.

With the help of the solution (3.42) of Eq. (3.41), we are now in a position to find a particular solution of Eq. (3.36) for an *arbitrary, time-dependent force*. Using (3.41), one can see at once that

$$x_p(t) = \int dt' G(t - t') k(t') \tag{3.43}$$

is a solution of Eq. (3.36). Thus, the problem of finding a particular, physically sensible solution is reduced to calculating the integral (3.43) for the given force $k(t)$.

3.4.2 The periodically driven, damped oscillator

We now consider a particular case of the damped, driven oscillator in which the driving force is periodic with frequency Ω. In Eq. (3.36), we then have $k(t) = k \sin \Omega t$, and the equation of motion reads

$$\ddot{x} + 2\gamma \dot{x} + \omega^2 x = k \sin \Omega t. \tag{3.44}$$

From (3.43) and (3.42), we obtain the particular solution,

$$x_p(t) = \frac{k}{\bar{\omega}} \int_{-\infty}^{t} e^{-\gamma(t-t')} \sin \bar{\omega}(t - t') \sin \Omega t' \, dt'.$$

After computing the integral (E), we have

$$x_p(t) = \frac{\omega^2 - \Omega^2}{(\Omega^2 - \omega^2)^2 + 4\gamma^2 \Omega^2} k \sin \Omega t - \frac{2\gamma \Omega}{(\Omega^2 - \omega^2)^2 + 4\gamma^2 \Omega^2} k \cos \Omega t. \tag{3.45}$$

This particular solution is independent from the initial values. It consists of two parts. The first term on the right hand side of (3.45) is the particle's reaction: the reaction is in phase with the driving force (both $\propto \sin \Omega t$). The second term, proportional to γ, is the absorptive part: it is shifted in phase by $\pi/2$ ($\propto \cos \Omega t$) with respect to the driving force, so the damping effect is delayed. The influence of the damping force can also be seen in the energy balance. Omitting for the moment the external driving force – we just want to focus on the system's energy loss – i.e. multiplying

$$\ddot{x} + 2\gamma\dot{x} + \omega^2 x = 0 \tag{3.46}$$

by $\dot{x}$, we get

$$\frac{d}{dt}\frac{1}{2}\left(\dot{x}^2 + \omega^2 x^2\right) = -2\gamma\dot{x}^2. \tag{3.47}$$

The term on the left is the time derivative of the energy of the free[9] oscillator at time t:

$$E(t) = \frac{m}{2}\left(\dot{x}^2 + \omega^2 x^2\right).$$

Integrating (3.47) gives the energy loss,

$$E(t) - E(0) = -2m\gamma\int_0^t \dot{x}^2 dt' < 0,$$

if the system is not driven.

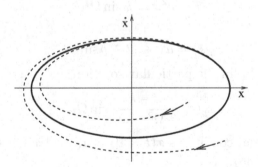

Figure 3.9: The driven, damped oscillator: all trajectories converge asymptotically onto the attracting ellipse.

Asymptotically (i.e. after a very long time), in the complete solution $x(t) = x_h(t) + x_p(t)$, the homogeneous solution (3.39) is vanishingly

---

[9]'Free' means no friction. Otherwise friction occurs, for instance, due to the interaction of the particle with the surrounding fluid, thus transferring energy into the fluid's degrees of freedom – which far outnumber the particle's degrees of freedom.

small, implying the loss of virtually all the information that the initial conditions had imparted to the system. Only the particular solution (3.45) remains, and, as we have seen, this is independent of the initial values. Therefore, for particular values of ω, γ, and Ω, all orbits tend to the same asymptotic state $x_p(t)$, even if the initial values x_0 and $\dot{x}_0$ are different. The trajectories in the phase plane are composed also of a reactive and a damping part. Plotting $\dot{x}_p$ versus x_p gives an ellipse that is independent of the initial values, while the $x_h(t)$ term contributes an exponentially decreasing amount. Depending on whether the initial point in phase space lies inside or outside the ellipse, one finds either one of two situations shown in Fig. 3.9. Asymptotically, all trajectories converge onto the same **attractor**.

Resonance catastrophe

If the oscillations are not damped, $\gamma = 0$, the amplitude in (3.45) diverges if the frequency of the driving force is near the frequency of the oscillator. When $\Omega \to \omega$ in (3.44), a so-called **resonance catastrophe** occurs.

The homogeneous solution of Newton's equation for driven, undamped motion,

$$\ddot{x} + \omega^2 x = k \sin \Omega t, \tag{3.48}$$

is

$$x_h(t) = a \cos \omega t + b \sin \omega t.$$

From (3.45), we obtain the particular solution

$$x_p(t) = \frac{-k}{\Omega^2 - \omega^2} \sin \Omega t.$$

Enforcing the initial conditions $x(t = 0) = x_0$ and $\dot{x}(t = 0) = 0$, yields the solution of (3.48), *viz.*:

$$x(t) = x_0 \cos \omega t + \frac{k}{\Omega^2 - \omega^2} \left(\frac{\Omega}{\omega} \sin \omega t - \sin \Omega t \right).$$

With $\Omega = \omega + \varepsilon$, in the limit $\varepsilon \to 0$, this becomes

$$x(t) = x_0 \cos \omega t + \frac{k}{2\omega} \left(\frac{1}{\omega} \sin \omega t - t \cos \omega t \right). \tag{3.49}$$

Because of the term linear in t, the displacement x increases without bound. The motion is not periodic! We will make use of this result later, in Section 3.6.

In summary, the equation of motion for the linear, driven, damped oscillator can be solved analytically even though no conserved quantity is apparent. For the driven, damped pendulum, the situation is entirely different due to the nonlinearity ($\sin x$) in the equation of motion. In this case, the nonexistence of energy conservation is reflected by the strong dependence of the (numerical) solutions on initial conditions. This feature makes predicting the future state of the pendulum impossible (see the final remarks of Section 4.4).

3.5 Stability of motion

Experimental investigation of a physical process demands that the process itself and the measurements of it must be repeated. Statistical techniques are then usually applied in evaluating the data. But every repetition of a process should have identical initial conditions. One therefore seeks to:

i) *Keep the external conditions* under which the process is observed (e.g. friction, driving force) *constant*; and,

ii) *Start the system's* dynamical variables (e.g. x_0 and $\dot{x}_0$) *at the same initial value* each time.

Technical capability limits the degree to which either of these can be achieved. Even when external conditions are regulated with utmost stringency, perfect accuracy cannot be achieved. A discrepancy in initial conditions will always exist, however small. If a small discrepancy in initial conditions evolves to a bounded discrepancy in the observed states, then the motion is called **stable**. In this section, we focus on the evolution of discrepancies in initial values and assume that the external conditions (forces) are constant.

3.5.1 Two examples

Consider the motion of a particle obeying some one-dimensional equation of motion. The variable $x(t)$ denotes the evolution during the experiment's first run, for which the initial values are

$$x(0) = x_0, \qquad v(0) = v_0. \tag{3.50}$$

On the experiment's second run, the initial vales $\bar{x}(0)$, $\bar{v}(0)$ are in the vicinity of x_0, v_0. We may write

$$\bar{x}(0) = x_0 + \delta x_0, \qquad \bar{v}(0) = v_0 + \delta v_0, \tag{3.51}$$

where the *discrepancies* δx_0 and δv_0 are due to technical inaccuracy in the preparation of the initial state. The corresponding evolution on the second run is denoted $\bar{x}(t)$. How do the differences between the two solutions

$$\delta x := \bar{x}(t) - x(t) \quad \text{and} \quad \delta v := \bar{v}(t) - v(t) \tag{3.52}$$

evolve with time?

For a harmonically bound particle, using (3.25) and (3.26), we obtain

$$\delta x\,(t) \;\;=\;\; \delta x_0 \cos\omega t + (\delta v_0/\omega)\sin\omega t \tag{3.53}$$
$$\delta v\,(t) \;\;=\;\; \delta v_0 \cos\omega t - \omega\delta x_0 \sin\omega t. \tag{3.54}$$

The discrepancy is of the same order of magnitude *for all time*. Indeed, we have

$$(\delta x)^2 + \frac{1}{\omega^2}(\delta v)^2 = (\delta x_0)^2 + \frac{1}{\omega^2}(\delta v_0)^2. \tag{3.55}$$

The discrepancy is bounded. One consequence is that in physical reality, two identical oscillators with slightly different initial positions and velocities always stay in time.

The situation is quite different for the plane pendulum, Eq. (3.20). A numerical calculation of the evolution of a rectangular discrepancy area $\{x_0 - \delta x \le x \le x_0 + \delta x, \; v_0 - \delta v \le v \le v_0 + \delta v\}$ around the point (x_0, v_0) is presented in Fig. 3.10. The area changes its shape with time[10]. The unit of time is the period T corresponding of the initial condition (x_0, v_0). The area moves along the dashed trajectory in the phase plane. The energy ε at every point in the rectangle is smaller than $\varepsilon = 1.45$. Note that the energy limit for oscillatory motion (rather than revolution in complete circles) is $\varepsilon = 2$. The trend is evident. After sufficiently many periods, the area of discrepancy is spread across the whole trajectory. This means that two identical pendulums with slightly different initial values eventually go completely out of synchronization.

Linear stability analysis offers a first answer to whether a system is stable (i.e. whether discrepancy remains bounded).

---

[10]But its magnitude remains the same all the time (cf. Subsection 13.5).

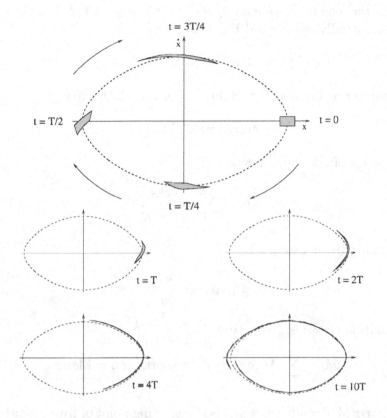

Figure 3.10: The temporal development of plane pendulums starting from different initial values within the shaded area.

3.5.2 Linear stability analysis

General features

A particles motion is determined by a system of ordinary first order differential equations (cf. Subsection 2.4.1),

$$\dot{\mathbf{u}} = \mathbf{G}(\mathbf{u}), \tag{3.56}$$

where $\mathbf{u}$ now denotes the vector $(\mathbf{r}, \mathbf{v})$, and $\mathbf{G}$ depends on the system considered. For Eqs. (2.8), $\mathbf{G} = (\mathbf{v}, \mathbf{F}/m)$. Consider a second solution $\mathbf{u} + \delta\mathbf{u}$ that is initially adjacent to $\mathbf{u}(t)$. It obeys the dynamical equations

$$\dot{\mathbf{u}} + \delta\dot{\mathbf{u}} = \mathbf{G}(\mathbf{u} + \delta\mathbf{u}). \tag{3.57}$$

What is the long time behavior of the discrepancy $\delta\mathbf{u}(t)$? To answer this at least partially, we expand Eq. (3.57) in $\delta\mathbf{u}$,

$$\dot{\mathbf{u}} + \delta\dot{\mathbf{u}} = \mathbf{G}(\mathbf{u}) + (\delta\mathbf{u}\,\boldsymbol{\nabla}_u)\,\mathbf{G}\,|_u + \cdots ,$$

and keep terms linear in $\delta\mathbf{u}$. Subtracting Eq. (3.56) gives

$$\delta\dot{\mathbf{u}} = (\delta\mathbf{u}\,\boldsymbol{\nabla}_u)\,\mathbf{G}\,|_u$$

or, in terms of the components,

$$\delta\dot{u}_i = \sum_j \delta u_j \frac{\partial G_i}{\partial u_j}.$$

Introducing the notation

$$M_{ij}(\mathbf{u}) := \frac{\partial G_i}{\partial u_j}, \tag{3.58}$$

the equation for $\delta\mathbf{u}$ is given by

$$\delta\dot{u}_i = \sum_j M_{ij}\delta u_j \quad \text{or, in short,} \quad \delta\dot{\mathbf{u}} = \mathsf{M}\delta\mathbf{u}. \tag{3.59}$$

The matrix M of coefficients M_{ij} is the starting point of **linear stability analysis**. It is important to note that although this system of equations is linear in δu_k, M generally depends on time, via $\mathbf{u}$. In this general case, the method is rather involved (see e.g. [LaSalle/Lefschetz]).

We discuss here only the case $\mathsf{M}(\mathbf{u})$ independent of time t, i.e. $\mathbf{u}$ is independent of time, $\dot{\mathbf{u}} = 0$. This means $\mathbf{u}$ has to be (close to) a **stationary solution** of (3.56). Such solutions are also called **fixed points** of the system of differential equations. Since M is time independent, Eq. (3.59) has the solution,

$$\delta\mathbf{u} = \exp(\mathsf{M}t)\,\delta\mathbf{u}_0, \tag{3.60}$$

where $\exp(\mathsf{M}t)$ is defined by the Taylor expansion, $\exp(\mathsf{M}t) = 1 + \mathsf{M}t + \frac{1}{2}(\mathsf{M}t)^2 + \cdots$, and $\delta\mathbf{u}_0$ is the initial discrepancy. The solution (3.60) becomes more useful if one introduces the eigenvalues λ_i and the eigenvectors $\mathbf{w}_i$ of M,

$$\mathsf{M}\mathbf{w}_i = \lambda_i\mathbf{w}_i. \tag{3.61}$$

Since the real matrix M in general is not symmetric, the eigenvalues and the eigenvectors are, in general, complex. In the complex case, $\mathbf{w}_i^*$ is also an eigenvector and belongs to the eigenvalue λ_i^*. One must bear this in mind when expanding $\delta\mathbf{u}_0$ in the linearly independent (but not necessarily orthogonal) eigenvectors $\mathbf{w}_i$,

$$\delta\mathbf{u}_0 = \sum_i c_i \mathbf{w}_i. \tag{3.62}$$

Since $\delta\mathbf{u}_0$ is real, the coefficients c_i, to be determined from (3.62), are in general complex. Because of $M^n\mathbf{w}_i = \lambda_i^n\mathbf{w}_i$, it follows from (3.60) that

$$\delta\mathbf{u} = \exp(Mt)\sum_i c_i\mathbf{w}_i = \sum_i c_i \exp(\lambda_i t)\,\mathbf{w}_i. \tag{3.63}$$

Without loss of generality, we assume λ_1 to have the largest real part. Then the dominant behavior for $t \to \infty$ is

$$\delta\mathbf{u} \simeq c_1 \exp(\lambda_1 t)\,\mathbf{w}_1.$$

Hence, if only one of the eigenvalues has a positive real part, the initial discrepancy increases without bound.

Let us now consider the phase plane (x, v) of a one-dimensional system. From the coupled system of differential equations

$$\dot{x} = v$$
$$\dot{v} = F(x, v)/m,$$

in accordance with Eq. (3.58), one obtains the matrix

$$M = \begin{pmatrix} 0 & 1 \\ \dfrac{1}{m}\dfrac{\partial F}{\partial x} & \dfrac{1}{m}\dfrac{\partial F}{\partial v} \end{pmatrix}. \tag{3.64}$$

There are two eigenvalues, λ_1 and λ_2, together with the corresponding linearly independent eigenvectors $\mathbf{w}_1$ and $\mathbf{w}_2$, which are time independent for stationary points. Expressed in terms of these quantities, we have

$$\delta\mathbf{u}(t) = c_1 \exp(\lambda_1 t)\,\mathbf{w}_1 + c_2 \exp(\lambda_2 t)\,\mathbf{w}_2. \tag{3.65}$$

The eigenvalues λ_i are either both real or complex conjugate (since M is a real matrix). Letting $t \to \infty$, one has the following cases (cf. Fig. 3.11):

a) λ_1 and $\lambda_2 > 0$: $|\delta u|$ increases without limit. u is a *totally unstable point (repeller)*.

b) $\lambda_1 > 0$ and $\lambda_2 < 0$: for $c_1 \neq 0$ the discrepancy $|\delta u|$ increases, for $c_1 = 0$ it vanishes: $|\delta u| \rightarrow 0$. u is a *saddle point* or *hyperbolic point*.

c) λ_1 and $\lambda_2 < 0$: $|\delta u| \rightarrow 0$. u is a *stable point (attractor)*.

d) λ_1 and λ_2 are imaginary $(\lambda_2 = \lambda_1^*)$: $|\delta u|$ stays bounded. u is an *elliptic point*.

e) λ_1 and λ_2 are complex conjugate and Re $\lambda_i < 0$: $|\delta u| \rightarrow 0$. u is the *attractor* of a spiralling flow of δu.

f) λ_1 and λ_2 are complex conjugate and Re $\lambda_i > 0$: $|\delta u| \rightarrow \infty$. u is the *repeller* of a spiralling flow of δu.

We present three concrete examples of this general mathematical treatment, focusing on the three cases of a particle's motion considered in the beginning of this chapter.

Particular examples

i) The harmonic oscillator: $\ddot{x} + \omega^2 x = 0$

Inserting the force
$$F = -m\omega^2 x$$
into Eq. (3.64) yields the matrix

$$M = \begin{pmatrix} 0 & 1 \\ -\omega^2 & 0 \end{pmatrix}.$$

Linear stability analysis can be performed globally (i.e. without referring to fixed points), since M is independent of $u = (x, v)$ and therefore of t. For *all* points in the phase plane, the discrepancy equations are
$$\delta \dot{x} = \delta v, \qquad \delta \dot{v} = -\omega^2 \delta x.$$

This is a special property of the harmonic oscillator. Setting $\det(\lambda 1 - M) = 0$ yields the eigenvalues

$$\lambda_{1,2} = \pm i\omega.$$

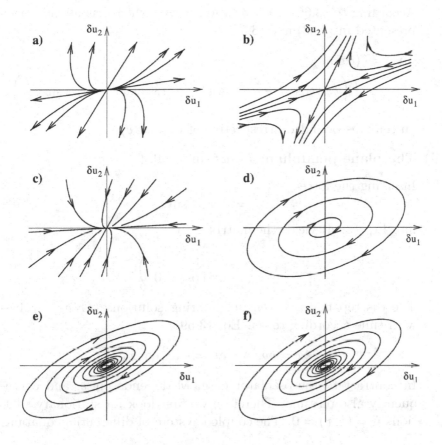

Figure 3.11: The kinds of behavior in the vicinity of a fixed point.

This is case (d) above and in Fig. 3.11. The respective eigenvectors are

$$\mathbf{w}_1 = \frac{1}{\sqrt{1+\omega^2}}\begin{pmatrix} 1 \\ i\omega \end{pmatrix} \quad \text{and} \quad \mathbf{w}_2 = \frac{1}{\sqrt{1+\omega^2}}\begin{pmatrix} 1 \\ -i\omega \end{pmatrix};$$

they form a complex conjugate pair. The initial discrepancy is

$$\delta\mathbf{u}_0 = (\delta x_0, \delta v_0) = 2\,\text{Re}(c_1\mathbf{w}_1) = (2\,\text{Re}\,c_1, -2\omega\,\text{Im}\,c_1)/\left(1+\omega^2\right)^{1/2}.$$

Solving for c_1 gives

$$c_1 = (\delta x_0 - i\delta v_0/\omega)(1+\omega^2)^{1/2}.$$

According to (3.65), the evolution of $\delta\mathbf{u}$, which has already been presented in (3.53) and (3.54), is

$$
\begin{aligned}
\delta\mathbf{u} &= (\delta x, \delta v) \\
&= \left(\left(\delta x_0 \cos\omega t + \frac{\delta v_0}{\omega} \sin\omega t \right), (\delta v_0 \cos\omega t - \delta x_0 \omega \sin\omega t) \right).
\end{aligned}
$$

$\delta\mathbf{u}$ remains bounded, irrespective of $\mathbf{u} = (x, v)$.

ii) **The plane pendulum:** $\ddot{x} + \omega^2 \sin x = 0$, $(\phi \to x)$

Inserting the force

$$
F = -m\omega^2 \sin x
$$

into Eq. (3.64) yields the matrix

$$
\mathsf{M} = \begin{pmatrix} 0 & 1 \\ -\omega^2 \cos x & 0 \end{pmatrix}.
$$

The discrepancy between neighboring solutions $(\delta x, \delta v)$ evolves with time according to (cf. Eq. (3.59))

$$
\delta\dot{x} = \delta v, \qquad \delta\dot{v} = -\omega^2 \cos x \, \delta x.
$$

In contrast to the oscillator case, M depends on x, and consequently also on time. Therefore we first look for stationary solutions $\dot{\mathbf{u}} = (\dot{x}, \dot{v}) = \mathbf{0}$. The coupled system of differential equations

$$
\dot{x} = v, \qquad \dot{v} = -\omega^2 \sin x
$$

has vanishing time derivatives $(\dot{x}, \dot{v})$ at the stationary points

$$
(x, v) = (0, 0) \qquad \text{and} \qquad (x, v) = (\pi, 0).
$$

At the stationary point $(0, 0)$, the matrix M is

$$
\mathsf{M} = \begin{pmatrix} 0 & 1 \\ -\omega^2 & 0 \end{pmatrix},
$$

and has eigenvalues

$$
\lambda_{1,2} = \pm i\omega.
$$

In a neighborhood of this point, the discrepancies do not grow (cf. Fig. 3.11 (d); see also the oscillator case). The size of this neighborhood cannot be determined from linear stability analysis.

At the second stationary point $(\pi, 0)$,

$$M = \begin{pmatrix} 0 & 1 \\ \omega^2 & 0 \end{pmatrix},$$

and the eigenvalues are

$$\lambda_1 = \omega \qquad \text{and} \qquad \lambda_2 = -\omega.$$

The future motion of the pendulum for displacements δx_0 from the vertical position $x = \pi$ (the point mass is above the pivot) depends on the direction of the initial velocity δv_0. If δv_0 has the same sign as δx_0, the eigenvector is $\mathbf{w}_1$, and the point mass goes away from the equilibrium position and 'falls'. If the sign of δv_0 is opposite to δx_0, the eigenvector is $\mathbf{w}_2$, and the point mass tends to the equilibrium position (**E**). We have a saddle point (Fig. 3.11 (b)).

iii) **The damped oscillator:** $\ddot{x} + 2\gamma\dot{x} + \omega^2 x = 0$

Inserting the force

$$F = -2m\gamma v - \omega^2 x$$

into Eq. (3.64) yields

$$M = \begin{pmatrix} 0 & 1 \\ -\omega^2 & -2\gamma \end{pmatrix}.$$

Since M is independent of (x, v), linear stability analysis can be applied globally. The eigenvalues are $(\gamma < \omega)$:

$$\lambda_{1,2} = -\gamma \pm i\sqrt{\omega^2 - \gamma^2}.$$

Since $\operatorname{Re} \lambda_i < 0$, $i = 1, 2$, the differences in the initial values decay exponentially. It is shown in Fig. 3.11 (e). Damping causes the oscillations of the particle to decay to the equilibrium point $(0, 0)$.

3.6 Anharmonic one-dimensional motion

Newton's equation for harmonic one-dimensional motion is solvable. If the forces – or potentials – are not harmonic, then the motion is still periodic (if it is bounded), in accordance with the general considerations in Section 3.2 (see, in particular, Eq. (3.10)). However, in this case, it might not be easy to determine a solution that is manifestly periodic. We give an example of such a case here.

A simple nonlinear system

We consider **Duffing's oscillator**, whose equation of motion,

$$\ddot{x} + \omega_0^2 x + \mu x^3 = 0, \tag{3.66}$$

is a nonlinear differential equation. Equation (3.66) arises, for instance, if one expands the sine term in the equation for the pendulum to third order, $\sin x = x - x^3/6$. One reason for doing this is to study the onset of nonlinear behavior. The potential corresponding to the force in Duffing's oscillator (fixing the mass by $m = 1$) is

$$V = \frac{1}{2}\omega_0^2 x^2 + \frac{\mu}{4}x^4. \tag{3.67}$$

From energy conservation, $E = \dot{x}^2/2 + V$, one can calculate $x(t)$ without approximations. The solution is an elliptic integral, which is periodic. In the limit $\mu \to 0$ – in a power series expansion of $x(t)$ in μ, only the μ^0-contribution remains – one obtains the well known solution to the harmonic oscillator with natural frequency ω_0.

Although the exact solution is readily obtainable here, it is instructive to solve the problem using an approximate method, too, since such a method may be applicable to other nonlinearities that are less amenable to exact analysis. One may choose, for instance, a Fourier series ansatz for $x(t)$ in the framework of harmonic analysis. Another possibility is a perturbation expansion. In the perturbation method, a solution is sought for an amended problem consisting only of the exactly solvable (preferably major) contribution to the interaction potential, while the remaining (smaller) contribution (e.g. the x^4-contribution in Eq. (3.67)) is treated as a 'perturbation'. The perturbation method uses the 'perturbing potential' to generate corrections to the exact solutions of the simplified problem in a systematic way. For the perturbation method to be useful, the perturbing potential must be small compared to the non-perturbing potential. In the following, a perturbation expansion is carried out for Duffing's oscillator.

We begin with a power series expansion of $x(t)$ in the parameter μ,

$$x(t) = x_0(t) + \mu x_1(t) + \mu^2 x_2(t) + \dots . \tag{3.68}$$

Inserting this expansion into the equation of motion (3.66), and collecting the contributions with the same power of μ, we regain a power series in μ whose coefficients contain the functions $x_n(t)$ and their time derivatives. Since μ can vary arbitrarily, the coefficients of each power μ^n, $n = 0, 1, 2, \ldots$, must equal zero. This gives the following ordinary differential equations:

$$\mu^0 \quad : \quad \ddot{x}_0 + \omega_0^2 x_0 = 0 \tag{3.69}$$

$$\mu^1 \quad : \quad \ddot{x}_1 + \omega_0^2 x_1 = -x_0^3 \tag{3.70}$$

$$\mu^2 \quad : \quad \ddot{x}_2 + \omega_0^2 x_2 = -3x_0^2 x_1 \tag{3.71}$$

etc.

The solution of the first equation is

$$x_0(t) = C \sin(\omega_0 t + \alpha),$$

where C and α are the integration constants. Inserting this solution into the second equation (3.70) yields

$$\begin{aligned}
\ddot{x}_1 + \omega_0^2 x_1 &= -C^3 \sin^3(\omega_0 t + \alpha) \\
&= -\frac{3}{4} C^3 \sin(\omega_0 t + \alpha) + \frac{1}{4} C^3 \sin(3\omega_0 t + 3\alpha). \tag{3.72}
\end{aligned}$$

We recognize this as simply an equation for a driven oscillator with natural frequency ω_0, and we have already computed the solution (see Eqs. (3.48) and (3.49) with $\omega \to \omega_0$ and $\Omega \to \omega$; E):

$$x_1(t) = \frac{3t}{8\omega_0} C^3 \cos(\omega_0 t + \alpha) - \frac{1}{32\omega_0^2} C^3 \sin(3\omega_0 t + 3\alpha) + C_1 \sin(\omega_0 t + \alpha_1),$$

$$\tag{3.73}$$

where C_1 and α_1 are, again, constants of integration.

The term linear in t appears because the driving term on the right hand side of Eq. (3.72) is proportional to $\sin(\omega_0 t + \alpha)$. This linear term increases without bound with time – the system seems to be heading for a resonance catastrophe. However, we know that the solution is periodic. The apparent contradiction is caused by the approximation[11]. In

---

[11] A simple example for an analogous situation is the expansion of $\sin(\omega t + \varepsilon t)$ in ε (actually εt):

$$\sin(\omega t + \varepsilon t) = \sin \omega t + \varepsilon t \cos \omega t - (1/2)(\varepsilon t)^2 \sin \omega t - \ldots \,.$$

The series does not converge uniformly in t. For moderate values of time t, where t is measured in units of the period $\tau = 2\pi/\omega$, the expansion can be useful.

celestial mechanics, where perturbation calculations of this kind originated, an unbounded term like this is called a **secular term**. Since
in celestial time scale τ is enormous, deviations from the true solution
remain unnoticeable in many situations. For longer times, one must use
a method that leads directly to periodic solutions.

Manifestly periodic solutions

As we have seen, the perturbation method produces a secular term
which originates from the zeroth order term with frequency ω_0. Now,
the frequency ω_0 determines the curvature of the potential close to the
minimum (cf. Section 3.2); however, the true frequency depends on the
behavior of the potential between the turning points (Fig. 3.4), as well
as on the energy E of the system, $\omega = \omega(E) = 2\pi/\tau$ ($\tau = \tau(E)$, see
(3.10)). Therefore, the contributions calculated from the power series
expansion about $\mu = 0$ should produce two functional dependencies:
the functional form of the actual solution we are seeking, as well as the
frequency shift. For this reason, frequency corrections, too, have to be
extracted from the power series, and summed to give a **renormalized
frequency** ω. The remaining terms in the power series sum up to the
true solution[12], which is periodic with frequency ω.

Accordingly, in addition to the power series expansion (3.68) for
$x(t)$, we introduce also the – yet to be determined – frequency ω, which
depends on μ. Since, for $\mu = 0$, the frequency ω should reduce to ω_0,
we write for ω^2 the power series in μ as

$$\omega^2 = \omega_0^2 + \mu w_1 + \mu^2 w_2 + \ldots , \tag{3.74}$$

where the coefficients w_n are to be determined. We express ω_0^2 in Eq.
(3.66) in terms of the exact frequency ω^2,

$$\ddot{x} + (\omega^2 - \mu w_1 - \mu^2 w_2 - \ldots)x = -\mu x^3, \tag{3.75}$$

and, together with the expansion (3.68) for $x(t)$, we have now the following system of equations for the coefficients of the powers μ^n:

$$\mu^0 : \qquad \ddot{x}_0 + \omega^2 x_0 = 0 \tag{3.76}$$

$$\mu^1 : \qquad \ddot{x}_1 + \omega^2 x_1 = -x_0^3 + w_1 x_0 \tag{3.77}$$

---

[12]In our case, instead of the solution $x(t) = x_0 \sin \omega_0 t$, a function *and* a frequency
depending on μ will exist, and the solution will be of the form $x(t) = f_\mu(\omega(\mu)t)$.
$f_\mu(\phi)$ is periodic in ϕ. In the limit $\mu \to 0$, we have $\omega(\mu) \to \omega_0$ and $f_\mu(\phi) \to x_0 \sin \phi$.

$$\mu^2 \quad : \qquad \ddot{x}_2 + \omega^2 x_2 = -3x_0^2 x_1 + w_2 x_0 + w_1 x_1 \qquad (3.78)$$

etc.

Note the differences between these and Eqs. (3.69), ... ! As above, the solution to Eq. (3.76) is

$$x_0(t) = C\sin(\omega t + \alpha).$$

Inserting this into Eq. (3.77) yields an equation for x_1,

$$\ddot{x}_1 + \omega^2 x_1 = -\frac{3}{4}C^3 \sin(\omega t + \alpha) + \frac{1}{4}C^3 \sin(3\omega t + 3\alpha) + w_1 C \sin(\omega t + \alpha).$$
$$(3.79)$$

Here, unlike the case of Eq. (3.72), the additional term proportional to w_1 enables us to avoid secular terms arising from the $\sin(\omega t + \alpha)$ driving term. In order to get rid of the $\sin(\omega t + \alpha)$ contributions, we simply choose

$$w_1 = 3C^2/4 \qquad (3.80)$$

and get the solution

$$x_1(t) = -\frac{1}{32\omega^2}C^3 \sin(3\omega t + 3\alpha) + C_1 \sin(\omega t + \alpha_1), \qquad (3.81)$$

with

$$\omega^2 = \omega_0^2 + \frac{3}{4}\mu C^2 + O(\mu^2). \qquad (3.82)$$

We have enough constants of integration of the zeroth order equation to satisfy the initial conditions, so we are free to set $C_1 = 0$, whence the solution in first order reduces to

$$x(t) = C\sin(\omega t + \alpha) - \frac{\mu}{32\omega^2}C^3 \sin(3\omega t + 3\alpha) + O(\mu^2). \qquad (3.83)$$

Choosing initial values

$$x(t=0) = x_0 \qquad \text{and} \qquad \dot{x}(t=0) = 0, \qquad (3.84)$$

fixes the integration constants α and C from the equations:

$$x_0 = C\sin\alpha - \frac{\mu}{32\omega^2}C^3 \sin 3\alpha, \qquad 0 = \omega C\cos\alpha - \frac{3\mu}{32\omega}C^3 \cos 3\alpha.$$

A solution to the second equation is $\alpha = \pi/2$, so that C follows from

$$x_0 = C + \frac{\mu}{32\omega^2}C^3 + O(\mu^2).$$

Conversely, expressing C in terms of x_0 (ansatz: $C = c_0 + \mu c_1 + \ldots$), we find

$$C = x_0 - \frac{\mu}{32\omega^2} x_0^3 + O(\mu^2). \tag{3.85}$$

Inserting this into (3.82) and (3.83) yields the solution – to first order in μ – given the initial conditions (3.84):

$$x(t) = x_0 \cos \omega t - \frac{\mu}{32\omega^2} x_0^3 (\cos \omega t - \cos 3\omega t) + O(\mu^2), \tag{3.86}$$

and

$$\omega^2 = \omega_0^2 + \frac{3}{4} \mu x_0^2 + O(\mu^2). \tag{3.87}$$

The result for the renormalized frequency ω^2 agrees with that one obtains by expanding the period τ given in Eq. (3.10) in powers of μ (E). The solution (3.86) is periodic, and ω depends on the amplitude (i.e. the distance between the turning points) as expected. This is true also for higher orders. In the additional harmonic terms that arise here, only odd multiples of the frequency ω appear: $3\omega, 5\omega, \ldots$. To determine the quality of the expansion (e.g. convergence properties) requires separate investigations.

Problems and examples

1. Discuss the motion of a particle along an inclined track. Solve the equation of motion and find the trajectories in the phase plane. When determining the time dependence, why is it advantageous to use an inclined track instead of a freely falling body?

2. Using conservation of energy (3.17), find the solution for motion on an inclined track.

3. A particle of mass m moves in a potential $V(x)$, starting from $x = 0$, towards the turning point x_u, $0 < x_u < a$. The potential in the interval $[0, a]$ is given by:

   a) $V(x) = V_0 \dfrac{x}{a}$;

   b) $V(x) = V_0 \left(1 - \left(\dfrac{x}{a} - 1 \right)^2 \right)$;

in both cases $V_0 > 0$. The energy E of the particle is slightly less than V_0: $E = V_0 \left(1 - \delta^2\right)$. How long does it take the particle to move from $x = 0$ to $x = x_u$? Discuss case (b) in the limit $\delta \to 0$.

4. Integrate the angular velocity $\dot{\phi}$ given in Eq. (3.22), and for energy value $\varepsilon = 2$, determine how long it takes the pendulum to arrive at $\phi = \pi$. (Solution: $\phi = -\pi + 4 \arctan(e^t)$.)

5. Prove that $dI/dt = 0$ for I defined in Eq. (3.29) by using conservation of energy (3.17).

6. Calculate the particular solution

$$x_p(t) = \frac{k}{\overline{\omega}} \int_{-\infty}^{t} e^{-\gamma(t-t')} \sin \overline{\omega}(t - t') \sin \Omega t' \, dt'.$$

The result is given in (3.45).

7. Derive the resonance solution, (3.49), for the driven oscillator.

8. Determine $\delta \mathbf{u}(t) = (\delta x, \delta v)$ for the plane pendulum in the vicinity of the unstable stationary point $(\pi, 0)$, and discuss the motion in the vicinity of this saddle point.

9. Solve Eq. (3.72).

10. Using a power series expansion, calculate, to first order in μ, the period τ of the motion in the potential given by Eq. (3.67) from the expression in Eq. (3.10). Compare the result with (3.87).

4

Encountering peculiar motion in two dimensions

To solve the equations of motion for a particle restricted to move in two-dimensional configuration space, even if the system is conservative, one requires at least one further integral of the motion in addition to the energy, $E = T(\dot{x}, \dot{y}) + V(x, y)$.

But if no more *isolating* (i.e. 'usable') integrals exist, and the system is nonlinear, chaotic behavior may appear.

4.1 The two-dimensional harmonic oscillator

We consider first the equations of motion of a two-dimensional anisotropic harmonic oscillator,

$$\ddot{x} + \omega_1^2 x = 0$$

$$\ddot{y} + \omega_2^2 y = 0. \tag{4.1}$$

The equations are not coupled (in general they may be coupled), meaning that the motion *separates* in the two one-dimensional motions, along the x-axis and y-axis, respectively. Consequently, the energies of the motion in the x-direction and in the y-direction are conserved independently:

$$\frac{m}{2}\left(\dot{x}^2 + \omega_1^2 x^2\right) = E_1 \tag{4.2}$$

$$\frac{m}{2}\left(\dot{y}^2 + \omega_2^2 y^2\right) = E_2. \tag{4.3}$$

(See remark (ii) on Page 31.) In the following, instead of considering E_1 and E_2 separately to be our independent conserved quantities, we choose E_2, together with the total energy $E = T + V$,

$$E = E_1 + E_2 = \frac{m}{2}\left(\dot{x}^2 + \dot{y}^2 + \omega_1^2 x^2 + \omega_2^2 y^2\right). \tag{4.4}$$

From Eqs. (4.3) and (4.4), $\dot{x}$ (=: v_x) and $\dot{y}$ (=: v_y) can be written explicitly as functions of x and y: The system is integrable. According to the general considerations of Subsection 2.4.1, a third conserved quantity must exist. However, as we show in Section 4.3, this quantity is 'useless' (specifically, *nonisolating*) in general. It can be expressed explicitly only for certain frequency ratios ω_1/ω_2. But the two independent conserved quantities we have already are sufficient to solve the equations of motion.

Since for the motion of the particle only the ratio ω_1/ω_2 is relevant, we choose the time scale such that $\omega_1 t \to t$, and set

$$\omega := \omega_2/\omega_1.$$

Eqs. (4.1) now read

$$\begin{aligned} \ddot{x} + x &= 0 \\ \ddot{y} + \omega^2 y &= 0. \end{aligned} \tag{4.5}$$

The solution of these equations of motion for the initial values x_0, y_0, v_{0x}, and v_{0y} are (cf. Eq. (3.25)):

$$\begin{aligned} x(t) &= x_0 \cos t + v_{0x} \sin t \\ y(t) &= y_0 \cos \omega t + (v_{0y}/\omega) \sin \omega t. \end{aligned} \tag{4.6}$$

The motion in configuration consists of **Lissajous's figures**. Their intricacy depends on ω. Three examples are presented on the left hand side of Fig. 4.2. If ω is rational, the orbit is closed; while for ω irrational, the motion never repeats, and eventually an entire region is covered by the orbit.

In four-dimensional phase space, with coordinates $\dot{x}, \dot{y}, x$, and y, trajectories are confined to the two-dimensional intersections of two (hyper)surfaces: the three-dimensional surface given by conservation of the total energy (4.4),

$$\dot{x}^2 + \dot{y}^2 + x^2 + \omega^2 y^2 = 2E/m, \tag{4.7}$$

and the surface given by the energy of the x- or y-component. We choose the energy (4.3) of the y-component,

$$\dot{y}^2 + \omega^2 y^2 = 2E_2/m, \tag{4.8}$$

i.e. a cylinder along the x-axis with an elliptical cross section. To fix a particular (one-dimensional) curve in phase space, we require a third conserved quantity that would define a third surface. The trajectory would then be given as the curve of intersection of these three surfaces.

Since a four-dimensional space is not easy to imagine, we use a conservation law to eliminate one coordinate of phase space, thus reducing the number of dimensions from four to three. From (4.7), let us express $\dot{x}$ in terms of the remaining variables:

$$\dot{x} = \sqrt{2E/m - \dot{y}^2 - x^2 - \omega^2 y^2}. \tag{4.9}$$

We are then left with a three-dimensional space with coordinates $x, y,$ and $\dot{y}$. In this reduced phase space, the energy E_2 defines a tube with elliptical cross-section. The trajectory is the curve of intersection of that tube with the surface given by the third, independent conserved quantity[1] (as shown for the isotropic oscillator in Fig. 4.3). However, as already mentioned in the introduction, a third conserved quantity cannot be given for arbitrary values of ω.

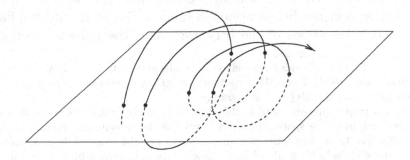

Figure 4.1: The sequence of intersection points of the trajectory in the Poincaré surface.

---

[1] E_1 can be expressed in terms of E and E_2, and therefore cannot be used!

A particular technique of representing the trajectory of bounded motion in four-dimensional phase space is due to Poincaré[2]. In the reduced phase space $(x, y, \dot{y})$, a surface is chosen, called **Poincaré surface** or **Poincaré section**, that partitions the phase space in such a way that the trajectory[3] repeatedly intersects it. The sequence of the **intersection points** or **return points** is notionally given by a **Poincaré map**, which is a mapping of a subset of the Poincaré section into itself. The Poincaré map is usually not known explicitly and its values are normally calculated numerically. A schematic view is presented in Fig. 4.1. This representation of the dynamics is particularly useful when the equations of motion can be solved only numerically.

In the case of the oscillator, a suitable choice for Poincaré section is the plane $x = 0$ (observe the symmetry under the transformation $x \to -x$). The intersection of the trajectory with the Poincaré section generates a sequence of return points[4] $(y, \dot{y})$. Now, for the oscillator defined by Eq. (4.1), the curve of intersection of the Poincaré section and the surface of constant energy E_2, (4.8), is the ellipse $\dot{y}^2 + \omega^2 y^2 = 2E_2/m$ in the plane $x = 0$. Therefore every trajectory is mapped onto a sequence of points on that ellipse. Figure 4.2 shows a trajectory's orbit and corresponding sequences of the Poincaré map for three values of ω. (Note that for fixed ω, varying the energy E does not affect the shape of the orbit or the number of intersections.) Figures 4.2 (a) and (b) correspond to the rational values $\omega = 1/2$ and $1/5$, while Fig. 4.2 (c) shows the orbit and the intersections for the 'not so rational' value $\omega = 0.679$. We can see from these figures that there is a one-to-one correspondence between intersections of the orbit with the line $x = 0$ and intersections of the trajectory with the Poincaré section.

---

[2]Poincaré H., Les Méthodes Nouvelles de la Mécanique Céleste (New Methods of Celestial Mechanics), Tome III: Invariants intégraux. Solutions périodiques du deuxième genre, Paris 1899 (see Bibliography).

The return map (Poincaré map) faithfully reflects many properties of the flow of the trajectories in phase space. In many ways the discrete-time dynamics of are easier to analyze than the continuous-time dynamics of the original flow. Poincaré exploited these ideas in his study of homoclinic orbits in the three-body problem. A **homoclinic orbit** is an orbit which starting from a saddle point (i.e. a hyperbolic fixed point) retuns to this point (see e.g. [Tabor]).

[3]One has to take care that the surface is not parallel to any part of the trajectory.

[4]As a rule, one only records intersections in which the surface is crossed in a certain direction; e.g., in our case, only points for which $\dot{x} > 0$.

For irrational values of ω, the orbit appears to *fill completely* an entire region[5] of the two-dimensional configuration space. In phase space, however, only the ellipse – the curve of intersection of the Poincaré section with the E_2-tube – is covered.

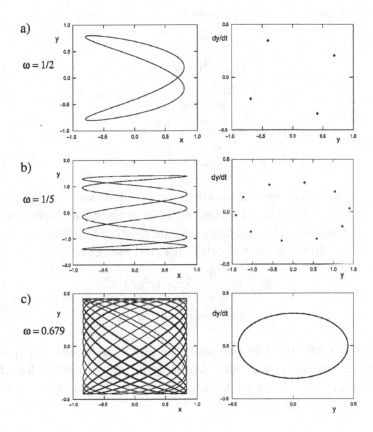

Figure 4.2: The two-dimensional harmonic oscillator: orbits and Poincaré maps for various values of ω.

> The fact *that the points of intersection are all situated on a certain curve is a consequence of the existence and the form of the conserved quantity* E_2. As one can see from Fig. 4.2,

---

[5]The size of this region is determined by the total energy (E).

in general, the orbit $y = y(x)$ is not unique; rather, *branches exist*, the number of these depending on the value of ω.

The two-dimensional isotropic harmonic oscillator

In the isotropic oscillator, whose equations of motion are

$$\begin{aligned} \ddot{x} + \omega^2 x &= 0 \\ \ddot{y} + \omega^2 y &= 0, \end{aligned} \tag{4.10}$$

with solution

$$\begin{aligned} x(t) &= x_0 \cos\omega t + \frac{v_{0x}}{\omega} \sin\omega t \\ y(t) &= y_0 \cos\omega t + \frac{v_{0y}}{\omega} \sin\omega t, \end{aligned} \tag{4.11}$$

the frequencies of the motion in the x and y direction are equal. Now, the parameter ω could be absorbed equally in the time dependences of the two coordinates. This explains why all isotropic oscillators have the same features. In addition to both conserved energies,

$$E_1 = \frac{m}{2} \left(\dot{x}^2 + \omega^2 x^2 \right) \tag{4.12}$$

$$E_2 = \frac{m}{2} \left(\dot{y}^2 + \omega^2 y^2 \right), \tag{4.13}$$

the isotropic oscillator has one more independent conserved quantity that can be explicitly written. Multiplying the equation for the x-component in Eq. (4.10) by y, and the y-component by x, and subtracting one from the other, yields

$$(x\ddot{y} - y\ddot{x}) = \frac{d}{dt} (x\dot{y} - y\dot{x}) = 0.$$

Since $(x\dot{y} - y\dot{x})$ is just the *z-component of the angular momentum*, we see that

$$L_z = m (x\dot{y} - y\dot{x}) = const =: L \tag{4.14}$$

is a conserved quantity. Because the three conserved quantities E_1, E_2, and L are independent[6], we have the maximal number of independent

---

[6]The quantities are independent (cf. section 14.1), if the differentials

$$dE_1 = m\omega^2 x dx + m\dot{x} d\dot{x}$$

conserved quantities. Each conserved quantity forms a three-dimensional hypersurface in the four-dimensional phase space (with coordinates $(x, y, \dot{x}, \dot{y})$). The curve of the intersection of this three hypersurfaces is the trajectory of the particle. The points on the curve have coordinates $(x(t), y(t), \dot{x}(t), \dot{y}(t))$, which contain information about the particle's position and velocity at time t. The particle's orbit $y(x)$ is the projection of the trajectory onto configuration space (x, y).

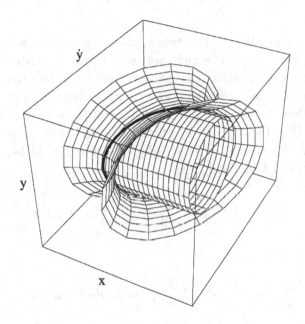

Figure 4.3: The trajectory of the isotropic oscillator in reduced phase space (cf. text).

To demonstrate these facts graphically, we eliminate again the variable $\dot{x}$, though this time using the conservation law (4.12),

$$\dot{x} = \pm\sqrt{\frac{2}{m}E_1 - \omega^2 x^2},$$

$$dE_2 = m\omega^2 y dy + m\dot{y}d\dot{y}$$

$$dL = m\dot{y}dx - m\dot{x}dy - my d\dot{x} + mx d\dot{y}$$

are linearly independent, i.e. if the matrix $(\partial I_k/\partial u_j)$ has rank 3 ($I_k = E_1, E_2, L$ and $u_j = x, y, \dot{x}, \dot{y}$). This is the case here (E).

and restrict ourselves to the reduced phase space; i.e. the three-dimensional projection of phase space with coordinates x, y, and $\dot{y}$. The second conservation law, Eq. (4.13), represents a tube with elliptical cross section parallel to the x-axis. The square of conservation of angular momentum, Eq. (4.14), also gives a two-dimensional surface: a hyperboloid consisting of one sheet, whose axis lies in the $(x, \dot{y})$-plane (E):

$$\frac{2E_2}{m}x^2 - \frac{2E_1}{m}y^2 + 2\frac{L}{m}\dot{y}x + \left(\frac{L}{m}\right)^2 = 0. \tag{4.15}$$

The curve of intersection of the two surfaces (note that we are now in the reduced phase space!), which actually turns out to be the curve formed where the surfaces touch, is the trajectory in this reduced phase space. The two surfaces and the trajectory are shown in Fig. 4.3. If we eliminate further the variable $\dot{y}$ with the aid of the second conservation law of energy (Eq. (4.13)), we obtain the orbit in the (x, y)-plane, which is the configuration space. To show this, we square Eq. (4.14), and using Eqs. (4.12) and (4.13), obtain

$$\left(\frac{L}{m}\right)^2 - \dot{x}^2y^2 - \dot{y}^2x^2 = \left(\frac{L}{m}\right)^2 - \frac{2E_2}{m}x^2 - \frac{2E_1}{m}y^2 + 2\omega^2x^2y^2 = -2\dot{x}\dot{y}xy,$$

which can be rewritten as

$$\frac{2E_2}{m}x^2 + \frac{2E_1}{m}y^2 - \left(\frac{L}{m}\right)^2 = 2xy(\dot{x}\dot{y} + \omega^2xy).$$

It is easy to see (differentiating with respect to time) that

$$I_4 = \dot{x}\dot{y} + \omega^2xy \tag{4.16}$$

is a conserved quantity; I_4 depends, however, on the other three conserved quantities (E), since[7]

$$(I_4)^2 = 4E_1E_2/m^2 - \omega^2(L/m)^2. \tag{4.17}$$

---

[7]The rank of the matrix corresponding to the differentials dE_1, dE_2, dL, and dI_4 (see the previous footnote and Section 14.1) is lower than four, since the determinant of the matrix vanishes.

Putting all this together, the three independent surfaces given by (4.12), (4.13), and (4.14) intersect to give a curve. The projection of this curve into the (x, y)-plane in reduced phase space gives the equation for the particle's orbit,

$$\frac{2E_2}{m}x^2 + \frac{2E_1}{m}y^2 - 2I_4(E_1, E_2, L)xy = \left(\frac{L}{m}\right)^2. \qquad (4.18)$$

Since (cf. (4.17))

$$4E_1E_2/m^2 - (I_4)^2 \geq 0,$$

the quadratic form in Eq. (4.18) is positive semidefinite. For $L \neq 0$, the orbit is an ellipse, whereas for $L = 0$, it is a line ($y = \pm\sqrt{E_2/E_1}x$).

There is no universal prescription for the construction of a conserved quantity. However, it turns out that conserved quantities are related to symmetries of the system. For example, the isotropic oscillator is invariant under (time independent) rotations of the coordinate system; consequently (as we see in Chapter 10), the angular momentum is constant in time, i.e. a conserved quantity. Therefore, in the analysis of dynamical systems, it is important to find symmetries (invariances) and to identify the corresponding conserved quantities. If we had used coordinates that naturally handle rotational symmetry – for instance, polar coordinates – the time independence of angular momentum would have emerged directly (E) (see also Chapter 5).

In any case, the two-dimensional harmonic oscillator is integrable. That this is not true in general for two-dimensional systems (and a fortiori not for three-dimensional ones), we demonstrate in the following.

4.2   The Hénon-Heiles system

The Hénon-Heiles system, though related to the isotropic oscillator, shows essentially new features (M. Hénon and C. Heiles, Astron.J.**69**,73-(1964)). The equations of motion are:

$$\begin{aligned} \ddot{x} &= -x - 2xy \\ \ddot{y} &= -y + y^2 - x^2. \end{aligned} \qquad (4.19)$$

The two equations are nonlinear and coupled. The Hénon-Heiles system originated in celestial dynamics, as a model for motion in an axially symmetric gravitational potential due to a galaxy. The system can also

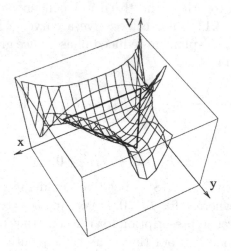

Figure 4.4: The potential of the Hénon-Heiles system.

be regarded as an extension of the isotropic harmonic oscillator to the anharmonic case (by setting $\omega = 1$ in (4.10)).

Multiplying the equations (4.19) by $\dot{x}$ and $\dot{y}$, respectively and adding the results shows that the total energy E

$$E = \frac{1}{2}(\dot{x}^2 + \dot{y}^2) + V(x,y), \qquad (4.20)$$

is conserved, where

$$V(x,y) = \frac{1}{2}(x^2 + y^2) + x^2 y - \frac{1}{3}y^3. \qquad (4.21)$$

No other integrals of the motion are known. The potential (4.21) is shown in Figs. 4.4 and 4.5. The border of the pit of the potential, where bounded motion occurs, is marked by the straight lines in Fig. 4.4. At the corners of this triangle there are three saddle points

$$(0,1), \qquad (\sqrt{3}/2, -1/2), \qquad \text{and} \quad (-\sqrt{3}/2, -1/2), \qquad (4.22)$$

where the potential has the value $V = 1/6$. At the point

$$(0,0) \qquad (4.23)$$

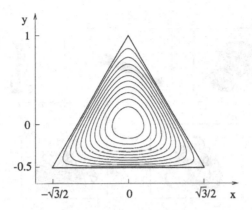

Figure 4.5: The equipotential lines in the Hénon-Heiles system.

is a local minimum with $V = 0$. Hence for energies in the range

$$0 \le E \le 1/6 \tag{4.24}$$

the motion is bounded if the starting point lies within the equilateral triangle shown in Fig. 4.5.

At the saddle points, as well as at the (local) minimum we have

$$\frac{\partial V}{\partial x} = \ddot{x} = \dot{v}_x = 0 \quad \text{and} \quad \frac{\partial V}{\partial y} = \ddot{y} = \dot{v}_y = 0.$$

Thus in phase space with coordinates $(x, y, \dot{x}, \dot{y})$ the points $(0, 1, 0, 0)$, $(\sqrt{3}/2, -1/2, 0, 0)$, $(-\sqrt{3}/2, -1/2, 0, 0)$, and $(0, 0, 0, 0)$ are the *stationary points* of the equations for the trajectories

$$\begin{aligned}
\dot{v}_x &= -x - 2xy \\
\dot{v}_y &= -y + y^2 - x^2 \\
\dot{x} &= v_x \\
\dot{y} &= v_y .
\end{aligned} \tag{4.25}$$

These four points are the *fixed points* of the flow (i.e. the set of all trajectories) in phase space; trajectories starting in these points stay there. As one may expect, linear stability analysis (introduced in Section 3.5) shows that the minimum of the potential, $(0, 0)$, is the only stable point (E).

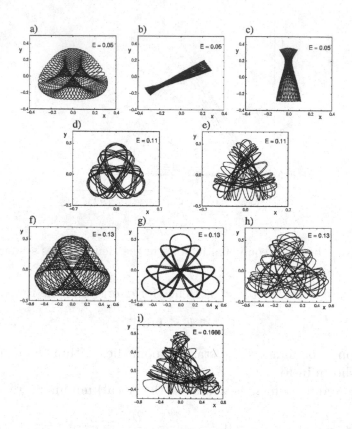

Figure 4.6: Various orbits in the Hénon-Heiles system.

In order to visualize the trajectories, we construct a Poincaré section. First, find an expression for any of the 4 coordinates, say $\dot{x}$, using conservation of energy, Eq. (4.20),

$$\dot{x} = \pm\sqrt{2\left(E - V(x,y)\right) - \dot{y}^2},$$

and choose the plane $x = 0$ as Poincaré section (as with the harmonic oscillator). Given the energy E, the region in the Poincaré section to which a trajectory returns is determined by the ranges of y and $\dot{y}$ when $\dot{x}$ is minimum, $\dot{x} = 0$. Therefore, the coordinates $(y, \dot{y})$ of returning trajectories obey the condition

$$E \geq \frac{1}{2}\dot{y}^2 + V(0, y). \tag{4.26}$$

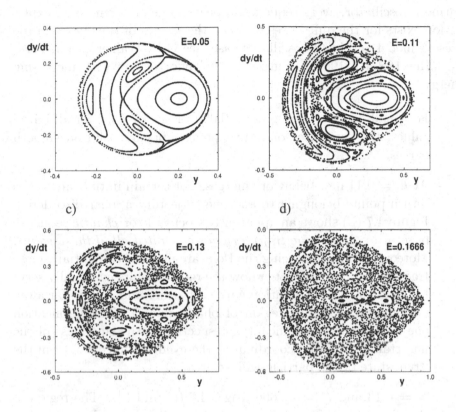

Figure 4.7: Poincaré maps for various energies (cf. text).

The *border line* of that area is given by the equation

$$\dot{y} = \pm\sqrt{2E - y^2 + \frac{2}{3}y^3},$$

where the values of y and E are restricted to the intervals $[-1/2, 1]$ and $[0, 1/6]$ respectively. The condition $\dot{y} = 0$ yields the maximal and minimal values of y for the chosen energy (cf. Figs. 4.7 a) to d)).

The sequence of return points $(y, \dot{y})$ in the Poincaré section $x = 0$ is obtained by numerical integration of the equations of motion. Figure 4.6 shows several orbits. For the smallest energy considered, $E = 0.05$, the orbits appear to be completely regular (Figs. 4.6 (a) – (c)). The same is true for the return points shown in Fig. 4.7 (a). As with the harmonic oscillator, every sequence appears to lie on a curve; an explanation exists for the differences between the shapes of the curves in the present model and in the oscillator case (see G.H. Walker and J. Ford, Phys.Rev.**188**,416(1969)). Surprising behavior occurs with increasing energy:

- For small energies E (e.g. $E = 0.05$ in Fig. 4.7 (a)), for all initial values, the sequences of return points appear to lie on smooth curves.

- At $E = 0.11$ new behavior emerges. For certain initial values, the return points belonging to any one trajectory appear disordered. Figure 4.7 (b) shows an apparently stochastic or *chaotic* sequence of return points – *one trajectory has generated all of these points.* Moreover, when producing the Poincaré maps of two initially close trajectories, one can watch how the return points eventually seem to lose all spatial correlation with each other and end up far away (within the accessible region of phase space[8]). The implication one may derive is significant: irrespective of the smallness of the uncertainty in initial coordinates, the eventual uncertainty in the 'true' state of the particle will be great.

- $E = 0.13$ and $E = 0.1666$, Figs. 4.7 (c) and (d). The region of chaotic motion becomes larger as energy is increased.

For increasing values of the energy, the size of the region of apparently chaotic motion, as well as the apparent degree of irregularity within a specific orbit (or trajectory), grows (cf. Figs. 4.6 (e), (h), and (i)). However, at any energy level, some initial conditions lead to apparently simple, regular motion (cf. Figs. 4.6 (d), (f), and (g)).

In their summary, Hénon and Heiles state that at energies smaller than a critical value of approximately $E_c = 0.11$, the sequences of return points, even when magnified, seem to lie exactly on curves. If several different initial values are chosen, the curves fill the region defined by Eq. (4.26) completely. At energies above the critical value, the size of the stochastic region – containing seemingly chaotic sequences of return points – increases linearly with energy. "... the situation could

---

[8]Though this part of phase space is of finite volume there is a way out so that the points nevertheless fall apart 'infinitely far' (see later).

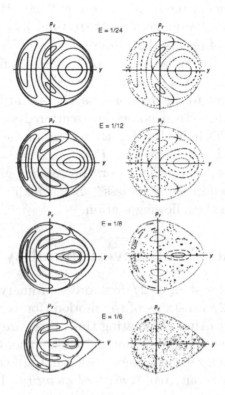

Figure 4.8: Poincaré maps calculated by F.G. Gustavson (left hand side) and the ones calculated numerically by Hénon and Heiles for increasing energy (from [Lichtenberg/Lieberman]).

be very roughly described by saying that the second integral exists for orbits below a "critical energy", and does not exist for orbits above that energy." Lichtenberg and Lieberman state, that the stochastic regions always exist, but for energies well below the critical value, they are thin and are not visible in numerically generated pictures of the entire region (see [Lichtenberg/Lieberman]).

Gustavson (F.G. Gustavson, Astron.J.**71**,670 (1966)), starting from the harmonic part of Eqs. (4.19), constructed a second integral using a perturbation method. This second integral of the motion prevents return points from scattering over the Poincaré section – the situation is analogous to the harmonic oscillator. At energies $E \leq E_c$, the Poincaré maps calculated by Gustavson agree very well with the results of Hénon and Heiles obtained by numerical integration, as can be seen in Fig. 4.8. The two sets of results show discrepancies in the chaotic region at energies $E > E_c$. The agreement at small energies may be understood with the help of the KAM theorem (as we shall see in Section 16.4).

Thus, in contrast to the harmonic oscillator – at least at sufficiently large energies – the return points are scattered stochastically within several regions. This indicates that a constant of the motion is missing, or that one exists but is ineffective. How can a constant of the motion, which should exist, in principle (for the reasons given in Subsection 2.4.1), become ineffective or useless? Some inkling of how this may happen is given in the following section.

4.3 A 'useless' conserved quantity

In an autonomous system of $2f$ first order ordinary differential equations, there exist $2f$ constants of the motion (for example, $2f$ independent initial values). After eliminating time, there are in principle $2f - 1$ conserved quantities (cf. Subsection 2.4.1). Some of these are known in advance and make a solution easier. But what are the other conserved quantities? It turns out that *conserved quantities* have varying quality: there are *isolating* and there are *nonisolating* ('useful' and 'useless') ones.

The difference between the two kinds of integrals can be demonstrated with the help of the two-dimensional harmonic oscillator (Section 4.1; see also [Balescu]). In the following, we set $m = 1$ and choose the notation $x_1 := x$, $x_2 := y$, $v_1 = v_x \, (= \dot{x}_1)$, $v_2 = v_y \, (= \dot{x}_2)$. From the outset, we know two integrals, the energies of the components of the motion in the 1-direction and 2-direction, Eqs. (4.2) and (4.3),

$$v_i^2 + \omega_i^2 x_i^2 = v_{0i}^2 + \omega_i^2 x_{0i}^2 = 2E_i, \qquad i = 1, 2 , \qquad (4.27)$$

where x_{0i} and v_{0i} are the initial values. The two conserved quantities have the form

$$G_i(x_i, v_i) - G_i(x_{0i}, v_{0i}) = 0, \qquad i = 1, 2 , \qquad (4.28)$$

where $G_i(\alpha, \beta) = \omega_i^2 \alpha^2 + \beta^2$ is an algebraic function. To find an independent third integral, which we know exists, we express the initial values x_{0i} and v_{0i}, $i = 1, 2$, as functions of the time-dependent positions and velocities $x_i(t)$ and $v_i(t)$, $i = 1, 2$ (see (2.10) and (3.33)):

$$
\begin{aligned}
x_{0i} &= x_i \cos \omega_i t - (v_i/\omega_i) \sin \omega_i t \\
v_{0i} &= v_i \cos \omega_i t + \omega_i x_i \sin \omega_i t.
\end{aligned}
\tag{4.29}
$$

This shows explicitly that the initial values serve as four constants of motion. From these relations, it follows that

$$
\cos \omega_i t = (v_i v_{0i} + \omega_i^2 x_i x_{0i})/2E_i
\tag{4.30}
$$

$$
\sin \omega_i t = \omega_i (x_i v_{0i} - v_i x_{0i})/2E_i.
\tag{4.31}
$$

Since

$$
t = (1/\omega_i) \arctan(\sin \omega_i t / \cos \omega_i t), \qquad i = 1, 2,
$$

the relations can be recast in the form

$$
t = \frac{1}{\omega_i} \arctan \left(\frac{\omega_i(x_i v_{0i} - v_i x_{0i})}{v_i v_{0i} + \omega_i^2 x_i x_{0i}} \right), \qquad i = 1, 2.
$$

Eliminating t, we obtain the time-independent (i.e. conservation) equation,

$$
\arctan \left(\frac{\omega_1(x_1 v_{01} - v_1 x_{01})}{v_1 v_{01} + \omega_1^2 x_1 x_{01}} \right) - \frac{\omega_1}{\omega_2} \arctan \left(\frac{\omega_2(x_2 v_{02} - v_2 x_{02})}{v_2 v_{02} + \omega_2^2 x_2 x_{02}} \right) = 0.
\tag{4.32}
$$

This equation defines a hypersurface in phase space, adding to those already known from Eq. (4.27). However, (4.32) does not (yet) have the form of Eq. (4.28).

Equation (4.32) is reducible to an algebraic equation for rational ratios

$$
\omega_1/\omega_2 = r/s,
\tag{4.33}
$$

where r and s are integers. For clearness, we write Eq. (4.32) in the form

$$
s \arctan z_1 - r \arctan z_2 = 0,
$$

where

$$
\begin{aligned}
z_1 &= \omega_1(x_1 v_{01} - v_1 x_{01})/\left(v_1 v_{01} + \omega_1^2 x_1 x_{01}\right) \\
z_2 &= \omega_2(x_2 v_{02} - v_2 x_{02})/\left(v_2 v_{02} + \omega_2^2 x_2 x_{02}\right).
\end{aligned}
$$

Applying the addition theorem for arctan-functions s-times and r-times, respectively, we arrive at the equation

$$\arctan f_1(z_1) - \arctan f_2(z_2) = \arctan \frac{f_1 - f_2}{1 + f_1 f_2} = 0,$$

or

$$f_1(z_1) - f_2(z_2) = 0. \tag{4.34}$$

The rational functions f_1 and f_2 can be written in the form

$$f_i = \frac{n_i(z_i)}{d_i(z_i)},$$

where n_i and d_i, $i = 1, 2$, are polynomials whose degrees depend on s and r, respectively. For general values of r and s, Eq. (4.34) hardly can be solved for any of the variables z_i, or even x_i and v_i.

In the case where $\omega_1 = \omega_2$, i.e. $r = s$, it is easy to express Eq. (4.32) in the form of Eq. (4.28). Indeed; in doing so, one regains for the oscillator the conservation laws for the angular momentum L and for the quantity I_4 (see Eqs. (4.16) and (4.17)), in the form of Eq. (4.28) (E); *viz.*:

$$L/I_4 - L_0/I_{04} = 0, \tag{4.35}$$

where the index 0 indicates that the variables assume their initial values; e.g. $L_0 = L(x_{01}, x_{02}, v_{01}, v_{02})$.

For even small values of r and s, however, Eq. (4.32) can no longer be reduced to the form of Eq. (4.28). Also, eliminating in Eq. (4.32) the variables v_i with the help of Eqs. (4.27) in order to obtain a relation between x_1 and x_2, one faces the problem, that due to the increasing degrees of the polynomials $n_i(z_i)$ and $d_i(z_i)$, there appears an increasing number of branches of the solution $x_2 = x_2(x_1)$. This feature can be inferred from the orbits in the (x_1, x_2)-plane. Figure 4.2 (a) shows an example of an orbit consisting of 4 branches. Along each branch, x_2 is a unique function of x_1, and in each branch, x_1 takes on all the values in its domain. As the number of branches grows with increasing r and s, the ambiguity in the relation between x_1 and x_2 increases (cf. Fig. 4.2 (b)).

Since the orbit is a projection of the trajectory onto configuration space, and the trajectory, in turn, is the curve of intersection of the hypersurfaces in phase space given by the conservation laws, the multiple branches of the orbits are produced by folds in the hypersurface given by Eq. (4.32). To give an example, we refer to the left hand side of Fig. 4.2 (a). The orbit shown is a cut through the hypersurface, which has four folds. For every value of x $(= x_1)$, there are four values of y $(= x_2)$. In any case, it remains true that for rational values of ω_2/ω_1, Eq. (4.32) is an algebraic equation for $x_2 = x_2(x_1)$.

If ω_2/ω_1 is irrational, the number of branches increases without bound, and any orbit fills the accessible part of configuration space (i.e. the region in configuration space allowed by energy; cf. the left hand picture in Fig. 4.2 (c) for $t \to \infty$). The equation obtained from (4.32) for $x_2 = x_2(x_1)$ is transcendental, the conservation law given in Eq. (4.32) cannot be written in the form of Eq. (4.28). Such a conserved quantity is called **nonisolating**; it is useless for the solution of the equations of motion, since neither x_1 can be given as a function of x_2 nor can x_2 be written as function of x_1. 'Nice' conserved quantities like the energies given in Eq. (4.27) are called **isolating integrals** of the motion.

4.4 Chaotic behavior

The Hénon-Heiles system admits irregular, chaotic solutions. The following properties of chaotic solutions characterize the temporal sequence of the return points in the Poincaré section:

i) The return points seem to scatter stochastically in (some subset of) the region defined by (4.26).

ii) Given two neighboring initial values on the Poincaré section, the two sequences of the return points eventually completely lose their spatial correlation.

Therefore we are led to the following

Working definition:

The **behavior** of a solution in a system **is called chaotic**, if the temporal evolution of the trajectory in phase space is completely irregular (stochastic) and depends sensitively on the initial values. Initially close trajectories eventually evolve completely differently.

The word 'sensitive' can be formulated more precise in simple terms. It means that the local distance between initially (at time $t = 0$) close trajectories always increases exponentially (cf. 4.38). Since it is the main characteristic feature of chaotic behavior, let us discuss this behavior of two trajectories – adjacent at some instant – in more detail.

If the accessible phase space is bounded in size, the tendency of two neighboring trajectories to diverge must mean that the trajectories are chaotic, in order to become arbitrarily distant (in a certain measure). When this happens, the topology of the manifold on which the trajectory lies must have certain properties. We have already had a hint of this in the above discussion on a useless integrals of the motion. If the trajectory is chaotic, this manifold has a **fractal structure**[9].

For a closer look at the evolution of the trajectory in phase space, we return to the representation of the equations of motion as an autonomous system; i.e., a representation of the form (cf. (3.56))

$$\dot{\mathbf{u}} = \mathbf{G}(\mathbf{u}) \quad \text{with} \quad \mathbf{u} = (\mathbf{r}, \mathbf{v}), \ \mathbf{G} = (\mathbf{v}, \mathbf{F}/m). \qquad (4.36)$$

The distance between two neighboring trajectories $\mathbf{u}(t)$ and $\mathbf{u}(t) + \delta\mathbf{u}(t)$ at time t is

$$|\delta\mathbf{u}(t)| = \sqrt{\sum_i (\delta u_i(t))^2}. \qquad (4.37)$$

---

[9] A standard example of a fractal set is the **Cantor set**, defined in the closed interval $[0, 1]$ of the real axis. The set is produced by the prescription:

Remove one third in the middle of the interval, i.e. remove the open interval $(1/3, 2/3)$ from the interval $[0, 1]$. In the next step from the remaining closed intervals $[0, 1/3]$ and $[2/3, 1]$ also the middle thirds are removed. And this procedure is repeated infinitely many times. The resulting set is the Cantor set.

The fractal character of the Cantor set can be seen when determining its size. If one measures the size using a stick of length $1/10$ and in next measurements a stick of length $1/100$ is used, the results differ. As a consequence one cannot attribute the dimension one to the set, if dimension means measurability by a standard length. One can however, determine the so-called **fractal dimension** or **Hausdorff dimension** d_H of the Cantor set. The result is: $d_H = 0.6309$. The difference to the dimension of the interval $[0, 1]$ ($d_H = 1$) which contains the Cantor set, reflects the fractal character of the latter.

Similar considerations one can apply to fractal sets in a plane ('coast lines' typically have Hausdorff dimension $d_H = 1.24$, Sierpinski's carpet $d_H = 1, 89$) and in 3-dimensional space (Sierpinski's sponge has $d_H = 2, 73$). For details we refer to the book by B. Mandelbrot, The Fractal Geometry of Nature, Freeman, New York 1977.

To first order, the variation of the distance with time is determined by linearizing equation Eq. (4.36) (cf. (3.59)):

$$\delta \dot{u}_i = \sum_j M_{ij}(\mathbf{u}) \, \delta u_j, \quad M_{ij}(\mathbf{u}) = \frac{\partial G_i}{\partial u_j}.$$

The eigenvectors $\mathbf{e}_i$ of the matrix M_{ij}, together with the corresponding eigenvalues λ_i, determine the behavior of $\delta\mathbf{u}$ in the time interval δt. The eigenvectors $\mathbf{e}_i$ for which $\lambda_i > 1$ determine the directions in which the distance between the two trajectories grows, while for eigenvectors whose corresponding eigenvalue $\lambda_i < 1$, the distance decreases.

For simplicity, we elaborate using only a two-dimensional example. Let the eigenvalues corresponding to $\mathbf{e}_1$ and $\mathbf{e}_2$ be $\lambda_1 > 1$ and $\lambda_2 < 1$, respectively. Then the evolution of two trajectories with time can be illustrated as follows. Connect the two initial points $\mathbf{u}_1(0)$ and $\mathbf{u}_2(0)$ of the trajectories by a curve $\mathcal{C}$. The eigenvectors and eigenvalues at the two initial points, as well as at the points on the curve $\mathcal{C}$ connecting them, are nearly equal. With passage of time, the connecting curve $\mathcal{C}$ will be stretched in the direction $\mathbf{e}_1$ and compressed in the direction $\mathbf{e}_2$. Stretching is only possible until the boundary of the accessible part of phase space is reached. What happens then? In the meantime, the eigenvectors and the eigenvalues have changed steadily at every point of $\mathcal{C}$, which, in turn, moves with time according to the time evolution of $\mathbf{u}(t)$. Now, at points sufficiently far apart, the local eigenvectors and eigenvalues differ considerably. The eigenvector directions and/or eigenvalues must change so that all points remain within the accessible region. Those parts of $\mathcal{C}$ that have arrived at the boundary begin to return to the interior; the curve folds over. With time, the length of $\mathcal{C}$ increases. Thus, the distance between the two initially close points $\mathbf{u}_1$ and $\mathbf{u}_2$ increases arbitrarily without the points crossing the boundary. The curve $\mathcal{C}$ stretches and folds with time. Such a behavior is sketched in Fig. 4.9, where five stages of the time evolution are shown. This is the idea behind what is going on with the two-dimensional *Poincaré maps* of dynamical systems such as the Hénon-Heiles system.

In a higher dimensional phase space, the situation is analogous. The manifold on which the trajectory lies can be imagined as infinitely often folded; therefore, the curve on which the return points of the Poincaré maps lie is also folded an infinite number of times. This explains the seemingly stochastic and chaotic sequence of the return points generated

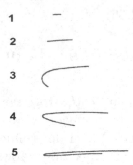

Figure 4.9: The time evolution of the curve $\mathcal{C}$ (cf. text).

by *one* trajectory. It also explains why sequences of return points generated by *two* trajectories originally close together lose their spatial correlation. The distance between two trajectories is initially $|\delta\mathbf{u}(t=0)|$. Locally, it increases strongly in the time interval δt, roughly as

$$|\delta\mathbf{u}(t)| \propto |\delta\mathbf{u}(t=0)|\, e^{\sigma t}, \qquad \sigma > 0, \qquad (4.38)$$

where σ is the largest (local) **Liapunov exponent**, defined by

$$\sigma\left(\delta\mathbf{u}\right) = \lim_{\tau\to\infty} \frac{1}{\tau}\ln\frac{|\delta\mathbf{u}(t+\tau)|}{|\delta\mathbf{u}(t)|}, \qquad \sigma = \max_{\delta\mathbf{u}}\sigma\left(\delta\mathbf{u}\right).$$

For a system to be chaotic, at least one of the Liapunov exponents has to be positive. The Liapunov exponents are closely related to the eigenvalues λ_i of the matrix M_{ij} above. (For further reading see [Lichtenberg/Lieberman].)

A simple picture of the process just described is the following mapping of the unit square[10] $0 \le x, y < 1$ into itself. The so-called **baker transformation** consists of two steps:

i) *Stretching* in one direction and a simultaneous *compression* in the other by a factor of two

$$x' = 2x, \qquad y' = y/2 \qquad \Rightarrow \qquad x' \in [0,2)\,, \; y' \in [0,1/2)\,;$$

---

[10]This transformation is a model of the Poincaré map of a two-dimensional system.

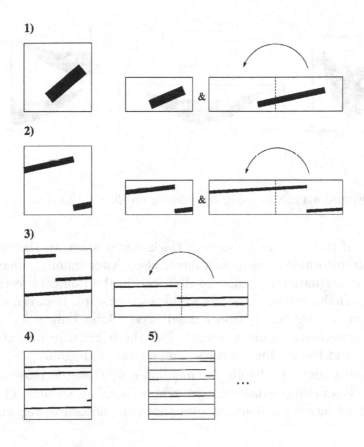

Figure 4.10: The fate of a 'thick line' under repeated baker transformations.

ii) *Folding*, i.e. clipping the interval $[1, 2)$ and putting it back on top of the squeezed original interval $[0, 1)$

$$x'' = x' \qquad y'' = y' \qquad \text{for} \qquad x' \in [0, 1)$$

$$x'' = x' - 1 \qquad y'' = y' + 1/2 \qquad \text{for} \qquad x' \in [1, 2).$$

The mapping preserves the magnitude of areas.

Figure 4.10 demonstrates what happens to a 'thick line' under repeated baker transformations. Shown here are the first four transformations of the initial state (1). As the transformation is applied more and more often, the line becomes distributed over the whole square.

Figure 4.11: Stretching and folding: on the road to chaos.

Moreover, if the 'line' had a certain thickness at start, in the process of being transformed it becomes thinner also. After infinitely many iterations, the situation is similar to that seen in the Poincaré section of the Hénon-Heiles system. Another well known example is shown in Fig. 4.11, where a cat's face is transformed once. After only a few iterations, the face becomes unrecognized. But the baker transformation is invertible, and the original pictures can be recovered again.

Without going into details, we mention only that the baker transformation is a *mixing* transformation, and is therefore *ergodic*: The line and the cat's face are distributed over the *entire* accessible region.

Remarks

Iterative maps

The temporal sequence of the return points in the Poincaré section can be considered as an **iterative map**, the Poincaré map. Due to the structure of the equations of motion of the Hénon-Heiles system, the **area is preserved**[11], i.e. if one lets different initial points filling a certain area evolve with time, the size of the area occupied by the points after every time step is the same as the initial area – but the shape changes (see Subsection 13.5). For nonintegrable systems, this mapping in the Poincaré section cannot be derived from the equations of motion. Therefore, one resorts to studying model maps. An example

---

[11]There exist also area nonpreserving iterative maps (see e.g. M. Hénon, *A two dimensional mapping with a strange attractor*, Commun. Math. Phys., **50**,69(1976)).

is the following (M. Hénon, Quart.Appl.Math., **27**, 291 (1969)):

$$x_{i+1} = x_i \cos \alpha - (y_i - x_i^2) \sin \alpha$$
$$y_{i+1} = x_i \sin \alpha + (y_i - x_i^2) \cos \alpha.$$

This iterative map shows features similar to those observed in the Poincaré sections of the Hénon-Heiles system.

When can chaos appear? When is one safe from it?

As we have seen, chaos can appear in four-dimensional phase space of a nonlinear two-dimensional system, but not in two-dimensional phase space of a one-dimensional system. In general, the following holds:

If the temporal evolution of n variables $\mathbf{u} = (u_1, \dots, u_n)$ is given by an autonomous system of n first order ordinary differential equations,

$$\dot{\mathbf{u}} = \mathbf{G}(\mathbf{u}), \tag{4.39}$$

where $\mathbf{G}$ is a nonlinear function of the u_i, the system can behave chaotically if the number of components is $n \geq 3$.

If the couplings between the components of $\mathbf{u}$ are of a general nature (e.g. the forces are not conservative), the system is integrable if $n - 1$ independent conserved quantities, i.e. $n - 1$ relations

$$I_j(\mathbf{u}) = 0, \qquad j = 1, \dots, n - 1,$$

are known. In this case there can be no chaotic behavior. For systems like the one given by Eqs. (2.8), fewer (namely $n/2$) conserved quantities are sufficient (cf. Section 14.1). Essentially, for the possibility of chaotic behavior to arise, the equations must be nonlinear and nonintegrable.

Dissipative systems, strange attractors

Consider once more the autonomous system in Eq. (4.39). Denoting the (generalized) divergence of the n-component vector field $\mathbf{G}(\mathbf{u})$ by

$$\boldsymbol{\nabla}_n \mathbf{G} = \left(\frac{\partial G_1}{\partial u_1} + \dots + \frac{\partial G_n}{\partial u_n} \right). \tag{4.40}$$

If

$$\boldsymbol{\nabla}_n \mathbf{G} = const < 0,$$

then the system is called **dissipative**[12]. In nonlinear, dissipative (e.g. damped) systems, the chaotic behavior is different from what we saw above. In these systems, the paths tend towards a so-called **strange attractor**. The strangeness of the attractor lies in its fractal structure (see footnote 9 above) and its **self-similarity**[13]. A strange attractor can be assigned a fractal dimension, whose value is between d and $d-1$, where d is the dimension of the embedding space.

Even for the driven, damped oscillator in one dimension (see Subsection 3.4.2), we found that, asymptotically, the trajectories are restricted to a submanifold – an ellipse – in phase space. (See the discussion of the solutions (3.39) and (3.45).) If the equation of motion is nonlinear, e.g. in the force term, $x \to \sin x$, we arrive at the *periodically driven, damped pendulum*

$$\ddot{x} + 2\gamma\dot{x} + \omega^2 \sin x = k \cos \Omega t.$$

Strange attractors and chaotic solutions may now appear.

One of the four parameters $(\gamma, \omega, k, \Omega)$ can be absorbed in a rescaling of time, e.g. $t' = \omega t$; again denoting t' by t, we get the equation

$$\ddot{x} + R\dot{x} + \sin x = F \cos 2\pi\eta t,$$

with the three parameters $R = 2\gamma/\omega$, $F = k/\omega^2$, and $\eta = \Omega/(2\pi\omega)$. Introducing the variable

$$\varphi(t) = 2\pi\eta t,$$

this equation can be recast in an autonomous system of first order ordinary differential equations:

$$\begin{aligned}
\dot{x} &= v \\
\dot{v} &= -Rv - \sin x + F \cos \varphi \\
\dot{\varphi} &= 2\pi\eta.
\end{aligned}$$

---

[12]The harmonic oscillator and the plane pendulum are Hamiltonian systems (see Chapter 13). In these systems the divergence of **G** vanishes as a consequence of Liouville's theorem about the 'incompressibility' of a given volume in phase space (see Subsection 13.5). In dissipative systems, an arbitrary volume of initial points in phase space shrinks as it evolves with time.

[13]A (infinitely large) object is self-similar, if arbitrarily small parts, when magnified, look similar to the original object. For a finite-sized object (e.g. an attractor), this criterion applies to the interior. An example of a self-similar object – at least for a few iterations of a self-similarity transformation (magnification) – is a rugged coast line, which for instance, may have bays and beaches, within bays and beaches, within bays and beaches, ...

The divergence (4.40) is equal to $-R$. Chaotic behavior occurs here (A.H. MacDonald and M. Plischke, Phys.Rev. **B27**, 201 (1983)), and since the system is dissipative, strange attractors exist, too, whose (Hausdorff-) dimension depends on R, F, and η, and is slightly larger than one.

Numerical integration

Given the sensitivity of chaotic systems to small discrepancies in initial conditions, the results of any numerical integration in the presence of chaotic dynamics have to be treated very carefully. Already for integrable equations, the accuracy of the numerical integration depends on factors such as the integration procedure and the step size. The reliability of numerical integration in chaotic systems still seems to be an unsettled question. But the numerical results can at least indicate that the system possesses interesting behavior.

4.5 Laplace's clock mechanism does not exist

The belief that the motion of particles obeying ordinary differential equations is under complete control was held universally until the 1960s. This belief was formulated in 1814, when Laplace published his "Philosophical Essay on Probabilities" (Essai philosophique sur les probabilités, 1814). For Laplace, the universe of macroscopic objects was like a giant clock. His statement about the predictability of the universe is famous (cited from [Peterson]):

> *Assume an intelligence that at a given moment knows all the forces that animate nature as well as the momentary positions of all the things of which the universe consists, and further that it is sufficiently powerful to perform a calculation based on these data. It would then include in the same formulation the motions of the largest bodies in the universe and those of the smallest atoms. To it, nothing would be uncertain. Both future and past would be present before his eyes.*

Given the initial conditions and sufficient computing facilities, the future should be predictable to any given precision, even if the effort may be enormous. This belief cannot be sustained anymore. It is an irony of history, that in these modern times, the enormous computing capability and theoretical knowledge in our possession have served to detect that the laws of classical mechanics do not allow prediction of an arbitrarily distant future. In chaotic systems, inevitable uncertainties in the initial conditions lead to large differences in the future states, a fact that poses a serious problem for the prediction of the future. Yet these systems are still deterministic.

Nonlinear equations of motions occur much more often than one would suspect if reading standard textbooks on mechanics. Chaos's emergence into prominence, therefore, is more accurately described as a revolution in physical thinking and reasoning. James Lighthill, in particular, brought this into focus in his article *The recently recognized failure of predictability in Newtonian dynamics* (Proc. R. Soc. Lond. A **407**, 35 (1986)):

> ... *We are deeply conscious today that the enthusiasm of our forebears for the marvellous achievements of Newtonian mechanics led them to make generalizations in this area of predictability which, indeed, we may have generally tended to believe before 1960, but which we now recognize were false. We collectively wish to apologize for having misled the general educated public by spreading ideas about the determinism of systems satisfying Newton's laws of motion that, after 1960, were to be proved incorrect. ...*

However, even today, what the exact implications of chaos are is not completely settled – the situation remains controversial. Even though, as just said, chaotic behavior is ubiquitous in classical model systems, apart from the unreliability of weather forecasts, it is hard to point out consequences in everyday experience. This has several causes. One of these is touched on at the end of this book. Regarding the others, we mention only that in real systems there exist much more and even different degrees of freedom than in the simplifying examples of this book. The opportunity for energy flow into such 'additional' degrees of freedom may change the behavior of that part of the system that when isolated, manifests chaotic behavior.

Problems and examples

1. Investigate the two-dimensional oscillator with bilinear coupling of the motion in the x and y directions (i.e. the potential energy is given by $V(x,y) = \omega_1^2 x^2 + \omega_2^2 y^2 + \lambda xy$).

2. Show that in the case of the two-dimensional isotropic oscillator, the conserved quantities E_1, E_2, and L (cf. Eqs. (4.12), (4.13), and (4.14)) are independent.

3. Deduce the relation given in Eq. (4.15) and show that it represents a hyperboloid of one sheet.

4. Show for the isotropic oscillator (4.10) that $I_4 = \dot{x}\dot{y} + \omega^2 xy$:

   i) Is time independent (i.e. conserved);

   ii) Is dependent on the conserved quantities E_1, E_2, and L.

5. What are the equations of motion for the isotropic oscillator in plane polar coordinates? Show that the energy is given by

$$E = \frac{m}{2}\dot{\rho}^2 + \frac{L^2}{2m\rho^2} + \frac{m\omega^2}{2}\rho^2.$$

   Discuss the solution, and comment on the implication of using polar coordinates rather than Cartesian coordinates.

6. Determine the stationary points for the motion in the potential given in Eq. (4.21). Apply linear stability analysis to the points.

7. Show that the relation given in Eq. (4.32) can be reduced to (4.35) for $\omega_1 = \omega_2$.

5

Motion in a central force

The central forces hold a preeminent position in mechanics. These forces produce a second conserved quantity, the angular momentum, which is a vector. This number of conserved quantities is sufficient to render the equation of motion for a particle in a central force field integrable. However, if an additional force breaks the spherical symmetry – for instance, if a homogeneous magnetic field is switched on – then integrability vanishes.

In Newton's "Principia" the 'centripetal forces', as he called them, are the central topic. In the light of his laws he investigates in Book I various problems. One issue is to find the force from a given orbit if the center of the force is known (the so-called *direct problem*). For example in the case of an elliptic orbit two completely different forces are found whether the center of the force is situated in the center of the ellipse (Prop. X) or in one of its foci (Prop. XI). On the other hand Newton solves also the *inverse problem* in Prop. XVII: Conic sections are the orbits for a force proportional to the inverse distance squared; the force is centered in a focus of the conic section[1].

---

[1]In 1710 Johann Bernoulli found faulty Newton's solution of the inverse problem. He criticized that Newton had not proved that conic sections, where a focus is the center of the force, are **necessarily** the orbits of a body attracted by an inverse square force. Solutions for the inverse problem were also presented by P. Varignon (1710) and by J. Keill (1708).

Bernoulli also criticized Newton's mathematical procedures applied to central forces in the "Principia", since, in his opinion, they lacked generality and could be used only if one knew the solution in advance (N. Guicciardini, *Johann Bernoulli, John Keill and the inverse problem of central forces*, Annals of Science **52** (6) 1995).

5.1   General features of the motion

5.1.1   Conserved quantities

We study the motion of a point mass in a central force (cf. Subsection 2.4.3)

$$\mathbf{F}(\mathbf{r}) = f(r)\,\mathbf{r}/r. \tag{5.1}$$

From Section 2.4, we know that for Newton's equation,

$$m\ddot{\mathbf{r}} = f(r)\mathbf{r}/r, \tag{5.2}$$

there are two conservation laws: conservation of energy and conservation of angular momentum[2].

Conservation of energy is a consequence of the existence of a potential. For the force $\mathbf{F}$ in Eq. (5.1), we have $\boldsymbol{\nabla} \times \mathbf{F} = \mathbf{0}$. Therefore every central force is conservative, which means that there exists a scalar potential $V(r)$, depending only on $r = |\mathbf{r}|$, such that

$$\mathbf{F}(\mathbf{r}) = -\boldsymbol{\nabla} V(r). \tag{5.3}$$

From this relation, we can construct $V(r)$ explicitly. Using Eq. (5.1) and multiplying by $\mathbf{r}$, we get (cf. Eq. (2.35))

$$f(r) = -\frac{dV}{dr}; \quad \text{i.e.,} \quad V(r) = -\int^{r} dr'\, f(r'). \tag{5.4}$$

After multiplying the equation of motion (5.2) by $\dot{\mathbf{r}}$ and using Eq. (5.3), one notes that the time derivative of the energy

$$E = \frac{m}{2}\dot{\mathbf{r}}^2 + V(r) \tag{5.5}$$

vanishes: *the total energy E is conserved.* The system is conservative.

Conservation of angular momentum is a consequence of the fact that the moment of a central force (5.1) taken about the origin of the force vanishes:

$$\mathbf{r} \times \mathbf{F}(\mathbf{r}) = \mathbf{0}.$$

---

[2]An interesting situation arises when the body - represented for the moment by a point - looses mass as is the case for some asteroids. This loss may be even not uniform; in the case of vaporization it increases in the vicinity of the sun.

The vector product of the equation of motion (5.2) and the radius vector **r** shows that the time derivative of the angular momentum vanishes (cf. Subsection 2.4.3). Therefore, the *angular momentum is constant in time*:

$$\mathbf{L} = m\mathbf{r} \times \dot{\mathbf{r}} = const. \tag{5.6}$$

Since the angular momentum, by definition, is perpendicular to the radius vector, **Lr** = 0, the *orbit* **r**(t) of a particle in a central force always lies *in a fixed plane* perpendicular to **L**. The position of the plane is determined by the initial values $\mathbf{L}_0 = m\mathbf{r}_0 \times \dot{\mathbf{r}}_0$ (= **L**). If we choose the coordinate system such that the direction of the angular momentum is parallel to the z-axis, i.e. $\mathbf{L} = L\mathbf{e}_z$, then the motion takes place only in a plane z = const. By an appropriate choice of origin, the motion occurs in the plane z = 0. In this plane, we introduce polar coordinates, so that in the whole space we have cylindrical coordinates (ρ, φ, z). For properties of cylindrical coordinates and some useful relations – which we use below – we refer the reader to Appendix A. Since the motion takes place only in the plane z = 0, the radius vector is just the vector ρ. Following the usual notation, from now on we denote ρ by **r**. This has to be kept in mind when applying relations in Appendix A. Because of Eq. (A.28), the equation of motion (5.2) in polar coordinates is

$$m(\ddot{r} - r\dot{\varphi}^2)\mathbf{e}_r + m(r\ddot{\varphi} + 2\dot{r}\dot{\varphi})\mathbf{e}_\varphi = f(r)\mathbf{e}_r, \tag{5.7}$$

or, separating the $\mathbf{e}_r$ and $\mathbf{e}_\varphi$ components, we have the system of two differential equations,

$$m(\ddot{r} - r\dot{\varphi}^2) = -\frac{dV}{dr} \tag{5.8}$$

$$m(r\ddot{\varphi} + 2\dot{r}\dot{\varphi}) = 0. \tag{5.9}$$

Inserting the velocity $\dot{\mathbf{r}}$ in polar coordinates (cf. Eq. (A.27))

$$\dot{\mathbf{r}} = \dot{r}\mathbf{e}_r + r\dot{\varphi}\mathbf{e}_\varphi \tag{5.10}$$

into Eq. (5.5), we obtain for the conservation energy,

$$E = \frac{m}{2}\left(\dot{r}^2 + r^2\dot{\varphi}^2\right) + V(r) = const. \tag{5.11}$$

Further, Eq. (5.10) inserted into Eq. (5.6) yields

$$\mathbf{L} = \mathbf{r} \times m\dot{\mathbf{r}} = r\mathbf{e}_r \times mr\dot{\varphi}\mathbf{e}_\varphi = mr^2\dot{\varphi}\mathbf{e}_z.$$

Conservation of angular momentum means conservation of the z-component of angular momentum,

$$L_z =: L = mr^2\dot\varphi = const, \qquad (5.12)$$

since both the other components of L vanish by our choice of coordinate system. (In the following, we denote by L the z-component and not the modulus of $\mathbf{L}$; $|\mathbf{L}| = |L_z| = |L|$.) So, from Eq. (5.12), the **angular velocity** $\dot\varphi$ is related to r by

$$\dot\varphi = \frac{L}{mr^2}, \qquad (5.13)$$

from which it follows that $\varphi(t)$ can only decrease or increase monotonically $(L < 0$ or $L > 0)$; i.e., *the particle moves along the orbit with a definite direction of rotation about the force center.* As we show later (see Chapter 10), conservation of angular momentum follows from the invariance of the potential $V(r)$ under time independent rotations.

The **areal law** (cf. Kepler's second law, Section 6.3) is a consequence of conservation of angular momentum. We show this as follows. Let $\dot A$ be the **areal velocity**, that is, the area swept out by the radius vector $\mathbf{r}(t)$ per unit time. The area dA swept out by the radius vector $\mathbf{r}$ during the infinitesimal time interval dt is

$$dA = \dot A\, dt = \frac{1}{2}\, r^2 d\varphi = \frac{1}{2}\, r^2\, \dot\varphi\, dt,$$

where we have used $r(t + dt)\, d\varphi = r(\varphi + d\varphi)\, d\varphi \approx r(\varphi)d\varphi = r(t)\, d\varphi.$

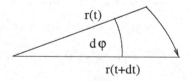

r(t)

dφ

r(t+dt)

Applying Eq. (5.12), one gets the areal law

$$\dot A = r^2\dot\varphi/2 = L/2m = const; \qquad (5.14)$$

the areal velocity $\dot A$ of a particle subject to a central force is constant.

5.1.2 The effective potential

Using conservation of angular momentum (Eq. (5.13)) to eliminate $\dot{\varphi}$ in Eq. (5.8), we find that the equation for radial motion depends only on r,

$$m\ddot{r} = -\frac{dV}{dr} + \frac{L^2}{mr^3} =: -\frac{dV_{eff}}{dr}, \qquad (5.15)$$

with the **effective potential** is given by

$$V_{eff} = V(r) + \frac{L^2}{2mr^2}. \qquad (5.16)$$

The additional term is the potential of the **centrifugal force**[3]

$$\mathbf{F}_{centr} = \frac{L^2}{mr^3}\mathbf{e}_r. \qquad (5.17)$$

Note that this force follows from the kinetic energy of the angular motion. The importance of the centrifugal force will be discussed later (cf. Subsection 8.5.3).

The solution of Eq. (5.15) can be found directly from the equation for conservation of energy, Eq. (5.11), by using Eq. (5.13) to eliminate $\dot{\varphi}$,

$$E = \frac{m}{2}\left(\dot{r}^2 + \frac{L^2}{m^2r^2}\right) + V(r) = \frac{m}{2}\dot{r}^2 + V_{eff}(r). \qquad (5.18)$$

For the radial velocity we get

$$\dot{r} = \frac{dr}{dt} = \pm\sqrt{\frac{2}{m}(E - V(r)) - \frac{L^2}{m^2r^2}} = \pm\sqrt{\frac{2}{m}(E - V_{eff}(r))} \qquad (5.19)$$

(the sign determines the direction of the radial motion). Integrating, one obtains the (implicit) solution

$$\int_{t_0}^{t} dt' = t - t_0 = \pm\int_{r_0}^{r} \frac{dr'}{\sqrt{\dfrac{2}{m}\left(E - V(r') - \dfrac{L^2}{2mr'^2}\right)}}. \qquad (5.20)$$

---

[3]Huygens had several communications on the 'vis centrifuga' (1669, 1673) but finally his *Tractatus de vi centrifuga* appeared only 1703 in the opuscula postuma, Leyden. He was the first to give the correct expression for the force. Concerning the cause of this force see however the footnote 24 on page 217.

The integral can be evaluated either analytically or numerically to give the radial motion $r(t)$. $\varphi(t)$ is found by inserting $r(t)$ into Eq. (5.13). *The equations of motion of a point mass subject to a central force are always integrable.*

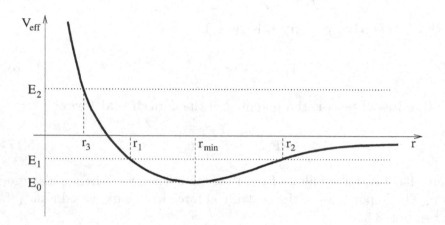

Figure 5.1: A typical form of the effective potential.

Discussion of the general features of motion in a central force is facilitated by considering the r-component of the motion as one-dimensional motion in the effective potential V_{eff}. For $V \propto -1/r$, the effective potential is shown in Fig. 5.1. The discussion of this one-dimensional motion is quite analogous to the discussion in Section 3.2. The only difference is that here the motion is restricted to $r \geq 0$. In particular, **totally bounded motion** ($E = E_1$ in Fig. 5.1) between the turning points r_1 (**internally bounded**) and r_2 (**externally bounded**) is *periodic* in r-*direction*. The *complete motion* (including motion in the φ-direction, too) is, in general, *not periodic!*

The **equation of the orbit** is obtained from Eq. (5.19) by using conservation of angular momentum, Eq. (5.13), to eliminate the time dependence in favor of φ, yielding

$$\frac{dr}{d\varphi} = \frac{dr}{dt}\frac{dt}{d\varphi} = \pm\frac{mr^2}{L}\sqrt{\frac{2}{m}\left(E - V(r)\right) - \frac{L^2}{m^2r^2}}. \qquad (5.21)$$

Integration gives the orbit in implicit form[4], $\varphi = \varphi(r)$,

$$\varphi - \varphi_0 = \pm \int_{r_0}^{r} \frac{L}{r'^2} \frac{dr'}{\sqrt{2m\left(E - V(r')\right) - \dfrac{L^2}{r'^2}}}. \tag{5.22}$$

From Eq. (5.21), it is evident that both $\dot{r}$ and $dr/d\varphi$ vanish *at the radial turning points*:

$$\dot{r} = \frac{dr}{d\varphi} = 0. \tag{5.23}$$

The choice of sign in Eq. (5.22) depends on whether $dr/d\varphi > 0$ or $dr/d\varphi < 0$ for the part of the orbit considered.

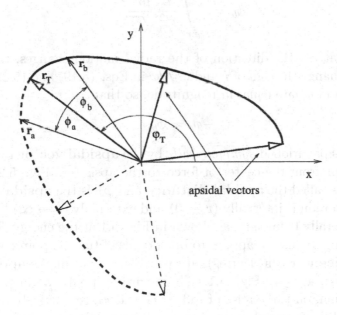

Figure 5.2: The symmetry of the orbit with respect to apsidal vectors.

---

[4]The constants of integration r_0 and φ_0 are the initial values $r(t = 0)$ and $\varphi(t = 0)$ if and only if, for the chosen values E and L, the initial values $\dot{r}(0)$ and $\dot{\varphi}(0)$ satisfy the conditions $E = E\left(r_0, \dot{r}(0), \dot{\varphi}(0)\right)$ (cf. (5.11)) and $L = L\left(r_0, \dot{\varphi}(0)\right)$ (cf. (5.12)). Otherwise r_0 and φ_0 are functions of the initial values $r(0)$, $\varphi(0)$, $\dot{r}(0)$, $\dot{\varphi}(0)$.

5.1.3   Properties of the orbits

Assuming $L \neq 0$, consider the orbit in the vicinity of a turning point r_T, which is also called an **apsis**. Starting on the orbit from a particular point $\mathbf{r}_a$ with polar coordinates $(r_a, \varphi_T + \phi_a)$, we move towards the apsis (r_T, φ_T) (see Fig. 5.2). According to (5.22), the angle $\varphi_T - (\varphi_T + \phi_a) = -\phi_a$ between the turning point and starting point is given by

$$-\phi_a = \int\limits_{r_a}^{r_T} \frac{L}{r'^2} \frac{dr'}{\sqrt{\cdots}} \, .$$

Continuing on until we reach the point $\mathbf{r}_b = (r_a, \varphi_T - \phi_b)$, again at distance r_a from the origin, the angle ϕ_b is given by

$$\phi_b = -\int\limits_{r_T}^{r_a} \frac{L}{r'^2} \frac{dr'}{\sqrt{\cdots}} \, .$$

Note that at r_T, the direction of the radial motion changes, therefore one must change the sign of $\dot{r}$ and $dr/d\varphi$ in Eqs. (5.19) and (5.22). The angles ϕ_a and ϕ_b are equal in magnitude, so that

$$-\phi_a = \phi_b.$$

The orbit is partitioned *symmetrically* by the **apsidal vector** $\mathbf{r}_T$, which is the vector from the center of force to the apsis (cf. Fig. 5.2). The length r_T is called the **apsidal distance** and φ_T is the **apsidal angle**.

    If the motion is internally $(r \to 0)$ and externally $(r \to \infty)$ bounded – that is, totally bounded – as shown in Fig. 5.1 at the energy E_1, then the foregoing discussion applies to both types of turning point: whether at the **pericenter** (inner apsis), at r_1 in Fig. 5.1; or at the **apocenter** (outer apsis), at r_2 in Fig. 5.1. These turning points lie in the plane of the motion, on two circles of radius r_1 and r_2, respectively. Because of Eq. (5.13), $\dot{\varphi}$ as well as $d\varphi/dr$ have always the same values at r_1 and r_2, respectively; therefore, the shape of the orbit in the vicinity of a pericenter or an apocenter is always the same. Since angular momentum is conserved, Eq. (5.12), $\dot{\varphi}$ and also $d\varphi/dr$ never change sign; hence, the orbit is always curved in one direction. The general shape of a bounded orbit[5] is shown in Fig. 5.4. Is the orbit **closed**? That is, is it true that

---

[5]An orbit as sketched in [Goldstein], Chapter 3, cannot appear.

when the polar angle φ increases by $2n\pi$, with n an integer, the radial distance returns to the initial value, $r(\varphi + 2n\pi) = r(\varphi)$? The answer depends on the particular form of the potential $V(r)$ (cf. Section 5.3).

Suppose a local minimum of V_{eff} occurs at distance r_{min} (cf. Fig. 5.1). In the case when the total energy E, denoted by E_0, is equal to V_{eff} at this local minimum (cf. Fig. 5.1), i.e. if

$$E_0 = V_{eff}(r_{min}),$$

the pericenter coalesces with the apocenter. Moreover, $\dot{r}$ (according to Eq. (5.19)) as well as $\ddot{r}$ vanish, because $dV_{eff}/dr|_{r_{min}} = 0$ (cf. Eq. (5.15)). The orbit is circular with radius r_{min}, and the particle moves with constant angular velocity

$$\omega := \dot{\varphi} = L/mr_{min}^2. \tag{5.24}$$

At energies higher than E_0, the radial motion is approximately an oscillation about r_{min}. (To see this, expand V_{eff} about r_{min} to the first nonvanishing order; see also Section 3.2.)

If the total energy is larger than the asymptotic value of the potential $V(r \to \infty)$ (this energy is denoted E_3 in Fig. 5.1), the particle motion is externally unbounded. There is only one apsis (at r_3 in Fig. 5.1). For appropriate initial conditions, the particle approaches the turning point and then runs off to infinity. Again, the orbit is symmetric with respect to the apsidal vector. Further properties of such an orbit will be investigated in the following Section, as well as in Chapter 7.

For the case $L = 0$, the equation of the orbit in polar coordinates follows from Eq. (5.12),

$$\varphi = const. \tag{5.25}$$

The orbit is a straight line through the center of the force (origin). In other words, *if $L = 0$, motion in a central force occurs along a straight line through the center of the force.*

In a central force, the motion is restricted to a fixed plane by conservation of angular momentum (more specifically, the direction of angular momentum is conserved). The motion is two-dimensional. The number of coordinates in the equation of motion is reduced to two in configuration space, which means four in phase space. The maximal number of independent conserved quantities is therefore three. To carry out the integration, one needs only two integrals of the motion; the energy

E and the conserved component of angular momentum L (perpendicular to the orbit) are two independent integrals. Thus *the equations of motion of a particle in a central force are always integrable.*

5.2   Motion in a $1/r$ potential

The gravitational force and the Coulomb force – the most important central forces in classical physics – in fact show the same r-dependence[6]:

$$\mathbf{F} = -km\mathbf{r}/r^3.$$

According to Eq. (5.4), the potential corresponding to this force is

$$V(r) = -\frac{km}{r}. \tag{5.26}$$

From the radial velocity (5.19), one finds

$$\dot{r} = \pm\sqrt{\frac{2}{m}\left(E - V_{eff}(r)\right)},$$

where

$$V_{eff}(r) = -\frac{km}{r} + \frac{L^2}{2mr^2}. \tag{5.27}$$

To obtain $r(t)$, and subsequently $\varphi(t)$, one has to use numerical methods or approximations. However, using Eq. (5.13), the orbit $r(\varphi)$ can be determined analytically.

The case $L \neq 0$, discussed presently, will be applied in the next chapter to planetary motion around the sun. In celestial mechanics, the apocenter and the pericenter are called **aphelion** and **perihelion**, respectively. We therefore introduce this terminology into the discussion now.

5.2.1   The case $L \neq 0$

For $k > 0$ the effective potential, Eq. (5.27), has form shown in Fig. 5.1; in particular, it has a minimum at $r_{\min}$. There are three different kinds of orbits. In Fig. 5.1, the energies of these orbits are denoted E_0, E_1, and E_2:

---

[6]Since we will consider mainly the gravitational force, it is advantageous to separate the mass m of the particle from the strength of the potential. The sign has been chosen so that for gravitational forces, $k > 0$.

i) Case $E = E_0$: The total energy is equal to the minimum of the effective potential $V_{eff}(r)$

$$E_0 = V_{eff}(r_{min}) = -\frac{k^2 m^3}{2L^2}, \qquad r_{min} = \frac{L^2}{m^2 k}. \qquad (5.28)$$

The orbit is circular ($\dot{r} = \ddot{r} \equiv 0$) with radius r_{min} and the particle moves with constant angular velocity (cf. Eq. (5.24))

$$\dot{\varphi} = \frac{L}{mr_{min}^2} = \frac{m^3 k^2}{L^3}. \qquad (5.29)$$

The kinetic energy of the radial motion vanishes. The effective force dV_{eff}/dr vanishes at $r = r_{min}$

$$\left[\frac{km}{r^2} - \frac{L^2}{mr^3} \right]_{r=r_{min}} = \frac{km}{r_{min}^2} - mr_{min}\dot{\varphi}^2 = 0.$$

This just means a balance exists between the centrifugal force ($mr_{min}\dot{\varphi}^2$) and the **centripetal force** (i.e. the gravitational force km/r_{min}^2; E). The angular velocity is restricted to

$$\dot{\varphi} = \sqrt{k/r_{min}^3} = const. \qquad (5.30)$$

ii) Case $E = E_1$: The orbit is still totally bounded between the perihelion r_1 and the aphelion r_2.

iii) Case $E = E_2$: The orbit is externally unbounded. The particle approaching the center arrives at the perihelion r_3 and goes away to infinity.

In all three cases, the angular velocity $\dot{\varphi}$ is either always positive or always negative.

The time dependence of the radius $r(t)$ cannot be calculated from Eq. (5.19) analytically[7] (further details are given in the next Chapter).

---

[7]Nevertheless, there exist excellent approximation methods. Many integrals in classical mechanics can be solved analytically to a very high degree of accuracy (i.e. nearly exact), using techniques of singular perturbation theory. Consult for example the books by A.H. Nayfeh, "Perturbation methods" and "Introduction to perturbation techniques" (see Bibliography).

But the orbit can be determined from the orbit equation[8] (5.22)

$$\varphi = \varphi_0 + \int_{r_0}^{r} \frac{L}{r'^2} \frac{dr'}{\sqrt{2m\left(E + \dfrac{km}{r'} - \dfrac{L^2}{2mr'^2}\right)}}.$$

Substituting $u' = 1/r'$ yields

$$\varphi = \varphi_0 - \int_{u_0}^{u} \frac{du'}{\sqrt{\dfrac{2mE}{L^2} + \dfrac{2m^2k}{L^2}u' - u'^2}}.$$

Consulting integral tables, one finds that

$$\int \frac{du}{\sqrt{\alpha + \beta u + \gamma u^2}} = \frac{1}{\sqrt{-\gamma}} \arccos\left(-\frac{\beta + 2\gamma u}{\sqrt{\beta^2 - 4\alpha\gamma}}\right).$$

Therefore, with $\alpha = 2mE/L^2$, $\beta = 2m^2k/L^2$, and $\gamma = -1$, the solution for φ is

$$\varphi = \varphi_K - \arccos \frac{\dfrac{L^2}{m^2|k|}\dfrac{1}{r} - \dfrac{k}{|k|}}{\sqrt{1 + \dfrac{2EL^2}{m^3k^2}}},$$

which may be inverted to give an expression for $1/r$,

$$\frac{1}{r} = \frac{m^2|k|}{L^2}\left(\frac{k}{|k|} + \sqrt{1 + \frac{2EL^2}{m^3k^2}}\cos(\varphi - \varphi_K)\right). \tag{5.31}$$

All constants of integration have been gathered in φ_K.

Relation (5.31) between r and φ is the representation of a conic section in polar coordinates. One focus is at the origin (i.e. the center of force). The **semilatus rectum**

$$p = \frac{L^2}{m^2|k|} \tag{5.32}$$

---

[8]The choice of sign in Eq. (5.22) fixes the orientation of the particle motion about the center of force.

and the **eccentricity**

$$\varepsilon = \sqrt{1 + \frac{2EL^2}{m^3 k^2}} = \sqrt{1 + \frac{E}{|E_0|}} \tag{5.33}$$

characterize the shape of the conic section. (ε is always real, since even for negative values of the energy, the condition $E \geq E_0$ holds; cf. Eq. (5.28) and Fig. 5.1.) The angle φ_K in the orbit equation

$$\frac{1}{r} = \frac{1}{p} \left(\mathrm{sgn}\,(k) + \varepsilon \cos(\varphi - \varphi_K) \right) \tag{5.34}$$

is the apsidal angle of the perihelion. After φ_K is fixed, every orbit for the $1/r$ potential is determined by the values p and ε, or L and E, in a unique manner. Each of the four conic sections

Hyperbola: $\quad \varepsilon > 1 \; (E > 0)$

Parabola: $\quad \varepsilon = 1 \; (E = 0)$

Ellipse: $\quad \varepsilon < 1 \; (E < 0)$

Circle: $\quad \varepsilon = 0 \; (E = E_0)$

may occur as an orbit.

The conic sections

In the following, we summarize some mathematical properties of conic sections. We choose the orientation of our coordinate system such that $\varphi_K = 0$ in Eq. (5.34); consequently, $r\,(\varphi)$ is given by

$$r = \frac{p}{\mathrm{sgn}\,(k) + \varepsilon \cos \varphi}. \tag{5.35}$$

Let us first consider an <u>attractive force</u>, $k > 0$, i.e. $r = p/\,(1 + \varepsilon \cos \varphi)$. The points of intersection of the conic section with the y-axis ($\varphi = \pi/2$ and $\varphi = 3\pi/2$, respectively) are at distance p from the origin. The origin is at a focus of the conic section; and is also the force center.

i) Ellipse $(\varepsilon < 1)$, circle $(\varepsilon = 0)$:

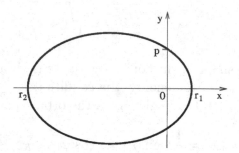

The perihelion at $\varphi = 0$ and the aphelion at $\varphi = \pi$ are at distance

$$r_1 = \frac{p}{1+\varepsilon}, \qquad r_2 = \frac{p}{1-\varepsilon}, \tag{5.36}$$

respectively, to the origin (for the circle, we have: $p = r_1 = r_2 = r$). Another complete set of parameters that fix the form of the ellipse consists of the semimajor axis

$$a = \frac{r_1 + r_2}{2} = \frac{p}{1 - \varepsilon^2} \tag{5.37}$$

and the semiminor axis

$$b = a\sqrt{1 - \varepsilon^2} = \frac{p}{\sqrt{1 - \varepsilon^2}} \tag{5.38}$$

(for the circle we have: $a = b = r$). Expressing the semilatus rectum p and the eccentricity ε in terms of a and b, we get

$$p = \frac{b^2}{a}, \qquad \varepsilon = \sqrt{1 - \frac{b^2}{a^2}}. \tag{5.39}$$

These conic sections (ellipse, circle) are the *closed* orbits in the $1/r$ potential.

ii) <u>Parabola</u> ($\varepsilon = 1$):

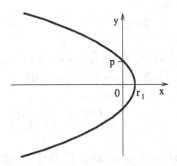

The only apsis (perihelion) with apsidal angle $\varphi = 0$ is at distance

$$r_1 = \frac{p}{2} \tag{5.40}$$

from the only focus (i.e. the origin). A parabolic orbit can only appear for $E = 0$. It is therefore of minor importance in physical systems.

iii) <u>Hyperbola</u> ($\varepsilon > 1$):

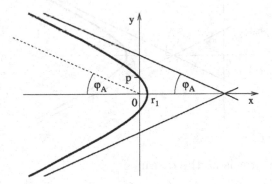

Equation (5.35) for the orbit only yields the branch of the hyperbola closest to the focus that contains the force center (i.e. the origin). The inequality $\cos\varphi \geq -1/\varepsilon$ holds; saturation of this inequality,

$$\cos\varphi_A = -1/\varepsilon, \tag{5.41}$$

determines the angles φ_A of the asymptotes ($r(\varphi_A) \to \infty$). At $\varphi = 0$, the distance of the perihelion from the origin is

$$r_1 = \frac{p}{1+\varepsilon}. \tag{5.42}$$

The equations for the asymptotes in Cartesian coordinates are

$$y = \mp x\sqrt{\varepsilon^2 - 1} \pm \frac{\varepsilon p}{\sqrt{\varepsilon^2 - 1}}.$$

From this relation, it is easy to deduce the distance of the focus (i.e. the center of the force) to each of the asymptotes (E)

$$s = \frac{p}{\sqrt{\varepsilon^2 - 1}}. \tag{5.43}$$

For a repulsive force, $k < 0$, Eq. (5.35) reads

$$r = \frac{p}{\varepsilon \cos\varphi - 1}, \tag{5.44}$$

and consequently, $\cos\varphi \geq 1/\varepsilon$. Only hyperbolic orbits[9] exist. Here, the origin (where the force center is located) is at the outer focus, close to the second, unphysical branch of the hyperbola.

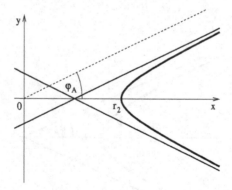

The pericenter is at the distance

$$r_{\min} = r_2 = p/(\varepsilon - 1) \tag{5.45}$$

from the force center. The equations of the asymptotes are the same as for $k > 0$; therefore, the distance from the force center to the asymptotes is again given by Eq. (5.43).

The hyperbolic orbits ($\varepsilon \geq 1$ and $E \geq 0$, respectively) are of interest in the problem of particle scattering in a $1/r$ potential (cf. Chapter 7).

---

[9]From the effective potential, one can immediately recognize the fact that only unbounded motion is possible.

5.2.2 Bounded motion for $L = 0$

The case $L = 0$ requires separate treatment. For $L \neq 0$, the centrifugal barrier – that is the dominance of the $1/r^2$ term of the effective potential as the particle approaches $r = 0$ (cf. Fig. 5.1) – prevents the particle from approaching the center of the potential. The singularity of the potential ($\propto 1/r$) remains inaccessible. For $L = 0$, as pointed out in Section 5.1, the motion is along a line through the origin (with equation $\varphi = const$). Let us consider this line to be the x-axis. Immediately upon inspecting the potential, it can be seen that the origin $x = r = 0$ is accessible. The only remaining question is the temporal dependence of the motion. A point mass initially at rest at $x = r_0$ (cf. Fig. 5.3) starts to move towards the origin $x = 0$. As it moves, its velocity and kinetic energy increase without bound (**E**). The point mass passes the origin at infinite velocity. Then the velocity decreases until the point $x = -r_0$ is reached, whereupon the point mass stops again. The point mass then accelerates in the opposite direction, passes the origin again, and returns to its original position. The cycle continues ad infinitum[10].

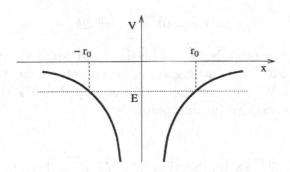

Figure 5.3: The situation for L=0 ($V_{eff} = V$).

Because of the singularity in the origin, the conservation of energy

---

[10]For bodies of finite dimensions, such as stars, the particle is reflected at the surface. If there would be a (small) tunnel through the center of the star, the attraction felt by the particle within the tunnel, for a homogeneous mass distribution, is given by a harmonic force (cf. Eq. (6.57); concerning its motion, see Problem 9 at the end of this Chapter).

equation only gives the particle motion until the singularity is almost reached. We restrict ourselves to studying the motion from $x = r_0$ to $x = 0$ (E). From conservation of energy, we have

$$\dot{r} = -\sqrt{\frac{2}{m}\left(-|E| + \frac{mk}{r}\right)} = -\sqrt{2k}\sqrt{\frac{1}{r} - \frac{1}{r_0}},$$

where for $\dot{x} = \dot{r} = 0$ we have

$$r = r_0 = \frac{mk}{|E|}.$$

Setting the constant of integration $t_0 = 0$ gives

$$t = -\sqrt{1/2k}\int\limits_r^{r_0} \frac{dr'}{\sqrt{\frac{1}{r'} - \frac{1}{r_0}}}.$$

Evaluating the integral yields

$$t(r) = \sqrt{\frac{r_0}{2k}}\left(\sqrt{r}\sqrt{r_0 - r} + r_0 \arctan \sqrt{r_0/r - 1}\right),$$

where t is in the interval $0 \le t \le T$, with

$$T := t(r = 0) = \frac{\pi}{2}\sqrt{r_0^3/2k}.$$

The period of the orbit is $4T$. This is the time which it takes the particle to go from one turning point to the other and back to the initial position. We see that the period is related to the distance $2r_0$ between two consecutive turning points by

$$T \sim r_0^{3/2}.$$

This is Kepler's 3rd law (cf. Section 6.3). As we will see presently, the exponent $3/2$ is characteristic for the $1/r$-dependence of the potential.

5.3   Motion in the potential $V(r) \propto 1/r^\alpha$

We now elaborate on the connection between the r-dependence of the potential $V(r)$ and the shape and period of the orbit. We assume a power law behavior for the r-dependence,

$$V(r) = -\frac{mk}{r^\alpha}.$$

Is there a connection between the properties of the motion and the exponent α? First, we present a compilation of values of α such that the integral in the orbit equation,

$$\varphi = \varphi_0 + \int_{r_0}^{r} \frac{L}{r'^2} \frac{dr'}{\sqrt{2m\left(E + \frac{mk}{r'^\alpha}\right) - \frac{L^2}{r'^2}}},$$

is expressible in terms of elementary functions. Elementary functions result if $\alpha = -2, 1, 2$. Certain other values of α lead to elliptic integrals.

Case $\alpha = -2$. Isotropic oscillator (see Section 4.1). In general, the orbits are ellipses, and the center of the force coincides with the center of the ellipse[11] (and not one of the foci). The period T is independent of the linear dimension a of the ellipse[12] (E): $T = 2\pi/\omega$.

Case $\alpha = 1$. Gravitational potential. In general, the orbits are ellipses and the center of the force is situated at one of the foci. The periods obey Kepler's third law: $T \sim a^{3/2}$ (see next Chapter).

Case $\alpha = 2$. $1/r^2$ potential. Motion in the effective potential V_{eff} is bounded only if $L^2 < 2m^2 k$. Since the orbit passes through the origin $r = 0$, intervals on either side of the origin must be considered separately (similar to the case $\alpha = 1$ when $L = 0$). With appropriately chosen constants of integration, the equation for the orbit is (E)

$$r = \frac{r_0}{\cosh(c\varphi)}, \qquad 0 < t < T,$$

where $c = \sqrt{2mk/L^2 - 1}$, $r_0 = c\sqrt{L/2m\,|E|}$, and $T = \pi r_0^2/2cL^2$. The orbit is a spiral, and the characteristic time of the motion is $T \sim r_0^2$.

---

[11]In his solution to Proposition X of Book I of the "Principia" ('*If a body revolves in an ellipse; it is proposed to find the law of the centripetal force tending to the centre of the ellipse*'), Newton shows that the force is proportional to the distance from the center.

[12] "Principia", Book I, Proposition X, Corollary 2.

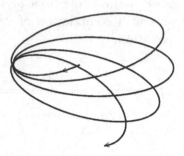

Figure 5.4: Part of a rosette-shaped orbit ($\alpha = 1.1$).

Do closed orbits – like the ellipses and circles – also exist for $\alpha \neq -2, 1$? The answer depends on the value of α. Consider values α in the vicinity of $\alpha = 1$. For $\alpha < 1$, the perihelion undergoes retrograde motion, and for $\alpha > 1$ the perihelion advances. In both cases, the result is a **rosette-shaped orbit**, in which the apsidal angle between two successive passes changes by a certain amount $\Delta\varphi$ (cf. Fig. 5.4).

5.3.1   Mechanical similarity

For potentials with a pure power-law,

$$V(r) = c\,r^{-\alpha},$$

a relation exists between a length-scale a that characterizes the size of the orbit (e.g. the distance between two consecutive radial turning points in Fig. 5.1) and the time T it takes a particle to move the distance[13] a. This relation is a consequence of the invariance of the equation of motion

$$m\ddot{\mathbf{r}} = -\boldsymbol{\nabla}V(r) = c\alpha\,\mathbf{r}/r^{\alpha+2}$$

under a simultaneous transformation of length and time scales. This invariance is called **mechanical similarity**. If one transforms the length scale according to

$$\mathbf{r}' = \lambda\mathbf{r}$$

---

[13]For example, in an elliptical orbit, the distance along the major axis is $2a$, and the corresponding time is half the period – the time it takes to go halfway around the ellipse.

and the time scale according to

$$t' = \mu t,$$

the equation of motion transforms to

$$(\mu^2/\lambda)m\ddot{\mathbf{r}}' = -\lambda^{(\alpha+1)}c\alpha\,\mathbf{r}'/r'^{\alpha+2}. \tag{5.46}$$

The new parameters (a', T') of the orbit follow from the scaling relations

$$a' = \lambda a, \qquad T' = \mu T.$$

The equation for the orbit remains invariant, if in Eq. (5.46) the factors depending on λ and/or μ cancel. This is achieved for

$$\mu = \lambda^{(\alpha+2)/2}. \tag{5.47}$$

Expressing now λ and μ in terms of a, a', T, and T',

$$\lambda = a'/a, \qquad \mu = T'/T,$$

one obtains from Eq. (5.47), $T'/T = (a'/a)^{(\alpha+2)/2}$. Hence the desired relation between T and a is

$$T \propto a^{(\alpha+2)/2}; \tag{5.48}$$

the constant $T'/a'^{(\alpha+2)/2}$ (which is not written here) takes its value in some fixed reference system. For $\alpha = -2, 1, 2$ we recover from Eq. (5.48) the relations between the sizes of the orbits and the periods given above.

5.4 The Runge-Lenz vector

For $\alpha = 1$, the perihelion does not advance. The orbits are always closed (ellipses). This fact is due to a further independent conservation law produced by the $1/r$ potential[14]. Taking the vector product of

---

[14]This occurs because there is a particular symmetry of the $1/r$ potential. A further conserved quantity also exists in the case of the isotropic harmonic oscillator, whose orbits are also ellipses (see Page 124).

both sides of the equation of motion, $m\ddot{\mathbf{r}} = -km\mathbf{r}/r^3$, with the angular momentum $\mathbf{L}$, yields the relation

$$\ddot{\mathbf{r}} \times \mathbf{L} = -k\frac{\mathbf{r} \times \mathbf{L}}{r^3} = -mk\frac{\mathbf{r} \times (\mathbf{r} \times \dot{\mathbf{r}})}{r^3} = -mk\left(\mathbf{r}\frac{r\dot{r}}{r^3} - \frac{\dot{\mathbf{r}}}{r}\right).$$

Since, for a central force, $\dot{\mathbf{L}} = \mathbf{0}$, we obtain

$$\frac{d}{dt}(\dot{\mathbf{r}} \times \mathbf{L}) = -mk\left(\mathbf{r}\frac{\dot{r}}{r^2} - \frac{\dot{\mathbf{r}}}{r}\right) = mk\frac{d}{dt}\frac{\mathbf{r}}{r},$$

where we have used the fact that $\mathbf{r}\dot{\mathbf{r}} = r\dot{r}$ (which follows from $dr^2/dt = d\mathbf{r}^2/dt$). Collecting the time derivatives, we see that the **Runge-Lenz vector**[15] $\mathbf{K}$,

$$\mathbf{K} = \dot{\mathbf{r}} \times \mathbf{L} - mk\frac{\mathbf{r}}{r}, \tag{5.49}$$

is constant in time:

$$\dot{\mathbf{K}} = 0.$$

This time independence is related to a degeneracy of the periodicity of the r and φ-component of the motion (cf. Chapter 16). For this reason, the orbits are closed.

Since, for motion in the $1/r$ potential, we have already 4 independent conserved quantities, and we know that there are at most five, it follows that not all of the components of $\mathbf{K}$ can be independent conserved quantities. There are two conditions that the vector $\mathbf{K}$ satisfies:

1.) Since $\mathbf{K}$ is perpendicular to $\mathbf{L}$ (i.e. $\mathbf{KL} = 0$), $\mathbf{K}$ must lie in the plane of the orbit. In polar coordinates, where the z-axis is given by the angular momentum, $\mathbf{L} = L\mathbf{e}_z$, we have

$$\mathbf{K} = (r\dot{\varphi}L - mk)\mathbf{e}_r - \dot{r}L\mathbf{e}_\varphi, \tag{5.50}$$

where we used Eq. (A.27) for $\dot{\mathbf{r}}$ and once again denote ρ by r.

2.) The length of $\mathbf{K}$ can be expressed in terms of the energy E and the angular momentum L, since from

$$\mathbf{K}^2 = (r\dot{\varphi}L - mk)^2 + (\dot{r}L)^2 = (\dot{r}^2 + r^2\dot{\varphi}^2)L^2 - 2r\dot{\varphi}Lmk + m^2k^2$$
$$= (2E/m + 2k/r)L^2 - 2kL^2/r + m^2k^2,$$

---

[15]That this quantity is conserved has been rediscovered several times. Also Laplace in his "Mécanique Céleste" (see Bibliography) showed it. Therefore the vector is also named Laplace-Runge-Lenz vector.

it follows that

$$\mathbf{K}^2 = m^2 k^2 \left(1 + \frac{2EL^2}{m^3 k^2}\right). \tag{5.51}$$

Comparison to Eq. (5.33) shows that the eccentricity ε is proportional to $K = |\mathbf{K}|$,

$$\varepsilon = K/mk. \tag{5.52}$$

The remaining independent conserved quantity related to $\mathbf{K}$ is the angle φ_K in the orbital plane,

$$\mathbf{K} = K \cos \varphi_K \mathbf{e}_x + K \sin \varphi_K \mathbf{e}_y = K \left(\cos(\varphi - \varphi_K)\mathbf{e}_r - \sin(\varphi - \varphi_K)\mathbf{e}_\varphi\right). \tag{5.53}$$

(The second part of this equation is obtained by expressing $\mathbf{e}_x$ and $\mathbf{e}_y$ in terms of $\mathbf{e}_r$ and $\mathbf{e}_\varphi$ with the help of Appendix A.) Combining Eq. (5.50) and Eq. (5.53), φ_K can be expressed in terms of the dynamical variables as follows:

$$\tan(\varphi - \varphi_K) = \frac{\dot{r}}{r\dot{\varphi} - mk/L}.$$

Thus, for the $1/r$ potential, in addition to the *energy* E and the *angular momentum* $\mathbf{L}$, we have a *fifth integral* $I_K(r, \varphi, \dot{r}, \dot{\varphi})$, *the direction of* $\mathbf{K}$,

$$I_K = \varphi_K = \varphi - \arctan\left(\frac{\dot{r}}{r\dot{\varphi} - mk/L}\right) = const. \tag{5.54}$$

Let us consider the motion in the plane fixed by the invariant direction of angular momentum. For this two-dimensional system, we have a set of three independent conserved quantities as functions of the phase space coordinates $(r, \dot{r}, \varphi, \dot{\varphi})$: the energy E

$$E = \frac{m}{2}\left(\dot{r}^2 + r^2\dot{\varphi}^2\right) - \frac{km}{r}, \tag{5.55}$$

the z-component of the angular momentum L

$$L = mr^2\dot{\varphi}, \tag{5.56}$$

and the angle φ_K. Since the number of conserved quantities is maximal, the orbit can be determined algebraically. One only has to eliminate $\dot{\varphi}$ and $\dot{r}$ in Eq. (5.54) with the help of (5.56) and (5.55) (E). A shorter method is the following. Consider the projection of the Runge-Lenz vector onto the radius vector (cf. Eqs. (5.49) and (5.53)),

$$
\begin{aligned}
\mathbf{rK} &= rK\cos(\varphi - \varphi_K) \\
&= \mathbf{r}\,(\dot{\mathbf{r}} \times \mathbf{L}) - mkr = \mathbf{L}\,(\mathbf{r} \times \dot{\mathbf{r}}) - mkr \\
&= \frac{L^2}{m} - mkr.
\end{aligned}
$$

This gives $r\,(mk + K\cos(\varphi - \varphi_K)) = \dfrac{L^2}{m}$, or

$$
r = \frac{L^2/m^2k}{1 + \dfrac{K}{mk}\cos(\varphi - \varphi_K)}, \tag{5.57}
$$

which is again a conic section in polar coordinates (cf. Eqs. (5.34), (5.32), and (5.52)). Eq. (5.57) also justifies the notation in Eq. (5.34). Moreover, from Eq. (5.57), we see that the point on the orbit that is closest to the origin (which is the focus, center of force) – the perihelion – has the angular coordinate φ_K. Thus the vector from the focus to the perihelion is parallel to the Runge-Lenz vector $\mathbf{K}$. (The two are in fact one and the same.)

The foregoing discussion of the Runge-Lenz vector is not restricted to ellipses. It also applies to the open orbits (parabola, hyperbola), where the Runge-Lenz vector is also directed from the focus to the perihelion. This fact is of importance in the scattering of particles in a $1/r$ potential (see Chapter 7).

We shall see later (cf. Chapter 10) that conserved quantities are linked to invariance properties. We note, finally, that the existence of the Runge-Lenz vector is not connected with a symmetry under transformation of the coordinate system in configuration space, but rather to a symmetry under transformation of the coordinates in phase space (F. Schweiger, Acta Phys.Austr.**17**,343(1964)).

Comparison to the isotropic oscillator

The three-dimensional isotropic harmonic oscillator potential is given by,

$$
V\,(r) = \frac{m\omega^2}{2}\,(x^2 + y^2 + z^2) = \frac{m\omega^2}{2}r^2. \tag{5.58}
$$

Since angular momentum is conserved, the orbits lie in a fixed plane perpendicular to the angular momentum. Consequently, this system is equivalent to a *two-dimensional isotropic oscillator*. One can see this as follows. Since the oscillator defined by Eq. (5.58) is isotropic, as in the case of the $1/r$ potential, one can choose the direction for the angular momentum to be along the z-axis, so that $\mathbf{r} = (x, y, 0)$. Now the potential $V(r)$ reduces to $V(\rho) = m\omega^2 \rho^2/2$, $\rho^2 = x^2 + y^2$.

In Section 4.1, for the two-dimensional isotropic oscillator, it was shown that one has the maximal number of three independent conserved quantities; namely, E_1, E_2, and L (see Eqs. (4.12), (4.13), and (4.14)). Furthermore, it was shown that the orbits are ellipses (cf. Eq. (4.18)), so the apocenter remains stationary. One may therefore ask: Which conserved quantities correspond to the Runge-Lenz vector and to φ_K in the $1/r$ potential? If one chooses the independent conserved quantities to be $E = E_1 + E_2$, L, and I_4 (cf. (4.16)), one can show (**E**) that for the oscillator, the (constant) angle of the main axis of the ellipse with respect to the x-axis is given by

$$\varphi_O = \frac{1}{2} \arcsin\left[\frac{mI_4}{\sqrt{E^2 - \omega^2 L^2}}\right],\qquad (5.59)$$

and that the conservation law, Eq. (4.16), in polar coordinates is

$$\varphi_O = \varphi + \frac{1}{2} \arctan \frac{\dot{\rho}}{\rho\left(\frac{E}{L} - \frac{L}{m\rho^2}\right)}.\qquad (5.60)$$

This corresponds to the conservation law (5.54) for the $1/r$ potential.

5.5 Integrability vanishes

The situation changes if a force with lower symmetry is superimposed onto the central force (which could be, say, the Coulomb force). The superimposed force may, for instance, define a preferred direction in space, as in the following case. We consider an additional, velocity-dependent force, $\frac{q}{c}(\dot{\mathbf{r}} \times \mathbf{B})$, $\mathbf{B} = const$, so that the equation of motion now reads:

$$m\ddot{\mathbf{r}} = f(r)\mathbf{r}/r + \frac{q}{c}(\dot{\mathbf{r}} \times \mathbf{B}).\qquad (5.61)$$

Such a force appears in electrodynamics as the Lorentz force; or, when $(q/c)\mathbf{B}$ is replaced by $2m\boldsymbol{\omega}$, as a pseudo-force in reference frames rotating with angular velocity $\boldsymbol{\omega}$ (see Subsection 8.5.3). In the Lorentz force case, q is the charge of the particle (c is the velocity of light). Multiplying the equation of motion (5.61) by $\dot{\mathbf{r}}$, one can easily see that the superimposed force does not contribute to the energy. The equation for *conservation of energy* still reads

$$E = \frac{m}{2}\dot{\mathbf{r}}^2 + V(r) = const, \qquad V(r) = -\int dr\, f(r). \qquad (5.62)$$

The energy E is unchanged by the presence of the field $\mathbf{B}$. Are there any more conserved quantities? The direction of $\mathbf{B}$ destroys the spherical symmetry of the central force. Therefore, the angular momentum with respect to the center of the force is not conserved anymore (cf. Chapter 10). But there is another conserved quantity. The vector product of Eq. (5.61) and $\mathbf{r}$ gives

$$m\left(\mathbf{r} \times \ddot{\mathbf{r}}\right) = \frac{d}{dt}m(\mathbf{r} \times \dot{\mathbf{r}}) = \frac{d}{dt}\mathbf{L} = \frac{q}{c}\left(\mathbf{r} \times (\dot{\mathbf{r}} \times \mathbf{B})\right) = \frac{q}{c}\left((\mathbf{r}\mathbf{B})\dot{\mathbf{r}} - (\mathbf{r}\dot{\mathbf{r}})\mathbf{B}\right).$$

Multiplying this equation by $\mathbf{B}$, we have

$$\begin{aligned}
\frac{d}{dt}\mathbf{L}\mathbf{B} &= \frac{q}{c}\mathbf{B}\left(\mathbf{r} \times (\dot{\mathbf{r}} \times \mathbf{B})\right) = \frac{q}{c}\left(\mathbf{B} \times \mathbf{r}\right)\left(\dot{\mathbf{r}} \times \mathbf{B}\right) \\
&= -\frac{q}{2c}\frac{d}{dt}(\mathbf{r} \times \mathbf{B})^2 = \frac{q}{2c}\frac{d}{dt}\left((\mathbf{r}\mathbf{B})^2 - \mathbf{r}^2\mathbf{B}^2\right),
\end{aligned}$$

which shows that even the component of the angular momentum parallel to the magnetic field is not conserved; and that there is a conserved quantity that is related to this component of the angular momentum. Collecting the time derivatives yields

$$I_{LB}B := m(\mathbf{r} \times \dot{\mathbf{r}})\mathbf{B} + \frac{q}{2c}\left(\mathbf{r} \times \mathbf{B}\right)^2 = const \qquad (5.63)$$

as a *second conserved quantity*. It turns out that the energy E and I_{LB} are, in general, the only ('useful') conserved quantities. Before proceeding any further, we consider the case $f(r) \equiv 0$.

5.5.1 The homogeneous magnetic field as the sole force

If the only force acting is a constant magnetic field (cf. Fig. 5.5), then the equation of motion reads

$$m\ddot{\mathbf{r}} = \frac{q}{c}\left(\dot{\mathbf{r}} \times \mathbf{B}\right). \qquad (5.64)$$

The energy, which reduces to the kinetic energy (cf. Eq. (5.62)),

$$E = \frac{m}{2}\dot{\mathbf{r}}^2,$$

(5.65)

is conserved together with the quantity I_{LB}, Eq. (5.63). A further conserved quantity is obtained by integrating the equation of motion (5.64):

$$\mathbf{u} := \dot{\mathbf{r}} - \frac{q}{mc}(\mathbf{r} \times \mathbf{B}) = const.$$

(5.66)

It turns out, that four out of the five conserved quantities given in Eqs. (5.65), (5.63), and (5.66) are independent (cf. Eq. (5.74)). This implies that the equations of motion are integrable.

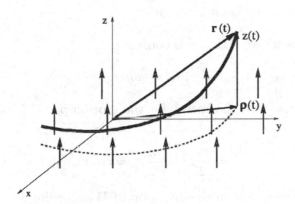

Figure 5.5: The orbit of a particle in a homogeneous magnetic field indicated by the arrows; the dashed line is its projection onto the (x, y)-plane.

Multiplying Eq. (5.66) by $\mathbf{B} = B\mathbf{e}_z$, one can easily see that the velocity of the particle in the direction of $\mathbf{B}$ is constant,

$$\dot{\mathbf{r}}\mathbf{e}_z = u\mathbf{e}_z = const.$$

Therefore the three-dimensional motion *separates* into a (simple) one-dimensional motion parallel to $\mathbf{B}$ and a two-dimensional motion perpendicular to $\mathbf{B}$. Accordingly, it is advantageous to split both $\mathbf{r}$ and $\dot{\mathbf{r}}$ into components parallel to $\mathbf{B}$, namely $z\mathbf{e}_z$ and $\dot{z}\mathbf{e}_z$, and into components perpendicular to $\mathbf{B}$, namely $\boldsymbol{\rho}$ and $\dot{\boldsymbol{\rho}}$ (see Fig. 5.5); *viz.*:

$$\mathbf{r} = z\mathbf{e}_z + \boldsymbol{\rho}, \qquad z = (\mathbf{r}\mathbf{B})/B, \quad \boldsymbol{\rho}\mathbf{B} = 0,$$

so that

$$\dot{\mathbf{r}} = \dot{z}\mathbf{e}_z + \dot{\boldsymbol{\rho}}, \qquad \dot{z} = (\dot{\mathbf{r}}\mathbf{B})/B, \quad \dot{\boldsymbol{\rho}}\mathbf{B} = 0.$$

Since

$$\begin{aligned}
\dot{\mathbf{r}} \times \mathbf{B} &= (\dot{\boldsymbol{\rho}} + \dot{z}\mathbf{e}_z) \times B\mathbf{e}_z \\
&= \dot{\boldsymbol{\rho}} \times B\mathbf{e}_z,
\end{aligned}$$

the equation of motion (5.64) splits into the system of differential equations

$$\begin{aligned}
m\ddot{\boldsymbol{\rho}} &= \frac{qB}{c}\dot{\boldsymbol{\rho}} \times \mathbf{e}_z \qquad\qquad (5.67) \\
m\ddot{z} &= 0.
\end{aligned}$$

The last equation shows that $\dot{z}$ is constant,

$$\dot{z} =: v_{\parallel} = const,$$

and therefore also that the energy for the motion parallel to $\mathbf{B}$ remains constant separately:

$$E_{\parallel} = \frac{m}{2}\dot{z}^2 = \frac{m}{2}v_{\parallel}^2 = const. \qquad (5.68)$$

Thus the particle moves in the direction of $\mathbf{B}$ according to

$$z = v_{\parallel}t + z_0.$$

The motion of the particle in the plane perpendicular to $\mathbf{B}$ is independent of its motion in the z-direction. Eqs. (5.65) and (5.68) imply that the corresponding energy E_{ρ} is also constant:

$$E_{\rho} = E - E_{\parallel} = \frac{m}{2}\dot{\boldsymbol{\rho}}^2 = const . \qquad (5.69)$$

Also, the conserved quantity I_{LB}, defined in Eq. (5.63), is determined only by the components of the dynamical variables perpendicular to $\mathbf{B}$. Since

$$(\mathbf{r} \times \mathbf{B}) = (\boldsymbol{\rho} \times \mathbf{B}) \qquad\qquad (5.70)$$

and

$$\mathbf{L}\mathbf{B} = m(\mathbf{r} \times \dot{\mathbf{r}})\mathbf{B} = m(\boldsymbol{\rho} \times \dot{\boldsymbol{\rho}})\mathbf{B} = L_B B,$$

where $\mathbf{L}_B = m(\boldsymbol{\rho} \times \dot{\boldsymbol{\rho}})$ is the component of angular momentum (with respect to the chosen origin) parallel to $\mathbf{B}$, we have

$$I_{LB}B = \mathbf{L}_B\mathbf{B} + \frac{q}{2c}\left(\boldsymbol{\rho} \times \mathbf{B}\right)^2 = const. \tag{5.71}$$

Moreover, Eqs. (5.66) and (5.70) also imply conservation of the component of $\mathbf{u}$ in the plane,

$$\mathbf{u}_\rho := \dot{\boldsymbol{\rho}} - \omega_Z\left(\boldsymbol{\rho} \times \mathbf{e}_z\right) = const. \tag{5.72}$$

The quantity

$$\omega_Z = \frac{qB}{mc} \tag{5.73}$$

is the so-called **cyclotron frequency**. We will presently see that ω_Z is the time scale of the circular motion in the plane perpendicular to the applied magnetic field.

The four conserved quantities (5.69), (5.71), and (5.72) are related to each other via

$$\frac{m}{2}\mathbf{u}_\rho^2 = E_\rho + \omega_Z I_{LB}. \tag{5.74}$$

This reflects the fact that for the planar motion, only three of these quantities can be independent. Proceeding as in derivation of the Runge-Lenz vector, the Cartesian components of $\mathbf{u}_\rho$ can be expressed as

$$\mathbf{u}_\rho = \begin{pmatrix} u_1 \\ u_2 \end{pmatrix} = u_\rho \begin{pmatrix} \cos\varphi_u \\ \sin\varphi_u \end{pmatrix},$$

while in polar coordinates (cf. Appendix A, Section A.3) one has

$$\mathbf{u}_\rho = \dot{\rho}\mathbf{e}_\rho + \rho\left(\dot{\varphi} + \omega_Z\right)\mathbf{e}_\varphi.$$

Similar to Eq. (5.54), the angle φ_u (cf. Fig. 5.6) is then established as an independent conserved quantity,

$$\varphi_u = \varphi + \arctan\frac{\rho\left(\dot{\varphi} + \omega_Z\right)}{\dot{\rho}} = const. \tag{5.75}$$

Since we now have the maximal number of three conserved quantities (E_ρ, I_{LB}, and φ_u), the orbit could be determined algebraically in the form $\rho = \rho\left(\varphi\right)$.

In polar coordinates, the conservation of the energy E_ρ, Eq. (5.69), takes the form

$$E_\rho = \frac{m}{2}\left(\dot\rho^2 + \rho^2\dot\varphi^2\right),$$

and the conserved quantity I_{LB} is written as

$$I_{LB} = m\rho^2\left(\dot\varphi + \frac{1}{2}\omega_Z\right). \tag{5.76}$$

This equation allows to eliminate the angular velocity $\dot\varphi$ in E_ρ in favor of the constant I_{LB}:

$$E_\rho = \frac{m}{2}\dot\rho^2 + \frac{I_{LB}^2}{2m\rho^2} + \frac{m}{2}\left(\frac{\omega_Z}{2}\right)^2\rho^2 - \frac{1}{2}\omega_Z I_{LB}. \tag{5.77}$$

Except for the last term, E_ρ here has the same form as the energy of a two-dimensional isotropic harmonic oscillator in polar coordinates (see Problem 5 on Page 99) with angular momentum $L_z = I_{LB}$ and oscillation frequency $\omega_Z/2$. Nevertheless, the orbits are quite different. The two systems also differ in the nature of the conserved quantities[16]. In particular, in the case of the magnetic field, we have the conservation of $\mathbf{u}_\rho$, whereas in the case of the harmonic oscillator, the energies of the motion in the x and y directions are conserved (cf. Section 4.1).

For the purpose of comparing the situation with the isotropic harmonic oscillator case, we now consider the planar motion of the system in Cartesian coordinates, where $\boldsymbol{\rho} = (x, y)$. In the plane, the equations of motion (5.64) reduce to the coupled differential equations

$$\begin{aligned}
\ddot{x} &= \omega_Z\dot{y} \\
\ddot{y} &= -\omega_Z\dot{x}.
\end{aligned} \tag{5.78}$$

The form of this system of equations appears to be quite different to that of the oscillator, Eqs. (4.10). However, differentiating with respect to time and rewriting, we obtain

$$\begin{aligned}
\frac{d^3}{dt^3}x + \omega_Z^2\frac{d}{dt}x &= 0 \\
\frac{d^3}{dt^3}y + \omega_Z^2\frac{d}{dt}y &= 0.
\end{aligned}$$

---

[16]The distinction vanishes in a reference frame rotating with a suitable angular velocity (see Subsection 16.1.2).

Integrating now with respect to time, the system of equations

$$\ddot{x} + \omega_Z^2 x = \omega_Z^2 a$$
$$\ddot{y} + \omega_Z^2 y = \omega_Z^2 b \tag{5.79}$$

resembles the isotropic oscillator system (4.10). The only difference is that here, the integration constants $\omega_Z^2 a$ and $\omega_Z^2 b$ appear as additional constant forces[17]. For this reason, the orbits of the two systems are fundamentally different. Here, the orbits always turn out to be circles (E).

We calculate the equation of the orbit from the conserved quantities E_ρ, I_{LB}, and $\mathbf{u}_\rho = (u_1, u_2)$ only. In Cartesian coordinates, Eqs. (5.69), (5.71), and (5.72) take the form

$$E_\rho = \frac{m}{2} \left(\dot{x}^2 + \dot{y}^2 \right), \tag{5.80}$$

$$I_{LB} = m \left((x\dot{y} - \dot{x}y) + \frac{1}{2}\omega_Z \left(x^2 + y^2 \right) \right), \tag{5.81}$$

and

$$\begin{pmatrix} u_1 \\ u_2 \end{pmatrix} = \begin{pmatrix} \dot{x} - \omega_Z y \\ \dot{y} + \omega_Z x \end{pmatrix}, \tag{5.82}$$

and satisfy Eq. (5.74); namely

$$\left(u_1^2 + u_2^2 \right) = 2E_\rho/m + 2\omega_Z I_{LB}/m. \tag{5.83}$$

By eliminating $\dot{x}$ and $\dot{y}$ in one of these conserved quantities with the help of the two other ones, we obtain the orbit. From Eq. (5.82) one can easily show[18] that

$$u_2 x - u_1 y = x\dot{y} - \dot{x}y + \omega_Z \left(x^2 + y^2 \right).$$

Using Eq. (5.81), we have

---

[17]The constants a and b are fixed by requiring that the solutions also satisfy Eqs. (5.78).

[18]Since $\mathbf{u}_\rho$ is conserved, Eq. (5.82) allows us to express the components (u_1, u_2) in terms of the initial values:

$$\begin{pmatrix} u_1 \\ u_2 \end{pmatrix} = \begin{pmatrix} v_x^0 - \omega_Z y^0 \\ v_y^0 + \omega_Z x^0 \end{pmatrix}.$$

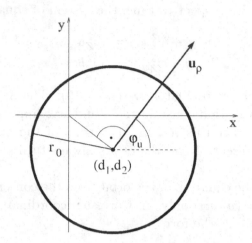

Figure 5.6: An orbit in a homogeneous magnetic field. The quantities are referred to in the text.

$$u_2 x - u_1 y = \frac{I_{LB}}{m} + \frac{1}{2}\omega_Z \left(x^2 + y^2\right).$$

This can be rewritten as (cf. Eq. (5.83))

$$\left(x - \frac{u_2}{\omega_Z}\right)^2 + \left(y + \frac{u_1}{\omega_Z}\right)^2 = \frac{1}{\omega_Z^2}\left(u_1^2 + u_2^2\right) - \frac{2I_{LB}}{m\omega_Z} = \frac{2E_\rho}{m\omega_Z^2}.$$

We recognize this as the equation of a circle in the (x, y)-plane (cf. Fig. 5.6),

$$(x - d_1)^2 + (y - d_2)^2 = r_0^2, \tag{5.84}$$

with center

$$(d_1, d_2) = \left(\frac{u_2}{\omega_Z}, -\frac{u_1}{\omega_Z}\right) \tag{5.85}$$

and radius

$$r_0 = \frac{\sqrt{2E_\rho/m}}{\omega_Z}. \tag{5.86}$$

The position of the center of the circle depends on the conserved quantity $\mathbf{u}_\rho$, and its radius on the conserved energy E_ρ. From (5.85) we see that $\mathbf{u}_\rho = (u_1, u_2)$ is perpendicular to the vector $\mathbf{d} = (d_1, d_2)$ from the origin to the center of the circle. Moreover, since Eq. (5.83) can be written as

$$\mathbf{d}^2 = r_0^2 + \frac{2I_{LB}}{m\omega_Z},$$

for $I_{LB} = 0$ the orbit crosses the z-axis $(0, 0, z)$. If $I_{LB} > 0$ or $I_{LB} < 0$, i.e. $\mathbf{d}^2 > r_0^2$ or $\mathbf{d}^2 < r_0^2$, the z-axis is either outside or inside the circular orbit. However, Eq. (5.83) gives a lower bound for I_{LB}; namely, $I_{LB} \geq -E_\rho/\omega_Z$.

Remarks

- The equation of motion (5.61), as well as Eq. (5.64) are invariant under *extended time reversal*

$$t \rightarrow -t, \quad \mathbf{B} \rightarrow -\mathbf{B}. \tag{5.87}$$

  Replacing time reversal by extended time reversal, the statements about the existence of constants of the motion (and therefore about conserved quantities) in Subsection 2.4.1 are applicable to the present system.

- The curvature of the orbit and the orientation of the motion along the orbit are fixed by magnitude and the direction of $\mathbf{B}$, respectively. In contrast to the central force problem, motion along all orbits can only proceed 'forward': the direction of the motion can only reverse if the magnetic field is reversed.

5.5.2 Addition of a central force

The symmetry properties of the two forces in Eq. (5.61) suggest the use of cylindrical coordinates (cf. Appendix A, Section A.3) for the equation of motion. Taking, as before, the z-axis to be parallel to $\mathbf{B}$, $\mathbf{B} = B\mathbf{e}_z$, Eq. (5.61) splits into (E)

$$\begin{aligned}
\ddot{\rho} - \rho\dot{\varphi}^2 &= f(r)\rho/r + \omega_Z \rho\dot{\varphi} \\
2\dot{\rho}\dot{\varphi} + \rho\ddot{\varphi} &= -\omega_Z\dot{\rho} \\
m\ddot{z} &= f(r)z/r,
\end{aligned} \tag{5.88}$$

where $r = \sqrt{\rho^2 + z^2}$. Of course, conservation of energy follows from these equations, too[19],

$$E = \frac{m}{2}\left(\dot{\rho}^2 + \rho^2\dot{\varphi}^2 + \dot{z}^2\right) + V\left(\sqrt{\rho^2 + z^2}\right) = const. \qquad (5.89)$$

Multiplying the second equation in the above system of differential equations by ρ, conservation of the quantity I_{LB} defined in Eq. (5.63) is recovered in the form Eq. (5.76); namely,

$$I_{LB} = m\rho^2\left(\dot{\varphi} + \frac{1}{2}\omega_Z\right) = const.$$

The value of I_{LB} determines where the orbit in the perpendicular plane is situated[20].

The system of differential equations (5.88) is integrable if one more isolating integral exists. Whether one exists or not was investigated by M. Robnik (J. Phys. A **14**,3195 (1981)), who numerically integrated the equations of motion for an attractive $1/r$ potential (Coulomb potential),

$$V(r) = -\frac{k}{r},$$

and considered the Poincaré maps for the trajectories. He concluded that a third, isolating, integral exists below a critical energy value that depends on B and I_{LB}. For energies above that value, the motion is in general chaotic. In Subsection 13.1.3, we study this system in a rotating frame of reference in detail and present numerical solutions.

5.5.3   Motion in the symmetry plane

At time $t = 0$, if one has $z(t = 0) = 0$ and $\dot{z}(t = 0) = 0$, then the motion remains in the plane perpendicular to $\mathbf{B}\ (= Be_z)$ containing the center

---

[19]To this end, one multiplies the three equations by $\dot{\rho}$, $\rho\dot{\varphi}$, and $\dot{z}$, respectively, and adds the resulting equations.

[20]As in the case $f(r) \equiv 0$, for a $1/r$ potential, one finds (R. Gajewski, Physica **47**,575(1970)):

  i) $I_{LB} > 0$: The intersection point of the z-axis with the orbital plane lies outside the (closed) orbit;

  ii) $I_{LB} = 0$: The intersection point lies on the orbit;

  iii) $I_{LB} < 0$: The intersection point lies within the orbit.

of the force (at $\mathbf{r} = \mathbf{0}$). For the initial value $z = 0$, the right hand side in the third equation of the system (5.88) vanishes, hence

$$m\ddot{z} = 0.$$

Integrating, we have

$$\dot{z} = const = 0,$$

because of the initial condition $\dot{z}(t = 0) = 0$. For the motion in this special plane, the first equation in the system (5.88) can now be written as (note that $r = \rho$)

$$m\left(\ddot{\rho} - \rho\dot{\varphi}^2\right) = f(\rho) + \frac{q}{c}B\rho\dot{\varphi}. \tag{5.90}$$

With the help of the conserved quantity Eq. (5.76), one can eliminate $\dot{\varphi}$

$$\dot{\varphi} = \frac{1}{m\rho^2}I_{LB} - \frac{1}{2}\omega_Z$$

from Eq. (5.90), thus showing that the radial motion obeys the equation

$$m\ddot{\rho} = f(\rho) + \frac{I_{LB}^2}{m\rho^3} - \frac{1}{4}m\omega_Z^2\rho. \tag{5.91}$$

This corresponds to one-dimensional motion in the effective potential

$$V_{eff}(\rho) = V(\rho) + \frac{I_{LB}^2}{2m\rho^2} + \frac{1}{8}m\omega_Z^2\rho^2. \tag{5.92}$$

The motion is integrable, since the energy, Eq. (5.89) (with $z = \dot{z} = 0$), is conserved. Therefore a class of orbits exists that are not chaotic for all parameter values. This is an example of coexistence of chaotic and nonchaotic regions in a system.

Summary

In general, a charged particle in a homogeneous magnetic field moves in a spiral. The angular velocity of the particle is the cyclotron frequency,

$$\boldsymbol{\omega}_Z = \frac{qB}{mc}\mathbf{e}_z.$$

The component of the motion perpendicular to the magnetic field lies on a circle whose center and radius depend on the initial values. The corresponding conserved quantity is no more the z-component of the angular momentum L_z anymore, but I_{LB}, given in Eq. (5.76).

This is in contrast the motion of a particle in the field of an attractive central force. The planar orbits about the center of force are in general bounded but not necessarily closed. The angular momentum $L_z = m\rho^2\dot\varphi$, taken with respect to the center of force (i.e. the origin), is constant.

When both fields are switched on at the same time (i.e. we have a charged particle in a static electric field $\mathbf{E} = k\mathbf{r}/r^3$ and a homogeneous magnetic field $\mathbf{B}$), competition between these two kinds of orbits generally brings about chaotic behavior. Remarkably, the two limiting cases $B \to 0$ and $k \to 0$, respectively, lead to integrable systems in which the maximal number of isolating integrals is known. In both cases, the orbits can be determined by pure algebra.

Problems and examples

1. Given a particle moving in an elliptic orbit. Derive the force, when the center of the force is

    a) the center of the ellipse (P. Varignon, Mém. Paris 1700, pp. 83-101)

    b) a focus of the ellipse (P. Varignon, Mém. Paris 1700, pp. 224-243)

2. For a particle in a central force, derive the equations of conservation of energy and angular momentum using cylindrical coordinates (cf. Section A.3).

3. Determine the equation for the orbit in a $1/r$ potential in Cartesian coordinates.

4. What can be inferred about the features of the motion from the effective potential of a three-dimensional isotropic oscillator? Find the conserved quantities. Solve the equations of motion.

5. **The isotropic harmonic oscillator in a plane.** Verify that the energy $E = E_1 + E_2$, Eqs. (4.12) and (4.13), the angular

momentum L (Eq. (4.14)), as well as I_4 (Eq. (4.16)), are given in plane polar coordinates by

$$E = \frac{m}{2}\left(\dot\rho^2 + \rho^2\dot\varphi^2\right) + \frac{m\omega}{2}\rho^2, \qquad L = m\rho^2\dot\varphi,$$

$$I_4 = \frac{L}{m}\left[\left(\frac{E}{L} - \frac{L}{m\rho^2}\right)\sin 2\varphi + \frac{\dot\rho}{\rho}\cos 2\varphi\right].$$

This is a maximal set of independent conserved quantities. From the equation for the orbits, Eq. (4.18), derive the polar angle of the semimajor axis φ_O, Eq. (5.59). Show that I_4 can be cast into the form of Eq. (5.60).

6. Starting from Eq. (5.35), calculate the equations for the asymptotes of the hyperbola, and hence verify Eq. (5.43).

7. Discuss the main features of the motion in the $1/r$ potential for $L = 0$. How long does it take a particle to reach the center of the force?

8. Determine the orbit of a particle in an attractive $1/r^2$ potential. What happens at $r = 0$?

9. Discuss the motion of a particle in the potential

$$V(r) = \begin{cases} -k(3 - r^2)/2 & r \le 1 \\ -k/r & r > 1. \end{cases}$$

10. Determine the equation for the orbit in the $1/r$ potential from the three conserved quantities given in Eqs. (5.55), (5.56), and (5.54).

11. For an attractive $1/r$ potential, express the conserved quantities of the motion in the plane perpendicular to $\mathbf{L}$ in terms of Cartesian coordinates, and determine from this the equation of the orbit.

12. **Motion in a homogeneous magnetic field.** The vector equation in three dimensions for the radius vector $\mathbf{r} = (x, y, z)$, in a shifted coordinate system (cf. Eq. (5.66)), is

$$\dot{\mathbf{r}} = -\left(\boldsymbol{\omega} \times \mathbf{r}\right),$$

with $\boldsymbol{\omega} = (\omega_1, \omega_2, \omega_3) = \omega_Z \mathbf{B}/B$, $\omega_Z = \dfrac{qB}{mc}$. In matrix notation, we have

$$\dot{\mathbf{r}} = \begin{pmatrix} 0 & \omega_3 & -\omega_2 \\ -\omega_3 & 0 & \omega_1 \\ \omega_2 & -\omega_1 & 0 \end{pmatrix} \mathbf{r}.$$

One can write this equation in the form

$$\dot{\mathbf{r}} = i(\boldsymbol{\omega}\mathbf{s})\,\mathbf{r},$$

where the three components of $\mathbf{s} = (\mathbf{s}_1, \mathbf{s}_2, \mathbf{s}_3)$ are the matrices

$$\mathbf{s}_1 = \begin{pmatrix} 0 & 0 & 0 \\ 0 & 0 & -i \\ 0 & i & 0 \end{pmatrix}, \quad \mathbf{s}_2 = \begin{pmatrix} 0 & 0 & i \\ 0 & 0 & 0 \\ -i & 0 & 0 \end{pmatrix}, \quad \mathbf{s}_3 = \begin{pmatrix} 0 & -i & 0 \\ i & 0 & 0 \\ 0 & 0 & 0 \end{pmatrix},$$

i.e. $i\boldsymbol{\omega}\mathbf{s} = i\sum_k \omega_k \mathbf{s}_k$. The solution of the equation can be easily found:

$$\mathbf{r}(t) = \exp\left[i(\boldsymbol{\omega}\mathbf{s})\,t\right]\mathbf{r}_0.$$

(Recall that for a matrix M, $\exp\mathsf{M} = 1 + \mathsf{M} + \frac{1}{2}\mathsf{M}^2 + \ldots$.) The matrices $\mathbf{s}_k$ have the properties,

$$\mathbf{s}_i\mathbf{s}_j - \mathbf{s}_j\mathbf{s}_i = i\varepsilon_{ijk}\mathbf{s}_k, \quad \mathbf{s}^2/2 = 1.$$

Use these properties to write down the components of $\mathbf{r}(t)$ explicitly.

13. Solve Eqs. (5.78) and verify the equation for the orbits, Eq. (5.84).

14. Show, for a charged particle in a homogeneous magnetic field $\mathbf{B} = B\mathbf{e}_z$, that the vector $\mathbf{r}_0 = (x, y) - (d_1, d_2)$ (cf. (5.85)) rotates with constant angular velocity ω_Z.

15. Show, for a charged particle in a homogeneous magnetic field $\mathbf{B} = B\mathbf{e}_z$, that the z-component of the angular momentum L_z taken with respect to the center of the circular orbit, is

$$L_z = -\frac{2E_\rho}{\omega_Z}.$$

6

Gravitational force between two bodies

So far we have considered *one* particle moving under the influence of an 'external' force. Now we want to apply the results of the last chapter to the planetary system, considering for the moment only the motion of a single planet around the sun. Therefore we have to show first that the motion of two interacting bodies can be related to the one of a single body in a field of force. In fact, under suitable assumptions about the force, the interaction of *two* point masses can be reduced to the motion of *one* point mass in a central field of force.

6.1   Two-body systems

Till now we ignored the source of the force. Indeed, the force must be due to bodies – or other physical objects – that we did not take into account. But in considering these bodies, questions about the interaction of two bodies inevitably arise: for example, whether the mutual forces are exerted instantaneously or with some delay ('retardation effect'). In classical nonrelativistic mechanics, the fields of force usually depend only on the radius vector. This implies that the interaction is instantaneous, since any movement of the source of the force has an immediate effect on other bodies within the range of the force. The theory of relativity demolished the view that forces act instantaneously: a force cannot propagate faster than the speed of light. If one wishes to take into account the finite speed of propagation of a particular in-

139

teraction, then one will arrive at a field theory of that interaction. For the electric and magnetic forces, the field theory is electrodynamics; for the gravitational interaction, it is Einstein's general theory of relativity. The common assumption in classical mechanics about the instantaneous propagation of forces is an approximation that is only justified by the small velocities (compared to the speed of light) of the bodies involved.

Let us consider two point masses m_1 and m_2 that exert forces on each other mutually; i.e. a so-called **interaction force** exists between them, as opposed to an external force:

$$m_1 \ddot{\mathbf{r}}_1 = \mathbf{F}_{12}$$
$$m_2 \ddot{\mathbf{r}}_2 = \mathbf{F}_{21}. \tag{6.1}$$

Here, $\mathbf{F}_{21} = \mathbf{F}_{21}(\mathbf{r}_2; \mathbf{r}_1)$ is the force exerted by m_1 on m_2, and vice versa for $\mathbf{F}_{12}$. According to *Newton's third law*, Eq. (2.6), the forces obey the condition

$$\mathbf{F}_{21} = -\mathbf{F}_{12}. \tag{6.2}$$

Since the forces $\mathbf{F}_{ij}$ only depend on radius vectors $\mathbf{r}_1$ and $\mathbf{r}_2$, and the sole distinct direction in the system comprising of the two point masses is the distance vector

$$\mathbf{r} = (\mathbf{r}_1 - \mathbf{r}_2) \tag{6.3}$$

between the point masses, the forces $\mathbf{F}_{21}$ and $\mathbf{F}_{12}$ can only be directed parallel to $\mathbf{r}$; i.e. $\mathbf{F}_{21}$ is parallel to $\mathbf{r}$. Therefore, $\mathbf{F}_{21}$ has the form

$$\mathbf{F}_{21} = f_{21}(\mathbf{r}_1, \mathbf{r}_2) \frac{\mathbf{r}}{|\mathbf{r}|}$$

(the vector character of the force is expressed by $\mathbf{r}$); and analogously, $\mathbf{F}_{12}$ has the form

$$\mathbf{F}_{12} = f_{12}(\mathbf{r}_1, \mathbf{r}_2) \frac{\mathbf{r}}{|\mathbf{r}|}.$$

Since, by assumption, the interaction of the two point masses is independent of the presence of other bodies or forces, the forces $\mathbf{F}_{ij}$ depend neither on the particles' location ('**homogeneity of (free) space**'), nor on the orientation of the vector $\mathbf{r}$ ('**isotropy of (free) space**'). Therefore the strength of the forces is a function only of the distance $r = |\mathbf{r}| = |\mathbf{r}_1 - \mathbf{r}_2|$:

$$f_{12}(\mathbf{r}_1, \mathbf{r}_2) = f_{12}(r)$$
$$f_{21}(\mathbf{r}_1, \mathbf{r}_2) = f_{21}(r). \tag{6.4}$$

Inserting these two expressions into Eq. (6.2), we find

$$f_{12}(r) = f_{21}(r) =: f(r),$$

where the indices are omitted in the last term because the order of the indices is immaterial ($f_{ij} = f_{ji}$). All in all, we have

$$\mathbf{F}_{21}(\mathbf{r}_1 - \mathbf{r}_2) = f(|\mathbf{r}_1 - \mathbf{r}_2|)\frac{\mathbf{r}_1 - \mathbf{r}_2}{|\mathbf{r}_1 - \mathbf{r}_2|}$$
$$= -\mathbf{F}_{12}(\mathbf{r}_2 - \mathbf{r}_1) = \mathbf{F}_{12}(\mathbf{r}_1 - \mathbf{r}_2) =: \mathbf{F}(\mathbf{r}_1 - \mathbf{r}_2). \quad (6.5)$$

The order of indices is also immaterial for $\mathbf{F}$; therefore, indices have been omitted here, too. In concise notation,

$$\mathbf{F}(\mathbf{r}) = f(r)\,\mathbf{r}/r = -\mathbf{F}(-\mathbf{r}). \quad (6.6)$$

But this relation implies that $\mathbf{F}(\mathbf{r})$ is a central force, and therefore conservative (cf. Section 5.1). Consequently, there exists a potential $V(r)$ such that

$$\mathbf{F}(\mathbf{r}) = -\boldsymbol{\nabla}V(r).$$

This **interaction potential** $V(r)$ of the two particles can be determined by integrating the force (cf. Eq. (5.4)):

$$V(r) = -\int^{r} dr'\, f(r').$$

The equations of motion (6.1) for the two point masses are now[1] (with $\boldsymbol{\nabla} =: \partial/\partial\mathbf{r} = \partial/\partial\mathbf{r}_1 = -\partial/\partial\mathbf{r}_2$)

$$m_1\ddot{\mathbf{r}}_1 = -\frac{\partial}{\partial\mathbf{r}_1}V(|\mathbf{r}_1 - \mathbf{r}_2|) \quad (6.7)$$

$$m_2\ddot{\mathbf{r}}_2 = -\frac{\partial}{\partial\mathbf{r}_2}V(|\mathbf{r}_1 - \mathbf{r}_2|). \quad (6.8)$$

We first look for conserved quantities in the system (6.7), (6.8). To this end we multiply the first equation by $\dot{\mathbf{r}}_1$ and the second by $\dot{\mathbf{r}}_2$.

---

[1] General features of two-body systems interacting via central forces are investigated in Section XI of the first book of the "Principia". Newton shows that the bodies describe similar figures about their common center of gravity, and about each other mutually (cf. Fig. 6.2).

Integrating the sum of the resulting equations over time we obtain the *conservation law of total energy*:

$$E_{tot} := \frac{1}{2}\left(m_1\dot{\mathbf{r}}_1^2 + m_2\dot{\mathbf{r}}_2^2\right) + V(|\mathbf{r}_1 - \mathbf{r}_2|) = const. \tag{6.9}$$

(Observe that $\dfrac{\partial V}{\partial \mathbf{r}_1}\dot{\mathbf{r}}_1 + \dfrac{\partial V}{\partial \mathbf{r}_2}\dot{\mathbf{r}}_2 = \dfrac{d}{dt}V(|\mathbf{r}_1 - \mathbf{r}_2|)$.) Hence, this two-particle system is *conservative*. Similarly, the sum of the cross products of Eqs. (6.7) and (6.8) with $\dot{\mathbf{r}}_1$ and $\dot{\mathbf{r}}_2$, respectively, leads to the *conservation law of total angular momentum*:

$$\mathbf{L}_{tot} := m_1\mathbf{r}_1 \times \dot{\mathbf{r}}_1 + m_2\mathbf{r}_2 \times \dot{\mathbf{r}}_2 = const. \tag{6.10}$$

Adding Eqs. (6.7) and (6.8) gives

$$m_1\ddot{\mathbf{r}}_1 + m_2\ddot{\mathbf{r}}_2 = 0, \tag{6.11}$$

which, when integrated, yields

$$m_1\dot{\mathbf{r}}_1 + m_2\dot{\mathbf{r}}_2 = const. \tag{6.12}$$

The total momentum is conserved.

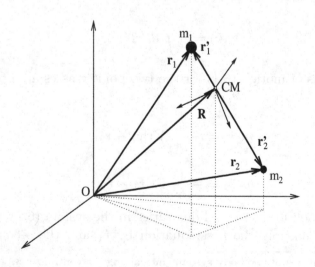

Figure 6.1: Center of mass coordinate and relative coordinates.

6.1.1 Center of mass coordinates and relative coordinates

In many applications, it is advantageous to introduce *center of mass coordinates* and *relative coordinates*. For this purpose, we define (cf. Fig. 6.1):

The center of mass: The coordinate $\mathbf{R}$ of the center of mass (CM in Fig. 6.1) of two point masses is given by the weighted average of their coordinates $\mathbf{r}_1$ and $\mathbf{r}_2$,

$$\mathbf{R} = \frac{m_1\mathbf{r}_1 + m_2\mathbf{r}_2}{M}, \tag{6.13}$$

where $M = m_1 + m_2$ is the total mass. The coordinate of the center of mass $\mathbf{R}$ lies on the straight line

$$\mathbf{t}(x) = (1 - x)\mathbf{r}_1 + x\mathbf{r}_2, \qquad 0 \leq x \leq 1$$

connecting the positions $\mathbf{r}_1$ and $\mathbf{r}_2$ of the point masses m_1 and m_2, respectively. (Note: if $x = m_2/M$, $\mathbf{t} - \mathbf{R}$.)

The relative coordinate: The relative position $\mathbf{r}$ of the two point masses is given by

$$\mathbf{r} = \mathbf{r}_1 - \mathbf{r}_2. \tag{6.14}$$

Expressing $\mathbf{r}_1$ and $\mathbf{r}_2$ in terms of $\mathbf{R}$ and $\mathbf{r}$, one obtains

$$\mathbf{r}_1 = \mathbf{R} + \frac{m_2}{M}\mathbf{r} \qquad \text{and} \qquad \mathbf{r}_2 = \mathbf{R} - \frac{m_1}{M}\mathbf{r}. \tag{6.15}$$

Introducing into Eqs. (6.7) and (6.8) $\mathbf{R}$ and $\mathbf{r}$ instead of $\mathbf{r}_1$ and $\mathbf{r}_2$, the equations of motion gain a far simpler form: they decouple. Equation (6.11) becomes simply

$$M\ddot{\mathbf{R}} = 0, \tag{6.16}$$

i.e. just the equation of motion of the center of mass. Multiplying Eqs. (6.7) and (6.8) by m_2 and m_1, respectively, and subtracting the resulting equations, yields[2] $m_1 m_2 (\ddot{\mathbf{r}}_1 - \ddot{\mathbf{r}}_2) = -(m_1 + m_2)\boldsymbol{\nabla} V(|\mathbf{r}_1 - \mathbf{r}_2|)$.

---

[2]Note that $\dfrac{\partial}{\partial \mathbf{r}_1} V = -\dfrac{\partial}{\partial \mathbf{r}_2} V = \dfrac{\partial}{\partial \mathbf{r}} V$.

Switching to the relative coordinate $\mathbf{r}$, one recognizes that this is just the equation of the relative motion of the two particles:

$$m\ddot{\mathbf{r}} = -\nabla V(r),\tag{6.17}$$

where

$$m = \frac{m_1 m_2}{m_1 + m_2}\tag{6.18}$$

is the **reduced mass** of the two particles. The two equations of motion (6.16) and (6.17) are independent: each depends only on either $\mathbf{R}$ or $\mathbf{r}$, respectively. Equation (6.16) shows that the *motion of the center of mass is a free motion* (the center of mass motion is conserved; see below). The **relative motion** of the particles under their mutual interaction potential V is equivalent to the motion of a *single particle* with reduced mass m at position $\mathbf{r}$ subject to a *central force* with potential $V(r)$.

The center of mass frame

A frame of reference that moves along with the center of mass (cf. Fig. 6.1), i.e. in which $\mathbf{R} = const =: \mathbf{0}$, is called the **center of mass frame**. The coordinates $\mathbf{r}'_1$ and $\mathbf{r}'_2$ of the point masses m_1 and m_2 with respect to the center of mass position $\mathbf{R}$ are given by

$$\begin{aligned}\mathbf{r}'_1 &= \mathbf{r}_1 - \mathbf{R} = \frac{m_2}{M}\left(\mathbf{r}_1 - \mathbf{r}_2\right)\\\mathbf{r}'_2 &= \mathbf{r}_2 - \mathbf{R} = -\frac{m_1}{M}\left(\mathbf{r}_1 - \mathbf{r}_2\right).\end{aligned}$$

Together with (6.13), these equations establish the coordinate transformation from the original system to the center of mass system: in particular, from coordinates $\mathbf{r}_1$ and $\mathbf{r}_2$ to coordinates $\mathbf{r}'_1$ and $\mathbf{r}'_2$ in the center of mass system. In terms of the relative coordinate $\mathbf{r}$ the radius vectors $\mathbf{r}'_1$ and $\mathbf{r}'_2$ are given by

$$\mathbf{r}'_1 = \frac{m_2}{M}\mathbf{r}\tag{6.19}$$

$$\mathbf{r}'_2 = -\frac{m_1}{M}\mathbf{r};\tag{6.20}$$

conversely, we have

$$\mathbf{r} = \mathbf{r}'_1 - \mathbf{r}'_2.$$

Here, $\mathbf{r} = \mathbf{0}$ defines the origin of some frame of reference which is not necessarily related to the physically distinguished frames shown in

Fig. 6.1. To draw conclusions from solutions of Eq. (6.17) about the motion of the particles in the center of mass frame, one has to use Eqs. (6.19) and (6.20).

6.1.2 Conserved quantities

Written in terms of $\mathbf{R}$ and $\mathbf{r}$, the conservation law in Eq. (6.9) reads

$$E_{tot} = \frac{1}{2}M\dot{\mathbf{R}}^2 + \frac{1}{2}m\dot{\mathbf{r}}^2 + V(r) = const. \tag{6.21}$$

Hence, the total energy is the sum of the kinetic energy of the center of mass and the energy of the relative motion. Proceeding as with Eq. (6.10), one can verify that conservation of the total angular momentum separates into the *angular momentum of the center of mass* and the *angular momentum of the relative motion*[3]:

$$\mathbf{L}_{tot} = M\mathbf{R} \times \dot{\mathbf{R}} + m\mathbf{r} \times \dot{\mathbf{r}} = const. \tag{6.22}$$

Of course, these conservation laws follow also from the equations of motion, (6.16) and (6.17). Multiplying Eq. (6.16) by $\dot{\mathbf{R}}$ and Eq. (6.17) by $\dot{\mathbf{r}}$ yields separate *conservation laws* for the *energy of the center of mass motion* and the *energy of the relative motion*:

$$E_S := \frac{1}{2}M\dot{\mathbf{R}}^2 = const \tag{6.23}$$

$$E := \frac{1}{2}m\dot{\mathbf{r}}^2 + V(r) = E_{tot} - E_S = const. \tag{6.24}$$

Taking the cross product of the equations of motion ((6.16) as well as (6.17)) and $\mathbf{R}$ and $\mathbf{r}$, respectively, leads to separate *conservation laws* for the *angular momentum of the center of mass motion* and *of the relative motion*, respectively:

$$\mathbf{L}_S := \mathbf{R} \times M\dot{\mathbf{R}} = const, \tag{6.25}$$

$$\mathbf{L} := \mathbf{r} \times m\dot{\mathbf{r}} = const, \tag{6.26}$$

since, due to $\mathbf{r} \times \nabla V(r) = 0$, the moment of the force vanishes. Thus the two-body problem separates into *two independent motions of two fictitious bodies*:

---

[3]We point out that the vectors $\mathbf{r}_1$, $\mathbf{r}_2$, and $\mathbf{R}$ refer to the (arbitrarily) chosen origin, whereas the relative vector $\mathbf{r}$ is independent of the choice.

i) Free uniform rectilinear motion of the center of mass (Eq. (6.16)), equivalent to the motion of a particle of mass M at point $\mathbf{R}$; and,

ii) Relative motion of the two particles, equivalent to the motion of a *single* point mass of reduced mass m (cf. Eq. (6.18)) in the potential $V(r)$ (cf. Eq. (6.17)).

Since the solution for the motion of the center of mass is trivial, the problem of the motion of *two interacting bodies* has been reduced to the problem of the *motion of one body in a central force*.

Considering now the center of mass motion, integration of Eq. (6.16) yields the center of mass velocity,

$$\dot{\mathbf{R}} = \mathbf{V} = \mathbf{V}_0 = const, \tag{6.27}$$

which can be interpreted as *conservation of the center of mass momentum*,

$$\mathbf{P} = M\mathbf{V} = M\mathbf{V}_0. \tag{6.28}$$

We therefore have three conserved quantities, given by the components of the velocity vector $\mathbf{V}_0$ (or, equivalently, the momentum vector $\mathbf{P}_0 = M\mathbf{V}_0$). A further integration of Eq. (6.27) yields the equation for the *orbit of the center of mass*,

$$\mathbf{R}(t) = \mathbf{V}_0 t + \mathbf{R}_0, \tag{6.29}$$

with $\mathbf{R}_0 = \mathbf{R}(t) - \mathbf{V}_0 t$ representing three more constants of the motion. Altogether we have six independent constants of the motion (given by $\mathbf{R}_0$ and $\mathbf{V}_0$), or five independent conserved quantities, since two conserved quantities can be extracted from Eq. (6.29) by eliminating time t. Instead of the latter two conserved quantities, one may take the conservation[4] of $\mathbf{L}_S$ (cf. Eq. (6.25)),

$$\mathbf{L}_S = M\mathbf{R} \times \mathbf{V} = M\mathbf{R}_0 \times \mathbf{V}_0.$$

However, since $\mathbf{L}_S$ and $\mathbf{P} = M\mathbf{V}$ are orthogonal, $\mathbf{L}_S$ only provides two more independent conserved quantities[5]. These can be used instead

---

[4]We note that the cross product of Eq. (6.27) with the vector $\mathbf{R} = \mathbf{R}_0 + \mathbf{V}_0 t$ yields also conservation of angular momentum of the center of mass, Eq. (6.25).

[5]Note also that the energy $E_S = \mathbf{P}^2/2M$ is a dependent conserved quantity because we have chosen $\mathbf{P}$ as independent quantity.

of the those obtained from Eq. (6.29). We have therefore found the complete set of five conserved quantities for the motion of the center of mass: these are the momentum $\mathbf{P}$ (cf. Eq. (6.28)), the modulus of the angular momentum (cf. Eq. (6.25)), $|\mathbf{L}_S|$, and the z-component of the angular momentum $L_{S,z}$.

The *relative motion* considered as motion in a central force is conveniently described in polar coordinates. The results of Section 5.1 concerning the motion of a particle in a central force can be applied here directly. In general, only four isolating integrals (conserved quantities) exist: the energy E (cf. Eq. (6.24)) and the three components of the angular momentum $\mathbf{L}$ of the relative motion (cf. Eq. (6.26)). The problem of the relative motion of two particles in the absence of an external force is therefore always integrable: conservation of energy and angular momentum yield a sufficient number of relations (integrals of the motion) to allow the velocities $\dot{r}$ and $\dot{\varphi}$ to be expressed as functions of r only (cf. Eqs. (5.19) and (5.13)):

$$\dot{r} = \dot{r}(r) = \sqrt{\frac{2}{m}\left(E - V(r)\right) - L^2/m^2 r^2}, \qquad (6.30)$$

$$\dot{\varphi} = \dot{\varphi}(r) = L/mr^2. \qquad (6.31)$$

In general, in the two-body problem in the absence of an external force, the nine known, independent conserved quantities give nine relations for the twelve coordinates in phase space.

6.2 The gravitational interaction

As early as 1665, Newton[6] concluded from Kepler's third law (see Section 6.3), that the strength of the gravitational force between two bodies is inversely proportional to the square of the distance between them[7].

---

[6]Compare "Let Newton be" (see Bibliography): Newton began developing his theory of the planetary orbits at about 1680.

[7]As already indicated in the last chapter, in Proposition XI of the first Book of the "Principia" ('*If a body revolves in an ellipse; it is required to find the law of the centripetal force tending to the focus of the ellipse*'), Newton deduces the $1/(\text{distance})^2$ law solely from the geometrical properties of an ellipse.

In the third Book of the "Principia", Proposition VII, he formulates the *universal law of gravitation* [Chandrasekhar] as follows: '*That there is a power of gravity to all bodies, proportional to the several quantities of matter* [i.e. the mass] *which they*

In modern form, the force between the sun (m_S) and a planet, e.g. the earth (m_E), is written as

$$|\mathbf{F}| = \frac{Gm_S m_E}{|\mathbf{r}_S - \mathbf{r}_E|^2}. \tag{6.32}$$

The corresponding potential is

$$V(\mathbf{r}) = -\frac{Gm_S m_E}{r}, \qquad \mathbf{r} = \mathbf{r}_S - \mathbf{r}_E.$$

This **universal law of gravitational interaction** is valid not only for planets, but also in general for the interaction of any pair of bodies having finite masses m_1 and m_2. This law has been repeatedly confirmed by experiments – beginning at around 1800 by H. Cavendish, then since the 1920s (L. Eötvös) with ever increasing accuracy[8]. Experiments confirm that the gravitational interaction between any two bodies of masses m_1 and m_2 is given by the potential

$$V(\mathbf{r}) = -\frac{Gm_1 m_2}{r}, \qquad \mathbf{r} = \mathbf{r}_1 - \mathbf{r}_2, \tag{6.33}$$

and the **universal value of the gravitational constant**[9]

$$G = (6.67384 \pm 0.00080) \times 10^{-8} g^{-1} cm^3 s^{-2}. \tag{6.34}$$

Note that the gravitational force is proportional to both masses, i.e. to m_1 as well as m_2; and these are the same masses that enter the left hand side of the respective equation of motion (Eqs. (6.7) and (6.8)); for example,

$$m_1 \ddot{\mathbf{r}}_1 = \mathbf{F}_{12} = -\frac{Gm_1 m_2}{|\mathbf{r}_1 - \mathbf{r}_2|^2} (\mathbf{r}_1 - \mathbf{r}_2).$$

---

contain.' In Proposition VIII, he writes: '*In two spheres gravitating each towards the other, if the matter in places on all sides round about and equidistant from the centres is similar* [i.e. the mass distribution inside the spheres is homogeneous], *the weight of either sphere towards the other will be inversely as the square of the distance between their centres.*' In Propositions LXXV and LXXVI in Book I, Newton proves that the force between spheres with homogeneous mass distribution depends on the distance between their respective centers.

[8]Further details are given at the end of this chapter.

[9]This is the *2010 CODATA recommended value* available since 2011 and recognized worldwide for use in all fields of science and technology.

Since G is asserted to be universal, this is a nontrivial statement! The mass appearing in the change of momentum (i.e. in the left hand side) is called **inertial mass**, and is a measure of the resistance to changes in the velocity of a body, whereas each of the masses in (6.33) is a **gravitational mass**; i.e., a mass that results from the weight (a measure of the gravitational force) of a body. Gravitational mass has two distinct aspects: it causes the force (**active gravitational mass**); and is subjected to the gravitational force due to the other mass (**passive gravitational mass**). The equivalence of active and passive gravitational mass is a consequence of Newton's third law. The equivalence (more precisely, the proportionality) of inertial and gravitational mass follows from experiment. Newton himself confirmed this equality, via experiments with a pendulum, where he demonstrated that the period of oscillation is independent of the mass. Previously, Galileo's free fall experiments showed that all objects fall at the same rate, regardless of their mass.

Now, consider a system consisting of two bodies of mass m_1 and m_2 interacting gravitationally via Eq. (6.33). If there are no external forces, then the motion of the center of mass (total mass $M = m_1 + m_2$) is easily found: it is uniform. On the other hand, the relative motion is nontrivial. The relative motion involves the reduced mass $m = m_1 m_2 / M$. Expressing m_1 and m_2 in terms of M and m, one obtains

$$m_1 = \frac{M}{2}\left(1 + \sqrt{1 - \frac{4m}{M}}\right), \qquad m_2 = \frac{M}{2}\left(1 - \sqrt{1 - \frac{4m}{M}}\right). \quad (6.35)$$

(Since $(m_1 - m_2)^2 \geq 0$, the condition $4m < M$ holds.) Because $m_1 m_2 = mM$, Eq. (6.33) can be also written as

$$V(r) = -G\frac{Mm}{r}. \quad (6.36)$$

Considering only the relative motion, $V(r)$ can be interpreted as an *external potential* acting and depending on the reduced mass[10],

$$V(r) = -k\frac{m}{r}, \qquad k = GM. \quad (6.37)$$

---

[10]Note the equivalence of the inertial reduced mass to the gravitational reduced mass.

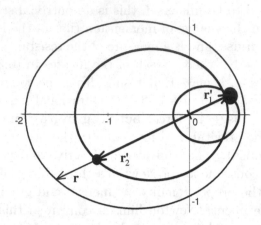

Figure 6.2: The orbits of two bodies (inner ellipses) and the orbit of the reduced mass (outer ellipse). The distance between the two bodies at a given time, $r_1 + r_2$, is equal the to the length of the radius vector $\mathbf{r}$ in the outer ellipse: $r = r_1 + r_2$.

The motion of this reduced mass corresponds completely to the motion of a particle in the $1/r$ potential, which we studied in the last chapter. The results from that chapter can be applied immediately here. The orbit of the reduced mass is an ellipse. One regains the two-particle view in the center of mass frame by noting that radius vectors of the two particles obey a simple proportionality with respect to the relative vector $\mathbf{r}$ (Eqs. (6.19) and (6.20)):

$$\mathbf{r}_1' = \frac{m_2}{M}\,\mathbf{r} \qquad \text{and} \qquad \mathbf{r}_2' = -\frac{m_1}{M}\,\mathbf{r}.$$

Thus one concludes that the orbits of the two particles form ellipses with a common focus, with the dimension of the respective ellipses corresponding to the mass ratios in the equations here for $\mathbf{r}_1'$ and $\mathbf{r}_2'$. The motion of the two particles is such that the relative vector $\mathbf{r} = \mathbf{r}_1' - \mathbf{r}_2'$ also sweeps out an ellipse, as shown in Fig. 6.2. If one of the masses is much larger than the other, e.g. $m_1 \gg m_2$, then the total mass M is approximately equal to m_1 and the reduced mass m is equal to m_2: $M \cong m_1$ and $m \cong m_2$. The position of the total mass nearly coincides with the center of gravity, which is at the origin for the motion of the

reduced mass:
$$\mathbf{r}_1 \simeq \mathbf{R}, \qquad \mathbf{r}_1' \simeq 0, \ \mathbf{r}_2' \simeq -\mathbf{r}.$$

6.3   Kepler's laws

The laws of planetary motion proposed by J. KEPLER[11] were immensely important for the development of dynamics. In deriving a number of consequences of Newton's law of gravitation, we have actually reversed the historical order, as this law is based on Kepler's laws and, indeed, the law of gravitation was published nearly a century later. In this section, we attempt to give a flavor of the magnitude of Kepler's achievements.

Since Kepler's laws follow easily from the law of gravitation, one may be tempted to hold a low opinion of Kepler's contribution. But this would be a mistaken attitude. In order to assess the significance of the laws, we first briefly outline the knowledge that was available in Kepler's time[12]. Table 6.1 shows some data of the nine planets in our solar system. In Kepler's time, only the first six planets were known. The heliocentric world view was beginning to challenge the geocentric world view. The catholic church permitted public support of the heliocentric system with reservations – for example, one was only allowed to present the heliocentric view as purely mathematical construct, with no basis in reality (which is why Galileo was put on trial). This is despite the fact that a century before Galileo, the heliocentric system had already been proposed by N. COPERNICUS[13], himself a catholic clergyman, in his work "De revolutionibus orbium coelestium" (published right after Copernicus's death[14]). Copernicus was of the opinion that the orbits of the planets around the sun have to be circular. Still existing discrepancies between the ideal, circular orbits and the observed orbits he removed by introducing epicycles – just as Ptolemy did in his geocentric

---

[11]Johannes Kepler (1571-1631), German astronomer and mathematician. A comprehensive biography of Johannes Kepler is that by M. Caspar (see the Bibliography).

[12]Historical details on Kepler's discovery may be gleaned from the books by A. Koestler and by I. Peterson. A comprehensive presentation of the history of astronomy is given by J. North (see the Bibliography).

[13]Nicolaus Copernicus (1473-1543), Polish astronomer and mathematician.

[14]An English translation "On the Revolutions of the Celestial Spheres" by C.G. Wallis appeared at Prometheus Books, Amherst, N.Y 1995.

system. However, Copernicus had to introduce only a smaller number of epicycles[15].

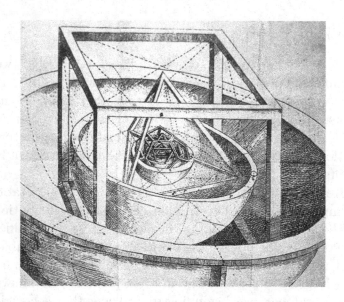

Figure 6.3: Kepler's model of the planetray orbits ("Mysterium Cosmographicum").

The laws are the results of Kepler's permanent quest for harmony in the proportions of the planetary orbits shown in Fig. 6.3[16]. He proposed his laws between 1605 and 1619. Kepler recognized that the sun actu-

---

[15]N. Copernicus, Commentariolus (a sketch of the "Revolutionibus" without formulae), manuscript written between 1507 and 1514. An English translation is given by E. Rosen in: "Three Copernican Treatises: The Commentariolus of Copernicus; The Letter against Werner; The Narratio Prima of Rheticus." Dover Publications, New York 2004 (1939).

However, in the "Revolutionibus" he had to introduce again some additional epicycles in order to account for the observed time dependence of the planets motion on the orbit.

[16]As early as 1596, in his "Mysterium Cosmographicum", Kepler had argued that in the Copernican system, the distances of the planets from the sun are given by nesting the five regular solids cocentrically, then imagining that each planet's sphere (the orbits were considered to be circles!) be placed outside one solid and inside the next one (see the famous Figure below).

The order of the bodies is determined by astrological considerations. For instance, the properties ascribed to Saturn in astrology are best reflected by the cube ("Mys-

ally caused planetary motion. To him, the heliocentric model was not
a mere aid to improving the accordance between mathematical model
and observation[17]. This was the prerequisite for deducing the first two
laws from the wealth of astronomical data inherited from the Danish
astronomer T. BRAHE[18]. As Kepler arrived in Prague[19], Brahe was occupied with the theory of the orbit of Mars showing the second largest
eccentricity (see Table 6.1). (With the observational tools at hand at
that time the deviation from a circular orbit could be determined only
for Mars.) Kepler entered the calculations the orbit. After laborious

terium Cosmographicum", Chapter IX). Concerning the orientation of the five Pla-
tonic solids in space Kepler's arguments given in Chapter XI are rather a matter of
taste. He apologizes right from the beginning: 'I shall have the physicists against
me in these chapters, because I have deduced ... from immaterial things and mathematical figures...' (translation by A.M. Duncan, The Secret of the Universe, Abaris
Books, Norwalk 1999) and refers in the following to the 'power and wisdom of God'.

[17]Right in the first chapter of Kepler's "Astronomia nova" Figure 2 (cf. below)
shows the orbit of Mars between 1580 and 1596 assuming a geocentric system. As
can be realized from the Figure this orbit is not in accord with the principle of
simplicity of planetary motion promoted by Kepler. In the comment to the Figure
Kepler states inter alia that "continuing this motion would result in a mess since the
series of windings continues without end and never returns to itself".

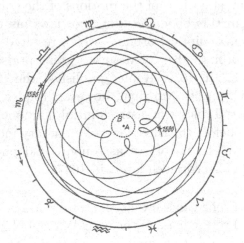

[18]Tycho Brahe (1546-1601), Danish astronomer.

[19]In 1601, Kepler was appointed – as a successor of Tycho Brahe – imperial mathematician at the court of the Hapsburg emperor, Rudolf II, in Prague. One of his
duties was to cast horoscopes. Due to his former experience he was already well-
versed in this art.

	mass ($\times m_E$)	mean distance ($\times 10^6$ km)	eccentricity ε	period T (years)
sun	332943	–	–	–
Mercury	0.0553	57.91	0.2056	0.2408
Venus	0.8150	108.21	0.0068	0.6152
earth	1	149.60	0.0162	1.00
Mars	0.1074	227.9	0.0934	1.881
Jupiter	317.8	778.4	0.0484	11.86
Saturn	95.16	1427	0.0526	29.42
Uranus	14.50	2883	0.0462	83.75
Neptune	17.20	4519	0.0066	163.7
(Pluto	0.0020	5960	0.2554	248.0)

Table 6.1: Some orbital elements of the planets (adapted from: H. Zimmermann and A. Weigert, Lexikon der Astronomie, 8th edition, Springer, Berlin 1999). Since 2006 Pluto does not belong to the planets anymore. It is now a 'dwarf planet.

and extensive calculations he presented in 1609 the "Astronomia nova" devoted to the "investigations of the motions of the star Mars". In particular one finds in this book the first two 'laws' as they were called later. In introducing elliptical orbits – which resulted in the abandonment of circular orbits – Kepler deserves great praise in overcoming traditional opinion. In 1619, Kepler published the third law in "De harmonice mundi". In his own time, however, Kepler was best known for his "Rudolphine Tables" (an astronomical handbook [Peterson]). It took quite a while before the three laws of planetary motion established Kepler's position in the history of astronomy and physics. Newton's "Principia" was published much later, in 1687.

From Table 6.1, one can see that the total mass of the solar system practically is equal to the mass of the sun. When looking at the motion of one planet (mass m_P), to a good approximation, the influence of the remaining planets can be ignored, and only this planet's interaction with the sun need be considered. (With some simplifying assumptions, we will study the influence of an additional planet in Chapters 14 and 16.) We consider the sun and the planets as point masses – for the moment, we justify this by pointing out that the interplanetary distances are enormous compared to the diameters of the planets. In Section 6.4, we investigate this assumption in more detail.

The relative motion between the planet m_P and the sun is determined by the potential (6.37). Taking into account the mass ratios, we have for the total and the reduced mass,

$$M \simeq m_S \quad \text{and} \quad m \simeq m_P,$$

respectively, and for the interaction strength,

$$k = GM \simeq Gm_S.$$

The sun practically coincides with the center of the force (center of mass). Bearing this in mind, we can now 'establish' **Kepler's laws** from the results derived in the previous chapter for motion in a $1/r$ potential.

The equation for the orbit, Eq. (5.31), for $E < 0$, corresponds to the

The first law:

Planets move in ellipses that have the sun at one focus.

Since $k > 0$, Eq. (5.35) takes the form

$$\frac{p}{r} = 1 + \varepsilon \cos \varphi. \tag{6.38}$$

Inserting Eqs. (5.32) and (5.33) for the semilatus rectum p and for the eccentricity ε, respectively, in the expression given in Eq. (5.37) for the semimajor axis of an ellipse, one obtains

$$a = \frac{p}{1 - \varepsilon^2} = -\frac{m_P k}{2E}. \tag{6.39}$$

Hence, a is determined solely by the energy (and not by the angular momentum L). Similarly, from Eq. (5.38) for the semiminor axis, we find that

$$b = a\sqrt{1 - \varepsilon^2} = \sqrt{ap} = \left(-\frac{L^2}{2m_P E} \right)^{1/2}, \tag{6.40}$$

so that b is proportional to the angular momentum L. If $L = 0$, i.e. $p = 0$ and $\varepsilon = 1$, the semiminor axis vanishes and, as we have shown already, the planet moves along a straight line (the major axis) through the center of mass (located approximately where the sun is).

A consequence of conservation of the angular momentum for central forces, Eq. (5.12), is the *areal law*, Eq. (5.14); i.e., the areal velocity,

$$\dot{A} = L/2m_P, \tag{6.41}$$

is constant. This corresponds to

The second law:

> *The radius vector from the sun to the planet sweeps out equal areas in equal times.*

The integral over $\dot{A}$, Eq. (6.41), for an entire period T, yields the area of the elliptical orbit:

$$A = \int_0^T \dot{A} dt = TL/2m_P.$$

Since the area A is also given by (see Eqs. (5.32) and (5.39))

$$A = \pi ab = \pi a\sqrt{ap} = \pi a^{3/2}\frac{L}{\sqrt{m_P^2 k}},$$

one obtains a relation between the period T and the length of the semi-major axis of the ellipse:

$$T = \frac{2\pi}{\sqrt{k}} a^{3/2} = \frac{2\pi}{\sqrt{GM}}a^{3/2}. \tag{6.42}$$

Since, for any planet, $M \cong m_S$, we have

$$T^2 \sim a^3.$$

This is

The third law:

> *The squares of the periods are proportional to the cubes of the mean distances from the sun.*

The validity of the third law can be checked from the periods and mean distances of the planets from the sun listed in Table 6.1.

The time dependence of the motion along the ellipse is harder to determine than the period (see, for example, [Goldstein]). Using Eqs. (5.33) and (6.39) for ε and a, Eq. (5.20) reads:

$$t = t_0 + \sqrt{\frac{1}{2k}} \int_{r_0}^{r} \frac{r' dr'}{\sqrt{r' - r'^2/2a - a(1 - \varepsilon^2)/2}}. \tag{6.43}$$

Taking t_0 as the time when the planet is at the perihelion, so that $r_0 = r(t_0) = p/(1 + \varepsilon) = a(1 - \varepsilon)$ (see Eq. (5.36)), $t - t_0$ measures the time elapsed since the planet was last at the perihelion. Replacing now r' by the **eccentric anomaly** ψ, defined by (see Fig. 6.4)

$$r' = a(1 - \varepsilon \cos \psi), \tag{6.44}$$

the integration can be performed analytically (**E**); the result is

$$t - t_0 = \left(a^3/k\right)^{1/2} (\psi - \varepsilon \sin \psi). \tag{6.45}$$

Using Eq. (6.42), one obtains **Kepler's equation,**

$$\mathcal{M}(t) := \frac{2\pi}{T}(t - t_0) = \psi - \varepsilon \sin \psi, \tag{6.46}$$

for the **mean anomaly**[20] $\mathcal{M}(t)$. This relates t to ψ. Of course, Kepler did not perform the integration. One can also deduce Eq. (6.46) via geometrical considerations (cf. e.g. [Whittaker]). Since the areal velocity $\dot{A}$ is constant, the area swept out by the radius vector in time t is proportional to t. Consequently, the ratio of the time elapsed, $(t - t_0)$, to the period T is equal to the ratio of the area of the region (PSO) (see Fig. 6.4) swept out by the radius vector in time $t - t_0$ and the total area of the ellipse; i.e.,

$$\frac{t - t_0}{T} = \frac{\text{Area}(PSO)}{\pi ab}.$$

Now, instead of the actual position O on the ellipse, consider the point

---

[20]The mean anomaly is the angle formed between a line drawn from the focus (sun) to the perihelion of the ellipse and a line from the sun to a hypothetical planet that has the same orbital period as the real planet, but has a constant angular speed. Kepler introduced Eq. (6.46) in his "Epitome astronomiae Copernicanae" (Books 1 to 3 appeared in 1618, Book 4 in 1620, and Books 5 to 7 in 1621).

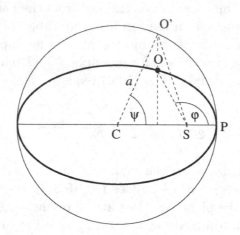

Figure 6.4: Definition of the eccentric anomaly ψ (C: center of ellipse, S: focus (sun), P: perihelion, O: actual planetary position, O': auxilliary position).

O' on the circle in Fig. 6.4, with corresponding area (PSO'). The ratio b/a must now appear on the right hand side,

$$\frac{t-t_0}{T} = \frac{b}{a}\frac{\text{Area}(PSO')}{\pi ab} = \frac{\text{Area}(PSO')}{\pi a^2}.$$

The area (PSO') is equal to the area of the sector (PCO') minus the area of the triangle (SCO') (observe that $SC = a\varepsilon$):

$$\begin{aligned}\text{Area}(PSO') &= \text{Area}\,(PCO') - \text{Area}\,(SCO') \\ &= \frac{a^2}{2}\psi - \frac{a^2\varepsilon}{2}\sin\psi.\end{aligned}$$

Putting all this together, we recover Eq. (6.46).

The eccentric anomaly $\psi(t)$, and subsequently $r(t)$, can be determined from Eq. (6.46) only numerically or approximately. Finding solutions to this equation is a basic task in astronomy. The connection between $\varphi(t)$ and the anomaly ψ follows from combining Eqs. (6.38) and (6.44), which gives (cf. Eq. (5.39))

$$(1+\varepsilon\cos\varphi)(1-\varepsilon\cos\psi) = p/a = 1-\varepsilon^2.$$

Using the relation $\cos\alpha = \left(1 - \tan^2(\alpha/2)\right)/\left(1 + \tan^2(\alpha/2)\right)$, we find

$$\tan\frac{\varphi}{2} = \sqrt{\frac{1+\varepsilon}{1-\varepsilon}}\,\tan\frac{\psi}{2}. \tag{6.47}$$

In his "Analytical Mechanics", J.L. de Lagrange presented approximate solutions of Eqs. (6.46) and (6.47) (see Bibliography).

The motion is not stable in configuration space. Since $T \sim a^{3/2}$, planets with close initial values tend to move away from each other. But the *motion is stable with respect to the orbits*: orbits initially close to each other remain close.

6.3.1 Beyond Kepler's laws

The planetary motion is not as regular (Keplerian) as it may seem at a first glance. Consider for instance the orbits. They are not really ellipses. Over long periods they can be described rather as rotating ellipses forming rosette-shaped patterns (cf. Fig. 5.4).

planet	$\dot\psi$
Mercury	5.75
Venus	2.05
earth	11.45
Mars	16.28
Jupiter	6.55
Saturn	19.50
Uranus	3.34
Neptune	0.36

Table 6.2: The observed perihelion precession of the planets (adapted from: E.M. Standish and J.G. Williams, *Orbital Ephemerids of the Sun, Moon, and Planets*, in Explanatory Supplement to the Astronomical Almanac, Seidelmann, P.K. (ed.). University Science Books)

One reason is that the planetary system consists of more than two players. Therefore the results of the former chapter apply only approximately. The Runge-Lenz-vector is not a conserved quantity anymore. The direction from the Sun to the perihelion of a planet's orbit does not stay constant, it processes with time. The observed **perihelion precession rates of the planets** are measured by an angular velocity

$\dot{\psi}$ of the major semi-axis. Table 6.2 shows the observed rates measured in *arc seconds/year*.

Another reason for deviations from Kepler's laws is the general theory of relativity implying deviations from the $1/r$-potential. This effect is particularly important for the planet closest to the Sun, Mercury. We will dwell a little bit more on this subject at the end of the chapter.

6.4   Gravitational potential of large bodies

Until now we considered the interacting bodies as point masses. But of course this is an idealization, valid only for large distances between the bodies. In more exact calculations one has to take into account the finite size of the bodies. Also Newton was aware of this. In Section XII of the first Book of the "Principia" he examined *the attractive forces of spherical bodies*, and in Section XIII *the attractive forces of bodies which are not of a spherical figure*.

To determine the force exerted by a body of finite size and total mass M on a point mass m from the law of gravitation, Eq. (6.33), we imagine that the body consists of a large number of point masses m_i on fixed sites $\mathbf{r}_i$. (This condition replaces the interaction between the constituents; see the beginning of Chapter 11.) The total mass of the body, $M = \sum_i m_i$, can be rewritten in terms of the **mass density**[21] $\rho(\mathbf{r})$,

$$\rho(\mathbf{r}) = \sum_i m_i \delta(\mathbf{r} - \mathbf{r}_i), \tag{6.48}$$

as

$$M = \int d^3x\, \rho(\mathbf{r}) \quad \left(= \sum_i m_i\right). \tag{6.49}$$

By the superposition principle of forces (stated on Page 17), the total potential is the sum of the potentials v_i of the single constituent point masses:

$$V(\mathbf{r}) = \sum_i v_i\left(\mathbf{r} - \mathbf{r}_i\right), \quad v_i\left(\mathbf{r} - \mathbf{r}_i\right) = -\frac{Gmm_i}{|\mathbf{r} - \mathbf{r}_i|}. \tag{6.50}$$

---

[21]The δ-function (cf. Appendix C.1) is the mathematical abstraction of a point particle of mass m_i, since the mass density of the particle must then be singular. The continuum description of a body is obtained by setting the singular nature of the particle density aside.

If we define the *reduced potential* as

$$V(\mathbf{r}) = -GmM\bar{V}(\mathbf{r}),$$

then using Eq. (6.50), we find

$$\bar{V}(\mathbf{r}) = \frac{1}{M}\sum_i \frac{m_i}{|\mathbf{r} - \mathbf{r}_i|};$$

or, expressed in terms of the mass density (6.48),

$$\bar{V}(\mathbf{r}) = \frac{1}{M}\int d^3x' \frac{\rho(\mathbf{r}')}{|\mathbf{r} - \mathbf{r}'|}. \tag{6.51}$$

This relation is now independent of the details of the body. In the continuum approach, the total potential is the integral over the contributions

$$v_{\mathbf{r}'}(\mathbf{r}) = -Gm\frac{\rho(\mathbf{r}')}{|\mathbf{r} - \mathbf{r}'|}$$

in the volume elements d^3x' at positions $\mathbf{r}'$:

$$V(\mathbf{r}) = \int d^3x'\, v_{\mathbf{r}'}(\mathbf{r}) = -Gm\int d^3x' \frac{\rho(\mathbf{r}')}{|\mathbf{r} - \mathbf{r}'|}. \tag{6.52}$$

What effects do the details of the mass distribution and the body's finite size have on the $\mathbf{r}$-dependence of the potential $\bar{V}(\mathbf{r})$?

6.4.1 The potential of a homogeneous sphere

Let us consider first a homogeneous sphere, i.e. one which has a constant mass density ρ_0. From the condition given in Eq. (6.49),

$$\int\limits_{r'<R} d^3x'\rho_0 = \frac{4\pi R^3}{3}\rho_0 = M,$$

(where $4\pi R^3/3$ is the volume of the sphere), we obtain the density

$$\rho_0 = \frac{3}{4\pi R^3}M. \tag{6.53}$$

Thus, Eq. (6.51) reduces to

$$\bar{V}(\mathbf{r}) = \frac{\rho_0}{M}\int\limits_{r'<R} d^3x' \frac{1}{|\mathbf{r} - \mathbf{r}'|}. \tag{6.54}$$

To calculate the integral, we align the coordinate system such that $\mathbf{r}$ is along the z-axis, i.e. $\mathbf{r} = (0,0,r)$, and introduce spherical coordinates for the components of $\mathbf{r}'$, i.e. $\mathbf{r}' = (r' \cos \varphi \sin \vartheta, r' \sin \varphi \sin \vartheta, r' \cos \vartheta)$ and $|\mathbf{r} - \mathbf{r}'| = (r^2 + r'^2 - 2rr' \cos \vartheta)^{1/2}$. For the integral in Eq. (6.54), we now have

$$\bar{V}(\mathbf{r}) = \frac{3}{4\pi R^3} \int_0^R r'^2 dr' \int_{-1}^1 d\cos\vartheta \int_0^{2\pi} d\varphi \, \frac{1}{(r^2 + r'^2 - 2rr' \cos \vartheta)^{1/2}}$$

$$= \frac{3}{2R^3} \int_0^R r'^2 dr' \int_{-1}^1 d\xi \frac{1}{(r^2 + r'^2 - 2rr'\xi)^{1/2}},$$

where the integration over φ is trivial, and $\xi = \cos \vartheta$. Due to the fact that

$$\int_{-1}^1 d\xi \, \frac{1}{(r^2 + r'^2 - 2rr'\xi)^{1/2}} = \begin{cases} 2/r & r' < r \\ 2/r' & r' > r \end{cases}, \qquad (6.55)$$

further calculation depends on whether the point of application of the force lies inside or outside the sphere.

i) $\underline{r < R}$:

Splitting the interval of integration into two parts (because space is divided into two regions, according to Eq. (6.55)),

$$\bar{V}(\mathbf{r}) = \frac{3}{R^3} \left[\int_0^r r'^2 dr' \frac{1}{r} + \int_r^R r'^2 \frac{1}{r'} dr' \right],$$

we find that the reduced potential is given by

$$\bar{V}(\mathbf{r}) = \frac{3}{R^3} \left[\frac{r^2}{3} + \frac{1}{2}(R^2 - r^2) \right];$$

i.e.

$$\bar{V}(r) = \frac{1}{2R^3}(3R^2 - r^2). \qquad (6.56)$$

The reduced potential depends only on $r = |\mathbf{r}|$. From the definition of the radial component of the force, $F_r = \mathbf{e}_r \mathbf{F}$, and using Eq. (A.19) along with Eq. (6.56), we find

$$F_r = -GmM\frac{d\bar{V}}{dr} = GmM\frac{r}{R^3}. \tag{6.57}$$

All the other spherical components of $\mathbf{F}$ vanish. In the interior of the sphere, the force is linear in r since the potential is harmonic.

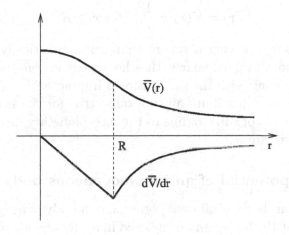

Figure 6.5: Gravitational potential and force of a sphere whose mass is distributed homogeneously.

ii) $\underline{r > R}$:

Clearly, we have also $r > r'$, so that

$$\bar{V}(\mathbf{r}) = \frac{3}{R^3}\frac{1}{r}\int\limits_{0}^{R} r'^2 dr' = \bar{V}(r) = \frac{1}{r} \tag{6.58}$$

and

$$F_r = \frac{GmM}{r^2}.$$

A sphere of homogeneous mass distribution acts on a point mass outside the sphere as if all of its mass M were located at the center of the sphere.

Figure 6.5 shows the gravitational potential $\bar{V}$ and its derivative $d\bar{V}/dr = -F_r/GmM$. At the surface of the sphere $(r = R)$, the potential as well as the force are continuous. These results were already known to Newton (see Footnote 7).

In a similar manner, one can prove that the potential of a sphere with a *spherical symmetric mass distribution* $\rho(\mathbf{r}) = \rho(r)$ is also given by (E)

$$\bar{V}(\mathbf{r}) = \bar{V}(r) = \frac{1}{r}, \quad \text{for } r \geq R.$$

This justifies our replacing a sphere that has a spherically symmetric mass distribution (in particular with a homogeneous one) with a point mass – when dealing with the gravitational interaction[22]. The assumption about the mass distribution is not quite true for the sun; nevertheless, the small ratio of sun's radius to the interplanetary distance allows such an approximation.

6.4.2   The potential of an inhomogeneous body

We now focus on the gravitational potential of a body of irregular shape. We assume that the body can be enclosed in a sphere with radius R (see sketch).

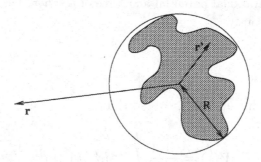

We are only interested in the potential outside the body (outside the enclosing sphere); that is, distances $r > R$. In this region, we have

---

[22]This property holds for the Coulomb interaction, too.

also $r > r'$, and we may expand the denominator in the integral in Eq. (6.51) in a power series of r'/r:

$$\frac{1}{|\mathbf{r} - \mathbf{r}'|} = \frac{1}{\sqrt{(\mathbf{r} - \mathbf{r}')^2}} = \frac{1}{r\sqrt{1 + (r'/r)^2 - 2\mathbf{r}\mathbf{r}'/r^2}}$$

$$= \frac{1}{r} + \mathbf{r}'\frac{\mathbf{r}}{r^3} + \frac{1}{2}\sum_{i,j} x_i' x_j' \left(\frac{3x_i x_j}{r^5} - \frac{\delta_{ij}}{r^3}\right) + \dots . \quad (6.59)$$

Since

$$\sum_{i,j} \delta_{ij} \left(\frac{3x_i x_j}{r^5} - \frac{\delta_{ij}}{r^3}\right) = 0,$$

the third term in the expansion (6.59) can be rewritten as the trace of the product of traceless tensors:

$$\frac{1}{2}\sum_{i,j} x_i' x_j' \left(\frac{3x_i x_j}{r^5} - \frac{\delta_{ij}}{r^3}\right) = \frac{1}{2}\sum_{i,j} \left(x_i' x_j' - \frac{1}{3}\delta_{ij}r'^2\right)\left(\frac{3x_i x_j}{r^5} - \frac{\delta_{ij}}{r^3}\right).$$

Inserting now the above power series expansion into Eq. (6.51), we get

$$\bar{V}(\mathbf{r}) - \frac{1}{M}\frac{1}{r}\int d^3x'\, \rho\,(\mathbf{r}') + \frac{1}{M}\frac{\mathbf{r}}{r^3}\int d^3x'\, \mathbf{r}'\rho\,(\mathbf{r}')$$

$$+\frac{1}{2M}\sum_{i,j}\left(\frac{3x_i x_j}{r^5} - \frac{\delta_{ij}}{r^3}\right)\int d^3x'\left(x_i' x_j' - \frac{1}{3}\delta_{ij}r'^2\right)\rho(\mathbf{r}') + \dots , \quad (6.60)$$

where the integrals extend over the volume of the body. The integral in the first term is the total mass M; the integral in the second term is the **dipole moment**, which also defines the *position of the center of mass* $\mathbf{r}_0$ according to (cf. also Eq. (6.13)),

$$\frac{1}{M}\int d^3x'\, \mathbf{r}'\rho(\mathbf{r}') =: \mathbf{r}_0. \quad (6.61)$$

The integral in the third term,

$$\int d^3x'\left(x_i' x_j' - \frac{1}{3}\delta_{ij}r'^2\right)\rho(\mathbf{r}') =: Q_{ij}, \quad (6.62)$$

is the traceless tensor of the **quadrupole moment** of the mass distribution. The behavior of the potential $\bar{V}(\mathbf{r})$ for sufficiently large values of r is

$$\bar{V}(\mathbf{r}) = \frac{1}{r} + \frac{\mathbf{r}\mathbf{r}_0}{r^3} + \mathcal{O}(1/r^3). \quad (6.63)$$

Choosing the origin at the center of mass $\mathbf{r}_0$ (by a translation of the coordinate system), the second term vanishes, and the lowest correction to the $1/r$ behavior is at least proportional to $1/r^3$. (In the case of a spherical symmetric body, $Q_{ij} \equiv 0$ (E) and only the $1/r$ term remains.) The quadrupole moment can not be removed by a transformation of the coordinate system. The force produced by this term is, in general, not a central force (E). If the corrections to the $1/r$ behavior of a central force are small (as for the sun and the planets), the mass distribution can be replaced by a point mass, to a good approximation.

Remarks

i) The potential of a point mass situated in $\mathbf{r}$, which acts on a mass distribution $\rho(\mathbf{r})$, is also given by $\bar{V}(\mathbf{r})$, Eq. (6.51) (active – passive gravitational mass).

ii) Potentials of point masses $V(r) \propto 1/r^\alpha$, $\alpha \neq 1$, can be discussed in a similar way. For a spherical symmetric distribution, the potential of a sphere is proportional to $1/r^\alpha$ (but only for $r \gg R$, even for a homogeneous mass distribution; (E)). The dipole moment can be set to zero by shifting the center of mass position to the origin.

6.5   On the validity of the gravitational law

An important feature of the *gravitational interaction* is that, in contrast to the Coulomb interaction, different contributions cannot cancel each other out – all masses have the same sign – so that the interaction *cannot be screened*. All bodies in the universe contribute a force at every point in space; most of them are negligible because these bodies are very far away. This property of the gravitational force plays an important role in several phenomena; for example, in the gravitational collapse of stars, as well as in the question of the stability of the universe (considering gravitational interaction only, no stable equilibrium configuration exists).

As we have seen already, there are, in general, deviations from the $1/r$ behavior of the gravitational potential (6.36) of finite size bodies, because of the shape and/or mass distribution of the bodies. However, one may still ask: apart from these effects, how sure can one be about the $1/r$ behavior of the elementary gravitational interaction? How well does the evidence from astronomical observations and from laboratory experiments prove the $1/r$ behavior? In answering these questions, one has to keep in mind that Newton's gravitational law is only a limiting case of Einstein's general theory of relativity. Therefore, for discrepancies to be truly controversial, they must contradict general relativity.

Evidence from astronomy

Kepler's second law applies to all central forces. So only Kepler's first law – orbits are elliptical – and the third law – giving the ratio of periods to the orbit size – refer specifically to gravity. Apropos the third law, planetary observations suggest that a potential $V \propto r^{-\alpha}$, $\alpha \neq 1$, is not very likely. Planetary orbits are safely elliptical and stable. Phenomena such as precession of planetary orbits (cf. Section 5.3), in particular that of Mercury, can be explained: even the classical theory of gravitation predicts the advance of Mercury's perihelion, caused by the presence of other planets in the solar system. Much smaller discrepancies do indeed exist in the precession of Mercury; however, these are explained by Einstein's general theory of relativity, in which no absolute time and space[23] exists. Einstein's theory predicts (correctly) that a correction to the $1/r$ potential, proportional to $1/r^3$, exists, and that this causes the observed advance of the perihelion of Mercury (see, for example, [Rindler]; the relativistic correction in the Coulomb interaction is $\propto 1/r^2$!). Astronomical observations clearly support the $1/r$ behavior of the gravitational potential. Nowadays, the gravitational law is tested down to the millimeter range (P. Binétruy, Europhysics news, **33**, 54 (2002)). In the context with the question of dark matter one possibility to avoid it is a deviation from the inverse square law. But investigations in the Solar system show that such a deviation can be only very tiny, so dark matter[24] is still needed (M. Sereno and Ph. Jetzer, MRNAS, 371

---

[23]Replacing Newton's absolute space ('proven' by the absolute rotation of Newton's pail; see Page 11) by the concept of the 'general covariance', leads to the *Thirring-Lense effect* in the general theory of relativity. In terms of classical concepts, this effect may be described as follows. *The rotation of a mass causes the surrounding inertial system in which the mass is at rest to rotate also. The effect on other masses is equivalent to a dipole field* ($V \propto k/r^2$; *the strength k is proportional to the velocity of rotation;* cf. [Rindler]). This supports Mach's explanation of Newton's pail experiment – the curvature of the water surface is due to this additional force.

[24]Dark matter accounts for the big difference between the mass calculated from the gravitational effects in the universe and the much smaller mass calculated from

(2), 626-632 (2006))

Evidence from laboratory experiments

The gravitational constant G was first determined by H. Cavendish in 1798, more than hundred years after the publication of the "Principia". Cavendish used a torsion balance[25] to determine the density of the earth. Precise torsion balance experiments by L. Eötvös (1922), as well as R.H. Dicke (1964), aimed at investigating the gravitational interaction between various bodies, showed that the constant G is independent of the material and the size of the bodies. Further experiments[26] have confirmed the $1/r$ law. They also improved the precision of the known value of the gravitational constant. The value found in these experiments is in agreement with the value (6.34).

Problems and examples

1. Solve Eq. (6.43) by introducing the eccentric anomaly ψ, Eq. (6.44).

2. The orbits of most planets in our solar system are nearly circular. For these planets, the energy E is only slightly larger than the energy E_0 in Eq. (5.28) for circular motion. In the effective potential $V_{eff}(r)$, Eq. (5.27), the motion of such planets is determined by the potential in the vicinity of $r_{\min}$, Eq. (5.28). Expand $V_{eff}(r)$ about $r_{\min}$ to second order. Solve the radial equation of motion in this approximation. Compare the oscillation frequency ('libration frequency') of the radial motion to the circular frequency Eq. (5.30). What is the shape of the orbit? Compare to the actual orbit.

3. Consider the equation of motion for a particle of mass m acted upon by earth's gravity: $\mathbf{F} = Gmm_E\mathbf{r}/r^3$. (Assume the earth is spherical.) Expand the radial component of the gravitational force

---

the luminosity of stellar and gas objects.

[25] A torsion balance consists of a rod hanging horizontally on a wire (or equivalent) undergoing harmonic motion (rotation) due to the harmonic restoring force of the wire.

[26] See, for example, J.H. Gundlach and S.M. Merkowitz, Phys.Rev.Lett., **85**, 2869 (2000).

in $z = r - r_E$ and determine in first order of z the solution $z(t)$ for the initial values $z(t = 0) = h$ and $\dot{z}(t = 0) = 0$. Compare this result (in the lowest order approximation) with the case of the homogeneous force $\mathbf{F} = m\mathbf{g}$, $\mathbf{g} = Gm_E\mathbf{r}_E/r_E^3$.

4. Investigate the motion of two point masses m_1 and m_2 attracting each other gravitationally in the homogeneous gravitational field of the earth; i.e., an additional force $\mathbf{F} = m_i\mathbf{g}$ acts on the two point masses. Give the equations of motion in center of mass coordinates and relative coordinates. Solve the equations and discuss the result.

5. Show that for a spherical mass distribution, $\rho(\mathbf{r}) = \Theta(R - r)\rho(r)$, the gravitational potential V for $r \geq R$ is given by $\bar{V}(r \geq R) = 1/r$ (cf. Eq. (6.51)).

6. Determine the potential $\bar{V}(r)$ of a sphere with radius R of homogeneous mass density ρ_0 for distances $r \geq R$, in which the contribution of each volume element is $v_{\mathbf{r}'}(\mathbf{r}) = \rho_0/|\mathbf{r} - \mathbf{r}'|^\alpha$. For a spherical mass distribution $\rho(r)$, use the equality(!) $\bar{V}(r \geq R) = 1/r^\alpha$ to show that $\alpha = 1$.

7. Show that for a homogeneous sphere, the quadrupole moment Q_{ij}, Eq. (6.62), vanishes.

8. Calculate the quadrupole moment of the earth and the corrections to the $1/r$ behavior of the potential. Consider the earth to be approximately an ellipsoid of revolution (semimajor axis $a = 6378.16\ km$, semiminor axis – which is the axis of rotation – $b = 6356.78\ km$) with a homogeneous mass distribution.

7

Collisions of particles. Scattering

7.1 Unbounded motion in a central force

The motion of a particle in a *central force* – that is, in a spherical symmetric potential $V(r)$ – is restricted to a fixed plane due to the conservation of angular momentum together with condition (2.34), $\mathbf{Lr} = 0$. The radial motion is equivalent to one-dimensional motion in the effective potential (5.16): $V_{eff}(r) = V(r) + L^2/2mr^2$. Once $r(t)$ has been found, the angular motion $\varphi(t)$ is easily determined from the conservation of angular momentum (5.13): $\dot{\varphi} = L/mr^2$.

In this chapter, we consider potentials that vanish as r goes to infinity,

$$V(r \to \infty) = 0. \tag{7.1}$$

This means the motion of a particle with energy $E > 0$ is unbounded, as discussed in Section 5.1. Figure 7.1 shows examples of $V_{eff}(r)$. Effective potentials tending to $-\infty$ as $r \to 0$ (see (2) and (4) in Fig. 7.1) must be looked at closely since the singularity at $r = 0$ may be part of the orbit. This complication arises because we are treating the particles as a point: if particles have finite size, the singularity in the potential lies outside the accessible region. Finite size can be represented by a 'steep' rise to $+\infty$ as $r \to 0$, with the rate of increase being a measure of the 'hardness' of the surface of the particle. A discontinuous jump to infinity corresponds to a perfectly hard surface. Thus a **hard sphere** (or a so-called hard core) of radius R is represented by replacing the

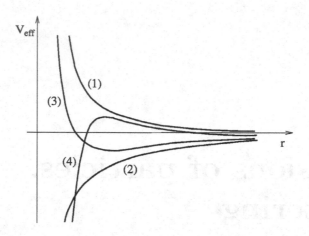

Figure 7.1: Various examples of effective potential. (1): $V_{eff}(r) = 1/r^2$; (2): $V_{eff}(r) = -1/r$; (3): $V_{eff}(r) = -1/r + 1/r^2$; (4): $V_{eff}(r) = -1/r + c(1/r^2 - 1/r^3)$, $c > 3$.

potential $V(r)$ in the interval $[0, R]$ by

$$V(r) = \infty \quad \text{for} \quad r \leq R. \tag{7.2}$$

If necessary, we exclude a singularity at $r = 0$ by assuming that the particles have finite size.

Consider an effective potential satisfying condition (7.1). For $E > 0$, radial motion $r(t)$ is externally unbounded. A particle approaching the center from $r = \infty$ with negative velocity $\dot{r} = -\sqrt{...}$, given in Eq. (5.19), turns around at $r_{\min}$ (where $\dot{r} = 0$) and heads back out to $r = \infty$ with positive velocity $\dot{r} = \sqrt{...}$. (In appropriate cases, $r_{\min}$ is fixed by condition (7.2).) For sufficiently large distances ($r \to \infty$), the particle will not be influenced perceptibly. Since $\dot{\varphi}$ ($\propto 1/r^2$) $\cong 0$, the angle φ will remain nearly constant. Hence, an *asymptotic angle* φ_A exists. As we saw in Section 5.1, every orbit is *symmetric* about the apsidal vector[1] giving the direction from the center of the force to the turning

---

[1]This also follows from the equation of the orbit (E),

$$\frac{L^2}{m}\left(\frac{d^2u}{d\varphi^2} + u\right) = -\frac{d}{du}V(1/u), \qquad u = 1/r,$$

point[2] whose polar coordinates are (r_{min}, φ_T). Since only one turning point exists, for $L \neq 0$ there is exactly a second asymptotic direction, related to the previous direction by reflection about the apsidal vector. The situation in a $1/r$ potential is shown in Fig. 7.2, where $\varphi_{1,2} = \varphi_K \pm \arccos(1/\varepsilon)$ (see Eq. (5.41)). In the context of scattering, the angle between the outgoing direction (angle φ_2) and the incident direction (angle $\varphi_1 - \pi$) is called the **scattering angle** Θ, so that (cf. Fig. 7.2)

$$\Theta = \varphi_2 - \varphi_1 + \pi. \tag{7.3}$$

From Eq. (5.22), we have the equation of the orbit for a particle incident from[3] $(r = \infty, \varphi = \varphi_1)$

$$\varphi - \varphi_1 = \frac{L}{\sqrt{2mE}} \int_{\infty}^{r} \frac{dr'}{r'^2} \frac{1}{\sqrt{1 - \dfrac{V(r')}{E} - \dfrac{L^2}{2mEr'^2}}}. \tag{7.4}$$

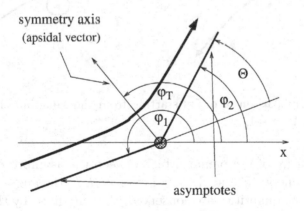

Figure 7.2: The angles characterizing the unbounded motion of a particle in a potential V(r).

which is symmetric with respect to the transformation $\varphi \to -\varphi$.

[2]Depending on the sign of the potential, for a hyperbolic orbit, the turning point is given by either Eq. (5.42) or Eq. (5.45), respectively.

[3]For a particular orbit, the sign in front of the integral in Eq. (5.22) is chosen according to the given situation. Here, the choice reflects the situation in Fig. (7.2).

In the derivation of Eq. (7.4), it was assumed that the relation between $r(t)$ and $\varphi(t)$ is unique. Consider, for instance, unbounded motion of a particle in the potential $V(r) = -k/r$. From the analytic treatment in Section 5.2, we know that the orbit in the $1/r$ potential is a hyperbola. The orbit does not intersect itself. This is not generally true if the potential has a different r-dependence; there may appear intersections of the orbit. The reason for the peculiar behavior of a particle in the $1/r$ potential is the presence of an additional conserved quantity. The Runge-Lenz vector, the vector from the focus to the pericenter (perihelion), is constant in time (see Section 5.4). However, for potentials of the form $1/r^{\alpha}$, where α is different from one, the orbits may form loops. This is illustrated in Fig. 7.3. For a $1/r^{1.8}$ potential, appropriate initial values produce looping orbits; after the particle completes a loop, φ changes by more than 2π. In the following, for simplicity, we consider only orbits that do not intersect themselves.

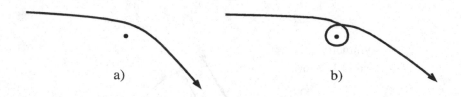

a)   b)

Figure 7.3: Comparison of orbits starting from the same initial values: a) $V = -1/r$; b) $V = -1/r^{1.8}$.

The curvature of the planar orbit between the asymptotes is determined by the energy E and the z-component L of the angular momentum. Since both quantities are conserved, they are fixed by their values before the particle enters the region where the potential is appreciable; i.e. before interacting with the potential. In that outer region, the particle can be considered to move freely. Rather than characterizing the state of the particle by its energy and angular momentum, it is convenient to specify the particle's **asymptotic velocity** v_{∞} and the **impact parameter** s. The impact parameter is defined as the perpendicular distance of the particle to the $L = 0$ orbit, that is the straight line between the source of the particle at $r = \infty$ and the scattering center (center of the potential). If the particle is incident along the negative x-axis, as in Fig. 7.4, asymptotically we have $v_x = v_\infty$ and $v_y = 0$. The angular momentum of the particle is $L = L_z = m\left(xv_y - yv_x\right) = -msv_\infty$, so that

$$|L| = msv_\infty, \qquad E = \frac{1}{2}mv_\infty^2. \tag{7.5}$$

In turn, the asymptotic velocity and the impact parameter are given by

$$v_\infty = \sqrt{\frac{2E}{m}}, \qquad s = \frac{|L|}{\sqrt{2mE}}. \tag{7.6}$$

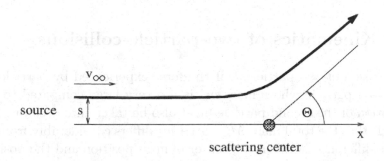

Figure 7.4: The 'scattering parameters' consist of the scattering angle Θ, impact parameter s, and asymptotic velocity v_∞.

For the choice of the coordinate axes shown in Fig. 7.4, the *angle of incidence* φ_1 is

$$\varphi_1 = \pi.$$

In order to calculate the scattering angle Θ $(= \varphi_2)$, due to the symmetry of the orbit, it suffices to extend the range of integration in Eq. (7.4) from $r = \infty$ (at $\varphi = \varphi_1$) to the closest approach at $(r_{\min}, \varphi_T)$. The angle φ_T is then the mean of the two asymptotic angles; expressed in terms of the scattering angle, φ_T is (cf. Eq. (7.3)) given by

$$\varphi_T = \frac{\varphi_1 + \varphi_2}{2} = \frac{\Theta + \pi}{2}. \tag{7.7}$$

By inserting the asymptotic velocity and the impact parameter into Eq.

(7.4), the scattering angle Θ can be calculated from the equation[4]

$$\varphi_T - \varphi_1 = \frac{\Theta - \pi}{2} = -s \int_{\infty}^{r_{min}} \frac{dr'}{r'^2} \frac{1}{\sqrt{1 - V(r')/E - \frac{s^2}{r'^2}}}, \qquad (7.8)$$

if r_{min} is known. In scattering experiments, many particles with a distribution of impact parameter s are fired at a target potential (or particle). The scattering angles of the deflected particles are recorded. From this data, the potential $V(r)$ can be deduced with the help of Eq. (7.8) (cf. Section 7.3 and (E)).

7.2   Kinematics of two-particle-collisions

In a collision of two particles[5], if the force experienced by particle m_1 is due to a particle whose mass m_2 is not very large compared to m_1, the motion of the second particle must also be taken into account. The reason is that the total mass $M = m_1 + m_2$ differs considerably from m_2 and the difference between the center of mass position and the position of the particle m_2 cannot be neglected anymore (cf. Section 6.1). At distances where the particles exert considerable influence on each other – i.e. in the **interaction region** (cf. Fig. 7.5) – the relative motion of the two particles in terms of relative coordinates is equivalent to the motion of one particle with reduced mass in the interaction potential. Hence the discussion in the preceding section applies to the relative motion of the particles in the interaction region. Concerning the asymptotic regions, we now ask: Given the momenta and the energy of the two particles before they enter the interaction region, how are these quantities related to the momenta and the energy, respectively, the particles have after leaving the interaction region?

---

[4]Now we have a minus sign in front of the integral because $s = -L / \sqrt{2mE}$ for an orbit as shown in Fig. 7.4.

[5]In classical mechanics, when considering two or more particles, we assume, that the particles can be distinguished, even if they have identical properties.

Moreover, if the particles have no further interaction, we have to conceive them to be of finite size, since the probability for point particles to scatter vanishes.

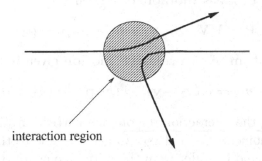

interaction region

Figure 7.5: Two particles colliding.

7.2.1 Elastic collisions of two particles

We investigate the motion of two interacting particles with masses m_1 and m_2. If the interaction depends only on the positions r_1 and r_2 of the particles, then the interaction force is a central force and the interaction potential $V(r)$, $r = r_1 - r_2$, is spherically symmetric (cf. Section 6.1). Consequently, the motion is effectively two-dimensional (due to the conservation of angular momentum)[6]. For the moment, we restrict ourselves to interaction potentials of *finite range* R; i.e.,

$$V(|r_1 - r_2|) = 0 \quad \text{for } |r_1 - r_2| > R.$$

Before entering the interaction region, $r > R$, the two particles move freely with constant momenta $p_i = m_i v_i$ $(i = 1, 2)$ towards each other. The total energy is simply the sum of the kinetic energies

$$E = p_1^2/2m_1 + p_2^2/2m_2. \tag{7.9}$$

Switching to center of mass coordinates and relative coordinates (cf. Eq. (6.21), $E = E_{tot}$, $V = \dot{R}$, and $v_i = \dot{r}_i$), the energy reads

$$E = MV^2/2 + m(v_1 - v_2)^2/2 = P^2/2M + q^2/2m, \tag{7.10}$$

---

[6]For the linear collision of two hard spheres Huygens gave the correct laws (*The laws of motion on the collision of bodies*, Phil. Trans. 1669; *Tractatus de motu corporum ex percussione*, 1703 (posthumous); the latter work includes also the proof of the laws). Descartes considerations about the collision of particles in his "Principia Philosophiae" contain some errors.

with the **center of mass momentum** given by[7]

$$\mathbf{P} = M\mathbf{V} = m_1\mathbf{v}_1 + m_2\mathbf{v}_2 = \mathbf{p}_1 + \mathbf{p}_2 \tag{7.11}$$

and the **momentum of the relative motion** given by

$$\mathbf{q} = m\mathbf{v} = m(\mathbf{v}_1 - \mathbf{v}_2) = (m_2\mathbf{p}_1 - m_1\mathbf{p}_2)/M. \tag{7.12}$$

After exiting the interaction region, the particles again move freely with constant momenta $\bar{\mathbf{p}}_1$ and $\bar{\mathbf{p}}_2$. In general, due to the interaction, the momenta $\bar{\mathbf{p}}_1$ and $\bar{\mathbf{p}}_2$ differ from the initial momenta $\mathbf{p}_1$ and $\mathbf{p}_2$. The change in the momenta depends on the interaction of the particles. The energy after interaction is

$$\bar{E} = \bar{\mathbf{p}}_1^2/2m_1 + \bar{\mathbf{p}}_2^2/2m_2 = \bar{\mathbf{P}}^2/2M + \bar{\mathbf{q}}^2/2m, \tag{7.13}$$

where the center of mass momentum and relative momentum are given by

$$\bar{\mathbf{P}} = \bar{\mathbf{p}}_1 + \bar{\mathbf{p}}_2 \quad \text{and} \quad \bar{\mathbf{q}} = (m_2\bar{\mathbf{p}}_1 - m_1\bar{\mathbf{p}}_2)/M, \tag{7.14}$$

respectively.

In an **elastic collision** of two particles, with interaction potential $V(r)$, two conservation laws relate the energy and momenta before and after the interaction; namely,

i) Conservation of total energy:

$$E = \bar{E}; \tag{7.15}$$

ii) Conservation of the center of mass momentum:

$$\mathbf{P} = \bar{\mathbf{P}}. \tag{7.16}$$

These conditions are deduced below.

For the relative motion, using Eqs. (7.10), (7.13), and (7.16), conservation of energy, Eq. (7.15), results in the condition

$$\mathbf{q}^2 = \bar{\mathbf{q}}^2. \tag{7.17}$$

Therefore, in an elastic collision, only the direction of the relative momentum – and consequently, the directions of the particle momenta –

---

[7]Note that Eq. (6.13) implies $M\dot{\mathbf{R}} = m_1\dot{\mathbf{r}}_1 + m_2\dot{\mathbf{r}}_2$.

may change[8]. The change is *not fixed by the kinematic conditions*: the change in the direction of the momenta, represented by the scattering angle Θ, *results from the interaction dynamics*[9].

In the interaction region, $|\mathbf{r}_1 - \mathbf{r}_2| \leq R$, the kinetic energy of the particles is converted partly into interaction energy V, and vice versa. The total energy $E = T + V$ of the two particles is conserved (we assume an elastic collision): condition (7.15) is valid. In concordance with the arguments in Section 6.1, the description of the collision separates into free motion of the center of mass and the motion of a *single fictitious particle* with reduced mass m in an *(external) potential* $V(r)$.

Since the center of mass moves freely, its momentum is equal to its value at the respective instants the interaction starts and stops having an effect on the particles. These values are just the center of mass momenta outside the interaction region, Eqs. (7.11) and (7.14). Thus, we have proven Eq. (7.16).

The relation between the scattering angle Θ and the potential is given by Eq. (7.8). The actual motion of the two particles about the center of mass is obtained from Eqs. (6.19) and (6.20),

$$\mathbf{r}_1(t) = \frac{m_2}{M}\mathbf{r}$$
$$\mathbf{r}_2(t) = -\frac{m_1}{M}\mathbf{r}.$$

(In the case of the $1/r$ potential, these define two hyperbolic orbits).

7.2.2 Kinematics of elastic collisions

In a collision scenario – or in a scattering experiment, when many particles collide – the initial momenta of the particles are fixed, and one observes the change in momenta after the particles have left the interaction region. To represent the collision process, one normally uses two *frames of reference* (cf. Chapter 8):

---

[8]For **inelastic collisions**, in which energy is channelled to other degrees of freedom such as heat or sound, we have: $\mathbf{q}^2/2m = \bar{\mathbf{q}}^2/2m + Q$, where Q is the energy transferred to these additional degrees of freedom.

[9]Conservation of energy and conservation of the momentum of the center of mass leave only one out of the four components of $\bar{\mathbf{p}}_1$ and $\bar{\mathbf{p}}_2$ undetermined (remember we have planar motion!). This undetermined component is related to the scattering angle.

i) The **center of mass frame**, in which the *center of mass is at rest*, $\mathbf{P} = 0$; and,

ii) The **laboratory frame**, in which one particle, the *projectile*, is shot towards the second particle, called the *target particle*, which is *at rest*.

We denote quantities in the laboratory frame with label L, while quantities in the center of mass frame shall have no extra label. In order to link the scattering angles θ_L and θ in the two reference frames, we investigate now the kinematics of the collision in these two frames.

   We consider first the elastic collision of two particles in the center of mass frame.

Kinematics in the center of mass frame

When the center of mass is at rest, we have $m_1 \mathbf{v}_1 + m_2 \mathbf{v}_2 = M\mathbf{V} = \mathbf{P} = 0$. The total momentum therefore vanishes at all times:

$$\mathbf{P} = \mathbf{p}_1 + \mathbf{p}_2 = \bar{\mathbf{p}}_1 + \bar{\mathbf{p}}_2 = 0. \qquad (7.18)$$

Hence, in the center of mass frame, the conditions

$$\mathbf{p}_1 = -\mathbf{p}_2 \qquad \text{and} \qquad \bar{\mathbf{p}}_1 = -\bar{\mathbf{p}}_2$$

for the momenta of the particles result. Inserting this into the definitions (7.12) and (7.14) of $\mathbf{q}$ and $\bar{\mathbf{q}}$, respectively, yields

$$\mathbf{q} = \mathbf{p}_1 = -\mathbf{p}_2 \qquad \text{and} \qquad \bar{\mathbf{q}} = \bar{\mathbf{p}}_1 = -\bar{\mathbf{p}}_2. \qquad (7.19)$$

Together with Eq. (7.17), it follows that the collision may change the directions of the momenta but not their magnitude as indicated in Fig. 7.6:

$$q = \bar{q} = p_i = \bar{p}_i. \qquad (7.20)$$

The change in the direction of the momentum of one particle, say m_1, is obtained from

$$\mathbf{p}_1 \bar{\mathbf{p}}_1 = \mathbf{q}\bar{\mathbf{q}} = q^2 \cos\theta. \qquad (7.21)$$

As stated above, the change θ is caused by the dynamics occurring within the region of nonvanishing potential $V(r)$.

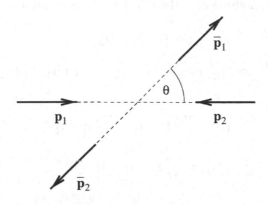

Figure 7.6: The collision in the center of mass frame.

Kinematics in the laboratory frame

Since the target particle (particle #2) is at rest in the laboratory frame, i.e. $\mathbf{v}_2^L = \mathbf{0}$, the center of mass moves with constant velocity $\mathbf{V}^L = (m_1/M)\,\mathbf{v}_1^L$, as can be seen from Eq. (7.11). Therefore, all velocities measured in the laboratory frame differ from their value in the center of mass frame by $\mathbf{V}^L$. However, the relative velocity between the particles is the same in both reference frames; therefore the relative momenta $\mathbf{q}$ and $\bar{\mathbf{q}}$ (Eqs. (7.12) and (7.14), respectively) are also independent of the frame. Hence the label L shall be omitted from the relative quantities $\mathbf{q}$ and $\bar{\mathbf{q}}$.

Suppose that *before the collision*, the target particle is at rest:

$$\mathbf{p}_2^L = \mathbf{0}.$$

Then the total momentum is due only to projectile particle (particle #1),

$$\mathbf{P} = \mathbf{p}_1^L,$$

and the relative momentum (7.12) is

$$\mathbf{q} = \frac{m_2}{M}\mathbf{p}_1^L = \frac{m_2}{M}\mathbf{P}. \tag{7.22}$$

Using conservation of total momentum (i.e. center of mass momentum),

$$\mathbf{P} = \mathbf{p}_1^L = \bar{\mathbf{p}}_1^L + \bar{\mathbf{p}}_2^L,$$

together with the definition of the relative momentum *after the collision*,

$$\bar{\mathbf{q}} = (m_2\bar{\mathbf{p}}_1^L - m_1\bar{\mathbf{p}}_2^L)/M,$$

the momenta $\bar{\mathbf{p}}_1^L$ and $\bar{\mathbf{p}}_2^L$ can be expressed in terms of $\mathbf{q}$ and $\bar{\mathbf{q}}$ using Eq. (7.22); *viz.*:

$$\bar{\mathbf{p}}_1^L = \frac{m_1}{M}\mathbf{P} + \bar{\mathbf{q}} = \frac{m_1}{m_2}\mathbf{q} + \bar{\mathbf{q}} \qquad (7.23)$$

$$\bar{\mathbf{p}}_2^L = \frac{m_2}{M}\mathbf{P} - \bar{\mathbf{q}} = \mathbf{q} - \bar{\mathbf{q}}. \qquad (7.24)$$

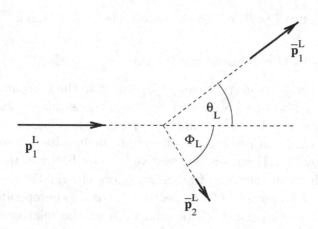

Figure 7.7: The collision viewed in the laboratory frame; θ_L and Φ_L are the scattering angles.

Since the target moves after the collision, we now have two *scattering angles* θ_L and Φ_L for the two particles with respect to the incident direction (see Fig. 7.7). θ_L is the angle between $\bar{\mathbf{p}}_1^L$ and $\mathbf{p}_1^L$, and Φ_L the angle between $\bar{\mathbf{p}}_2^L$ and $\mathbf{p}_1^L$. These two angles are not independent (see below).

Relating θ_L to θ

Using the fact that the two relative momenta $\mathbf{q}$ and $\bar{\mathbf{q}}$ are frame-independent, we establish the connection between the scattering angles

θ and θ_L. To this end, we consider the triangle formed by the vectors in Eq. (7.23). According to Eq. (7.21), θ is the angle between $\mathbf{q}$ and $\bar{\mathbf{q}}$, and due to Eq. (7.22), θ_L is also the angle between $\mathbf{q}$ and $\bar{\mathbf{p}}_1^L$. Hence, from Fig. 7.8, we conclude that

$$\tan \theta_L = \frac{|\bar{\mathbf{q}}| \sin \theta}{|\mathbf{q}m_1/m_2| + |\bar{\mathbf{q}}| \cos \theta}.$$

Since $\bar{q} = q$ (see Eq. (7.20)), the desired relation between θ and θ_L is

$$\tan \theta_L = \frac{\sin \theta}{m_1/m_2 + \cos \theta}. \tag{7.25}$$

Obviously, $\tan \theta_L$ is always smaller than $\tan \theta$. This means that the scattering angle in the laboratory frame is always smaller than in the center of mass frame,

$$\theta_L < \theta. \tag{7.26}$$

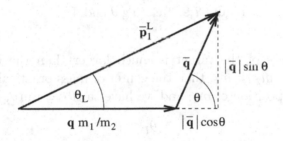

Figure 7.8: Relating the scattering angles θ and θ_L.

One can deduce the scattering angle Φ_L of the target particle in the laboratory frame from the triangle in Fig. 7.9, which is a graphical representation of Eq. (7.24). Since $\mathbf{q}$ and $\mathbf{p}_1^l$ are parallel (cf. Eq. (7.22)), Φ_L is also the angle between $\mathbf{q}$ and $\bar{\mathbf{p}}_2^L$, and due to $q = \bar{q}$ (cf. Eq. (7.20)), the triangle is isosceles. Therefore the angle between $\bar{\mathbf{p}}_2^L$ and $\bar{\mathbf{q}}$ is also Φ_L. Because the angle between $\mathbf{q}$ and $\bar{\mathbf{q}}$ is the scattering angle θ in the center of mass frame (cf. Eq. (7.21)), we have $\theta + 2\Phi_L = \pi$; i.e.,

$$\Phi_L = \frac{\pi - \theta}{2}. \tag{7.27}$$

Let us now discuss some limiting cases.

- If the mass of the incident particle, the projectile, is larger than the mass of the target, $m_1 > m_2$, then there is a maximal scattering angle in the laboratory frame (E),

$$\sin \theta_L^{\max} = m_2/m_1. \tag{7.28}$$

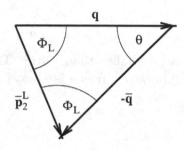

Figure 7.9: Relating θ and Φ_L.

- If the mass of the target is much larger than the mass of the projectile, $m_2 \gg m_1$, then the center of mass practically coincides with the position of m_2, and we have $\tan \theta_L \cong \tan \theta$; i.e.,

$$\theta_L \cong \theta. \tag{7.29}$$

The laboratory frame coincides with the center of mass frame. This is the case, for instance, in *potential scattering*, in which the center of the potential is fixed. (Formally, this is achieved in the limit $m_2 = \infty$; consequently $\theta_L = \theta = \Theta$.) From Eq. (7.27), we see that the sum of the two scattering angles θ_L and Φ_L is always larger than $\pi/2$ (the sum is $\pi/2 + \theta_L/2$).

- If the mass of the target and the mass of the projectile are equal, $m_2 = m_1$, we have

$$\tan \theta_L = \frac{\sin \theta}{1 + \cos \theta} = \tan \frac{\theta}{2};$$

that is,

$$\theta_L = \theta/2. \tag{7.30}$$

Consequently, the maximal scattering angle $\theta_{L,\text{max}}$ in the laboratory frame is 90°: there is no back-scattering. Since, for $m_2 = m_1$, we have (cf. Eq. (7.27))

$$\theta_L + \Phi_L = \pi/2,$$

the angle between the momenta of the two particles after the collision is 90° in the laboratory frame.

7.3 Potential scattering

The interaction of two particles is determined by dynamics. As already stated several times, the interaction can, in turn, be considered as the motion of one particle in an external potential (cf. Section 6.1). This is why understanding the scattering of particles from a potential is also relevant to understanding the scattering of two particles[10].

There are two opposite cases to consider, depending on the type of potential – whether it is short-range or long-range. The limiting case of a short-range potential is the hard-core potential, defined by

$$V(r) = \begin{cases} \infty \\ 0 \end{cases} \text{ for } \begin{matrix} r < R \\ r \geq R \end{matrix}.$$

An example of hard-core scattering is the scattering of point particles from an infinitely hard sphere of radius R. In this case, the dynamics reduces to the laws of reflection at the surface of the sphere (E).

On the other hand, the limiting case of a long-range potential is any potential with infinite range (which does not fall off too fast when approaching infinity, e.g. does not decay exponentially). Examples are the gravitational and Coulomb potential; both have the form

$$V(r) = -k/r.$$

Although for the gravitational force the coupling $k = Gm_1m_2$ is weak, the force has *infinite range*; moreover, gravity cannot be screened. For these reasons, scattering experiments investigating only the effect of

---

[10]The following considerations are due to E. Rutherford (*The Scattering of α and β rays by Matter and the Structure of the Atom*, Philos. Mag., Series 6, vol. 21, 1911, pp. 669-688).

gravitational interaction are difficult to perform. In contrast, the **Coulomb interaction**, $k = -q_1 q_2$, between two particles with charges q_1 and q_2, is much stronger than the gravitational interaction between their masses m_1 and m_2[11]. Since both positive and negative charges exist,

i) The $1/r$ interaction can be attractive or repulsive; and,

ii) Screening is possible – therefore two-particle systems can be created in a laboratory.

In our treatment of scattering from (or, actually, within) a $1/r$ potential, the Coulomb rather than the gravitational interaction is foremost in our minds.

In scattering experiments, particularly in particle physics, it is often difficult or impossible to know the impact parameter s precisely. In practice, the properties of the interaction potential are investigated by firing a beam of particles at a target. (In a *colliding-beam* experiment, two beams collide.) The impact parameter thereby has some distribution of values. Via the concept of the scattering cross section, the number of particles detected at different directions after the collision (normalized to the number of incident particles) provides information about the interaction potential.

7.3.1   The scattering cross section

The concept of scattering cross section facilitates the description of particle scattering. The **differential (scattering) cross section** $\sigma(\Omega)$ is defined by

$$\sigma(\Omega)\, d\Omega = I_s(\Omega, d\Omega)/I_{inc}, \qquad (7.31)$$

where

---

[11]For the interaction between an electron and a proton ($q_1 = -e$ and $q_2 = e$, where e is the elementary charge), the strengths of the couplings are as follows.

$$\begin{aligned}
\text{Gravitation: } Gm_e m_p &= 6.673 \times 10^{-8} \times 9.11 \times 10^{-28} \times 1.67 \times 10^{-24} g\ cm^3\ sec^{-2} \\
&= 1.015 \times 10^{-58} g\ cm^3\ sec^{-2};
\end{aligned}$$

$$\begin{aligned}
\text{Coulomb: } e^2 &= \left(4.8 \times 10^{-10}\right)^2 g\ cm^3\ sec^{-2} \\
&= 2.3 \times 10^{-19} g\ cm^3\ sec^{-2}.
\end{aligned}$$

The Coulomb force is therefore stronger by a factor 2.27×10^{39}!

- $I_s(\Omega, d\Omega)$ is the number of the particles scattered into the solid angle element $d\Omega$ at the solid angle Ω per unit time; and,

- I_{inc} is the incident flux, the number of particles traversing a unit area per unit time.

The cross section σ has therefore the physical dimension of area.

The scattering of particles from a *central force* is spatially symmetric with respect to the incident beam. The element of solid angle,

$$d\Omega = \sin\Theta\, d\Theta\, d\Phi,$$

can be replaced by the conic element (see Fig. 7.10),

$$d\Omega \to 2\pi\sin\Theta\, d\Theta,$$

where Θ is the angle between the asymptotic direction of motion of the scattered particle and the axis of the incident beam (in the laboratory frame)[12]. The quantity $I_s(\Theta, d\Theta)$ is then the number of particles scattered per unit time into this conic element; so instead of Eq. (7.31), we can write

$$2\pi\sigma(\Theta)\sin\Theta\, d\Theta = I_s(\Theta, d\Theta)/I_{inc}. \tag{7.32}$$

Determination of $I_s(\Theta, d\Theta)/I_{inc}$

We still have to find an explicit expression for $\sigma(\Theta)$ in terms of the parameters of the system, e.g. E and s. This expression can be obtained using the fact that in elastic scattering, the number of particles entering the potential has to be equal the number of particle leaving it.

The orbits of the particles, and thus the scattering angle Θ, are functions of the parameters s and the energy E (see, for example, Eq. (7.8)). The number of particles entering the potential through an imaginary annulus s and $s + ds$ (see Fig. 7.10) must equal the number of particles $I_s(\Theta, d\Theta)$ that leave the potential through the conic element between Θ and $\Theta - d\Theta$ per unit time. Let Θ be the scattering angle of a particle entering asymptotically at an impact parameter s, then the

---

[12]In the notation of the previous section, Θ is the scattering angle in the laboratory frame, $\Theta = \theta_L$. In the scattering from a potential (corresponding to an infinite mass of the target particle), $\theta_L = \theta$ (the scattering angle in the center of mass frame), so that $\Theta = \theta$, too.

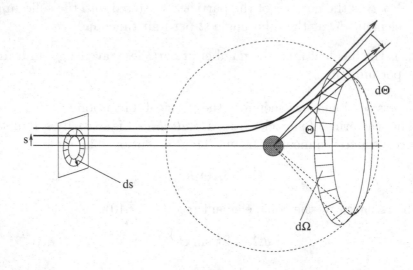

Figure 7.10: Scattering of particles from a potential.

angle $\Theta - d\Theta$ is associated with a particle whose impact parameters is $s+ds$. Since the area of the annulus is $2\pi s\,ds$ (note that the corresponding solid angle is $-2\pi \sin\Theta\,d\Theta$), the balance in the number of particles can be written as

$$I_s(\Theta, d\Theta) = I_{inc}2\pi s\,ds. \tag{7.33}$$

In this derivation, we have assumed that the relation between s and Θ is unique for the energy E considered: different values of s are associated with different values of Θ. Inserting this equation into the definition of the cross section σ, we obtain

$$2\pi\sigma(\Theta)\sin\Theta\,|d\Theta| = \frac{I_{inc}2\pi s\,|ds|}{I_{inc}} = 2\pi s\,|ds|.$$

The absolute value takes care of the opposite sign of the increments ds and $d\Theta$ (the number of particles has to be positive). Thus the scattering cross section is given by

$$\sigma(\Theta) = \frac{s}{\sin\Theta}\left|\frac{ds}{d\Theta}\right|. \tag{7.34}$$

For fixed angle Θ, the cross section still depends on the energy, so strictly speaking, $\sigma(\Theta) = \sigma(\Theta, E)$.

The **total scattering cross section** σ_{tot} is obtained from the differential cross section σ by integrating over the entire solid angle:

$$\sigma_{tot} = \int \sigma(\Omega)\, d\Omega. \tag{7.35}$$

For a cylindrically symmetric scattering process, this reduces to

$$\sigma_{tot} = 2\pi \int_0^\pi \sigma(\Theta) \sin\Theta\, d\Theta. \tag{7.36}$$

In the case of a finite-range potential, the total cross section is finite and gives the effective area of the potential[13]. For example, when point masses are incident onto a hard sphere, σ_{tot} is the cross section of the sphere – only particles incident within that area are deflected (**E**).

7.3.2 Scattering in the $1/r$ potential

In Section 5.2.1, we found that the orbits of particles with energy $E > 0$ in the potential $V(r) = -km/r$ are hyperbolas. The equation of the orbit in polar coordinates is (cf. Eq. (5.34))

$$r = \frac{p}{\mathrm{sgn}(k) + \varepsilon \cos(\varphi - \varphi_K)}, \tag{7.37}$$

where the semilatus rectum p and the eccentricity ε are related to the energy E and to the angular momentum L via Eqs. (5.32) and (5.33). Since, for a hyperbola, $\varepsilon > 1$, the denominator can vanish for certain values of φ. These angles are the asymptotic angles φ_1 and φ_2. In the following, we restrict ourselves to a repulsive potential, i.e. $k < 0$. This specifies the equation of the orbit as that given in Eq. (5.44):

$$r = \frac{p}{\varepsilon \cos(\varphi - \varphi_K) - 1}.$$

The angles of the asymptotes φ_A, $A = 1, 2$, are the solutions of the equation

$$\cos(\varphi_A - \varphi_K) = 1/\varepsilon. \tag{7.38}$$

Comparing the orbits for different values of ε and p, one observes that the asymptotes have the following property (see also Section 5.2). The

---

[13]This is actually the definition of a finite-range potential.

closer the turning point $(r_{\min} = p/(\varepsilon - 1), \varphi_{\min} = \varphi_K)$ to the center of the force (origin) – i.e. the deeper the particle enters the potential – the stronger is the deviation from the asymptotic incident direction, and the smaller is the angle between the asymptotes (between the incident and the deflected direction; cf. Fig. 7.11 (a)).

In a scattering experiment, the incident direction (the beam axis) is fixed. Only the impact parameter s – consequently the angle $\varphi_K = \varphi_{\min}$ and the direction of the scattered particle – varies (see Fig. 7.11 (b)). We already know the impact parameter as a function of ε and p from Section 5.2, since s is the distance between the center of the force and the asymptote. From Eq. (5.43), we have

$$s = \frac{p}{\sqrt{\varepsilon^2 - 1}}. \tag{7.39}$$

A particle is deflected symmetrically with respect to φ_K from its in-

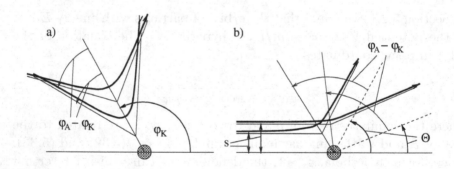

Figure 7.11: The asymptotes of two orbits at (a) fixed angle φ_K; and at (b) fixed incident direction.

cident direction φ_1. Thus the scattering angle (see Eq. (7.7) and Fig. 7.11 (b)) is

$$\Theta = \pi - 2(\varphi_1 - \varphi_K).$$

Consequently, we have from Eq. (7.38),

$$\cos(\varphi_1 - \varphi_K) = \cos\left(\frac{\pi - \Theta}{2}\right) = \sin\frac{\Theta}{2} = 1/\varepsilon.$$

Hence, Eq. (7.39) yields the relation between the impact parameter and the scattering angle:

$$s = p \tan \frac{\Theta}{2}. \tag{7.40}$$

Now we only have to express p given in Eq. (5.32) in terms of s and v_∞ (or E). Since the potential is extremely long-ranged ($R = \infty$), the asymptotic velocity v_∞ is difficult to define. But if we consider some point which is 'sufficiently' distant from the center of the force, we can let $v \simeq v_\infty$ and $E \simeq T$, so that again (cf. Eq. (7.5)), $E \simeq mv_\infty^2/2$ and $L = mv_\infty s$. If we insert these expressions into Eqs. (5.32) and (5.33), we find for the semilatus rectum p and the eccentricity ε,

$$p = \frac{v_\infty^2 s^2}{|k|} \tag{7.41}$$

$$\varepsilon = \sqrt{1 + v_\infty^4 s^2/k^2}. \tag{7.42}$$

For Eq. (7.40), we then have

$$s = \frac{|k|}{v_\infty^2} \cot \frac{\Theta}{2} = \frac{m|k|}{2E} \cot \frac{\Theta}{2}, \tag{7.43}$$

which finally yields

$$\left| \frac{ds}{d\Theta} \right| = \frac{|k|}{2v_\infty^2} \frac{1}{\sin^2 \Theta/2}. \tag{7.44}$$

From Eq. (7.34) we obtain for the differential cross section,

$$\sigma(\Theta) = \frac{k^2}{4v_\infty^4} \frac{1}{\sin^4 \Theta/2} = \frac{m^2 k^2}{16E^2} \frac{1}{\sin^4 \Theta/2}. \tag{7.45}$$

This is the famous **Rutherford scattering formula** (E. Rutherford, 1871-1937) for scattering of particles from a Coulomb potential[14]. The differential cross section diverges in the forward direction ($\Theta = 0$) because of the infinite range of the potential. This is true for all potentials of infinite range. For increasing values of the energy, the particles are preferentially scattered close to the forward direction, since $\sigma(\Theta)$ tends to zero, except for $\Theta \cong 0$. Divergence of $\sigma(\Theta)$ in the forward direction implies the divergence of the total cross section (7.36):

$$\sigma_{tot} \propto \int_0^\pi \frac{1}{\sin^4 \Theta/2} \sin \Theta \, d\Theta = 4 \int_0^1 \frac{1}{\sin^3 \Theta/2} d\sin \Theta/2 = \infty.$$

---

[14]See Footnote 10.

Remarks

i) The Rutherford scattering cross section does not depend on the sign of k; it is therefore equal for the attractive case as well as the repulsive case of an interaction.

ii) If the potential is strongly attractive ($1/r^\alpha$, $\alpha \geq 2$), the orbit can pass close to the singularity at the center of the force. In this case, problems in determining the scattering angle occur. A more realistic potential with a hard core avoids this problem.

Problems and examples

1. Derive the differential equation for the orbit of a particle in a central force using the variable $u = u(\varphi) = 1/r$.

2. Let $m_{\text{projectile}} > m_{\text{target}}$. Calculate the maximal scattering angle, defined in Eq. (7.28), in the laboratory frame and the center of mass frame.

3. Deduce the relation between s and Θ, given in Eq. (7.40), from Eq. (7.8).

4. Point masses are scattered by a fixed hard sphere (radius R). Calculate the differential and the total cross section. Hint: To establish $s(\Theta)$ it is necessary to investigate the reflection of a particle at the surface of the sphere (circle).

5. **Reconstruction of a potential $V(r)$ from the cross section.** If the differential scattering cross section $\sigma(\Theta)$ is known experimentally, the Θ-dependence in $s(\Theta)$ can be determined from $\sigma(\Theta)$, and subsequently one can draw conclusions about the potential ([Landau/Lifschitz], vol. I).

   The first relation is obtained from Eq. (7.34). Since the impact parameter has to be zero for back-scattered particles, $s(\Theta = \pi) = 0$, integrating Eq. (7.34) yields

$$2\pi \int_{\Theta}^{\pi} \sigma(\Theta') \sin \Theta' \, d\Theta' = \pi \, s^2(\Theta), \qquad (7.46)$$

where the modulus in Eq. (7.34) has been taken into account. Thus $s(\Theta)$, as well as $\Theta(s)$, can be determined from the scattering data. The second relation we need is Eq. (7.8),

$$\frac{\pi - \Theta}{2} = \int_{\infty}^{r_{min}} \frac{1}{r'^2} \frac{dr'}{\left((1 - V(r')/E)/s^2(\Theta) - 1/r'^2\right)^{1/2}}.$$

Inserting in the denominator the calculated impact parameter $s(\Theta)$ (cf. Eq. (7.46)), one obtains an integral equation for the potential $V(r)$. Now, $V(r)$ is present here in a very indirect way: it appears only in the integrand; also, the upper limit r_{min} of the integral depends on $V(r)$ via an integral. However, rewriting this equation (see loc.cit.), one can obtain an integral equation more suitable for the determination of $V(r)$; namely,

$$W = \exp\left(\frac{1}{\pi} \int_{rW}^{\infty} \frac{\Theta(s)}{\sqrt{s^2 - r^2 W^2}} ds\right), \quad W = \sqrt{1 - \frac{V}{E}}.$$

6. The frame of reference in the discussion of the potential scattering was the laboratory frame; i.e. in the notation of Section 7.2.2, $\Theta = \theta_L$ and σ is to be replaced by σ_L. (We continue to denote by σ the cross section in the center of mass frame.) When the masses of the colliding particles are similar enough (e.g. in colliding-beam experiments), the dynamics is considered in the center of mass frame. What is the connection between the cross sections in the two frames?

We deduce this relation using the fact that the number of particles scattered into a given element of solid angle is independent of the frame:

$$2\pi I_{inc}\, \sigma(\theta) \sin\theta\, |d\theta| = 2\pi I_{inc}\, \sigma_L(\theta_L) \sin\theta_L\, |d\theta_L|, \qquad \theta = \theta(\theta_L).$$

Therefore

$$\sigma_L(\theta_L) = \sigma(\theta)\frac{\sin\theta}{\sin\theta_L}\left|\frac{d\theta}{d\theta_L}\right| = \sigma(\theta)\frac{d\cos\theta}{d\cos\theta_L}.$$

From Eq. (7.25), one finds

$$\cos\theta_L = \left(\frac{m_1}{m_2} + \cos\theta\right)\bigg/\left[\left(\frac{m_1}{m_2}\right)^2 + 2\frac{m_1}{m_2}\cos\theta + 1\right]^{1/2}.$$

Hence one can calculate $|d\cos\theta/d\cos\theta_L|$. One then obtains,

$$\sigma_L(\theta_L) = \sigma(\theta) \left[\left(\frac{m_1}{m_2}\right)^2 + 2\frac{m_1}{m_2}\cos\theta + 1 \right]^{3/2} \bigg/ \left(1 + \frac{m_1}{m_2}\cos\theta\right).$$

If $m_2 \gg m_1$, the scattering angles are equal, $\theta_L = \theta$ (see Eq. (7.29)), and the two cross sections are also equal,

$$\sigma_L(\theta_L) = \sigma(\theta) = \sigma(\theta_L).$$

If $m_2 = m_1$, we have $\theta_L = \theta/2$ (cf. Eq. (7.30)); then

$$\sigma_L(\theta_L) = 4\sigma(\theta)\cos\frac{\theta}{2} = 4\sigma(2\theta_L)\cos\theta_L.$$

Even if the cross section in one of the frames were constant, it would display an angular dependence in the other frame.

The total cross sections in both frames must, of course, be equal:

$$\int \sigma_L(\theta_L)\,d\cos\theta_L = \int \sigma(\theta)d\cos\theta.$$

8

Changing the frame of reference

In our treatment of physical processes so far, we have always tacitly chosen some particular **point of reference**. The positions of physicals events – for instance, the instantaneous location of a point particle – have been referred to a point of reference that coincides with the origin of the coordinate system. Until now, we have not given much thought to how the choice may influence the result of an observation or the mathematical form of physical laws. Only when considering collisions and scattering, we distinguished between kinematics in the laboratory frame and in the center of mass frame – and found that the scattering angles and also the differential cross sections differ in these two frames. We give two further examples.

In Kepler's laws, the sun is clearly the point of reference. As seen from the sun, the planets move on elliptical (nearly circular) orbits without changing their direction of motion. Observing the motion of the planets from earth gives a completely different picture. In particular, retrograde motion of planets is observed. The ancient as well the medieval descriptions of the orbits in the geocentric system required numerous cycles and epicycles. The second example is an airplane performing a dive. To an observer in the plane, the objects in the plane float weightlessly, whereas an observer on the ground would see the objects as falling together with the plane: zero gravity is 'simulated' inside the plane.

To fix the positions and time intervals in a physical process, a **frame**

of reference is used. Conceptually, a frame of reference comprises of a point of reference[1], together with devices for determining the position of some event with respect to the point of reference, and appropriately installed clocks for determining the time of occurrence of the event at any position[2]. The mathematical counterpart of such a reference frame is a coordinate system (origin and axes) and a time scale, which together allow us to describe, for example, the motion of a particle at any time – its position $\mathbf{r}$ at time t, as well as its rate of change with time by $\dot{\mathbf{r}}$ and $\ddot{\mathbf{r}}$. In any particular reference frame, different coordinate systems may be chosen (e.g. Cartesian or spherical coordinates).

When relating the observations of physical quantities in two different frames of reference, one has to state how to compare them: what do the views in the two systems have in common? A *fundamental assumption* of nonrelativistic, classical physics enters here: that *the distance between two physically distinguished positions* – e.g. between two bodies – *is independent of the reference frame*[3]. However, changing the reference frame may also affect physical quantities, as in the example of particle collision the scattering angle or the velocity observed in the center of mass frame and in the laboratory frame. A change of coordinate system within a given frame, has, of course, no influence on the physical processes; only the (mathematical) description is affected by the choice[4]. The coordinate system is often considered to be the refer-

---

[1]The specific choice of frame is linked to some other physical fact. For instance, for the motion (translational or rotational) of a single body, it is indispensable to think of other bodies (in general, physical objects). Motion can only occur with respect to other bodies (physical objects): a body alone cannot move. There does not exist Newtonian absolute, abstract space.

[2]A theorist's conception of such a measuring device may be represented by a 3-dimensional lattice of infinitely long, extremely thin rigid rods, each with scales and synchronized clocks at regular intervals. A nice picture of such a device is shown in the book by E.F. Taylor and J.A. Wheeler, Space-time Physics, Freeman & Comp., San Francisco, 1966.

[3]Here, Newton's concept of absolute time is essential, since the distance between two such points is measured in all reference frames at the same (absolute) time.

[4]We frequently choose a Cartesian coordinate system. But it may be advantageous to use different (e.g. curvilinear) coordinates.

The statement that the orbit of a particle is an ellipse is independent of the coordinates chosen in a given frame of reference; but the equation for the ellipse is quite different when using Cartesian or polar coordinates. Changing the frame of reference – for example, going over to a rotating frame – the orbit might not appear like an ellipse anymore.

ence frame, but strictly speaking one should distinguish between these two notions. Of course, a change of reference frame is accompanied by a coordinate transformation (but not, in general, vice versa). In a certain sense, one may say that the concept of a reference frame emphasizes the physical world view, while the concept of coordinate systems emphasizes the mathematical point of view.

8.1 Inertial frames

In view of the possible dependence of physical laws on the frame of reference, it is natural to ask: given a particular frame of reference, do there exist 'equivalent' reference frames, in which a *preferred* physical law is the same as in that frame?

The fundamental law in nonrelativistic mechanics is Newton's equation of motion for a particle influenced by a force: $m\ddot{\mathbf{r}} = \mathbf{F}$. In Newton's first law, free motion, i.e. $\mathbf{F} = \mathbf{0}$, is selected as the basic case, relative to which – by the second law – the presence of forces can be assessed. Thus the distinguished role of uniform and rectilinear motion is established. In former times the view was advanced that every motion needs a cause. Also to maintain free motion of a body there has to be a permanent push. Later on, instead of this view, free motion was explained by the property of **inertia**[5] of a body.

Returning to the question of reference frames: Indeed, an important class of reference frames is selected by the requirement that in each of

---

[5]Kepler coined the notion of inertia, but his inertia brought a body to rest (Epitome Astronomiae Copernicanae, Linz 1620, Book IV, Second Part, Section II: On the causes of the planetary motion; see also Kepler's note 76 in his Somnium (The dream), opus posthumum, Sagan 1634).

Descartes formulated the first version of the principle of inertia in the **first law of nature**: 'Each thing when left to itself continues in the same state; hence any thing, which started moving goes on moving' and he added the **second law of nature**: 'Each motion by itself is rectilinear; ...'(Principia Philosophiae, Amsterdam 1644, Second Part, 37).

Also Galileo had a notion of inertia, but not as clear as Descartes's. So, for instance, in his "Dialogo ... sopra i due massimi sistemi del mondo ..." (Dialogue concerning the two chief World Systems: Ptolemaic and Copernican; Florence 1632) the Second day Simplicio and Salviati discuss whether the motion of a ball on a horizontal plane will be perpetual. Another aspect of that topic in the "Dialogue" is touched in footnote 16.

the frames, motion – in the absence of a force – obeys the equation

$$m\ddot{\mathbf{r}} = \mathbf{0}. \tag{8.1}$$

In each frame free motion is uniform and rectilinear. Such frames are called **inertial frames**. In 1886, L. Lange proposed[6] a procedure for finding out whether a given reference frame is an inertial frame:

> *A reference frame in which a mass point thrown from the same point in three different (non-coplanar) directions follows rectilinear paths each time it is thrown, is called inertial frame.*

8.2   Changing the inertial frame

We consider two frames of reference, I and I'. Each frame includes an origin, O and O', respectively, and a set of orthogonal basis vectors[7], $\{\mathbf{e}_i\}$ and $\{\mathbf{e}'_i\}$ $(i = 1, 2, 3)$, respectively. In each frame the basis vectors generate the coordinate axes of a Cartesian coordinate system. According to Appendix (A.2) any radius vector $\mathbf{r}$ in I is a linear combination of the basis vectors[8]

$$\mathbf{r} = x_i\mathbf{e}_i,$$

where the coefficients $x_i = \mathbf{r}\mathbf{e}_i$ are the Cartesian components of the radius vector $\mathbf{r}$. Similarly, for a radius vector $\mathbf{r}'$ in I' holds

$$\mathbf{r}' = x'_i\mathbf{e}'_i.$$

Considering, the velocity vector and acceleration vector in a frame of reference, for instance $\dot{\mathbf{r}}$ and $\ddot{\mathbf{r}}$, we have (cf. Appendix A)

$$\dot{\mathbf{r}} = \dot{x}_i\mathbf{e}_i \qquad \text{and} \qquad \ddot{\mathbf{r}} = \ddot{x}_i\mathbf{e}_i. \tag{8.2}$$

---

[6]Cited by M. v. Laue in his book "Die spezielle Relativitätstheorie" (Vieweg, Braunschweig 1955).

[7]The basis vectors are kind of a mathematical image of measuring rods of unit length.

[8]From now on, unless otherwise noted, we use Einstein's summation convention: i.e., one sums over repeated indices:

$$x_i\mathbf{e}_i := \sum_{i=1}^{3} x_i\mathbf{e}_i.$$

(Since the time derivatives are taken within the frame of reference there is no time dependence of the basis vectors.)

Assume that the vectors $\mathbf{r}$ and $\mathbf{r}'$ are the respective radius vectors of a particle. If the origins O and O' of the two frames differ, then $\mathbf{r}$ is different from $\mathbf{r}'$. If, however, one considers two distinct physical points in the two frames of reference (at the same time), i.e. the points $\mathbf{r}_1$ and $\mathbf{r}_2$ in I correspond to $\mathbf{r}'_1$ and $\mathbf{r}'_2$ in I', the vectors $\mathbf{d}$ and $\mathbf{d}'$ between these two points are the same:

$$\mathbf{d} = \mathbf{r}_2 - \mathbf{r}_1 = \mathbf{r}'_2 - \mathbf{r}'_1 = \mathbf{d}'. \tag{8.3}$$

When changing the reference frame it may be important whether a vector depends on the origin or not.

If we suppose now that I and I' are inertial frames, then the motion of a free particle is described by an equation of motion in the form (8.1). Thus, projecting the acceleration vector $\ddot{\mathbf{r}}$ in I onto the basis vectors, the components of Newton's equation of free motion (8.1) in the inertial frame I are

$$m\ddot{x}_i = 0, \qquad i = 1, 2, 3. \tag{8.4}$$

Similarly, in the inertial frame I' free motion in terms of Cartesian components of $\ddot{\mathbf{r}}' = \ddot{x}'_i \mathbf{e}'_i$ is described by

$$m\ddot{x}'_i = 0, \qquad i = 1, 2, 3.$$

In consideration of the equivalence of inertial frames, we ask: What is the class of transformations leading from $\ddot{x}_i = 0$ to $\ddot{x}'_i = 0$? Which transformations leave (the form of) Newton's equation of motion invariant?

Remark

As customary, we often designate also the three-tuple of components by the symbol $\mathbf{r}$,

$$\mathbf{r} = (x, y, z) = (x_1, x_2, x_3),$$

but this may sometimes lead to confusion[9].

---

[9]Evidently, relation (8.3) is not fulfilled by $\mathbf{d}$ defined as three-tuple; we have

$$\mathbf{d} = (d_1, d_2, d_3) \neq (d'_1, d'_2, d'_3) = \mathbf{d}'.$$

However, the quadratic form $d_1^2 + d_2^2 + d_3^2$, the distance between the points, is invariant:

$$d_i d_i = d'_i d'_i.$$

8.3 Linear transformations of the coordinates

Equations (8.4) are invariant under general linear transformations[10]

$$x_i' = D_{ik}x_k + s_i, \qquad (8.5)$$

with D_{ik} and s_i denoting constants, since Newton's equation – written in terms of the new components x_i' – retains its form[11]:

$$m\ddot{x}_i' = 0, \qquad i = 1, 2, 3.$$

The class of *linear transformations* (8.5) is too general for our purposes. We consider only transformations that map the Cartesian components (x_1, x_2, x_3) of a vector $\mathbf{r}$ in the system I onto the components (x_1', x_2', x_3') in another system I' (in the following I and I' are not necessarily inertial systems). If this change of basis and origin does not affect a vector, as in the case of the vector $\mathbf{d}$ above (see (8.3); it only affects its components), we have the condition

$$\mathbf{d} = d_i\mathbf{e}_i = d_i'\mathbf{e}_i'. \qquad (8.6)$$

Since the coordinate system is transformed and $\mathbf{d}$ is fixed, the components of $\mathbf{d}$ are said to be **transformed passively**[12]. The condition expressed in Eq. (8.6) places a restriction on the coefficients D_{ik}, as is shown in Appendix B.

8.3.1 Translation of the coordinate system

Setting $D_{ik} = \delta_{ik}$ and focussing on the parameters s_i, Eq. (8.5) reduces to the simple transformation

$$x_i' = x_i + s_i, \qquad s_i = const.$$

The component representation of a vector emphasizes the transformational aspect. This point of view of vectors (and tensors) is taken in the theory of relativity. There, transformations that leave a particular quadratic form invariant, are fundamental.

Occasionally, one has to state in which sense the concept 'vector' is used, $\mathbf{d} = d_i\mathbf{e}_i$ or $\mathbf{d} = (d_1, d_2, d_3)$. Whenever the distinction is important, we will stress this explicitly.

[10]Recall that $D_{ik}x_k := \sum_{k=1}^{3} D_{ik}x_k$.

[11]This can be seen by taking the second time derivative of Eq. (8.5) and multiplying Eq. (8.4) by D_{ik}.

[12]If the coordinate system is fixed and the vector is transformed, then the transformation is called an **active transformation**.

Interpreting this transformation in terms of coordinate systems I and I', the origin O' differs from the origin O by a constant vector, $\mathbf{s} = s_i \mathbf{e}_i$. This **translation** affects the radius vectors,

$$\mathbf{r}' = \mathbf{r} + \mathbf{s}, \tag{8.7}$$

while leaving vectors that do not depend on the origin – such as $\mathbf{d}$ (cf. (8.3)) or a force $\mathbf{F}$ – invariant. Since $\dot{\mathbf{s}} = 0$, the velocity and acceleration remain invariant,

$$\dot{\mathbf{r}}' = \dot{\mathbf{r}}, \qquad \ddot{\mathbf{r}}' = \ddot{\mathbf{r}}.$$

(This is because velocity and acceleration are defined as (infinitesimal) differences of vectors, which remain invariant.) Hence, *Newton's equation of motion (8.1) is invariant under time-independent translations of the origin.*

8.3.2 Rotation of the coordinate system

Next we consider linear transformations (8.5), with $s_i = 0$, $i = 1, 2, 3$:

$$x'_j = D_{ji} x_i, \qquad D_{ji} = const. \tag{8.8}$$

Again, we consider Eq. (8.8) to define a relation between the components of the radius vector of a particular point in space in the two coordinate systems I and I' with common origin ($O = O'$). Since the origin remains unchanged, and the radius vector is directed towards a physically distinct point whose position is independent of the choice of the coordinate system, we have

$$x_i \mathbf{e}_i = x'_i \mathbf{e}'_i;$$

therefore the length of the radius vector remains unchanged,

$$x_i x_i = x'_i x'_i. \tag{8.9}$$

Consequently, as shown in Appendix B (Section B.1), the constants D_{ij} are the elements of an orthogonal matrix $\mathbf{D}$,

$$\mathbf{D}^T \mathbf{D} = 1, \quad \text{i.e. } \mathbf{D}^{-1} = \mathbf{D}^T, \tag{8.10}$$

which implies that we can write Eq. (8.8) in the form[13]

$$\mathbf{r}' = \mathbf{D} \mathbf{r}. \tag{8.11}$$

---

[13]The behavior (8.8) under rotations of the coordinate system characterizes a vector; for a vector quantity $\mathbf{V}$ (e.g. a force $\mathbf{F}$), $V'_j = D_{ji} V_i$ holds (cf. Section B.2).

Here $\mathbf{r}$ and $\mathbf{r}'$ are to be taken as three-tuples.

The orthogonality condition (8.10) gives only six independent conditions for the nine components of D. Therefore, three quantities can be chosen freely. The transformation is basically[14] a **rotation** of the basis $\{\mathbf{e}_i'\}$ with respect to the basis $\{\mathbf{e}_i\}$.

The three rotational degrees of freedom can be parametrized, for example, by the unit vector $\mathbf{n}$ pointing along the rotation axis (corresponding to two quantities), together with the angle of rotation ϕ about this axis. The matrix D has a particularly simple form for rotations about a coordinate axis; say, the z-axis. If the rotation angle is denoted by ϕ, then (cf. Eq. (B.13))

$$\mathsf{D}_z(\varphi) = \begin{pmatrix} \cos\varphi & \sin\varphi & 0 \\ -\sin\varphi & \cos\varphi & 0 \\ 0 & 0 & 1 \end{pmatrix}. \tag{8.12}$$

For arbitrary rotation axis, $\mathbf{r}' = \mathsf{D}\mathbf{r}$ can be written in the form[15]

$$\mathbf{r}' = \mathbf{r}\cos\phi + \mathbf{n}(\mathbf{nr})(1 - \cos\phi) - (\mathbf{n}\times\mathbf{r})\sin\phi. \tag{8.13}$$

In this equation, if $\mathbf{r}$ and $\mathbf{r}'$ are considered as the three-tuples of the respective components of the same vector represented in two coordinate systems, I and I', then the rotation of a coordinate system is called a *passive rotation*. In this case the basis vectors are rotated by the angle ϕ about the axis $\mathbf{n}$. But $\mathbf{r}$ and $\mathbf{r}'$ can also taken to be vectors (not just the components); then the transformation (8.13) is an *active* rotation of the vector $\mathbf{r}$ into the vector $\mathbf{r}'$ while keeping the basis fixed (cf. also Fig. 8.2).

---

[14]Actually, the transformation (8.8) may include also reflection or permutation of the coordinates (basis vectors). However, we do not consider such transformations here.

[15]One can easily proof the following properties of the transformation given in Eq. (8.13):

i) $\phi = 0$ is the identity transformation $\mathbf{1}$;

ii) If $\mathbf{r} = \mathbf{n}$, then $\mathbf{r}' = \mathbf{n}$ too. The rotation axis does not change under the transformation;

iii) Rotation by $-\phi$ is the inverse rotation;

iv) $\mathbf{r}'^2 = \mathbf{r}^2$.

Euler's angles

Another choice for the parameters of a rotation are the three **Euler's angles** φ, ϑ, and ψ. We adopt the following definition. The transformation from I to I' is achieved by the three rotations shown in Fig. 8.1.

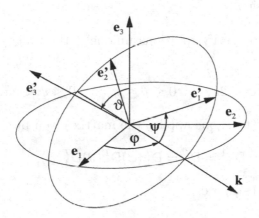

Figure 8.1: Euler's angles.

The first rotation is about the z-axis (in system I) by angle φ (cf. Eq. (8.12)):

$$D_1(\varphi) = \begin{pmatrix} \cos\varphi & \sin\varphi & 0 \\ -\sin\varphi & \cos\varphi & 0 \\ 0 & 0 & 1 \end{pmatrix}. \tag{8.14}$$

The second rotation is about the new x-axis, shown as **k** in Fig. 8.1, by angle ϑ:

$$D_2(\vartheta) = \begin{pmatrix} 1 & 0 & 0 \\ 0 & \cos\vartheta & \sin\vartheta \\ 0 & -\sin\vartheta & \cos\vartheta \end{pmatrix}. \tag{8.15}$$

The third rotation is about the new z-axis, by angle ψ:

$$D_3(\psi) = \begin{pmatrix} \cos\psi & \sin\psi & 0 \\ -\sin\psi & \cos\psi & 0 \\ 0 & 0 & 1 \end{pmatrix}. \tag{8.16}$$

The total rotation is equivalent to the product matrix (note the order of multiplication!)

$$D = D(\psi, \vartheta, \varphi) = D_3(\psi)D_2(\vartheta)D_1(\varphi) =$$

$$\begin{pmatrix} \cos\psi\cos\varphi - \sin\psi\cos\vartheta\sin\varphi & \cos\psi\sin\varphi + \sin\psi\cos\vartheta\cos\varphi & \sin\psi\sin\vartheta \\ -\sin\psi\cos\varphi - \cos\psi\cos\vartheta\sin\varphi & -\sin\psi\sin\varphi + \cos\psi\cos\vartheta\cos\varphi & \cos\psi\sin\vartheta \\ \sin\vartheta\sin\varphi & -\sin\vartheta\cos\varphi & \cos\vartheta \end{pmatrix},$$

(8.17)

where, in order to specify rotations uniquely, the ranges of the angles are fixed by

$$0 \le \varphi < 2\pi, \qquad 0 \le \vartheta \le \pi, \qquad 0 \le \psi < 2\pi. \tag{8.18}$$

To return from I' to I, one applies the matrix given by

$$D^{-1} = D^T = D_1^T(\varphi)D_2^T(\vartheta)D_3^T(\psi), \tag{8.19}$$

with $D_1^T(\varphi) = D_1(-\varphi)$ etc.

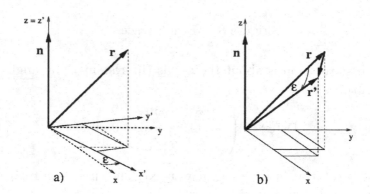

Figure 8.2: Infinitesimal rotations about the z-axis: a) passive rotation, b) active rotation.

Infinitesimal rotations

For a very small rotation angle $\phi = \varepsilon$, to first order in ε, Eq. (8.13) reduces to

$$\mathbf{r}' = \mathbf{r} - \varepsilon(\mathbf{n} \times \mathbf{r}). \tag{8.20}$$

The inverse of this relation is

$$\mathbf{r} = \mathbf{r}' + \varepsilon(\mathbf{n} \times \mathbf{r}'). \tag{8.21}$$

Since the triple product $\mathbf{r}(\mathbf{n} \times \mathbf{r})$ vanishes in the order of ε chosen, we have

$$(\mathbf{r}')^2 = \mathbf{r}^2 \quad (+\, \mathcal{O}\left(\varepsilon^2\right)).$$

We consider an infinitesimal rotation of the radius vector about the z-axis, specifying $\mathbf{n} = (0,0,1)$ in Eq. (8.20),

$$\begin{aligned} x' &= x + \varepsilon y \\ y' &= y - \varepsilon x \\ z' &= z. \end{aligned} \tag{8.22}$$

Figure 8.2 (a) shows passive rotation of the coordinate system, whereas in Fig. 8.2 (b) active rotation of the radius vector is represented. Both transformations correspond to the same relation (8.22) between the components of the radius vector. Note that while the angle of rotation ε is the same in both Figs. 8.2 (a) and (b), the direction of rotation of the respective objects (the x and y-axis in (a) and the vector in (b)) is different.

8.4 The Galilean group

We found that the equation of motion (8.1) remains invariant if the following time-independent transformations between the two frames (coordinate systems) I and I' are applied:

  i) Translations of the coordinate system

$$\mathbf{r}' = \mathbf{r} + \mathbf{s}, \qquad \dot{\mathbf{s}} = 0; \tag{8.23}$$

 ii) Rotations of the coordinate system

$$\mathbf{r}' = \mathbf{D}\mathbf{r}, \qquad \dot{\mathbf{D}} = 0. \tag{8.24}$$

Invariance of physical systems under these two transformations is often referred to as the **homogeneity** and **isotropy** of (free) space, respectively.

Two further classes of transformations between frames I and I' that leave Eq. (8.1) invariant, are the following:

iii) Velocity transformations or boosts with constant velocity (also called **special Galilean transformations**[16]),

$$\mathbf{r}' = \mathbf{r} + \mathbf{w}t, \qquad \dot{\mathbf{w}} = \mathbf{0}; \tag{8.25}$$

iv) Translations of the time origin ('**homogeneity of time**'),

$$t' = t + \tau. \tag{8.26}$$

In both cases force-free motion in frame I, $m\ddot{\mathbf{r}} = \mathbf{0}$, is also force-free in frame I', $m\ddot{\mathbf{r}}' = \mathbf{0}$.

Hence, the most general transformation leaving $m\ddot{\mathbf{r}} = \mathbf{0}$ invariant is the **Galilean transformation**

$$\mathbf{r}' = \mathbf{Dr} + \mathbf{w}t + \mathbf{s} \tag{8.27}$$
$$t' = t + \tau. \tag{8.28}$$

Transformations with $\det \mathbf{D} = 1$ form the **proper Galilean group**, with ten parameters $(\mathbf{n}, \phi, \mathbf{s}, \mathbf{w}, \tau)$ (**E**). All reference frames that are 'reachable' from a given inertial frame by application of a Galilean transformation are inertial frames. It is easy to see that the inverse transformation is (remember: $\mathbf{D}^{-1} = \mathbf{D}^T$)

$$\mathbf{r} = \mathbf{D}^T(\mathbf{r}' - \mathbf{w}(t' - \tau) - \mathbf{s}), \qquad t = t' - \tau,$$

or, written in the form above,

$$\mathbf{r} = \mathbf{D}^T\mathbf{r}' + \mathbf{w}'t' + \mathbf{s}',$$
$$t = t' + \tau',$$

with

$$\mathbf{w}' = -\mathbf{D}^T\mathbf{w}, \qquad \mathbf{s}' = -\mathbf{D}^T(\mathbf{s} - \mathbf{w}s), \qquad \tau' = -\tau.$$

---

[16]In his "Dialogue" Galileo presents the dialogue of three persons. Four days Sagredo, Simplicio (rather simple-minded), and Salviati (representing Galileo's views) discuss mainly on the motion on earth and in the planetary system. On the second day they also talk about a stone falling down from the top of a mast on a sailing ship. Will it come down at the foot of the mast or in some distance from it depending on the speed of the ship, since the ship is perceived to have advanced during the time of fall? For Simplicio the stone will be left behind, whereas Salviati asserts that the stone will come down at the foot of the mast, whether the ship is standing still or moving. (What is the view of an observer on the shore?)

The group of Galilean transformations is closely connected to the concepts of absolute time and absolute space in Newtonian mechanics. In Section 10.4, the consequences of the invariance of a dynamical system with respect to Galilean transformations will be elucidated[17].

8.4.1 Transformation of forces

We now investigate the effects of a Galilean transformation applied to a force field. A time-independent force is only affected by rotations. In particular, the *components* of a constant (homogeneous) force $\mathbf{F}$ transform as follows (cf. Eq. (8.8)):

$$F_i' = D_{ik}F_k, \qquad (8.29)$$

or, in matrix notation,

$$\mathbf{F}' = \mathbf{DF}.$$

Note that, once again, the change involves only the three-tuple of components, not the vector.

Suppose now that the force depends on the radius vector: it is a *vector field* $\mathbf{F} = \mathbf{F}(\mathbf{r})$. Then, in a Galilean transformation – in particular, in a rotation – one has to take into account the fact that the components of $\mathbf{r}$ are transformed too. Therefore, to obtain the right three-tuples for the force in the mathematical representation, we must use

$$\mathbf{F}'(\mathbf{r}') = \mathbf{DF}(\mathbf{r}), \quad \mathbf{r}' = \mathbf{Dr}. \qquad (8.30)$$

Thus, if the force depends only on the radius vector in the reference frame I – $\mathbf{F} = \mathbf{F}(\mathbf{r})$ for a single particle, or $\mathbf{F} = \mathbf{F}(\mathbf{r}_1, \mathbf{r}_2)$ in a two particle system – it becomes, in general, velocity and time-dependent in I'.

Which forces remain invariant under Galilean transformations (8.27)?

---

[17]If another differential equation were considered fundamental – in the sense that its validity establishes the meaning of inertial frame – and if one demands that equation should be invariant when changing to an 'equivalent' reference frame, then the invariance group may differ from the Galilean group. This is the case, for example, with Maxwell's equations: their invariance group is the Lorentz group of transformations. Lorentz transformations change the coordinates *and* time simultaneously. They are important in the special theory of relativity.

a) <u>External forces</u>

Let $\mathbf{F} = \mathbf{F}(\mathbf{r})$ be an external force and suppose a Galilean transformation is applied to it. Since $\mathbf{F}$ is a vector field, the components of $\mathbf{F}$ and the components of $\mathbf{r}$ are both transformed; but the components of $\mathbf{F}$ are only transformed by rotations (cf. Eq. (8.30)). We therefore have

$$\mathbf{F'}(\mathbf{r'}) = \mathbf{D}\mathbf{F}(\mathbf{r}) = \mathbf{D}\mathbf{F}(\mathbf{D}^T\mathbf{r'} + \mathbf{w'}t' + \mathbf{s'}). \qquad (8.31)$$

Consequently, only a constant external force (i.e. independent of space and time) is invariant under Galilean transformations. Otherwise the force appears, for example, time-dependent in the inertial system I'. This expresses the fact that, in general, an external force field establishes a physically distinguishable frame of reference (e.g. where the source of the field is at rest) and thus destroys Galilean invariance.

b) <u>Two particle forces</u>

Let $\mathbf{F} = \mathbf{F}(\mathbf{r}_1, \mathbf{r}_2)$ be the force between two particles at the positions $\mathbf{r}_1$ and $\mathbf{r}_2$, respectively. Expressed in terms of center of mass and relative coordinates, $\mathbf{R}$ and $\mathbf{r}$, respectively (cf. Section 6.1), $\mathbf{F}(\mathbf{r}_1, \mathbf{r}_2) = \mathbf{F}(\mathbf{R}, \mathbf{r})$, and using

$$\mathbf{r}'_i = \mathbf{D}\mathbf{r}_i + \mathbf{w}t + \mathbf{s}, \qquad i = 1, 2 \ ,$$

one finds

$$\mathbf{R'} = \frac{m_1\mathbf{r}'_1 + m_2\mathbf{r}'_2}{m_1 + m_2} = \mathbf{D}\mathbf{R} + \mathbf{w}t + \mathbf{s}$$

$$\mathbf{r'} = \mathbf{r}'_1 - \mathbf{r}'_2 = \mathbf{D}(\mathbf{r}_1 - \mathbf{r}_2) = \mathbf{D}\mathbf{r};$$

i.e. $\mathbf{R}$ is transformed like a radius vector and $\mathbf{r}$ is only affected by the rotation. Inserting this into

$$\mathbf{F'}(\mathbf{r}'_1, \mathbf{r}'_2) = \mathbf{D}\mathbf{F}(\mathbf{r}_1, \mathbf{r}_2) = \mathbf{D}\mathbf{F}(\mathbf{R}, \mathbf{r})$$

yields

$$\mathbf{F'}(\mathbf{R'}, \mathbf{r'}) = \mathbf{D}\mathbf{F}(\mathbf{D}^T\mathbf{R'} + \mathbf{w'}t' + \mathbf{s'}, \mathbf{D}^T\mathbf{r'}). \qquad (8.32)$$

If $\mathbf{F}$ is *independent of* $\mathbf{R}$, then

$$\mathbf{F'}(\mathbf{r'}) = \mathbf{D}\mathbf{F}(\mathbf{D}^T\mathbf{r'}). \qquad (8.33)$$

The force in I' still only depends on the relative positions of the particles. If the force is a central force $\mathbf{F}(\mathbf{r}) = f(r)\mathbf{r}$ in reference frame I, then

$$\mathbf{F}'(\mathbf{r}') = \mathsf{D}f(|\mathsf{D}^T\mathbf{r}'|)\mathsf{D}^T\mathbf{r}' = f(r')\mathbf{r}'$$

$(|\mathsf{D}^T\mathbf{r}'| = r' = r)$ implies that $\mathbf{F}'$ is also a central force.

Remark

If the coordinate systems of two inertial frames differ only in the position of the origin and/or the orientation of the coordinate axes (or in the time origin), the difference between them can be removed by a time-independent translation and/or rotation of the coordinate system in one of the frames (or by a translation of the zero of time). Such frames are actually completely equivalent, so they can be considered to be identical. Hence, the only criterion distinguishing one inertial frame from another is their uniform relative motion. Therefore to answer the question, "How does the choice of the inertial frame influence the representation of a physical process?", we only need consider the class of inertial frames that are related to each other by a special Galilean transformation (boost).

8.5   Transformations to non-inertial frames

We now consider the behavior of Newton's equation when we change from an inertial reference frame I to a non-inertial frame, i.e. a *nonuniformly moving frame*, K. We will consider two cases for K: an accelerated frame, such as that of a rocket while its motors are on; and a uniformly rotating reference frame, such as the surface of the earth.

8.5.1   Accelerated frames of reference

Suppose the velocity of reference frame K relative to inertial frame I changes in time, but only in magnitude, not direction. (Therefore the relative orientation of the coordinate systems does not change with time.) Then the radius vectors from the respective origins to the same point (e.g. at which a point mass is located) differ according to

$$\mathbf{r}' = \mathbf{r} + \mathbf{s}(t), \tag{8.34}$$

where $\mathbf{s}(t)$ is the vector between the origins at time t. Starting from the equation of motion in the inertial frame, we are interested in seeing how the equation of motion appears in the non-inertial frame K.

The time derivatives of the spatial vectors, appearing as velocity and acceleration, are related by[18]

$$\dot{\mathbf{r}}' = \dot{\mathbf{r}} + \dot{\mathbf{s}} \qquad \text{and} \qquad \ddot{\mathbf{r}}' = \ddot{\mathbf{r}} + \ddot{\mathbf{s}}.$$

We ensure K is an **accelerated frame of reference** by

$$\ddot{\mathbf{s}} \neq 0.$$

In frame I, if the equation of motion is $m\ddot{\mathbf{r}} = \mathbf{F}(\mathbf{r})$, then in frame K we have

$$m\ddot{\mathbf{r}}' = \mathbf{F}(\mathbf{r}' - \mathbf{s}(t)) + m\ddot{\mathbf{s}}. \tag{8.35}$$

The term $m\ddot{\mathbf{s}}$ appearing in K can be interpreted as a force,

$$\mathbf{F}_{acc} = m\ddot{\mathbf{s}}.$$

Such an additional force due to nonuniform relative motion of two reference frames is called a **pseudo-force**. If $\ddot{\mathbf{s}} = const$, the pseudo-force is homogeneous and time-independent.

Consider the transformation to center of mass coordinates and to relative coordinates for a system of two particles (cf. Section 6.1). For the transformed Newton's equation for the relative motion of the two particles, it is easy to show that the pseudo-force term is absent in the accelerated frame of reference. Furthermore, in two-particle systems where $\mathbf{F}$ is an internal force (i.e. an interaction force between the particles), the force field depends only on the relative coordinate $\mathbf{F} = \mathbf{F}(\mathbf{r}_1 - \mathbf{r}_2)$; then the force in K also depends only on $\mathbf{r}_1' - \mathbf{r}_2'$. Therefore, the pseudo-force term is absent in the equation for the relative motion:

$$m\ddot{\mathbf{r}}' = \mathbf{F}(\mathbf{r}'), \qquad \mathbf{r}' = \mathbf{r}_1' - \mathbf{r}_2'.$$

In the relative motion, the difference between the two frames of reference is not noticeable. In the center of mass frame (with coordinate $\mathbf{R}$), however, there is an effect, since

$$\ddot{\mathbf{R}}' = \ddot{\mathbf{s}}.$$

(If there is no external force, in the inertial frame I one has $\ddot{\mathbf{R}} = 0$.)

---

[18]Due to Newton's absolute time, the time is the same in the two frames.

8.5.2   Rotating frames of reference

Transformation of the basis vectors

Let the motion of K with respect to I be a *time-dependent* rotation. The transformation is of the form given by Eq. (8.8), but now the rotation matrix depends on time:

$$x'_i = D_{ik}(t)x_k. \tag{8.36}$$

The relation derived in Section B.1 of Appendix B is also valid for time-dependent rotations. In particular, the basis vectors transform according to the Eqs. (B.3) and (B.10):

$$\mathbf{e}'_i = D_{ik}\mathbf{e}_k \qquad \text{and} \qquad \mathbf{e}_k = D_{jk}\mathbf{e}'_j. \tag{8.37}$$

Since the vectors $\mathbf{e}_i$ in the inertial frame I are time independent, we get for the temporal change of the basis vectors $\mathbf{e}'_i$,

$$\dot{\mathbf{e}}'_i = \frac{d}{dt}\left(D_{ik}\mathbf{e}_k\right) = \dot{D}_{ik}\mathbf{e}_k = \dot{D}_{ik}D_{jk}\mathbf{e}'_j,$$

or, in simpler notation,

$$\dot{\mathbf{e}}'_i = \Omega_{ij}(t)\mathbf{e}'_j \qquad \text{where} \qquad \Omega_{ij} := \dot{D}_{ik}D_{jk}. \tag{8.38}$$

The matrix $\Omega = \dot{\mathbf{D}}\mathbf{D}^T$ describes the change of the basis due to the rotation *in the rotating frame!* Multiplying Eq. (8.38) by $\mathbf{e}'_k$, we find the elements of Ω:

$$\Omega_{ik} = \dot{\mathbf{e}}'_i\mathbf{e}'_k. \tag{8.39}$$

The derivative of the orthogonality condition for the rotation matrix (B.6), $D_{ik}D_{jk} = \delta_{ij}$, with respect to time is

$$\dot{D}_{ik}D_{jk} + D_{ik}\dot{D}_{jk} = 0,$$

and from the definition of Ω_{ij} in Eq. (8.38), we obtain

$$\Omega_{ij} + \Omega_{ji} = 0. \tag{8.40}$$

Hence, Ω is an *antisymmetric matrix*. It therefore has only *three nonvanishing, independent elements*, and can be represented in the following form:

$$\Omega_{ij} = \varepsilon_{ijk}\omega'_k. \tag{8.41}$$

The explicit matrix form is

$$\Omega = \begin{pmatrix} 0 & \omega_3' & -\omega_2' \\ -\omega_3' & 0 & \omega_1' \\ \omega_2' & -\omega_1' & 0 \end{pmatrix}. \tag{8.42}$$

Using Eq. (8.41), one can express the ω_k' in terms of Ω multiplying this equation by ε_{ijl}:

$$\omega_k' = \frac{1}{2}\varepsilon_{kij}\Omega_{ij}. \tag{8.43}$$

In order to show that

$$\boldsymbol{\omega} = \omega_i' \mathbf{e}_i' \tag{8.44}$$

is the *angular velocity* vector of the rotating frame (here represented in the basis $\mathbf{e}_i'$), we have to show that: (i) $\boldsymbol{\omega}$ is a vector; and (ii) $\boldsymbol{\omega} = \boldsymbol{\omega}(t)$ is the instantaneous rotation axis about which K rotates with angular velocity $\omega(t)$. One recognizes that $\boldsymbol{\omega}$ is a vector, from its behavior under rotations[19]. To show that $\boldsymbol{\omega}$ is equal to the current vector $\dot{\phi}\mathbf{n}$ of the angular velocity of K, we first observe that the basis vectors transform according to[20]

$$\dot{\mathbf{e}}_i' = \dot{\phi}\left(\mathbf{n} \times \mathbf{e}_i'\right). \tag{8.45}$$

---

[19]Consider a rotation $\bar{D}$ of a coordinate system in the frame K. Such a rotation applied to Ω causes a transformation of the three-tuple $(\omega_1, \omega_2, \omega_3)$ according to Eq. (8.8). Proof: Eq. (8.39) implies that the matrix elements of Ω under the rotation $\bar{D}$ transform according to:

$$\Omega_{mn} = \bar{D}_{mj}\bar{D}_{nl}\bar{\Omega}_{jl}.$$

Inserting this into $\omega_k' = \frac{1}{2}\varepsilon_{kmn}\Omega_{mn}$ and multiplying the resulting relation by $\bar{D}_{ki}$ $(= (\bar{D}^T)_{ik}!)$, one has

$$\bar{D}_{ki}\omega_k' = \frac{1}{2}\varepsilon_{kmn}\bar{D}_{ki}\bar{D}_{mj}\bar{D}_{nl}\bar{\Omega}_{jl}.$$

Since the ε-tensor is an invariant tensor, we have $\bar{D}_{ki}\bar{D}_{mj}\bar{D}_{nl}\varepsilon_{kmn} = \varepsilon_{ijl}$, and it follows that

$$\bar{D}_{ki}\omega_k' = \frac{1}{2}\varepsilon_{ijl}\bar{\Omega}_{jl} = \bar{\omega}_i \qquad \text{or} \qquad \omega_j' = \bar{D}_{ji}\bar{\omega}_i, \quad \text{respectively.}$$

This shows that ω transforms like a vector. In the new frame, we again have the relation (8.43) between $\bar{\omega}$ and $\bar{\Omega}$

$$\bar{\omega}_i = \frac{1}{2}\varepsilon_{ijl}\bar{\Omega}_{jl}.$$

[20]This relation follows from the representation (8.21) of an infinitesimal rotation

We then express relations (8.38) in terms of $\boldsymbol{\omega}$ by multiplying the equation

$$\dot{\mathbf{e}}'_i = \Omega_{ij}\mathbf{e}'_j = \varepsilon_{ijk}\omega'_k\mathbf{e}'_j$$

by the i'th component of $\mathbf{e}'_j$, $(\mathbf{e}'_j)_i = \delta_{ij}$. Since

$$(\mathbf{e}'_j)_i\dot{\mathbf{e}}'_i = \dot{\mathbf{e}}'_j = \mathbf{e}'_k\varepsilon_{kli}\omega'_l(\mathbf{e}'_j)_i = \mathbf{e}'_k(\boldsymbol{\omega}\times\mathbf{e}'_j)_k = \boldsymbol{\omega}\times\mathbf{e}'_j,$$

we obtain

$$\dot{\mathbf{e}}'_j = \boldsymbol{\omega}\times\mathbf{e}'_j. \tag{8.46}$$

Comparison with (8.45) shows that

$$\dot{\phi}\mathbf{n} = \boldsymbol{\omega}; \tag{8.47}$$

hence $\boldsymbol{\omega}$ is the angular velocity of the rotating reference frame. Only the time derivatives of the components ω'_j in Eq. (8.44) contribute to the time derivative of $\boldsymbol{\omega}$, which is the angular acceleration $\dot{\boldsymbol{\omega}}$,

$$\dot{\boldsymbol{\omega}} = \dot{\omega}'_i\mathbf{e}'_i + \omega'_i\dot{\mathbf{e}}'_i = \dot{\omega}'_i\mathbf{e}'_i. \tag{8.48}$$

This can be seen by multiplying Eq. (8.46) by ω'_j, whence $\omega'_j\dot{\mathbf{e}}'_j = \boldsymbol{\omega}\times\boldsymbol{\omega} = 0$.

The angular velocity expressed in terms of Euler's angles

Expressing $\boldsymbol{\omega}$ in terms of Euler's angles is more involved. It would be very tedious to use relation (8.43) together with the definition (8.38) of Ω,

$$\omega'_k = \frac{1}{2}\varepsilon_{kij}\Omega_{ij} = \frac{1}{2}\varepsilon_{kij}\dot{D}_{il}D_{jl} = \frac{1}{2}\varepsilon_{kij}\dot{D}_{il}(\mathsf{D}^T)_{lj},$$

to calculate the components of the angular velocity $\boldsymbol{\omega}$ as functions of Euler's angles by inserting D_{ij} from Eq. (8.17). It is easier to express Ω

---

if one sets $\mathbf{r} = \mathbf{r}' + d\mathbf{r}'$ and $\varepsilon\to d\phi$. Dividing by dt gives:

$$\dot{\mathbf{r}}' = \dot{\phi}(\mathbf{n}\times\mathbf{r}').$$

Thus we find for each basis vector $(\mathbf{r}' \to \mathbf{e}'_i)$

$$\dot{\mathbf{e}}'_i = \dot{\phi}(\mathbf{n}\times\mathbf{e}'_i).$$

in terms of the product of the rotations (8.14), (8.15), (8.16) and only then to use relation (8.43) between $\boldsymbol{\omega}$ and Ω. From

$$
\begin{aligned}
\Omega \;=\; \dot{\mathsf{D}}\mathsf{D}^T &= \left(\frac{d}{dt}\left(\mathsf{D}_3(\psi)\mathsf{D}_2(\vartheta)\mathsf{D}_1(\varphi) \right) \right) \mathsf{D}_1^T(\varphi)\mathsf{D}_2^T(\vartheta)\mathsf{D}_3^T(\psi) \\
&= \dot{\mathsf{D}}_3\mathsf{D}_2\mathsf{D}_1\mathsf{D}_1^T\mathsf{D}_2^T\mathsf{D}_3^T + \mathsf{D}_3\dot{\mathsf{D}}_2\mathsf{D}_1\mathsf{D}_1^T\mathsf{D}_2^T\mathsf{D}_3^T + \mathsf{D}_3\mathsf{D}_2\dot{\mathsf{D}}_1\mathsf{D}_1^T\mathsf{D}_2^T\mathsf{D}_3^T \\
&= \dot{\mathsf{D}}_3\mathsf{D}_3^T + \mathsf{D}_3\dot{\mathsf{D}}_2\mathsf{D}_2^T\mathsf{D}_3^T + \mathsf{D}_3\mathsf{D}_2\dot{\mathsf{D}}_1\mathsf{D}_1^T\mathsf{D}_2^T\mathsf{D}_3^T \,,
\end{aligned}
$$

we get

$$
\Omega = \Omega_3(\psi) + \mathsf{D}_3\Omega_2(\vartheta)\mathsf{D}_3^T + \mathsf{D}_3\mathsf{D}_2\Omega_1(\varphi)\mathsf{D}_2^T\mathsf{D}_3^T , \tag{8.49}
$$

where the matrices $\Omega_i = \dot{\mathsf{D}}_i\mathsf{D}_i^T$ are given by rotations D_i about the three axes of rotation. The matrices Ω_i operate in different coordinate systems: $\Omega_1(\varphi)$ in the inertial frame; $\Omega_2(\vartheta)$ in the intermediate system; $\Omega_3(\psi)$ and Ω in the rotating frame K. Therefore, in Eq. (8.49), $\Omega_1(\varphi)$ and $\Omega_2(\vartheta)$ are transformed by the rotation matrices to the rotating frame K. According to Eq. (8.43), the matrices Ω and Ω_i are related to the angular velocities $\boldsymbol{\omega}$ and $\boldsymbol{\omega}_i$ in the respective systems by multiplying Ω_{jk} in Eq. (8.49) by $\frac{1}{2}\varepsilon_{ijk}$. This yields for the i'th components of the angular velocities the following relation in the rotating frame

$$
\omega_i' = (\boldsymbol{\omega}_3(\psi))_i' + (\mathsf{D}_3\,\boldsymbol{\omega}_2(\vartheta))_i' + (\mathsf{D}_3\mathsf{D}_2\,\boldsymbol{\omega}_1(\varphi))_i' . \tag{8.50}
$$

This relation implies that the vectors of the angular velocities, considered in the frame K, are added:

$$
\boldsymbol{\omega} = \boldsymbol{\omega}_3(\psi) + \boldsymbol{\omega}_2(\vartheta) + \boldsymbol{\omega}_1(\varphi). \tag{8.51}
$$

Since $\boldsymbol{\omega}_1(\varphi)$ in the inertial frame has the components $(0,0,\dot{\varphi})$, $\boldsymbol{\omega}_2(\vartheta)$ in the intermediate system is oriented along the x-axis and thus has the components $(\dot{\vartheta},0,0)$; and finally $\boldsymbol{\omega}_3(\psi) = (0,0,\dot{\psi})$ in the rotating frame, by construction. Now the components of the angular velocity in the rotating frame K ($\boldsymbol{\omega} \to \boldsymbol{\omega}'!$) follow from Eqs. (8.50), (8.15), and (8.16):

$$
\begin{aligned}
\boldsymbol{\omega}' &= \begin{pmatrix} 0 \\ 0 \\ \dot{\psi} \end{pmatrix} + \mathsf{D}_3(\psi)\begin{pmatrix} \dot{\vartheta} \\ 0 \\ 0 \end{pmatrix} + \mathsf{D}_3(\psi)\mathsf{D}_2(\vartheta)\begin{pmatrix} 0 \\ 0 \\ \dot{\varphi} \end{pmatrix} \\[2mm]
&= \begin{pmatrix} 0 \\ 0 \\ \dot{\psi} \end{pmatrix} + \begin{pmatrix} \dot{\vartheta}\cos\psi \\ -\dot{\vartheta}\sin\psi \\ 0 \end{pmatrix} + \begin{pmatrix} \dot{\varphi}\cos\psi\sin\vartheta \\ \dot{\varphi}\sin\psi\sin\vartheta \\ \dot{\varphi}\cos\vartheta \end{pmatrix}
\end{aligned}
$$

and thus $\boldsymbol{\omega}'$ expressed in terms of Euler's angles reads

$$\boldsymbol{\omega}' = \begin{pmatrix} \dot{\varphi}\cos\psi\sin\vartheta + \dot{\vartheta}\cos\psi \\ \dot{\varphi}\sin\psi\sin\vartheta - \dot{\vartheta}\sin\psi \\ \dot{\varphi}\cos\vartheta + \dot{\psi} \end{pmatrix}. \tag{8.52}$$

Time derivatives of vectors in the rotating frame

We consider a physical vector quantity $\mathbf{A} = \mathbf{A}(t)$ from two viewpoints: the inertial frame I and the rotating frame K. We represent the given vector $\mathbf{A}$ in terms of both bases (cf. Eq. (8.6)); thus we have

$$\mathbf{A}(t) = A_i(t)\mathbf{e}_i = A_i'(t)\mathbf{e}_i'(t). \tag{8.53}$$

Inspecting the time derivative in both frames and using Eq. (8.46) to yield

$$\frac{d\mathbf{A}}{dt} = \dot{A}_i\mathbf{e}_i = \dot{A}_i'\mathbf{e}_i' + A_i'\left(\boldsymbol{\omega}\times\mathbf{e}_i'\right) = \dot{A}_i'\mathbf{e}_i' + \left(\boldsymbol{\omega}\times\mathbf{A}\right),$$

and denoting the time derivative of $\mathbf{A}$ in the rotating frame[21] K by

$$\left(\frac{d\mathbf{A}}{dt}\right)' := \dot{A}_i'\mathbf{e}_i', \tag{8.54}$$

one obtains the desired relation between the time derivatives of the vector $\mathbf{A}$ in each frame[22]:

$$\frac{d\mathbf{A}}{dt} = \left(\frac{d\mathbf{A}}{dt}\right)' + \left(\boldsymbol{\omega}\times\mathbf{A}\right). \tag{8.55}$$

The $(\boldsymbol{\omega}\times\mathbf{A})$ term results from the rotation of the basis of the rotating frame. Setting $\mathbf{A} = \mathbf{r}$, the relation between the velocity in the I and K frames, respectively, is

$$\mathbf{v} = \mathbf{v}' + \left(\boldsymbol{\omega}\times\mathbf{r}\right), \tag{8.56}$$

---

[21] Absolute time enters here again!

[22] This relation is also obtained from the transformation law for the components of $\mathbf{A}$,

$$A_i' = D_{ik}(t)A_k,$$

by taking the derivative with respect to time.

where $\mathbf{v} = \dfrac{d\mathbf{r}}{dt}$ and

$$\mathbf{v}' := \left(\frac{d\mathbf{r}}{dt}\right)'. \tag{8.57}$$

If $\mathbf{A} = \boldsymbol{\omega}$ then

$$\frac{d\boldsymbol{\omega}}{dt} = \left(\frac{d\boldsymbol{\omega}}{dt}\right)' =: \dot{\boldsymbol{\omega}}, \tag{8.58}$$

so the time derivative of $\boldsymbol{\omega}$ in the two systems is equal (since $\boldsymbol{\omega}$ lies in the rotation axis); this is simply the result given already in Eq. (8.48).

Due to Eq. (8.58), a further time derivative of $\mathbf{A}$ takes the form

$$
\begin{aligned}
\frac{d^2\mathbf{A}}{dt^2} &= \frac{d}{dt}\left(\left(\frac{d\mathbf{A}}{dt}\right)' + (\boldsymbol{\omega} \times \mathbf{A})\right)' + \boldsymbol{\omega} \times \left(\left(\frac{d\mathbf{A}}{dt}\right)' + (\boldsymbol{\omega} \times \mathbf{A})\right) \\
&= \left(\frac{d^2\mathbf{A}}{dt^2}\right)' + \dot{\boldsymbol{\omega}} \times \mathbf{A} + 2\boldsymbol{\omega} \times \left(\frac{d\mathbf{A}}{dt}\right)' + \boldsymbol{\omega} \times \boldsymbol{\omega} \times \mathbf{A}, \quad (8.59)
\end{aligned}
$$

so that the second time derivative in the rotating frame K has three additional terms. If the angular velocity is constant ($\dot{\boldsymbol{\omega}} = 0$), the first extra term vanishes.

8.5.3   Motion in a rotating frame

In the following, we only consider frames of reference K which rotate with *constant* angular velocity $\boldsymbol{\omega}$ (with respect to the inertial frame I),

$$\dot{\boldsymbol{\omega}} = \mathbf{0},$$

with the origins of the two frames I and K coinciding.

From Eq. (8.59), the relation between the accelerations $\ddot{\mathbf{r}}$ and $\ddot{\mathbf{r}}' := (d^2\mathbf{r}/dt^2)'$ in the two frames is given by[23]

$$\ddot{\mathbf{r}} = \ddot{\mathbf{r}}' + 2\boldsymbol{\omega} \times \mathbf{v}' + \boldsymbol{\omega} \times \boldsymbol{\omega} \times \mathbf{r}', \tag{8.60}$$

and hence, from the equation of motion in the inertial frame, $m\ddot{\mathbf{r}} = \mathbf{F}$, the law of motion in the rotating frame follows; namely,

$$m\ddot{\mathbf{r}}' = \mathbf{F}' - 2m(\boldsymbol{\omega} \times \mathbf{v}') - m\boldsymbol{\omega} \times \boldsymbol{\omega} \times \mathbf{r}', \tag{8.61}$$

---

[23] Again, $\mathbf{r}' = \mathbf{r}$; in order to stress that the components of $\mathbf{r}$ should be taken in K we write $\mathbf{r}'$.

where $\mathbf{F}'$ is the force in K. Its components in K are given by

$$\mathbf{F}'\left(\mathbf{r}',t\right) = \mathsf{D}\left(t\right)\mathbf{F}\left(\mathbf{r}\right).$$

Equation (8.61) shows two new contributions acting as forces. These pseudo-forces are the *centrifugal force* and the *Coriolis force*.

- The **centrifugal force**[24] in K is given by

$$\mathbf{F}_{cent} = -m\boldsymbol{\omega} \times \boldsymbol{\omega} \times \mathbf{r}'. \tag{8.62}$$

If $\mathbf{v}' = \mathbf{0}$, i.e. no translational motion in the rotating frame K, it is the only pseudo-force present. The centrifugal force is directed outward (from the rotation axis), perpendicular to the rotation axis $\boldsymbol{\omega}$, and its absolute value is $m\omega^2 r \sin\theta$, where θ is the angle between $\mathbf{r}'$ $(=\mathbf{r})$ and $\boldsymbol{\omega}$.

- The **Coriolis force**[25] is given by

$$\mathbf{F}_{Cor} = -2m(\boldsymbol{\omega} \times \mathbf{v}'). \tag{8.63}$$

It is experienced by a body moving with non-zero velocity in K. The acceleration $(-2(\boldsymbol{\omega} \times \mathbf{v}')$ is called *Coriolis acceleration*) is perpendicular to $\boldsymbol{\omega}$ and the velocity $\mathbf{v}'$ in K.

---

[24]According to Hertz ("Die Prinzipien der Mechanik"), one encounters a '*logical imperfection*' in applying Newton's third law to a body forced by a rope to move in a circle, if, as customary, one considers the centrifugal force as a counter force to the attracting force of the rope. The centrifugal force actually is a part of the change of momentum on the left side of Eq. (2.4) (it is a manifestation of the inertia of the body). Being rather a reaction to the applied force, the centrifugal acceleration should not be considered as a force appearing in Newton's third law.

[25]Gaspard Gustave DE CORIOLIS (1792-1843), French scientist.

The effect was already anticipated before Coriolis published his *Mémoire sur les équations du mouvement relatif des systèmes de corps*. Journal de l'école Polytechnique, Vol 15, 142, 1835.

In the context of collisions one should mention also his work on billiards: "Théorie mathématique des effets du Jeu de Billard", Paris 1835.

Motion in a homogeneous magnetic field considered in a rotating frame

In an inertial frame, let the motion of a point mass be subject to a force together with a constant magnetic field (see Section 5.5). The equation of motion (cf. Eq. (5.61)) reads

$$m\ddot{\mathbf{r}} = \mathbf{F}(\mathbf{r}) + \frac{q}{c}\dot{\mathbf{r}} \times \mathbf{B}. \tag{8.64}$$

To see how this equation appears in a rotating reference frame, we transform according to Eq. (8.59):

$$
\begin{aligned}
m\ddot{\mathbf{r}}' &= \mathbf{F}'(\mathbf{r}') + \frac{q}{c}\left(\dot{\mathbf{r}}' + \boldsymbol{\omega} \times \mathbf{r}'\right) \times \mathbf{B}' + 2m\dot{\mathbf{r}}' \times \boldsymbol{\omega} - m\boldsymbol{\omega} \times \boldsymbol{\omega} \times \mathbf{r}' \\
&= \mathbf{F}'(\mathbf{r}') + \frac{q}{c}\dot{\mathbf{r}}' \times \mathbf{B}' + \frac{q}{c}\left(\boldsymbol{\omega} \times \mathbf{r}'\right) \times \mathbf{B}' + 2m\dot{\mathbf{r}}' \times \boldsymbol{\omega} - m\boldsymbol{\omega} \times \boldsymbol{\omega} \times \mathbf{r}'.
\end{aligned}
$$

We now choose the rotation axis $\hat{\boldsymbol{\omega}}$ ($\boldsymbol{\omega} = \omega\hat{\boldsymbol{\omega}}$, $|\hat{\boldsymbol{\omega}}| = 1$) to be parallel to $\mathbf{B}$. Therefore, $\mathbf{B}$ is not changed by the rotation, and we have[26] $\mathbf{B}' = \mathbf{B}$. If, in addition,

$$\boldsymbol{\omega} = \boldsymbol{\omega}_L = -\frac{q}{2mc}\mathbf{B}, \tag{8.65}$$

where ω_L is the **Larmor frequency**, the Coriolis term in the equation of motion vanishes, but there is still another term present. The equation of motion in the frame rotating with angular velocity $\boldsymbol{\omega}_L$ is (E)

$$m\ddot{\mathbf{r}}' = \mathbf{F}'(\mathbf{r}') + \frac{q^2}{4mc^2}\mathbf{B} \times \mathbf{B} \times \mathbf{r}'. \tag{8.66}$$

The angular momentum in this frame, $\mathbf{L}' = m\mathbf{r}' \times \dot{\mathbf{r}}'$, is given by (E)

$$\mathbf{L}'\mathbf{B} = \mathbf{L}\mathbf{B} + \frac{q}{2c}(\mathbf{r} \times \mathbf{B})^2 = BI_{LB}. \tag{8.67}$$

This is exactly the conserved quantity found in Section 5.5. Now we can also understand the meaning of I_{LB}: it is the component of the angular momentum parallel to the magnetic field in a reference frame rotating with the Larmor frequency $\boldsymbol{\omega}_L$.

In Subsection 13.1.3, we consider the equation of motion (8.64) for a particle in a central force field, $\propto 1/r^2$, in a rotating frame. Numerical integration shows that the motion is in general chaotic.

---

[26] Another way to see this is to use Eq. (8.55). According to this equation, $d\mathbf{B}/dt = (d\mathbf{B}/dt)'$, and since $d\mathbf{B}/dt = \mathbf{0}$, it follows that $\mathbf{B}' = \mathbf{B}$.

The Coriolis force

The earth rotates with angular velocity

$$\omega_e = \frac{2\pi}{T_{day}} = \frac{2\pi}{24 \times 3600} \sec^{-1} = 7.3 \times 10^{-5} \sec^{-1}$$

with respect to the fixed stars. The centrifugal acceleration, which is dependent on the geographic latitude, is at most 0.3% of the gravitational acceleration (r is of the order of the radius of the earth) and is therefore neglected. The Coriolis force is responsible for effects such as the direction of rotation of cyclones (counterclockwise on the northern hemisphere and clockwise on the southern hemisphere – looking down from space), as well as the direction of flow of the Gulf stream.

We consider a body falling near the earth's surface, i.e. the gravitational acceleration is considered to be constant. We investigate the magnitude of the deviation from the vertical incurred by the body due to the Coriolis acceleration. Since, in the following, all dynamical quantities are taken in the rotating reference frame K located at the earth's surface, we omit the dashes from the velocity and acceleration vectors. In this rotating frame, the force of gravity has the simple form $\mathbf{F}' = m\mathbf{g}$. Thus the equation of motion – neglecting the centrifugal force – is given by

$$\ddot{\mathbf{r}} = \mathbf{g} - 2(\boldsymbol{\omega}_e \times \dot{\mathbf{r}}). \tag{8.68}$$

Choosing the coordinates as shown in Fig. 8.3, the components of $\mathbf{g}$ and $\boldsymbol{\omega}_e$ (in K) are

$$\mathbf{g} = (0, 0, -g), \qquad \boldsymbol{\omega}_e = (0, -\omega_e \sin\theta, \omega_e \cos\theta). \tag{8.69}$$

The initial values are taken to be

$$\mathbf{r}(0) = (0, 0, h) \qquad \text{and} \qquad \dot{\mathbf{r}}(0) = (0, 0, 0). \tag{8.70}$$

Equation (8.68) can be solved exactly (see e.g. [Sommerfeld]). Rather than laboring through the exact solution, it is more instructive to consider the following approximation procedure, which relies on the smallness of ω_e. (Hence, the approximation is good only for small time intervals.)

We start from Eq. (8.68), written in the form

$$\dot{\mathbf{v}} + 2(\boldsymbol{\omega}_e \times \mathbf{v}) = \mathbf{g}. \tag{8.71}$$

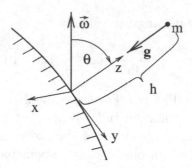

Figure 8.3: The coordinate system at the surface of the earth.

Since one can see that $\mathbf{v}(t)$ depends on w_e in the rotating frame, we set

$$\mathbf{v}(t) = \mathbf{v}_0(t) + w_e \mathbf{v}_1(t) + w_e^2 \mathbf{v}_2(t) + \dots \, , \qquad (8.72)$$

where the $\mathbf{v}_i$ are independent of w_e. Inserting this into the equation of motion, we obtain

$$\left(\dot{\mathbf{v}}_0 + w_e \dot{\mathbf{v}}_1 + w_e^2 \dot{\mathbf{v}}_2 + \dots \right) + 2\boldsymbol{w}_e \times \left(\mathbf{v}_0 + w_e \mathbf{v}_1 + w_e^2 \mathbf{v}_2 + \dots \right) = \mathbf{g}. \quad (8.73)$$

Comparing the contributions of like powers in w_e yields

$$
\begin{array}{rl}
w_e^0 \ : & \dot{\mathbf{v}}_0 = \mathbf{g} \\
w_e^1 \ : & w_e \dot{\mathbf{v}}_1 + 2(\boldsymbol{w}_e \times \mathbf{v}_0) = 0 \\
w_e^2 \ : & w_e^2 \dot{\mathbf{v}}_2 + 2w_e(\boldsymbol{w}_e \times \mathbf{v}_1) = 0
\end{array}
\qquad (8.74)
$$

$$\vdots$$

The solution of the first equation for the initial values given in Eq. (8.70) is

$$\mathbf{v}_0 = \mathbf{g}t. \qquad (8.75)$$

This is just the free fall solution, which, inserted into the second equation, yields

$$w_e \dot{\mathbf{v}}_1 = -2(\boldsymbol{w}_e \times \mathbf{g})t.$$

Since the choice (8.69) gives $\boldsymbol{w}_e \times \mathbf{g} = (w_e g \sin\theta, 0, 0)$, it follows that the x-component is

$$\dot{v}_{1x} = -2gt \sin\theta$$

and hence[27]

$$v_{1x} = -gt^2 \sin\theta. \tag{8.76}$$

At this order of ω_e, the velocity $\mathbf{v}$ is then

$$\mathbf{v}(t) = \begin{pmatrix} -\omega_e g t^2 \sin\theta \\ 0 \\ -gt \end{pmatrix}, \tag{8.77}$$

and a further integration, using the initial values (8.70), yields the radius vector

$$\mathbf{r}(t) = \begin{pmatrix} -\frac{1}{3}\omega_e g t^3 \sin\theta \\ 0 \\ -\frac{g}{2}t^2 + h \end{pmatrix}. \tag{8.78}$$

The z-component is the law of the free fall. In addition, however, a *displacement to the east* arises due the Coriolis force. The magnitude of the displacement depends also on the geographic latitude; it vanishes at the poles. The body arrives at the earth's surface ($z = 0$) at time $t = \sqrt{2h/g}$, having suffered a displacement in the x-direction of

$$\frac{1}{3}\omega_e \sqrt{(2h)^3/g}\sin\theta. \tag{8.79}$$

At the next order ($\mathcal{O}\left(\omega_e^2\right)$) of accuracy, a displacement in y-direction also appears. The orbit deviates towards the equator (**E**).

The Coriolis force is also responsible for the precession of the plane of oscillation of a pendulum (see (**E**) in the next chapter). In 1851, this effect was used by Jean Bernard Léon FOUCAULT (1819-1868, French physicist) to demonstrate the rotation of the earth.

Problems and examples

1. Express the elements of the rotation matrix **D** in terms of the rotation axis **n** and the rotation angle ϕ appearing in Eq. (8.13).

2. Show that the ε-tensor is invariant under rotations of the coordinate system.

---

[27]The constants of integration are set to zero, since $\mathbf{v}_0\,(t)$ takes care of the initial values.

3. Verify the group properties of the Galilean transformations (8.27) and (8.28).

4. What are the components of $\boldsymbol{\omega}$ (cf. Eq. (8.52)) expressed in terms of Euler's angles in the inertial frame I?

5. For the motion in a constant magnetic field **B**, derive the equation of motion and the angular momentum in the frame rotating with angular velocity $\boldsymbol{\omega}_L$, given in Eq. (8.65).

6. Calculate the deviation from the vertical suffered by a body in free fall due to the Coriolis force to second order in the rotation frequency ω_e.

9

Lagrangian mechanics

At the end of the 18th century, an important stage in the development of classical mechanics was achieved. In 1788 Lagrange presented mechanics in a uniform, clear and very elegant manner in his "Mécanique analytique". Lagrange's book is the first comprehensive representation of *analytical* mechanics. It is an edifice of ideas: the building blocks are logical related to each other. The representation of physical processes by mathematical theories and models is unified and elegant. This "Analytical Mechanics" is a 'phenomenal achievement in the economy of thought[1]' (E. Mach, "The Science of Mechanics").

The treatment in this chapter differs from the historical development: Lagrange's work is based on D'Alembert's principle[2] and on the concept of virtual displacements; we consider neither in this book. Instead, after a brief account of the calculus of variations, Lagrangian mechanics is introduced by identifying Newton's equation of motion with the Euler-Lagrange equations for the Lagrangian.

9.1   Constrained motion

In many problems in classical mechanics, we are given not only the forces, but also certain **constraints** that the orbits must obey. For example, in the case of the pendulum, the mass must remain a constant

---

[1]In the sense of 'based on a minimal set of assumptions'.

[2]Jean-Baptiste le Rond D'ALEMBERT (1717-1783). He proposed his Principle in the "Traité de dynamique" (Treatise of dynamics; Paris 1743).

distance away from the pivot point. One distinguishes various kinds of constraints. A constraint depending only on the position $\mathbf{r}(t)$ of the particle and on time t, i.e. of the form

$$f\left(\mathbf{r}(t), t\right) = 0, \tag{9.1}$$

is called a **holonomic** constraint. (In the case of the pendulum, the holonomic constraint is $\mathbf{r}^2(t) - l^2 = 0$.) Constraints that cannot be written in this form, for example, those that

- Contain velocities or differentials only of the radius vectors (e.g.: $\sum g_i(\mathbf{r})dx_i = 0 \Rightarrow \sum g_i(\mathbf{r})\dot{x}_i dt = 0$); or,

- Appear only as inequalities,

are **non-holonomic**. Examples of systems with non-holonomic constraints include:

- The motion of a particle restricted to the interior of a sphere, $|\mathbf{r(t)}| \leq R$;

- A wheel (radius R) rolling on a surface. Here the constraint $v = R\dot{\phi}$ relates the angular velocity $\dot{\phi}$ of the wheel to its linear velocity v; since this condition can not be integrated without knowing the motion of the wheel, the constraint is non-holonomic.

Time independent constraints are called **scleronomic**, while time dependent ones are called **rheonomic**.

In the following, we only consider holonomic constraints for the motion of a single particle. The generalization to several constraints, or to more than one particle, ultimately follows the same logic. As we have seen, the solutions of the equations of motion for a given set of forces are unique functions of the initial positions and velocities. This implies that, in general, given some arbitrary constraint of the form of Eq. (9.1), the solutions of the equation

$$m\ddot{\mathbf{r}} = \mathbf{F}$$

do not satisfy the constraint, even if the initial values satisfy it. To fix this problem, an extra force must appear in the equation of motion, the so-called **force of constraint Z**,

$$m\ddot{\mathbf{r}} = \mathbf{F} + \mathbf{Z}, \tag{9.2}$$

which guarantees that the present solutions observe the constraint. Since the motion is determined by $\mathbf{r}$, $\dot{\mathbf{r}}$, and t, these quantities constitute the most general set of variables for the force of constraint,

$$\mathbf{Z} = \mathbf{Z}(\mathbf{r}, \dot{\mathbf{r}}, t), \tag{9.3}$$

even if the force $\mathbf{F}$ depends only on $\mathbf{r}$. From another point of view, for a particle with position $\mathbf{r}$ and velocity $\dot{\mathbf{r}}$, $\mathbf{Z}$ has to be such that at any time t, all components of the force $\mathbf{F}$ that are perpendicular to the prescribed orbit – and that could knock the particle sideways – are balanced by the force of constraint.

The constraint $f(\mathbf{r}, t) = 0$ defines a surface in configuration space. Every solution of the equation of motion must lie on this surface. As argued above, $\mathbf{Z}$ has to be perpendicular to the surface in order to prevent the particle from leaving the surface, if $\mathbf{F}$ has components orthogonal to the surface. Hence, $\mathbf{Z}$ is proportional to the surface normal ∇f:

$$\mathbf{Z} = \lambda(\mathbf{r}, \dot{\mathbf{r}}, t)\nabla f(\mathbf{r}, t), \tag{9.4}$$

where λ in general depends on the variables shown in Eq. (9.3). The function λ is to be determined[3] from Eqs. (9.2) and (9.1). The constraint plays the role of a constant of the motion (or conserved quantity; see the following examples).

What effect does a constraint have on the energy of an otherwise conservative system? To answer this question, we multiply the equation of motion,

$$m\ddot{\mathbf{r}} = \mathbf{F} + \lambda(\mathbf{r}, \dot{\mathbf{r}}, t)\nabla f(\mathbf{r}, t), \tag{9.5}$$

by $\dot{\mathbf{r}}$:

$$\frac{d}{dt}\frac{1}{2}m\dot{\mathbf{r}}^2 = -\nabla V(\mathbf{r})\dot{\mathbf{r}} + \lambda\dot{\mathbf{r}}\nabla f(\mathbf{r}, t).$$

Observing that (since $\dfrac{df}{dt} = \dfrac{\partial f}{\partial \mathbf{r}}\dfrac{d\mathbf{r}}{dt} + \dfrac{\partial f}{\partial t} = 0$) the last term can be rewritten as $\dot{\mathbf{r}}\nabla f = -\dfrac{\partial f}{\partial t}$, we get

$$\frac{d}{dt}\left(\frac{1}{2}m\dot{\mathbf{r}}^2 + V(r)\right) = -\lambda\frac{\partial f}{\partial t}. \tag{9.6}$$

---

[3]One can avoid determining $\lambda(\mathbf{r}, \dot{\mathbf{r}}, t)$ if it is possible – e.g. by the choice of coordinates – to take into account only the components of the force that are tangential to the prescribed orbit (as in the problem of the inclined track and the plane pendulum. See Chapter 3.)

Hence, *if the constraint does not depend explicitly on time, the total energy is still conserved.* The constraint does not contribute to the energy; and the equation of the constraint

$$f(\mathbf{r}) = 0$$

has the *form of a further conservation law*[4].

Two examples shall illustrate this point.

Motion on an inclined track

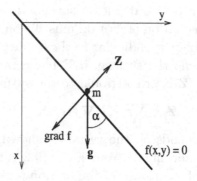

Figure 9.1: Motion on an inclined track.

A point mass in the earth's gravitational field is constrained to move without friction on a track inclined at angle α with respect to the direction of gravitational acceleration $\mathbf{g}$ (see Fig. 9.1; cf. also Section 3.3.1). The constraint is the equation for a straight line,

$$f(x,y) = x \tan \alpha - y = 0. \tag{9.7}$$

We use three different methods to obtain the result.

1) Inserting the normal vector,

$$\boldsymbol{\nabla} f = \begin{pmatrix} \tan \alpha \\ -1 \end{pmatrix} = \frac{1}{\cos \alpha} \begin{pmatrix} \sin \alpha \\ -\cos \alpha \end{pmatrix},$$

---

[4]Thus when $V(\mathbf{r}) = 0$ the kinetic energy is conserved. This has already observed by Euler in the second volume of his "Mechanica ... analytice exposita" (St. Petersburg 1736; see Bibliography)

into Eq. (9.5) yields the system of equations of motion:

$$m\ddot{x} = \lambda \tan\alpha + mg$$
$$m\ddot{y} = -\lambda.$$

The energy of this system (cf. Eq. (9.6)) is given by

$$E = \frac{m}{2}\left(\dot{x}^2 + \dot{y}^2\right) - mgx. \qquad (9.8)$$

Expressing $\ddot{x}$ and $\ddot{y}$ in the second time derivative of Eq. (9.7), $\ddot{x}\tan\alpha = \ddot{y}$, with the help of the above equations of motion, we find for λ

$$\lambda = -mg\sin\alpha\cos\alpha.$$

The force of constraint (cf. Eq. (9.4)),

$$\mathbf{Z} = mg\sin\alpha \begin{pmatrix} -\sin\alpha \\ \cos\alpha \end{pmatrix},$$

is perpendicular to the track, and is directed opposite to the vector normal ∇f. $\mathbf{Z}$ balances the component of the gravitational force parallel to ∇f (E). The 'appropriate' system of equations of motion for the system is now:

$$\ddot{x} = g\cos^2\alpha$$
$$\ddot{y} = g\sin\alpha\cos\alpha.$$

Integrating these equations yields

$$x = \frac{g}{2}t^2\cos^2\alpha + v_x^0 t + x_0$$
$$y = x\tan\alpha.$$

2) A more direct way of finding the system of equations of motion here avoids having to determine the force of constraint. In writing the equations of motion for a particle moving along the straight line given by Eq. (9.7), we can immediately take into account only the component of gravity parallel to the track:

$$F_{\parallel} = mg\cos\alpha.$$

Projecting this force onto the x-axis,

$$F_{\parallel,x} = mg\cos^2\alpha,$$

leads to the above result for the motion in the x-direction

$$m\ddot{x} = mg\cos^2\alpha.$$

Similarly, the motion in the y-direction is obtained.

3) The energy, together with the given constraint, constitute sufficiently many conserved quantities to allow direct integration of
the equations of motion. From the constraint, we have $\dot{y} = \dot{x}\tan\alpha$,
so that the energy is

$$E = \frac{m}{2}\dot{x}^2\frac{1}{\cos^2\alpha} - mgx.$$

This allows us to express $\dot{x}$ as a function of x:

$$\dot{x} = \cos\alpha\sqrt{2E/m + 2gx}.$$

The integration then yields the above result (**E**).

The spherical pendulum

We consider the spherical pendulum, defined as a point mass constrained to move on the surface of a sphere (radius l) in the earth's
gravitational field $\mathbf{g} = (0, 0, -g)$. According to Eq. (9.4), the constraint

$$f(\mathbf{r}) = \mathbf{r}^2 - l^2 = 0 \tag{9.9}$$

gives rise to the force of constraint

$$\mathbf{Z} = 2\lambda\mathbf{r},$$

leading to the following equation of motion (cf. Eq. (9.2)):

$$m\ddot{\mathbf{r}} = m\mathbf{g} + 2\lambda\mathbf{r}. \tag{9.10}$$

It follows from Eq. (9.9) that $\dfrac{d^2 f}{dt^2} = \dfrac{d}{dt}(2\mathbf{r}\dot{\mathbf{r}}) = 2\dot{\mathbf{r}}^2 + 2\mathbf{r}\ddot{\mathbf{r}} = 0$: i.e., that

$$\ddot{\mathbf{r}}\mathbf{r} = -\dot{\mathbf{r}}^2.$$

Multiplying Eq. (9.10) by $\mathbf{r}$ to give $m\ddot{\mathbf{r}}\mathbf{r} = mgr + 2\lambda\mathbf{r}^2$, then substituting into the left hand side for $\ddot{\mathbf{r}}\mathbf{r}$ the expression just found, gives

$$-m\dot{\mathbf{r}}^2 - mgr = 2\lambda\mathbf{r}^2 = 2\lambda l^2.$$

Thus, for λ, we have

$$\lambda(\mathbf{r}, \dot{\mathbf{r}}) = -\frac{m}{2l^2}\left(\dot{\mathbf{r}}^2 + gr\right)$$

(note that $\lambda = \lambda(\mathbf{r}, \dot{\mathbf{r}})$). Hence the force of constraint is given by

$$\mathbf{Z} = -\frac{m}{l^2}(\dot{\mathbf{r}}^2 + gr)\mathbf{r}. \tag{9.11}$$

So, the equation of motion for the spherical pendulum reads

$$m\ddot{\mathbf{r}} = m\mathbf{g} - \frac{m}{l^2}(\dot{\mathbf{r}}^2 + gr)\mathbf{r}. \tag{9.12}$$

The first term in the force of constraint $\mathbf{Z}$ ($\propto \dot{\mathbf{r}}^2\mathbf{r}$) is the **centripetal force**, an attractive force that balances the centrifugal force. The balance can be seen if one considers a circle in a vertical plane in spherical coordinates $\dot{\mathbf{r}} = \dot{r}\mathbf{e}_r + r\dot{\vartheta}\mathbf{e}_\vartheta + r\dot{\varphi}\sin\vartheta\mathbf{e}_\varphi$ (see Eq. (A.20)). Since $\varphi = const$ and $\dot{r} = \dot{\varphi} = 0$, the velocity is $\dot{\mathbf{r}} = l\dot{\vartheta}\mathbf{e}_\vartheta$. Hence, the first term in Eq. (9.11),

$$-\frac{m}{l^2}\dot{\mathbf{r}}^2\mathbf{r} = -ml\dot{\vartheta}^2\mathbf{e}_r,$$

has opposite sign to the centrifugal force $ml\dot{\vartheta}^2\mathbf{e}_r$. The second term in Eq. (9.11) simply cancels the component of the gravitational force acting along the normal to the sphere's surface at the point considered:

$$-m(g\mathbf{e}_r)\mathbf{e}_r \quad \left(= -\frac{m}{l^2}(gr)\mathbf{r}\right).$$

The equation of motion (9.12) is nonlinear, and, moreover, it depends on the velocity $\dot{\mathbf{r}}$. The velocity dependence can be removed using conservation of energy,

$$E = \frac{1}{2}m\dot{\mathbf{r}}^2 - mgr. \tag{9.13}$$

(The constraint (9.9) is independent of time; therefore the right hand side in Eq. (9.6) vanishes.) The equation of motion then reads

$$m\ddot{\mathbf{r}} = m\mathbf{g} - \frac{m}{l^2}\left(\frac{2E}{m} + 3gr\right)\mathbf{r}. \tag{9.14}$$

The nonlinearity in $\mathbf{r}$, however, is still present. Nevertheless, no chaotic behavior appears, because the number of isolating conserved quantities suffices.

Multiplying Eq. (9.12) by $\mathbf{r}$ leads to

$$m\left(\ddot{\mathbf{r}}\mathbf{r} + \dot{\mathbf{r}}^2\mathbf{r}^2/l^2\right) = mg\mathbf{r}\left(1 - \mathbf{r}^2/l^2\right).$$

Note that solutions $\mathbf{r}(t)$ obeying the constraint (9.9), satisfy automatically each side of this equation. Here, again, the constraint acts like a conserved quantity. Multiplying Eq. (9.12) by $\dot{\mathbf{r}}$, as already shown, yields the conservation of energy, Eq. (9.13). And finally taking the cross product of Eq. (9.12) with $\mathbf{r}$ gives

$$m\mathbf{r} \times \ddot{\mathbf{r}} = m\mathbf{r} \times \mathbf{g}.$$

A further multiplication by $\mathbf{g}$,

$$\frac{d}{dt}m(\mathbf{r} \times \dot{\mathbf{r}})\mathbf{g} = \frac{d}{dt}\mathbf{Lg} = 0,$$

yields conservation of the component of angular momentum parallel to $\mathbf{g}$:

$$L_g := \mathbf{Lg}/g = ml^2 \sin^2 \vartheta \dot{\varphi} = const. \tag{9.15}$$

The three conserved quantities, (9.9), (9.13), and (9.15), guarantee the integrability of the system; we can directly calculate the solution from them. In spherical coordinates (see again Eq. (A.20)), with the negative z-axis aligned with $\mathbf{g}$, we can use the constraint (9.9) and conservation of L_g, Eq. (9.15), to express the energy (9.13) only in terms of $\dot{\vartheta}$ and ϑ:

$$E = \frac{ml^2}{2}\left(\dot{\vartheta}^2 + \left(\frac{L_g}{ml^2}\right)^2 \frac{1}{\sin^2 \vartheta}\right) + mgl\cos\vartheta.$$

This can be immediately written in the form

$$\dot{\vartheta} = \sqrt{\frac{2E}{ml^2} - 2\frac{g}{l}\cos\vartheta - \left(\frac{L_g}{ml^2}\right)^2 \frac{1}{\sin^2 \vartheta}}. \tag{9.16}$$

Substituting $\zeta = \cos\vartheta$, the integral of this equation may be written as

$$t = \int \frac{d\zeta}{\sqrt{(2E/ml^2)(1 - \zeta^2) - 2(g/l)(\zeta - \zeta^3) - (L_g/ml^2)^2}}. \tag{9.17}$$

This integral leads to an elliptic function for the solution, $t = t(\vartheta)$ (see, for example, [Whittaker]).

The motion of the pendulum is described by a simple differential equation, when spherical coordinates are simultaneously introduced into the constraint as well as the equation of motion (9.12); by taking advantage of Eqs. (9.9) and (9.15), one finds (E)

$$\ddot{\vartheta} - \left(\frac{L_g}{ml^2}\right)^2 \frac{\cos\vartheta}{\sin^3\vartheta} = \frac{g}{l}\sin\vartheta, \tag{9.18}$$

with $L_g/ml^2 = \sin^2\vartheta\dot{\varphi} = const$. This second order ordinary differential equation is integrable. Its solution is given in Eq. (9.17). For $\dot{\varphi} = 0$, i.e. $L_g = 0$, the equation reduces to that of the plane pendulum (Section 3.3.2), as expected.

To proceed similar to method (2) for the track, and to write down the equation of motion for a particle moving on the surface of a sphere directly, is difficult. The force of constraint considered above is already quite a challenge. For the case $\dot{\varphi} = 0$ (plane pendulum), this method has been anticipated in Section 3.3.2. A more elegant derivation of Eq. (9.18) will be presented in Subsection 9.3.5.

The two examples considered above suggest that solution methods that require determination of the constraint forces can be laborious. The question therefore arises whether there exists a method that allows more direct application of the constraint. If, for example, the equations of motion could be regarded as the equations for the minimum of some quantity, then the constraints could be taken into account by applying the well known method of Lagrange multipliers. In fact, such a method exists. But before continuing, we must introduce some mathematics.

9.2 Calculus of variations

The **brachistochrone problem**, formulated in 1696 by Johann BERNOULLI[5], marks the beginning of the calculus of variations. The problem is as follows (see Fig. 9.2):

---

[5]Johann Bernoulli (1667-1748), Swiss mathematician. Together with his brother Jacob (1655-1705) he founded a very important center of mathematics and mechanics in Basel. His son Daniel (1700-1782) carried on this tradition.

Given two points A and B in a vertical plane, and a point mass M able to move under its own weight, find the path from A to B that minimizes the travel time.

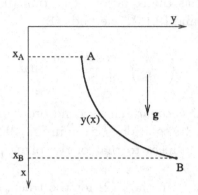

Figure 9.2: The brachistochrone problem.

Since the path $y = y(x)$ to be found for the particle represents a time independent constraint $f(x, y) = 0$, according to Eq. (9.6), the equation for conservation of energy for the point mass initially at rest in x_A reads

$$\frac{M}{2}v^2 - Mgx = -Mgx_A;$$

hence the velocity of the particle is given by

$$v = \sqrt{2g(x - x_A)}. \tag{9.19}$$

On the other hand, v is the velocity along the yet unknown curve $\mathcal{C}$, whose equation is $y = y(x)$, and which has element of arclength $ds = \sqrt{(dx)^2 + (dy)^2} = dx\sqrt{1 + (dy/dx)^2}$,

$$v = \frac{ds}{dt} = \frac{dx}{dt}\sqrt{1 + \left(\frac{dy}{dx}\right)^2}. \tag{9.20}$$

According to Eqs. (9.19) and (9.20), the time taken for the particle to move from x_A to x_B along the curve C is

$$T = \int_{t(x_A),C}^{t(x_B)} dt = \int_{x_A,C}^{x_B} \frac{dt}{dx} dx = \int_{x_A}^{x_B} \frac{1}{v} \sqrt{1 + \left(\frac{dy}{dx}\right)^2} dx$$

$$= \int_{x_A}^{x_B} \frac{1}{\sqrt{2g(x - x_A)}} \sqrt{1 + \left(\frac{dy}{dx}\right)^2} dx.$$

The travel time T is a **functional**[6] of the path $y(x)$

$$T = \int_{x_A}^{x_B} f(y, y'; x) dx, \tag{9.21}$$

where $y' = \dfrac{dy}{dx}$. (Note: in general, we allow f to depend also on y.) For this brachistochrone (meaning 'shortest time') problem, the integrand

$$f(y, y'; x) = \frac{\sqrt{1 + y'^2}}{\sqrt{2g(x - x_A)}} \tag{9.22}$$

is independent of y. Each curve $y(x)$ between the points A and B (with $y_A = y(x_A)$ and $y_B = y(x_B)$) is assigned a real number T. The problem now is to look for the path $y = y(x)$ for which the time T taken for the point mass M to move from A to B is minimal. We shall complete the problem later; first we present the general method.

9.2.1 The Euler-Lagrange equation

The following *fundamental lemma* is basic to the calculus of variations:

Let $g(x)$ be a continuous function. If any continuously differentiable function $\eta(x)$, with

$$\eta(x_1) = \eta(x_2) = 0, \tag{9.23}$$

---

[6]I.e. the function $y(x)$ is mapped on a real number (by the integral).

satisfies

$$\int_{x_1}^{x_2} g(x)\eta(x)dx = 0, \qquad (9.24)$$

then $g(x)$ vanishes identically in the interval $x_1 \leq x \leq x_2$.

This lemma enables us to find an equation for the function $y(x)$ that satisfies the boundary conditions

$$y_1 = y(x_1) \qquad \text{and} \qquad y_2 = y(x_2) \qquad (9.25)$$

and at the same time minimizes (more precisely, extremizes) the functional

$$I = I[y] = \int_{x_1}^{x_2} f(y, y'; x)dx. \qquad (9.26)$$

To find this function $y(x)$, we set

$$y(x, \alpha) = y(x) + \alpha\eta(x),$$

where $\eta(x)$ satisfies the boundary conditions in Eq. (9.23) but is otherwise arbitrary (consequently, $y(x, \alpha)$ assumes the boundary values y_1 and y_2 given by (9.25), for any α). We consider the integral

$$I(\alpha) = \int_{x_1}^{x_2} f\left(y(x, \alpha), y'(x, \alpha); x\right) dx$$

as a function of α. Since $\alpha = 0$ will yield the $y(x)$ for which the function $I(\alpha)$ is minimal, it must be true that

$$\left.\frac{dI}{d\alpha}\right|_{\alpha=0} = \int_{x_1}^{x_2} \left\{\frac{\partial f}{\partial y}\frac{dy}{d\alpha} + \frac{\partial f}{\partial y'}\frac{dy'}{d\alpha}\right\} dx \Bigg|_{\alpha=0} = 0.$$

From the definition of $y(x, \alpha)$, we have $\dfrac{dy}{d\alpha} = \eta$ and $\dfrac{dy'}{d\alpha} = \dfrac{d}{dx}\dfrac{dy}{d\alpha} = \dfrac{d}{dx}\eta(x)$; integration by parts gives

$$\frac{dI}{d\alpha} = \int_{x_1}^{x_2} \left\{\frac{\partial f}{\partial y}\eta - \left(\frac{d}{dx}\frac{\partial f}{\partial y'}\right)\eta\right\} dx + \frac{\partial f}{\partial y'}\eta\Bigg|_{x_1}^{x_2}.$$

The last term vanishes because of the assumption in Eq. (9.23). For $\alpha = 0$, we have $y(x, 0) = y(x)$, and hence

$$\left. \frac{dI}{d\alpha} \right|_{\alpha=0} = \int_{x_1}^{x_2} \left\{ \frac{\partial f}{\partial y} - \frac{d}{dx} \frac{\partial f}{\partial y'} \right\} \eta(x) dx.$$

By virtue of the fundamental lemma, Eq. (9.24), the function $y(x)$ is a solution of the **Euler-Lagrange differential equation**[7]

$$\frac{\partial f}{\partial y} - \frac{d}{dx} \frac{\partial f}{\partial y'} = 0. \tag{9.27}$$

Written out explicitly (recall $f = f(y, y'; x)$), this equation reads

$$\frac{\partial f}{\partial y} - \frac{\partial^2 f}{\partial y' \partial y} y' - \frac{\partial^2 f}{\partial y'^2} y'' - \frac{\partial^2 f}{\partial y' \partial x} = 0. \tag{9.28}$$

This second order ordinary differential equation for the function $y(x)$ has the form

$$g_1(y, y', x) \frac{d^2 y}{dx^2} + g_2(y, y', x) \frac{dy}{dx} + g_3(y, y', x) = 0.$$

The two arbitrary constants in the general solution are fixed by the boundary conditions in Eq. (9.25). Given a solution of Eq. (9.27), the value of I, Eq. (9.26), is extremal but not necessarily minimal.

Two particular cases of $f(y, y'; x)$

i) f does not depend on y,

$$f = f(x, y'). \tag{9.29}$$

---

[7]This equation was developed when Lagrange starting in 1754 tackled the *tautochrone problem* (to find the curve on which a particle will fall to a fixed point in a time independent of the starting point; Huygens showed already 1659 that the solution is a cycloid. Lagrange's solution was published not until 1765 in the Berlin Academy proceedings). He communicated his results to Euler. Earlier, in 1733, Euler had begun with his activities in the calculus of variations (as he called it later). He immediately adopted Lagrange's method and published 1766 his *Elementa calculi variationum* in the St. Petersburg Academy transactions [Thiele].

Since $\partial f/\partial y = 0$, the Euler-Lagrange equation implies that $\dfrac{d}{dx}\dfrac{\partial f}{\partial y'} = 0$, and therefore

$$\frac{\partial f}{\partial y'} = const \qquad (9.30)$$

is an integral of Eq. (9.27).

ii) f does not depend on x explicitly,

$$f = f(y, y'). \qquad (9.31)$$

Regarding y as the independent variable, one has

$$y' = dy/dx = 1/(dx/dy) =: 1/x'$$

and (note that $x' = dx/dy$!)

$$\int f(y, y')dx = \int f\left(y, 1/x'\right) x'dy =: \int F(x', y)dy.$$

Since F is independent of x, according to (i), we have

$$\frac{\partial F}{\partial x'} = \frac{\partial}{\partial x'} f\left(y, 1/x'\right) x' = const;$$

i.e.,

$$f\left(y, 1/x'\right) + x'\frac{\partial f}{\partial (1/x')}\left(-\frac{1}{x'^2}\right) = const,$$

which may be rewritten in terms of y and y' as

$$f(y, y') - \frac{\partial f}{\partial y'}y' = const. \qquad (9.32)$$

This is an integral of (9.27).

Now we return to the brachistochrone problem. The function f, Eq. (9.22), in the brachistochrone problem is independent of y. This corresponds to case (i) above. Hence, according to Eq. (9.30),

$$\frac{\partial f}{\partial y'} = \frac{y'}{\sqrt{1+y'^2}\sqrt{2g(x - x_A)}} = const =: \frac{1}{\sqrt{2gc_1}},$$

and therefore

$$y' = \sqrt{\frac{x - x_A}{c_1 - (x - x_A)}}. \tag{9.33}$$

The integral of this equation for $x \leq c_1$ is

$$y = c_2 - \sqrt{c_1(x - x_A) - (x - x_A)^2} - c_1 \arcsin \sqrt{\frac{c_1 - (x - x_A)}{c_1}}.$$

If the starting point A has the coordinates $x_A = 0$ and $y_A = y(x_A) = 0$, then $c_2 = \frac{\pi}{2} c_1$ and

$$\begin{aligned} y &= \frac{\pi}{2} c_1 - \sqrt{c_1 x - x^2} - c_1 \arcsin \sqrt{\frac{c_1 - x}{c_1}} \qquad (9.34) \\ &= c_1 \arccos \sqrt{\frac{c_1 - x}{c_1}} - \sqrt{c_1 x - x^2}. \end{aligned}$$

The constant c_1 is determined by the coordinates $(x_B, y_B = y(x_B))$ of B. The equation for c_1 is, in general, transcendental. The slope $y'(x_A = 0)$ at A (cf. Eq. (9.33)) is

$$y'(0) = 0;$$

the point mass initially moves in the direction of gravity.

It may be easier to envisage the curve using the following parameterization. Let

$$c_1 = 2a.$$

Substituting this into Eq. (9.34), we find that the resulting equation is satisfied by

$$x = a(1 - \cos t), \qquad y = a(t - \sin t). \tag{9.35}$$

This shows that the curve of the brachistochrone is a **cycloid** (i.e. the path traced by a particular point on a circle of radius a that is rolling along a straight line).

Finally, we only state the generalization of the Euler-Lagrange problem to a function f that depends on the functions $y_1(x), \ldots, y_n(x)$. We look for the extremum of the functional

$$I = \int_{x_1}^{x_2} f(y_1, \ldots, y_n; y_1', \ldots, y_n'; x) dx \tag{9.36}$$

with boundary conditions

$$\alpha_i = y_i(x_1), \quad \beta_i = y_i(x_2), \qquad i = 1, \ldots, n. \tag{9.37}$$

Repeating the above steps for these n functions leads to a system of Euler-Lagrange differential equations for f

$$\frac{\partial f}{\partial y_i} - \frac{d}{dx}\frac{\partial f}{\partial y_i'} = 0, \qquad i = 1, \ldots, n. \tag{9.38}$$

9.2.2 Transforming the variables

In applications to mechanics, it is important to consider the behavior of the variational method for a function $f\left(\{y\}; \{y'\}; x\right)$ under transformations of the variables – which, here, are functions – $y_i(x)$ to some other variables $q_i(x)$:

$$q_i(x) = q_i(y_1(x), \ldots, y_n(x)) =: q_i(\{y\}), \qquad i = 1, \ldots, n.$$

From the inverse transformation, $y_i = y_i(\{q\})$, the derivatives y_i' can be expressed in terms of the derivatives q_j',

$$y_i' = \sum_k \frac{\partial y_i}{\partial q_k} q_k' = y_i'(\{q\}, \{q'\}).$$

Thus the function $h\left(\{q\}; \{q'\}; x\right) = f\left(y\left(\{q\}\right); y'\left(\{q\}, \{q'\}\right); x\right)$ is found. Since the functional I (which, by definition, is real-valued – for instance, the travel time T between A and B) must be independent of the choice of variables (polar or Cartesian coordinates, say), we must have

$$I = \int_{x_1}^{x_2} f\left(\{y\}; \{y'\}; x\right) dx = \int_{x_1}^{x_2} h\left(\{q\}; \{q'\}; x\right) dx, \tag{9.39}$$

for which the transformed boundary conditions for each of the q_i (cf. Eq. (9.37)) are denoted as

$$\bar{\alpha}_i = q_i(\{\alpha\}) \quad \text{and} \quad \bar{\beta}_i = q_i(\{\beta\}).$$

The Euler-Lagrange equations with respect to the $\{q_i\}$ follow immediately from Eq. (9.39),

$$\frac{\partial h}{\partial q_i} - \frac{d}{dx}\frac{\partial h}{\partial q_i'} = 0, \qquad i = 1, \ldots, n. \tag{9.40}$$

Thus the Euler-Lagrange equations for the function $h\left(\{q\};\{q'\};x\right)$ are of the same form as before; i.e. the prescription for finding the function that extremizes the functional is the same in the new variables.

A frequently used transformation is parametrization. Let us consider, for example, a variational problem for a function f of three Cartesian coordinates x, y, z:

$$I = \int_{x_1}^{x_2} f(y, z; y', z'; x)dx. \qquad (9.41)$$

The solution $y = y(x)$ and $z = z(x)$ is a curve in Euclidean space. This unsymmetrical representation of the integral may be undesirable. One may introduce a parameter t for all three coordinates[8]

$$x = x(t), \qquad y = y(t), \qquad z = z(t). \qquad (9.42)$$

Since

$$t_1 = t(x_1) \qquad \text{and} \qquad t_2 = t(x_2), \qquad (9.43)$$

the boundary values are

$$x_i = x(t_i), \quad y_i = y(t_i), \quad z_i = z(t_i), \qquad i = 1, 2.$$

Instead of y' and z', we now have derivatives with respect to t,

$$y' = \dot{y}/\dot{x}, \quad z' = \dot{z}/\dot{x} \qquad \text{with} \quad \dot{x} = \frac{dx}{dt}, \text{ etc.}$$

The integral (9.41) now takes the form

$$\int_{x_1}^{x_2} f(y, z; y', z'; x)dx = \int_{t_1}^{t_2} f(y, z; \dot{y}/\dot{x}, \dot{z}/\dot{x}; x)\dot{x}dt.$$

If we define

$$h(x, y, z; \dot{x}, \dot{y}, \dot{z}) = f(y, z; \dot{y}/\dot{x}, \dot{z}/\dot{x}; x)\dot{x},$$

---

[8]Since there are many parametric representations $(x(t), y(t), z(t))$ of a function $(y(x), z(x))$, one must make sure that the value of the integral is independent of the parameterization. This condition places a restriction on the possible forms of the function f.

we get finally

$$\int_{x_1}^{x_2} f(y, z; y', z'; x) dx = \int_{t_1}^{t_2} h(x, y, z; \dot{x}, \dot{y}, \dot{z}) dt. \qquad (9.44)$$

All three coordinates now appear as dependent variables. This can be useful when it reflects the symmetry of the problem given. The Euler-Lagrange equations for h are:

$$\frac{\partial h}{\partial x} - \frac{d}{dt}\frac{\partial h}{\partial \dot{x}} = 0, \qquad \frac{\partial h}{\partial y} - \frac{d}{dt}\frac{\partial h}{\partial \dot{y}} = 0, \qquad \frac{\partial h}{\partial z} - \frac{d}{dt}\frac{\partial h}{\partial \dot{z}} = 0. \qquad (9.45)$$

Since h depends on t only implicitly (cf. Eq. (9.44)), one can integrate the general Eqs. (9.45) once. Because $\partial h/\partial t = 0$, we have

$$\begin{aligned}
\frac{d}{dt}h &= \frac{\partial h}{\partial x}\dot{x} + \frac{\partial h}{\partial y}\dot{y} + \frac{\partial h}{\partial z}\dot{z} + \frac{\partial h}{\partial \dot{x}}\ddot{x} + \frac{\partial h}{\partial \dot{y}}\ddot{y} + \frac{\partial h}{\partial \dot{z}}\ddot{z} \\
&= \dot{x}\frac{d}{dt}\frac{\partial h}{\partial \dot{x}} + \dot{y}\frac{d}{dt}\frac{\partial h}{\partial \dot{y}} + \dot{z}\frac{d}{dt}\frac{\partial h}{\partial \dot{z}} + \frac{\partial h}{\partial \dot{x}}\ddot{x} + \frac{\partial h}{\partial \dot{y}}\ddot{y} + \frac{\partial h}{\partial \dot{z}}\ddot{z} \\
&= \frac{d}{dt}\left(\frac{\partial h}{\partial \dot{x}}\dot{x} + \frac{\partial h}{\partial \dot{y}}\dot{y} + \frac{\partial h}{\partial \dot{z}}\dot{z}\right),
\end{aligned}$$

where, in the second step, we made use of Eqs. (9.45). This shows that

$$h - \left(\frac{\partial h}{\partial \dot{x}}\dot{x} + \frac{\partial h}{\partial \dot{y}}\dot{y} + \frac{\partial h}{\partial \dot{z}}\dot{z}\right) = const \qquad (9.46)$$

is an integral of the Euler-Lagrange equations (9.45) (cf. also Eq. (9.32)).

9.2.3   Constraints

If constraints are given as part of the variational problem, these can be taken into account in a simple way. The procedure is demonstrated in the following problem.

Find the curve through the points in space $P_1(x_1, y_1, z_1)$ and $P_2(x_2, y_2, z_2)$ such that[9]

$$I = \int_{x_1}^{x_2} f(y, z; y', z'; x) dx \qquad (9.47)$$

---

[9]The curve is given e.g. by $y = y(x)$ and $z = z(x)$; the boundary conditions are $y_i = y(x_i)$, $z_i = z(x_i)$, $i = 1, 2$.

is extremal (minimal). Moreover, the curve must lie on the surface

$$G(x, y, z) = 0. \tag{9.48}$$

(Of course, the given points P_1 and P_2 must also lie on this surface.)

As with numerous everyday optimization problems, the method of **Lagrange multipliers** can be used here, too. Thus, one considers the functional

$$\bar{I} = \int_{x_1}^{x_2} h(y, z; y', z'; x) dx,$$

with

$$h(y, z; y', z'; x) = f(y, z; y', z'; x) + \lambda(x) G(x, y, z),$$

where $\lambda(x)$ is a Lagrange multiplier. Since the boundary values for y and z must lie on the constraining surface given by Eq. (9.48), these are still valid for the adapted Lagrangian multipliers problem. From the Euler-Lagrange equations for $\bar{I}$, we get the system of differential equations,

$$\frac{\partial h}{\partial y} - \frac{d}{dx}\frac{\partial h}{\partial y'} = \frac{\partial f}{\partial y} - \frac{d}{dx}\frac{\partial f}{\partial y'} + \lambda(x)\frac{\partial G}{\partial y} = 0$$

$$\frac{\partial h}{\partial z} - \frac{d}{dx}\frac{\partial h}{\partial z'} = \frac{\partial f}{\partial z} - \frac{d}{dx}\frac{\partial f}{\partial z'} + \lambda(x)\frac{\partial G}{\partial z} = 0. \tag{9.49}$$

Together with Eq. (9.48), we have now three equations for the functions $y(x)$, $z(x)$, and $\lambda(x)$. The generalization to several variables or several constraints is easy.

Another way of treating constraints, namely, *transformation to coordinates that are more adapted to the constraint*, is demonstrated in the following example.

> *Find the **geodesic** (i.e. the line of shortest distance) between two points $P_1(a, 0, 0)$ and $P_2(-a, 0, \pi)$ on the cylindrical surface given by $x^2 + y^2 - a^2 = 0$.*

After parametrizing the coordinates, *viz.*, $x = x(t)$, $y = y(t)$, and $z = z(t)$, the element of arclength is

$$ds = \sqrt{(dx)^2 + (dy)^2 + (dz)^2} = \sqrt{\left(\frac{dx}{dt}\right)^2 + \left(\frac{dy}{dt}\right)^2 + \left(\frac{dz}{dt}\right)^2}\, dt.$$

Consequently, for the arclength between P_1 and P_2, we have

$$L = \int_{s_1}^{s_2} ds = \int_{t_1}^{t_2} \sqrt{\left(\frac{dx}{dt}\right)^2 + \left(\frac{dy}{dt}\right)^2 + \left(\frac{dz}{dt}\right)^2} \, dt,$$

where the positions s_i on the curve of integration at parameter values t_i correspond to the points P_i. The cylinder $x^2 + y^2 = a^2$ is parametrized according to

$$x = a\cos t, \quad y = a\sin t, \quad z = z(t),$$

so that only the function $z(t)$ remains to be determined, and we need not care about the constraint anymore. At P_1, the value of the parameter t is $t = t_1 = 0$, and at P_2, it is $t = t_2 = \pi$. Since $dx/dt = -a\sin t$, and $dy/dt = a\cos t$, we now have the condition that the length

$$L = \int_0^\pi \sqrt{a^2 + \dot{z}^2} \, dt$$

is extremal. Because the integrand is independent of z, the Euler-Lagrange equation for $z(t)$ reduces to $\frac{\partial}{\partial \dot{z}}\sqrt{a^2 + \dot{z}^2} = const$ (cf. Eq. (9.30)); or,

$$\frac{\dot{z}}{\sqrt{a^2 + \dot{z}^2}} = const =: \frac{1}{\sqrt{c}}.$$

Now it follows that $\dot{z} = \dfrac{a}{\sqrt{c-1}}$; furthermore,

$$z = \frac{a}{\sqrt{c-1}} t + z_0.$$

The boundary conditions at $t_1 = 0$ and $t_2 = \pi$ require that $z_0 = 0$ and $a/\sqrt{c-1} = 1$ (whence $c = a^2 + 1$). The solution is a screw curve, shown in Fig. 9.3:

$$\begin{aligned} x &= a\cos t \\ y &= a\sin t \\ z &= t \, . \end{aligned}$$

The arclength between P_1 and P_2 is

$$L = \int_0^\pi \sqrt{a^2 + 1} \, dt = \pi\sqrt{a^2 + 1}.$$

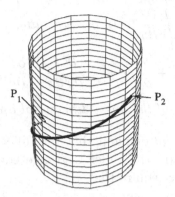

Figure 9.3: The geodesic on the surface of a cylinder (cf. text).

This method often proves to be particularly convenient when taking into account constraints. The ease with which constraints are handled or transformations of variables are performed, suggests that it is favorable to use a variational formulation of mechanics to start with, rather than the basic approach using Newton's equation.

9.3 The Lagrangian

9.3.1 The inverse problem in the calculus of variations

The inverse problem in the calculus of variations[10] may be stated as follows.

Given a second order ordinary differential equation,

$$y'' + \Phi(x, y, y') = 0,$$

does there exist a variational problem,

$$\int_{x_1}^{x_2} f(y, y'; x)dx = extreme\ value,$$

---

[10] Only since the beginning of the 20th century there has been an appreciable activity by numerous authors in the study of the inverse problem.

such that the solutions of the Euler-Lagrange differential equations $\dfrac{\partial f}{\partial y} - \dfrac{d}{dx}\dfrac{\partial f}{\partial y'} = 0$, *or, explicitly,*

$$\frac{\partial^2 f}{\partial y'^2}y'' + \frac{\partial^2 f}{\partial y'\partial y}y' + \frac{\partial^2 f}{\partial y'\partial x} - \frac{\partial f}{\partial y} = 0,$$

are also solutions of the given differential equation?

In general, one *cannot* demand that the differential equation $y'' + \Phi(x, y, y') = 0$ is already the Euler-Lagrange equation (9.28) for the functional; one can only require that

$$My'' + M\Phi = 0 \tag{9.50}$$

is the Euler-Lagrange equation, where one allows for a multiplying function $M = M(x, y, y')$. Comparing with Eq. (9.28), we see that

$$\frac{\partial^2 f}{\partial y'^2} = M \quad \text{and} \quad \frac{\partial^2 f}{\partial y'\partial y}y' + \frac{\partial^2 f}{\partial y'\partial x} - \frac{\partial f}{\partial y} = M\Phi. \tag{9.51}$$

Differentiating the second equation with respect to y' and making use of the first equation, one obtains a linear partial differential equation for the factor M:

$$\frac{\partial M}{\partial x} + y'\frac{\partial M}{\partial y} - \Phi\frac{\partial M}{\partial y'} - \frac{\partial \Phi}{\partial y'}M = 0. \tag{9.52}$$

The solution of this differential equation contains an arbitrary function, so that there are infinitely many solutions of the inverse problem (G. Darboux, about 1890). In order to regain uniqueness, one needs to impose still another condition on the function $f(y, y'; x)$ (see below).

We apply the inverse problem to Newton's equation, to see what kind of function $f(y, y'; x)$ emerges.

9.3.2   The inverse problem for Newton's equation of motion

Let us consider the equation for the one-dimensional motion of a particle in a potential $V(x)$:

$$m\ddot{x} = -\frac{d}{dx}V(x). \tag{9.53}$$

We ask:

Is there a function $L(x, \dot{x}, t)$, together with a variational problem

$$\int_{t_1}^{t_2} L(x, \dot{x}, t)dt = extreme\ value, \qquad (9.54)$$

such that the Euler-Lagrange differential equation,

$$\frac{\partial L}{\partial x} - \frac{d}{dt}\frac{\partial L}{\partial \dot{x}} = 0, \qquad (9.55)$$

has the same solutions as the equation of motion, Eq. (9.53); or, even more desirable, takes the same form as the equation of motion?

We focus on the second part of the question, whether for some choice of $L(x, \dot{x}, t)$, the Euler-Lagrange equation can take the same form as Newton's equation of motion. Since only the second term in the Euler-Lagrange equation can contain $\ddot{x}$, by comparing it with Newton's equation, we immediately deduce the two equations

$$\frac{d}{dt}\frac{\partial L}{\partial \dot{x}} = m\ddot{x} \qquad (9.56)$$

$$\frac{\partial L}{\partial x} = -\frac{\partial V}{\partial x}. \qquad (9.57)$$

The solution of Eq. (9.57) is

$$L = T(\dot{x}) - V(x), \qquad (9.58)$$

with some unknown function $T(\dot{x})$. This function $T(\dot{x})$ can be determined from Eq. (9.56), which, after inserting $L = T(\dot{x}) - V(x)$, reads

$$\frac{d}{dt}\frac{\partial T}{\partial \dot{x}} = m\ddot{x}.$$

Integration over t yields

$$\frac{\partial T}{\partial \dot{x}} = m\dot{x} \quad (+\ const),$$

and a further integration over $\dot{x}$ shows that T is the kinetic energy of the particle:

$$T = \frac{1}{2}m\dot{x}^2 \quad (+\ const). \qquad (9.59)$$

Hence, the **Lagrangian**[11] $L(x, \dot{x}, t)$,

$$L(x, \dot{x}, t) = \frac{1}{2}m\dot{x}^2 - V(x) = T - V, \qquad (9.60)$$

is a solution of our inverse problem. Since L does not depend on time explicitly, according to Eq. (9.32),

$$L - \frac{\partial L}{\partial \dot{x}}\dot{x} = const \qquad (9.61)$$

is a first integral of the Euler-Lagrange equation. Because $\dfrac{\partial L}{\partial \dot{x}} = 2T$, we have

$$L - \frac{\partial L}{\partial \dot{x}}\dot{x} = L - 2T = -(T + V) = -E = const,$$

so that the total energy is conserved:

$$E = T + V = const. \qquad (9.62)$$

Even if V also depends on time, the Lagrangian is $L(x, \dot{x}, t) = T - V(x, t)$, and the Euler-Lagrange equations take the form of Eq. (9.55). However, in this case, the energy $T + V$ is not conserved anymore (since Eq. (9.61) is then not a solution of the Euler-Lagrange equation).

The derivative of the Lagrangian (9.60) with respect to $\dot{x}$, is the momentum p_x,

$$\frac{\partial L}{\partial \dot{x}} = m\dot{x} = p_x. \qquad (9.63)$$

Provided that $\partial L/\partial x = 0$, i.e. L does not depend on x explicitly – x is said to be a *cyclic coordinate* – it follows from Eq. (9.55) that the *momentum p_x conjugate to x* is a conserved quantity.

Remark

Since the solution to the inverse variational problem is not unique, to ensure uniqueness, we must impose the requirement that $L = T - V$, where T and V are the kinetic and potential energy, respectively. For example, the equation of motion

$$m\ddot{x} + kx = 0,$$

---

[11]This function was introduced 1788 by Lagrange in his "Analytical Mechanics". Its first applications have a touch of intricateness.

is the Euler-Lagrange equation for the following function L:

$$L = \frac{m}{2}\dot{x}^2 - \frac{k}{2}x^2.$$

However, other functions L' exist whose Euler-Lagrange equations yield the same equation of motion but are not of the form $L' = T - V$. An example is

$$L' = \frac{m}{3}\dot{x}^4 + 2km\dot{x}^2 x^2 - k^2 x^4$$

(here $M = m\dot{x}^2 + kx^2$; cf. Eq. (9.50)).

9.3.3 The Lagrangian for a single particle

We apply the above procedure to the motion of a particle in three-dimensional space. Identifying the x, y, and z-components of the equation of motion,

$$m\ddot{\mathbf{r}} = -\frac{\partial}{\partial \mathbf{r}} V(\mathbf{r}), \qquad \mathbf{r} = (x, y, z), \qquad (9.64)$$

with the Euler-Lagrange equations

$$\frac{\partial L}{\partial x} - \frac{d}{dt}\frac{\partial L}{\partial \dot{x}} = 0, \quad \text{etc.},$$

for a Lagrangian $L(x, y, z, \dot{x}, \dot{y}, \dot{z}, t) = L(\mathbf{r}, \dot{\mathbf{r}}, t)$, we obtain (E)

$$L(\mathbf{r}, \dot{\mathbf{r}}, t) = \frac{m}{2}\dot{\mathbf{r}}^2 - V(\mathbf{r}). \qquad (9.65)$$

From now on, as customary, the Euler-Lagrange equation is written in the form

$$\frac{d}{dt}\frac{\partial L}{\partial \dot{\mathbf{r}}} - \frac{\partial L}{\partial \mathbf{r}} = 0, \qquad (9.66)$$

and is called simply **Lagrange's equation**. It yields the equation of motion (9.64).

The derivative $\partial L/\partial \dot{\mathbf{r}}$ appearing in Eq. (9.66) is called the **(canonically) conjugate momentum**[12] or **canonical momentum p**,

$$\mathbf{p} = \frac{\partial L}{\partial \dot{\mathbf{r}}}. \qquad (9.67)$$

---

[12]The momentum is conjugate to the radius vector $\mathbf{r}$.

For the Lagrangian (9.65), the canonical momentum

$$\frac{\partial L}{\partial \dot{\mathbf{r}}} = m\dot{\mathbf{r}}$$

coincides with the momentum that is already familiar to us.

Finding the Lagrangian for the motion of a charged particle in a homogeneous magnetic field, as discussed in Section 5.5, is somewhat more difficult. For a magnetic field, to obtain the equation of motion (5.64) from the Lagrangian, instead of the field $\mathbf{B}$, one must introduce the vector potential $\mathbf{A}$ connected with $\mathbf{B}$ via $\mathbf{B} = \nabla \times \mathbf{A}$. In the Lagrangian,

$$L = \frac{m}{2}\dot{\mathbf{r}}^2 + \frac{q}{c}\dot{\mathbf{r}}\mathbf{A} - V(\mathbf{r}),$$

$\mathbf{A}$ is coupled to the velocity $\dot{\mathbf{r}}$. (See text books on electrodynamics, and also, for example, [Goldstein].) The coupling of the magnetic field (or rather its potential) to the velocity alters the relation between the canonical momentum and the velocity. This relation is now

$$\mathbf{p}_{can} = \frac{\partial L}{\partial \dot{\mathbf{r}}} = m\dot{\mathbf{r}} + \frac{q}{c}\mathbf{A}. \qquad (9.68)$$

In the case of a homogeneous magnetic field, the vector potential can be taken as[13]

$$\mathbf{A} = -\frac{1}{2}\mathbf{r} \times \mathbf{B}.$$

It is easy to see that the equations of motion (5.61) are just Lagrange's equations for the Lagrangian,

$$L = \frac{m}{2}\dot{\mathbf{r}}^2 - \frac{q}{2c}\dot{\mathbf{r}}\left(\mathbf{r} \times \mathbf{B}\right) - V(\mathbf{r}). \qquad (9.69)$$

The canonical momentum is then

$$\mathbf{p}_{can} = m\dot{\mathbf{r}} - \frac{q}{2c}\left(\mathbf{r} \times \mathbf{B}\right). \qquad (9.70)$$

Defining the *canonical angular momentum* by

$$\mathbf{L}_{can} := \mathbf{r} \times \mathbf{p}_{can} \qquad (9.71)$$

---

[13]Since the vector potential is not unique, it is only defined up to a gauge function (cf. textbooks on Electrodynamics).

we find that its projection on the magnetic field is given by

$$\mathbf{L}_{can}\mathbf{B} = m\left(\mathbf{r} \times \dot{\mathbf{r}}\right)\mathbf{B} - \frac{q}{2c}\left[\mathbf{r} \times \left(\mathbf{r} \times \mathbf{B}\right)\right]\mathbf{B}$$
$$= m\left(\mathbf{r} \times \dot{\mathbf{r}}\right)\mathbf{B} + \frac{q}{2c}\left(\mathbf{r} \times \mathbf{B}\right)^2. \tag{9.72}$$

Therefore, the conserved quantity appearing in Section 5.5 is nothing but the component of the canonical angular momentum parallel to the magnetic field $\mathbf{B}$:

$$I_{LB} = \mathbf{L}_{can}\mathbf{B}/B. \tag{9.73}$$

9.3.4 Hamilton's principle

The integral over L is called the **action integral** S:

$$S = \int_{t_1}^{t_2} L(\mathbf{r}, \dot{\mathbf{r}}, t)dt. \tag{9.74}$$

It is extremal when the variables appearing in L are solutions of Lagrange's equations (cf. Eq. (9.54)). The condition

$$S = extremum,$$

is written as

$$\delta S = 0 \tag{9.75}$$

(meaning that the variation of S vanishes), and can be interpreted as follows. Of all the conceivable paths $\mathbf{r}(t)$ with fixed endpoints $\mathbf{r}(t_i)$, $i = 1, 2$, only those are realized physically whose action integral S is minimal (extremal). That is, only paths that are solutions of Lagrange's equations (9.66) – and therefore also of Newton's equations of motion – are physical. This statement is called **Hamilton's principle**[14].

---

[14]William Rowan HAMILTON (1805-1865), Irish physicist, astronomer, and mathematician.

W.R. Hamilton, *On a General Method in Dynamics* (First and second essay), Philosophical Transaction of the Royal Society, Part II (1834) pp. 247; Part I (1835) pp. 95.

With the principle of least action, Hamilton completed ideas originally advanced 1744 by Pierre-Louis Moreau de MAUPERTUIS (1698-1759), published in "Essai de cosmologie" (Amsterdam, 1750), and more clearly by Euler in "Methodus inveniendi lineas curvas" (Lausanne & Geneva, 1744).

Formulation of Hamilton's principle in terms of functional derivatives

Analogous to the derivative dx/dt of a function $x(t)$, the functional derivative $\delta S/\delta x(t)$ of a functional $S[x(t')] = \int L(x(t'))\, dt'$ is defined by

$$\frac{\delta S[x(t')]}{\delta x(t)} = \lim_{\varepsilon \to 0} \frac{S[x(t') + \varepsilon\delta(t - t')] - S[x(t')]}{\varepsilon}.$$

We consider the functional derivative of the action integral with respect to the path, $\dfrac{\delta S}{\delta x(t)}$. Because $\dfrac{\delta x(t')}{\delta x(t)} = \delta(t' - t)$, one obtains

$$\frac{\delta S}{\delta x(t)} = \int_{t_1}^{t_2} \frac{\delta L(x, \dot{x}, t')}{\delta x(t)}\, dt' = \int_{t_1}^{t_2} \left(\frac{\partial L}{\partial x} \frac{\delta x(t')}{\delta x(t)} + \frac{\partial L}{\partial \dot{x}} \frac{\delta \dot{x}(t')}{\delta x(t)} \right) dt'$$

$$= \int_{t_1}^{t_2} \left(\frac{\partial L}{\partial x} - \frac{d}{dt'} \frac{\partial L}{\partial \dot{x}} \right) \delta(t' - t) dt' = \frac{\partial L}{\partial x} - \frac{d}{dt} \frac{\partial L}{\partial \dot{x}},$$

where the identity $\dfrac{\delta \dot{x}(t')}{\delta x(t)} = \dfrac{d}{dt'} \dfrac{\delta x(t')}{\delta x(t)}$ has been used. For the solution of Lagrange's equation, the functional derivative of the action integral vanishes, $\dfrac{\delta S}{\delta x(t)} = 0$. Or, putting it another way, S is extremal.

9.3.5   The Lagrangian in generalized coordinates

It is often advantageous to change from Cartesian coordinates to other coordinates that are more suitable for the symmetry of the problem (and/or constraint). We therefore review the procedure of Subsection 9.2.2 for the Lagrangian case.

Let $L(\mathbf{r}, \dot{\mathbf{r}}, t)$ be the Lagrangian, in Cartesian coordinates, for the motion of a particle. A transformation to non-Cartesian coordinates (e.g. spherical or cylindrical coordinates),

$$q_i = q_i(x, y, z), \qquad i = 1, 2, 3,$$

can be carried out according to Subsection 9.2.2. Since

$$x = x(\{q\}), \quad y = y(\{q\}), \quad z = z(\{q\}), \qquad \{q\} = (q_1, q_2, q_3),$$

the Lagrangian $L(\mathbf{r}, \dot{\mathbf{r}}, t)$ is transformed to a *new* function $L(\{q\}, \{\dot{q}\}, t)$ of the new variables (cf. (9.39)), where, for convenience, we did not change the name of the function: the functions are distinguished by their arguments. In the new coordinates q_i, Lagrange's equations for the new Lagrangian $L = L(\{q\}, \{\dot{q}\}, t)$ have the same form (cf. (9.40)), i.e.

$$\frac{d}{dt}\frac{\partial L}{\partial \dot{q}_i} - \frac{\partial L}{\partial q_i} = 0, \qquad i = 1, 2, 3, \tag{9.76}$$

as in the original system of coordinates. The **generalized momenta** p_i conjugate to the coordinates q_i are defined as

$$p_i = \frac{\partial L}{\partial \dot{q}_i}. \tag{9.77}$$

Again, we have the following. If L is independent of q_i, $\partial L/\partial q_i = 0$ (q_i is a **cyclic coordinate**), then Eq. (9.76) implies that the *conjugate momentum is a constant of the motion*,

$$p_i = \frac{\partial L}{\partial \dot{q}_i} = const. \tag{9.78}$$

We now present two particular examples: cylindrical coordinates and spherical coordinates.

Cylindrical coordinates

Representing the velocity vector $\dot{\mathbf{r}}$ in the local basis of the vectors $\mathbf{e}_\rho, \mathbf{e}_\varphi$, and $\mathbf{e}_z$ (cf. Eq. (A.27)),

$$\dot{\mathbf{r}} = \dot{\rho}\mathbf{e}_\rho + \rho\dot{\varphi}\mathbf{e}_\varphi + \dot{z}\mathbf{e}_z,$$

one obtains for the Lagrangian of a particle in the potential $V(\mathbf{r})$,

$$L(\rho, \varphi, z, \dot{\rho}, \dot{\varphi}, \dot{z}) = \frac{1}{2}m(\dot{\rho}^2 + \rho^2\dot{\varphi}^2 + \dot{z}^2) - V(\rho, \varphi, z). \tag{9.79}$$

Hence, Lagrange's equations (9.76) yield the equations of motion in cylindrical coordinates:

$$\frac{d}{dt}\frac{\partial L}{\partial \dot{\rho}} - \frac{\partial L}{\partial \rho} = 0 \qquad \rightarrow \qquad m\ddot{\rho} - m\rho\dot{\varphi}^2 + \frac{\partial V}{\partial \rho} = 0 \tag{9.80}$$

$$\frac{d}{dt}\frac{\partial L}{\partial \dot{\varphi}} - \frac{\partial L}{\partial \varphi} = 0 \qquad \rightarrow \qquad m\frac{d}{dt}(\rho^2\dot{\varphi}) + \frac{\partial V}{\partial \varphi} = 0 \tag{9.81}$$

$$\frac{d}{dt}\frac{\partial L}{\partial \dot{z}} - \frac{\partial L}{\partial z} = 0 \qquad \rightarrow \qquad m\ddot{z} + \frac{\partial V}{\partial z} = 0. \tag{9.82}$$

This is clearly easier than expressing Newton's equation in terms of cylindrical coordinates directly.

Cylindrical coordinates are particularly advantageous when the potential (or constraint) has cylindrical symmetry; that is, when the potential does not depend on φ: $V = V(\rho, z)$. Since, in this case, $\partial V/\partial \varphi = 0$, from Eq. (9.81) one has[15]

$$p_\varphi = \frac{\partial L}{\partial \dot\varphi} = m\rho^2\dot\varphi = const = L_z. \qquad (9.83)$$

Inserting this result into Eq. (9.80), one obtains

$$m\ddot\rho - \frac{L_z^2}{m\rho^3} + \frac{\partial V}{\partial \rho} = 0.$$

Together with

$$m\ddot z + \frac{\partial V}{\partial z} = 0,$$

these are the equations of motion for a particle in a cylindrically symmetric potential.

For the motion of a charged particle in a homogeneous magnetic field **B** acting together with a central force – expressed in terms of a spherically symmetric potential $V(r)$ – cylindrical coordinates are the natural choice, since they reflect the symmetry of the system. The z-axis is chosen along the direction of **B**, $\mathbf{B} = B\mathbf{e}_z$, then $V(r) = V(\sqrt{\rho^2 + z^2}) = V(\rho, z)$. The Lagrangian (9.69) reads

$$L(\rho, \varphi, z, \dot\rho, \dot\varphi, \dot z) = \frac{1}{2}m(\dot\rho^2 + \rho^2\dot\varphi^2 + \dot z^2) + \frac{qB}{2c}\rho^2\dot\varphi - V(\rho, z), \quad (9.84)$$

---

[15]The identity of p_φ and L_z follows already from the generally valid relation between an angle of rotation φ about an axis **n** and its conjugate momentum p_φ.

Let $\mathbf{r} = \mathbf{r}(\varphi, ...)$; from $\dot{\mathbf{r}} = \dfrac{\partial \mathbf{r}}{\partial \varphi}\dot\varphi + ...$, we have $\partial\dot{\mathbf{r}}/\partial\dot\varphi = \partial\mathbf{r}/\partial\varphi$, and hence

$$p_\varphi = \frac{\partial L}{\partial \dot\varphi} = \frac{\partial T}{\partial \dot\varphi} = \frac{\partial T}{\partial \dot{\mathbf{r}}}\frac{\partial \dot{\mathbf{r}}}{\partial \dot\varphi} = m\dot{\mathbf{r}}\cdot\frac{\partial \mathbf{r}}{\partial \varphi}.$$

For an infinitesimal (active) rotation of the radius vector (cf. Eq. (8.20) with $d\mathbf{r} = \mathbf{r}' - \mathbf{r}$ and $-\varepsilon = d\varphi$),

$$\frac{\partial \mathbf{r}}{\partial \varphi} = \mathbf{n} \times \mathbf{r},$$

and thus

$$p_\varphi = m\dot{\mathbf{r}}(\mathbf{n} \times \mathbf{r}) = m\mathbf{n}(\mathbf{r} \times \dot{\mathbf{r}}) = \mathbf{nL}.$$

and Lagrange's equations are identical to the equations of motion (5.88). The magnetic field only affects the conserved canonical momentum p_φ,

$$p_\varphi = m\rho^2 \left(\dot\varphi + \frac{1}{2}\omega_Z \right) = const, \quad \omega_Z = \frac{qB}{mc}. \quad (9.85)$$

In Lagrangian mechanics, conservation of p_φ is a direct consequence of the fact that φ is a cyclic variable – the momentum p_φ conjugate to φ is constant. Expressing also the conserved quantity $I_{LB} = [m(\mathbf{r} \times \dot{\mathbf{r}})\mathbf{B} + (q/2c)(\mathbf{r} \times \mathbf{B})^2]/B$ (cf. Eq. (5.76)) in terms of cylindrical coordinates, we see that

$$p_\varphi = I_{LB}.$$

Hence, p_φ also equals the canonical angular momentum $L_{can,z}$ (cf. Eq. (9.72)).

Spherical coordinates

According to Eq. (A.20), the velocity in spherical coordinates is given by

$$\dot{\mathbf{r}} = \dot{r}\mathbf{e}_r + r\sin\vartheta\,\dot\varphi\mathbf{e}_\varphi + r\dot\vartheta\mathbf{e}_\vartheta.$$

Inserting this into the Lagrangian (9.65),

$$L(r,\vartheta,\varphi,\dot{r},\dot\vartheta,\dot\varphi) = \frac{1}{2}m(\dot{r}^2 + r^2\dot\vartheta^2 + r^2\sin^2\vartheta\,\dot\varphi^2) - V(r,\varphi,\vartheta), \quad (9.86)$$

and one obtains the equations of motion in spherical coordinates,

$$\frac{d}{dt}\frac{\partial L}{\partial \dot{r}} - \frac{\partial L}{\partial r} = 0 \quad \rightarrow \quad m\ddot{r} - mr\dot\vartheta^2 - mr\sin^2\vartheta\,\dot\varphi^2 + \frac{\partial V}{\partial r} = 0 \quad (9.87)$$

$$\frac{d}{dt}\frac{\partial L}{\partial \dot\vartheta} - \frac{\partial L}{\partial \vartheta} = 0 \quad \rightarrow \quad m\frac{d}{dt}(r^2\dot\vartheta) - mr^2\sin\vartheta\cos\vartheta\,\dot\varphi^2 + \frac{\partial V}{\partial \vartheta} = 0 \quad (9.88)$$

$$\frac{d}{dt}\frac{\partial L}{\partial \dot\varphi} - \frac{\partial L}{\partial \varphi} = 0 \quad \rightarrow \quad m\frac{d}{dt}(r^2\dot\varphi\sin^2\vartheta) + \frac{\partial V}{\partial \varphi} = 0.$$

$$(9.89)$$

Even if the potential is spherical symmetric, $V = V(r)$, the equations are not as simple as in the cylindrically symmetric case. This is why in spherically symmetric systems, too, one takes advantage of the conservation of the angular momentum by introducing cylindrical (actually polar) coordinates in the plane perpendicular to the direction of the angular momentum (cf. Section 5.1; in spherical coordinates $L_z = mr^2\dot\varphi\sin^2\vartheta$, see Eq. (9.89)).

The spherical pendulum

The potential energy of a particle attached to a pivot by a rigid massless rod of length l is $V = -m\mathbf{gr}$ (cf. Section 9.1). The presence of the spherically symmetric constraint (9.9), $\mathbf{r}^2 - l^2 = 0$, suggests that we introduce spherical coordinates into the Lagrangian,

$$L = \frac{m}{2}\dot{\mathbf{r}}^2 + m\mathbf{gr}. \tag{9.90}$$

The transformation is given by:

$$\begin{aligned}
x &= l\cos\varphi\sin\vartheta \\
y &= l\sin\varphi\sin\vartheta \\
z &= l\cos\vartheta.
\end{aligned}$$

In these coordinates,

$$\begin{aligned}
\dot{\mathbf{r}} &= l\dot\vartheta\mathbf{e}_\vartheta + l\dot\varphi\sin\vartheta\mathbf{e}_\varphi \\
\mathbf{gr} &= -gz = -gl\cos\vartheta,
\end{aligned}$$

and hence the Lagrangian takes the form:

$$L = \frac{ml^2}{2}\left(\dot\vartheta^2 + (\sin\vartheta\,\dot\varphi)^2\right) - mgl\cos\vartheta. \tag{9.91}$$

By dealing with the constraint in this way, we have eliminated the coordinate r and the velocity $\dot r$ from the Lagrangian. The number of variables has thus already been reduced right from the beginning.

Lagrange's equations

i) φ is cyclic, therefore

$$p_\varphi = \partial L/\partial\dot\varphi = ml^2\dot\varphi\sin^2\vartheta = const =: ml^2 c;$$

the z-component of the angular momentum is conserved.

ii) Written out explicitly, $\dfrac{d}{dt}\dfrac{\partial L}{\partial\dot\vartheta} - \dfrac{\partial L}{\partial\vartheta} = 0$ takes the form

$$\frac{d}{dt}(ml^2\dot\vartheta) - ml^2\sin\vartheta\cos\vartheta\,\dot\varphi^2 - mgl\sin\vartheta = 0.$$

Taking advantage of the constant angular momentum in the second term, one obtains an equation of motion for the variable ϑ only,

$$\ddot{\vartheta} - c^2 \cos\vartheta / \sin^3 \vartheta = \frac{g}{l} \sin \vartheta.$$

This derivation of the equation is much shorter than the one given in the beginning of the chapter (leading to Eq. (9.18)), which used the force of constraint as a starting point.

9.3.6 Further applications of the Lagrangian

Non-conservative forces

In a conservative system a potential V exists for the force, and the equation of motion is given by Lagrange's equation (9.66) for the Lagrangian $L = T - V$. If a force $\mathbf{F}$ is added that cannot be written as the gradient of a potential, then Lagrange's equation for $L = T - V$ is modified to

$$\frac{d}{dt}\frac{\partial L}{\partial \dot{\mathbf{r}}} - \frac{\partial L}{\partial \mathbf{r}} = \mathbf{F}. \tag{9.92}$$

Evidently this is the equation of motion. If $\mathbf{F}$ is conservative, $\mathbf{F} = -\partial W/\partial \mathbf{r}$, the modified equation follows from the Lagrangian $L' = T - V - W$.

In non-Cartesian coordinates (e.g. spherical or cylindrical coordinates), Lagrange's equations read

$$\frac{d}{dt}\frac{\partial L}{\partial \dot{q}_i} - \frac{\partial L}{\partial q_i} = Q_i, \qquad i = 1, 2, 3, \tag{9.93}$$

where the Q_i are projections of the non-conservative force $\mathbf{F}$, originally given in Cartesian components, onto the axes of the coordinates q_i (cf. also Appendix A),

$$Q_i = \mathbf{F}\frac{\partial \mathbf{r}}{\partial q_i}.$$

(Note that if $\mathbf{F} = -\dfrac{\partial W(\mathbf{r})}{\partial \mathbf{r}}$, then $Q_i = -\dfrac{\partial W}{\partial \mathbf{r}}\dfrac{\partial \mathbf{r}}{\partial q_i} = -\dfrac{\partial W}{\partial q_i}$.)

Non-holonomic constraints

A non-holonomic constraint of the kind

$$\sum_i^3 a_i dq_i + a dt = 0, \tag{9.94}$$

which is equivalent to

$$\sum_i^3 a_i \dot{q}_i + a = 0, \tag{9.95}$$

can be dealt with by the Lagrange's equations (for a proof see [Spiegel]) of the form,

$$\frac{d}{dt}\frac{\partial L}{\partial \dot{q}_i} - \frac{\partial L}{\partial q_i} = \lambda a_i. \tag{9.96}$$

The additional parameter λ, which is, in general, time dependent, now appears, but since the equations of motion (9.96) are supplemented by the equation of the constraint (9.95), there are as many equations as there are unknowns.

Holonomic constraints of the form

$$G(q_1, q_2, q_3, t) = 0$$

are special cases of this method. This follows from the form of the equation $dG = 0$; i.e.

$$\sum \frac{\partial G}{\partial q_i} dq_i + \frac{\partial G}{\partial t} dt = 0$$

has the form of Eq. (9.94) ($a_i = \partial G/\partial q_i$ and $a = \partial G/\partial t$). Using Eq. (9.96), the familiar Lagrange's equations for holonomic constraints, Eq. (9.49), follow immediately; namely,

$$\frac{d}{dt}\frac{\partial L}{\partial \dot{q}_i} - \frac{\partial L}{\partial q_i} = \lambda \frac{\partial G}{\partial q_i}.$$

9.3.7   Nonuniformly moving frames of reference

In passing from an inertial frame I to an *accelerated frame of reference* K (i.e. to a non-inertial frame), the form of Newton's equation changes. Therefore, we may expect that the Lagrangian L changes, too. We do not, however, expect the form of Lagrange's equation (9.66) to change (cf. Subsection 9.2.2).

The relation between the velocity $\mathbf{v}$ observed in an inertial frame I and the velocity observed in a frame K – moving with velocity $\mathbf{W}(t)$ with respect to I – is[16] (cf. Subsection 8.5.1)

$$\mathbf{v}' = \mathbf{v} + \mathbf{W}(t). \tag{9.97}$$

---

[16]This relation is not true if one does not accept absolute time (as in the theory of relativity).

Inserting this into the Lagrangian for the motion in the frame I,

$$L = \frac{m\mathbf{v}^2}{2} - V(\mathbf{r}),$$

we get the Lagrangian for the motion in the frame K,

$$L' = \frac{m\mathbf{v}'^2}{2} - m\mathbf{v}'\mathbf{W}(t) + \frac{m\mathbf{W}^2}{2} - V(\mathbf{r}).$$

(The vector $\mathbf{r}$ remains unchanged: $\mathbf{r} = \mathbf{r}'$.) The term $m\mathbf{W}^2/2$ is independent of the dynamical variables $\mathbf{r}$ and $\mathbf{v}'$. It therefore does not influence Lagrange's equations and can be omitted. Hence, the Lagrangian in K reduces to

$$L'(\mathbf{r}, \mathbf{v}', t) = \frac{m\mathbf{v}'^2}{2} - m\mathbf{v}'\mathbf{W}(t) - V(\mathbf{r}), \tag{9.98}$$

and Lagrange's equation $\dfrac{d}{dt}\dfrac{\partial L'}{\partial \mathbf{v}'} = \dfrac{\partial L'}{\partial \mathbf{r}}$ yields the equation of motion in K (cf. Eq. (8.35)),

$$m\dot{\mathbf{v}}' = -\frac{\partial V}{\partial \mathbf{r}} + m\dot{\mathbf{W}}(t). \tag{9.99}$$

Rotating frames of reference

If the frame of reference K is *rotating* with respect to the inertial frame I with *constant angular velocity* $\boldsymbol{\omega}$, the relation between the velocities in the two frames is (cf. Eq. (8.56))

$$\mathbf{v} = \mathbf{v}' + \boldsymbol{\omega} \times \mathbf{r}. \tag{9.100}$$

From the Lagrangian L',

$$L'(\mathbf{r}, \mathbf{v}', t) = \frac{m\mathbf{v}'^2}{2} + m\mathbf{v}'(\boldsymbol{\omega} \times \mathbf{r}) + \frac{m}{2}(\boldsymbol{\omega} \times \mathbf{r})^2 - V(\mathbf{r}), \tag{9.101}$$

one obtains the equation of motion (8.61) already found in Subsection 8.5.3 (E).

Consider now the motion of a charged particle in a homogeneous magnetic field $\mathbf{B}$, in a frame of reference rotating with Larmor's angular velocity (cf. Eq. (8.65)),

$$\boldsymbol{\omega}_L = -\frac{q}{2mc}\mathbf{B}.$$

Inserting into the Lagrangian (9.69) in the inertial frame, $L = (m/2)\,\mathbf{v}^2 - (q/2c)\,\mathbf{v}\,(\mathbf{r} \times \mathbf{B}) - V(\mathbf{r})$, Eq. (9.100) and $\boldsymbol{\omega} = \boldsymbol{\omega}_L$, one finds the Lagrangian in the rotating frame K:

$$L' = \frac{m\mathbf{v}'^2}{2} - \frac{q^2}{8mc^2}(\mathbf{r} \times \mathbf{B})^2 - V(\mathbf{r}). \qquad (9.102)$$

Observe that the anomalous velocity dependence is not present in the Lagrangian in the rotating frame. The magnetic field simply produces an additional potential in the rotating frame. Here, the canonical momentum is directly related to the velocity:

$$\mathbf{p}' = m\mathbf{v}'. \qquad (9.103)$$

Problems and examples

1. For the motion of a particle on an inclined track, determine the force of constraint and the force along the track.

2. Determine the motion of a particle on the inclined track using only the energy, Eq. (9.8), and the constraint given in Eq. (9.7).

3. A particle moves without friction in the vertical (x, y)-plane under the influence of gravity (acceleration vector $\mathbf{g} = (g, 0)$) along the parabola $x = ky^2$. At time $t = 0$, it is at the origin $(0, 0)$.

   a) Suppose the particle is not restricted to the parabola. Determine the limiting initial velocity for which the orbit of the particle is exactly the given parabola.

   b) Suppose the particle is constrained to move along the parabola. Find the force of constraint. Determine $\dot{y} = \dot{y}\,(y)$ and discuss the dependence of the solution on the initial velocity v_0.

4. Derive the equations of motion for the spherical pendulum in spherical coordinates.

5. Determine the stationary solutions of Eq. (9.18) for the spherical pendulum. One integral $\dot{\vartheta} = \dot{\vartheta}\,(\vartheta)$ (cf. Eq. (9.16)) is obtained from conservation of energy. Discuss the dependence of the turning points of $\dot{\vartheta}\,(\vartheta)$ on E and L_g. Carry out the stability analysis.

Determine the solutions for small displacements from the stable equilibrium position.

6. **Foucault's pendulum** is a spherical pendulum (length l) in the rotating frame of reference K at the earth's surface (cf. Subsection 8.5.3 and above). Its initial velocity is small. Neglecting the centrifugal term $m(\boldsymbol{\omega} \times \mathbf{r})^2/2$ in Eq. (9.101), the Lagrangian in K is

$$L'(\mathbf{r}, \mathbf{v}', t) = \frac{m\mathbf{v}'^2}{2} + m\mathbf{v}'(\boldsymbol{\omega} \times \mathbf{r}) + mg\mathbf{r}.$$

In addition, the motion of the point mass is constrained by

$$\mathbf{r}^2 - l^2 = 0.$$

Derive the equations of motion in suitable coordinates. Solve the equations of motion for small displacements from the equilibrium position.

7. Find the geodesic between two points on the surface of a cylinder (cf. Subsection 9.2.3) by:

   a) Direct elimination of, for instance, y from the constraint;

   b) Using Lagrange multipliers.

8. Find the Lagrangian for the isotropic harmonic oscillator and for motion in the $1/r$ potential in coordinates appropriate to the symmetry. Reduce the number of equations of motion using the first integrals of the cyclic coordinates.

9. A pearl (mass m) slides without friction on a circular wire (radius R) positioned vertically to the surface of the earth. The wire circle rotates with constant angular velocity ω about a vertical diameter.

   Determine:

   - The Lagrangian for the position $s(t) = R\,\vartheta(t)$ of the pearl on the circle (take the center of the circle as the origin) and find the equations of motion;

   - All stationary solutions (the number of solutions depends on $\alpha := g/\omega^2 R$; what does this parameter mean?);

- The stability of the stationary solutions as a function of α.

Solve the equations of motion for small displacements from the stationary positions.

10. Use L' given in Eq. (9.101) to derive the equations of motion in a rotating frame of reference. (Convince yourself that the equation of motion (8.66) follows from Eq. (9.102).)

10

Conservation laws and symmetries in many particle systems

Until now, we have investigated the dynamics of at most two particles. Henceforth, we focus on the dynamics of more than two particles. First, in this chapter, we will study general properties of systems of N interacting point masses.

10.1 Equations of motion for N point masses

In a system of N point masses, the particles may interact with each other, and may also be influenced by external forces. We suppose that the interactions between the particles consist only of **two-particle interactions** $\mathbf{F}_{\alpha\beta}(\mathbf{r}_\alpha, \mathbf{r}_\beta)$, where $\mathbf{F}_{\alpha\beta}$ is the force exerted by particle β on particle α. By *Newton's third law*, the force $\mathbf{F}_{\beta\alpha}$ exerted by particle α on particle β satisfies the condition

$$\mathbf{F}_{\beta\alpha} = -\mathbf{F}_{\alpha\beta}. \tag{10.1}$$

The maximal number of *degrees of freedom* f in the system (that is, the number of independent directions that the particles can move in) is the number $3N$ of the coordinates[1]. This number may be reduced by constraints placed on the system. According to the superposition principle of forces (cf. Page 17), the equations of motion for the N-particle system are

$$m_1 \ddot{\mathbf{r}}_1 = \mathbf{F}_1 + \sum_{\beta=1}^{N} \mathbf{F}_{1\beta}$$

$$m_2 \ddot{\mathbf{r}}_2 = \mathbf{F}_2 + \sum_{\beta=1}^{N} \mathbf{F}_{2\beta} \qquad (10.2)$$

$$\vdots$$

$$m_N \ddot{\mathbf{r}}_N = \mathbf{F}_N + \sum_{\beta=1}^{N} \mathbf{F}_{N\beta},$$

where $\mathbf{F}_\alpha = \mathbf{F}_\alpha(\mathbf{r}_\alpha)$ is the external force acting on particle α.

This is a system of $3N$ second order ordinary differential equations for the $3N$ coordinates of the N particles, denoted as

$$\{\mathbf{r}\} := (\mathbf{r}_1, \mathbf{r}_2, \dots, \mathbf{r}_N) = (x_1, \dots, x_f) =: \{x\}, \qquad f = 3N. \quad (10.3)$$

(For convenience, we occasionally label the coordinates consecutively from 1 to f.) As with the equations of motion for a single point mass, this system of differential equations possesses a unique solution[2] for given initial values of the variables x_i and v_i $(= \dot{x}_i)$, $i = 1, \dots, f$ (cf. Subsection 2.4.1). Let the solutions be denoted as

$$x_i = X_i(\{x_0\}, \{v_0\}; t), \qquad x_{0i} = x_i(t = 0)$$
$$v_i = W_i(\{x_0\}, \{v_0\}; t), \qquad v_{0i} = v_i(t = 0). \quad (10.4)$$

Since the equations of motion are reversible, i.e. invariant under *reversal of time* $t \to -t$ (cf. Subsection 2.4.1), we can start at the values $x_i(t), v_i(t)$, $i = 1, \dots, f$, and go backwards in time for duration t,

---

[1]Since we consider point masses, the particles have no internal degrees of freedom – such as rotational degrees of freedom that, say, a body of finite size possesses (see also Footnote 17 in Chapter 2).

[2]The system has a unique solution provided that the forces fulfill certain conditions. Here, we suppose that they do.

i.e. we let time $-t$ pass, arriving at the original initial values x_{0i} and v_{0i}, $i = 1, \ldots, f$. That is, we have the equations

$$
\begin{aligned}
x_{0i} &= X_i\left(\{x\}_t, \{v\}_t; -t\right) \\
v_{0i} &= W_i\left(\{x\}_t, \{v\}_t; -t\right), \qquad i = 1, \ldots, f,
\end{aligned}
\tag{10.5}
$$

where $\{x\}_t = (x_1(t), \ldots, x_f(t))$, and similarly for $\{v\}_t$. This implies there are $2f$ functions of the dynamical variables x_i, v_i, and time t that remain separately constant (x_{0i} and v_{0i}) along a trajectory ($\{x\}_t, \{v\}_t$) in the $2f$-dimensional phase space. These functions are $2f$ *constants of the motion* (or integrals). Using any one of these relations, we can eliminate the explicit time dependence from the remaining $(2f - 1)$ relations in Eqs. (10.5) and obtain $(2f - 1)$ time independent constants of motion: the $(2f - 1)$ *conserved quantities*,

$$
I_j\left(\{x\}_t, \{v\}_t\right) = i_j = const, \qquad j = 1, \ldots, 2f - 1.
\tag{10.6}
$$

Each of these relations can be considered to be a $(2f - 1)$-dimensional hypersurface in the $2f$-dimensional phase space. The trajectory of the many particle system is the curve formed by the intersection of these hypersurfaces. As in the single particle case, the existence of the conserved quantities does not imply that the conserved quantities are isolating, or 'useful' (cf. Section 4.3, in particular Eq. (4.28)). In the following, only an isolating conserved quantity (constant of the motion) is referred to as integral.

As we will see, for a conservative system (meaning *each* force acting in the system can be expressed as the gradient of a potential) of N particles, conservation of total energy holds. Hence, we are left with the problem of determining 'only' $2f - 2$ conserved quantities – or $2f - 1$ constants of the motion – which, of course, have to be independent. It is, in general, impossible to find all constants of the motion. (In Section 14.1, we show that f conserved quantities suffice.) But in many cases, the constants of the motion (conserved quantities, too) are related to symmetries and invariances of the system. In the following, we deduce the constants of motion for some rather general systems of N point masses.

Conservative N particle systems

The interaction force between two particles is directed along the straight line connecting them and depends only on the distance (see the discussion in Section 6.1). We may write[3]

$$\mathbf{F}_{\alpha\beta}(\mathbf{r}_\alpha, \mathbf{r}_\beta) = f_{\alpha\beta}(r_{\alpha\beta})(\mathbf{r}_\alpha - \mathbf{r}_\beta)/r_{\alpha\beta}, \qquad r_{\alpha\beta} = |\mathbf{r}_\alpha - \mathbf{r}_\beta|. \qquad (10.7)$$

Consequently, an interaction potential $V_{\alpha\beta}$ exists, such that (note: $\boldsymbol{\nabla}_\alpha = \partial/\partial\mathbf{r}_\alpha$)

$$\mathbf{F}_{\alpha\beta} = -\boldsymbol{\nabla}_\alpha V_{\alpha\beta}(r_{\alpha\beta}). \qquad (10.8)$$

Since $\mathbf{F}_{\alpha\beta} = -\mathbf{F}_{\beta\alpha}$, the functions $f_{\alpha\beta} = f_{\beta\alpha}$ are symmetric with respect to the particles α and β; therefore the relations

$$V_{\alpha\beta} = V_{\beta\alpha}, \qquad (V_{\alpha\alpha} = 0), \qquad (10.9)$$

hold. Inserting the expression (10.8) for $\mathbf{F}_{\alpha\beta}$ into the equations of motion, we have

$$m_\alpha \ddot{\mathbf{r}}_\alpha = \mathbf{F}_\alpha + \sum_{\beta=1}^{N} \mathbf{F}_{\alpha\beta} = \mathbf{F}_\alpha - \boldsymbol{\nabla}_\alpha \sum_{\beta=1}^{N} V_{\alpha\beta}(r_{\alpha\beta}), \qquad \alpha = 1, \ldots, N.$$
$$(10.10)$$

If also the external forces are conservative, i.e. if

$$\mathbf{F}_\alpha(\mathbf{r}) = -\boldsymbol{\nabla} U_\alpha(\mathbf{r}), \qquad (10.11)$$

then the system is *conservative*, since the total energy is conserved (see below, Subsection 10.2.3).

In the rest of this chapter, we elaborate on conservation laws and symmetries in many particle systems, taking two cases of external force as examples. The first case is the force of gravity in the vicinity of the surface of the earth,

$$\mathbf{F}_\alpha = m_\alpha \mathbf{g} \qquad \text{and} \qquad U_\alpha(\mathbf{r}) = -m_\alpha \mathbf{g}\mathbf{r}. \qquad (10.12)$$

In this approximation, the force of gravity is *homogeneous*, i.e. independent of the position of the particle. The second case is a **closed system**, meaning that no external forces exist,

$$\mathbf{F}_\alpha = 0, \qquad \alpha = 1, \ldots, N, \qquad (10.13)$$

---

[3]In Eq. (10.1), if one allows $\alpha = \beta$, then one has to include the condition $\mathbf{F}_{\alpha\alpha} = \mathbf{0}$. We do not allow of self-interactions of the point particles.

and the motion of the particles is determined solely by the interactions between them:

$$m_\alpha \ddot{\mathbf{r}}_\alpha = -\boldsymbol{\nabla}_\alpha \sum_{\beta=1}^{N} V_{\alpha\beta}(r_{\alpha\beta}), \qquad \alpha = 1, \ldots, N. \qquad (10.14)$$

When discussing properties of a many body system, it is advantageous to introduce the **center of mass coordinate**[4],

$$\mathbf{R} := \frac{\sum_{\alpha=1}^{N} m_\alpha \mathbf{r}_\alpha}{\sum_{\alpha=1}^{N} m_\alpha} = \frac{1}{M} \sum_{\alpha=1}^{N} m_\alpha \mathbf{r}_\alpha. \qquad (10.15)$$

The positions of the particles with respect to the center of mass are given by the **relative coordinates**

$$\mathbf{r}'_\alpha = \mathbf{r}_\alpha - \mathbf{R}. \qquad (10.16)$$

Because of Eq. (10.15), the weighted sum of the particles' relative coordinates vanishes:

$$\sum_{\alpha=1}^{N} m_\alpha \mathbf{r}'_\alpha = \mathbf{0}. \qquad (10.17)$$

Consequently, the weighted sums of the time derivatives of the particles' relative coordinates vanish

$$\sum_{\alpha=1}^{N} m_\alpha \dot{\mathbf{r}}'_\alpha = \sum_{\alpha=1}^{N} m_\alpha \ddot{\mathbf{r}}'_\alpha = \mathbf{0}. \qquad (10.18)$$

This is true also in the *center of mass frame*, in which $\mathbf{R} = 0$.

10.2 The conservation laws

10.2.1 The motion of the center of mass

Let us sum the N equations of motion (10.2) for the individual particles. The interaction forces do not contribute to the sum, because $\mathbf{F}_{\alpha\beta} =$

---

[4]One can show that the center of mass is located inside the convex hull of the set of points $\mathbf{r}_\alpha$ – or more colloquially, inside the largest polyhedron arising from connecting lines between the particles.

$-\mathbf{F}_{\beta\alpha}$. We therefore have

$$\sum_{\alpha=1}^{N} m_\alpha \ddot{\mathbf{r}}_\alpha = \sum_{\alpha=1}^{N} \mathbf{F}_\alpha. \tag{10.19}$$

Using Eqs. (10.15) and (10.19), the equation of *motion of the center of mass* is:

$$M\ddot{\mathbf{R}} = \dot{\mathbf{P}} = \sum_{\alpha=1}^{N} \mathbf{F}_\alpha =: \mathbf{F}, \tag{10.20}$$

where $\mathbf{P}$ is the *center of mass momentum*, defined as the total momentum of the particles,

$$\mathbf{P} := \sum_{\alpha=1}^{N} m_\alpha \dot{\mathbf{r}}_\alpha = M\dot{\mathbf{R}}. \tag{10.21}$$

The motion of the center of mass is determined solely by the sum of the external forces (acting at the center of mass).

(This is the first variant of the law of the center of mass motion.)
    If the external forces are fields,

$$\mathbf{F}_\alpha = \mathbf{F}_\alpha(\mathbf{r}_\alpha) = \mathbf{F}_\alpha(\mathbf{R} + \mathbf{r}'_\alpha), \tag{10.22}$$

then the total force $\mathbf{F}$ depends on the positions of all particles:

$$\mathbf{F} = \mathbf{F}(\mathbf{r}_1,\ldots,\mathbf{r}_N) = \mathbf{F}(\mathbf{R}; \mathbf{r}'_1,\ldots,\mathbf{r}'_N). \tag{10.23}$$

According to Eq. (10.20), the motion of the center of mass, $\mathbf{R}(t)$, also depends on the (instantaneous) positions $\mathbf{r}'_\alpha$ of the individual point masses m_α. Therefore, in general, the motion of the center of mass is coupled to the relative motion of the point masses; for example, as in the motion of interacting point masses in the gravitational field of another body. However, in the case of the homogeneous gravitational force – in the approximation given in Eq. (10.12) – the motion of the center of mass is independent of the positions of the point masses, since

$$\mathbf{F} = \sum_{\alpha=1}^{N} \mathbf{F}_\alpha = \sum_{\alpha=1}^{N} m_\alpha \mathbf{g} = M\mathbf{g}. \tag{10.24}$$

For the homogeneous gravitational force, then, it follows from Eq. (10.20)
that

$$\ddot{\mathbf{R}} = \mathbf{g}. \tag{10.25}$$

The solution,

$$\mathbf{R} = \frac{1}{2}\mathbf{g}t^2 + \mathbf{V}_0 t + \mathbf{R}_0, \tag{10.26}$$

shows that the center of mass is uniformly accelerated; $\mathbf{V}_0$ and $\mathbf{R}_0$ are
the initial values of the center of mass velocity and position, respectively,
and are therefore also constants of motion.

In a closed system (no external forces, $\mathbf{F}_\alpha = 0$), the law of *conservation of center of mass momentum* follows by integrating Eq. (10.20):

$$M\dot{\mathbf{R}} = \mathbf{P} = const = \mathbf{P}_0. \tag{10.27}$$

In a closed system, the total momentum is conserved.

Integrating over time once more yields

$$M\mathbf{R} - \mathbf{P}_0 t = M\mathbf{R}_0, \tag{10.28}$$

where $\mathbf{R}_0$ is the constant of integration.

In a closed system the center of mass moves freely.

(This is the second variant of the law of conservation of center of mass
motion.) Thus we have found the two constant vectors $\mathbf{P}_0$ and $\mathbf{R}_0$ (see
Eqs. (10.27) and (10.28)) as constants of the motion. As can be seen
from the derivation, these two statements of conservation of center of
mass momentum are also valid for the weaker condition $\mathbf{F} = \sum_{\alpha=1}^{N} \mathbf{F}_\alpha = 0$. The difference between this and the closed system case (all $\mathbf{F}_\alpha = 0$)
comes to light if one looks at the angular momentum.

10.2.2 Conservation of angular momentum

The angular momentum of each particle shall be defined relative to the
origin. The origin is usually some physically distinct point in the system;
for example, the center of mass or the center of an external force such
as gravity. Let us sum all the individual angular momenta. The time
derivative of the *total angular momentum*,

$$\mathbf{L}_{tot} = \sum_{\alpha=1}^{N} m_\alpha \mathbf{r}_\alpha \times \dot{\mathbf{r}}_\alpha, \tag{10.29}$$

is then[5]

$$\frac{d}{dt}\mathbf{L}_{tot} = \frac{d}{dt}\left(\sum_{\alpha=1}^{N} m_\alpha \mathbf{r}_\alpha \times \dot{\mathbf{r}}_\alpha\right) = \sum_{\alpha=1}^{N} m_\alpha \mathbf{r}_\alpha \times \ddot{\mathbf{r}}_\alpha$$

$$= \sum_{\alpha=1}^{N} (\mathbf{r}_\alpha \times \mathbf{F}_\alpha) + \sum_{\alpha,\beta=1}^{N} \mathbf{r}_\alpha \times \mathbf{F}_{\alpha\beta}$$

$$= \mathbf{N}_{tot} + \frac{1}{2}\sum_{\alpha,\beta=1}^{N} (\mathbf{r}_\alpha - \mathbf{r}_\beta) \times \mathbf{F}_{\alpha\beta},$$

where $\mathbf{N}_{tot}$ is the total moment of the external forces,

$$\mathbf{N}_{tot} = \sum_{\alpha=1}^{N} (\mathbf{r}_\alpha \times \mathbf{F}_\alpha). \tag{10.30}$$

Since the forces $\mathbf{F}_{\alpha\beta}$ act parallel to $\mathbf{r}_\alpha - \mathbf{r}_\beta$ according to Eq. (10.7), the cross product $(\mathbf{r}_\alpha - \mathbf{r}_\beta) \times \mathbf{F}_{\alpha\beta}$ vanishes, and the time derivative of the total angular momentum $\mathbf{L}_{tot}$ is equal to $\mathbf{N}_{tot}$,

$$\frac{d}{dt}\mathbf{L}_{tot} = \mathbf{N}_{tot}. \tag{10.31}$$

While the total momentum $\mathbf{P}$ is equal to the center of mass momentum (cf. Eq. (10.21)), the total angular momentum $\mathbf{L}_{tot}$ consists of two contributions:

$$\mathbf{L}_{tot} = \sum_{\alpha=1}^{N} m_\alpha(\mathbf{r}_\alpha \times \dot{\mathbf{r}}_\alpha) = \sum_{\alpha=1}^{N} m_\alpha(\mathbf{R} + \mathbf{r}'_\alpha) \times (\dot{\mathbf{R}} + \dot{\mathbf{r}}'_\alpha)$$

$$= M\mathbf{R} \times \dot{\mathbf{R}} + \sum_{\alpha=1}^{N} m_\alpha \mathbf{r}'_\alpha \times \dot{\mathbf{r}}'_\alpha$$

$$= \mathbf{R} \times \mathbf{P} + \sum_{\alpha=1}^{N} m_\alpha \mathbf{r}'_\alpha \times \dot{\mathbf{r}}'_\alpha = \mathbf{L}_S + \mathbf{L}. \tag{10.32}$$

---

[5]By first relabelling the indices, then using the condition in Eq. (10.1), one obtains

$$\sum_{\alpha,\beta} \mathbf{r}_\alpha \times \mathbf{F}_{\alpha\beta} = \frac{1}{2}\left(\sum_{\alpha,\beta} \mathbf{r}_\alpha \times \mathbf{F}_{\alpha\beta} + \sum_{\alpha,\beta} \mathbf{r}_\beta \times \mathbf{F}_{\beta\alpha}\right) = \frac{1}{2}\sum_{\alpha,\beta} (\mathbf{r}_\alpha - \mathbf{r}_\beta) \times \mathbf{F}_{\alpha\beta}.$$

Here, $\mathbf{L}_S$ is the *angular momentum of the center of mass* about the origin,

$$\mathbf{L}_S = \mathbf{R} \times \mathbf{P}, \qquad (10.33)$$

and $\mathbf{L}$ is the *total angular momentum of the relative motion* about the center of mass,

$$\mathbf{L} = \sum_{\alpha=1}^{N} m_\alpha \mathbf{r}'_\alpha \times \dot{\mathbf{r}}'_\alpha. \qquad (10.34)$$

The *moment of the external force*, Eq. (10.30), also splits into two contributions:

$$\begin{aligned}
\mathbf{N}_{tot} &= \sum_{\alpha=1}^{N} \mathbf{r}_\alpha \times \mathbf{F}_\alpha = \sum_{\alpha=1}^{N} (\mathbf{R} + \mathbf{r}'_\alpha) \times \mathbf{F}_\alpha \\
&= \mathbf{R} \times \mathbf{F} + \sum_{\alpha=1}^{N} \mathbf{r}'_\alpha \times \mathbf{F}_\alpha = \mathbf{N}_S + \mathbf{N}.
\end{aligned} \qquad (10.35)$$

Here,

$$\mathbf{N}_S = \mathbf{R} \times \mathbf{F} \qquad (10.36)$$

is the moment, about the origin, of the total force acting on the center of mass, and

$$\mathbf{N} = \sum_{\alpha=1}^{N} \mathbf{r}'_\alpha \times \mathbf{F}_\alpha \qquad (10.37)$$

is the sum of all individual moments, about the center of mass, of the external forces.

Taking the cross product of the equation of the center of mass motion (10.20) with $\mathbf{R}$ yields an equation for the angular momentum of the center of mass,

$$\frac{d}{dt}(M\mathbf{R} \times \dot{\mathbf{R}}) = \mathbf{R} \times \mathbf{F}; \quad \text{in short,} \quad \frac{d\mathbf{L}_S}{dt} = \mathbf{N}_S. \qquad (10.38)$$

Subtracting this from Eq. (10.31), we obtain an equation for the angular momentum of the relative motion only[6]

$$\frac{d}{dt} \sum_{\alpha=1}^{N} m_\alpha \mathbf{r}'_\alpha \times \dot{\mathbf{r}}'_\alpha = \sum_{\alpha=1}^{N} \mathbf{r}'_\alpha \times \mathbf{F}_\alpha; \quad \text{in short,} \quad \frac{d\mathbf{L}}{dt} = \mathbf{N}. \quad (10.39)$$

---

[6]Equation (10.39) can also be obtained directly by taking the angular momenta of the point masses relative to the center of mass; i.e., one evaluates the vector product of the equations of motion, Eq. (10.10), and $\mathbf{r}'_i$, then adds all resulting equations.

In general, the equations of motion for the angular momenta $\mathbf{L}_S$ and $\mathbf{L}$, (10.38) and (10.39), are coupled. This is because according to Eqs. (10.22, 10.23), the torques $\mathbf{N}_S$ and $\mathbf{N}$ are not independent. The equation of motion for the angular momentum $\mathbf{L}$ of the relative motion, Eq. (10.39), is the most important equation for finding the motion of a spinning top (see Subsection 11.3.2).

For the gravitational force in the approximation given in Eq. (10.12), using Eq. (10.17), the torque $\mathbf{N}$ (about the center of mass, Eq. (10.37)) acting on the particles is

$$\mathbf{N} = \sum_{\alpha=1}^{N} \mathbf{r}'_\alpha \times \mathbf{F}_\alpha = -\mathbf{g} \times \sum_{\alpha=1}^{N} m_\alpha \mathbf{r}'_\alpha = \mathbf{0}, \qquad (10.40)$$

so that

$$\mathbf{L} = const. \qquad (10.41)$$

For N point masses m_α ($\alpha = 1, \ldots, N$) subjected to gravity $m_\alpha \mathbf{g}$, the *angular momentum of the relative motion is a conserved quantity*. What about the torque applied by the external forces, Eq. (10.36), to the center of mass? Using Eq. (10.24), we get $\mathbf{N}_S = M\mathbf{R} \times \mathbf{g}$, so that the angular momentum of the center of mass varies with time according to

$$\frac{d\mathbf{L}_S}{dt} = M\mathbf{R} \times \mathbf{g}. \qquad (10.42)$$

Multiplying this equation by $\mathbf{g}$ shows that the component of $\mathbf{L}_S$ parallel to the gravitational force is conserved:

$$\mathbf{g}\mathbf{L}_S = const. \qquad (10.43)$$

(Compare this to the result for the spherical pendulum, Eq. (9.15) in Section 9.1.)

In a closed system ($\mathbf{F}_\alpha = \mathbf{0}$), we not only have $\mathbf{N} = \mathbf{0}$ but also $\mathbf{N}_{tot} = \mathbf{0}$. Therefore, *the total angular momentum is conserved*:

$$\mathbf{L}_{tot} = const. \qquad (10.44)$$

The angular momentum of the center of mass is already determined by the conservation of momentum, Eqs. (10.27) and (10.28),

$$\mathbf{L}_S = \mathbf{R} \times \mathbf{P} = (\mathbf{R}_0 + \frac{1}{M}\mathbf{P}_0 t) \times \mathbf{P}_0 = \mathbf{R}_0 \times \mathbf{P}_0 = const.$$

Thus, since $\mathbf{L}_{tot} = \mathbf{L}_S + \mathbf{L}$, we find that for a closed system, only the angular momentum of the relative motion,

$$\mathbf{L} = \sum_{\alpha=1}^{N} m_\alpha \mathbf{r}'_\alpha \times \dot{\mathbf{r}}'_\alpha = const, \tag{10.45}$$

provides us with an additional independent conserved quantity.

10.2.3 Conservation of energy

Taking the scalar product of the equation of motion (10.10) for particle α with $\dot{\mathbf{r}}_\alpha$ gives

$$\frac{1}{2}\frac{d}{dt}m_\alpha \dot{\mathbf{r}}_\alpha^2 = \dot{\mathbf{r}}_\alpha \mathbf{F}_\alpha - \dot{\mathbf{r}}_\alpha \nabla_\alpha \sum_{\substack{\beta=1 \\ \beta \neq \alpha}}^{N} V_{\alpha\beta}(r_{\alpha\beta}).$$

Summing over all particles yields

$$\frac{d}{dt}\sum_{\alpha-1}^{N}\frac{1}{2}m_\alpha \dot{\mathbf{r}}_\alpha^2 = \sum_{\alpha-1}^{N}\dot{\mathbf{r}}_\alpha \mathbf{F}_\alpha - \sum_{\substack{\alpha,\beta-1 \\ \beta \neq \alpha}}^{N}\dot{\mathbf{r}}_\alpha \nabla_\alpha V_{\alpha\beta}(r_{\alpha\beta})$$

$$= \sum_{\alpha=1}^{N}\dot{\mathbf{r}}_\alpha \mathbf{F}_\alpha - \frac{1}{2}\sum_{\substack{\alpha,\beta=1 \\ \beta \neq \alpha}}^{N}\left(\dot{\mathbf{r}}_\alpha \nabla_\alpha V_{\alpha\beta}(r_{\alpha\beta}) + \dot{\mathbf{r}}_\beta \nabla_\beta V_{\beta\alpha}(r_{\alpha\beta})\right).$$

Using the fact that $V_{\alpha\beta} = V_{\beta\alpha}$, Eq. (10.9), and the identity

$$\frac{d}{dt}V_{\alpha\beta}(r_{\alpha\beta}) = \frac{d}{dt}V_{\alpha\beta}(|\mathbf{r}_\alpha - \mathbf{r}_\beta|) = (\dot{\mathbf{r}}_\alpha \nabla_\alpha + \dot{\mathbf{r}}_\beta \nabla_\beta)V_{\alpha\beta}(r_{\alpha\beta}),$$

one finds

$$\frac{d}{dt}\left[\sum_{\alpha=1}^{N}\frac{1}{2}m_\alpha \dot{\mathbf{r}}_\alpha^2 + \frac{1}{2}\sum_{\substack{\alpha,\beta=1 \\ \beta \neq \alpha}}^{N}V_{\alpha\beta}(r_{\alpha\beta})\right] = \frac{d}{dt}(T+V) = \sum_{\alpha=1}^{N}\dot{\mathbf{r}}_\alpha \mathbf{F}_\alpha,$$

$$\tag{10.46}$$

where T is the total kinetic energy,

$$T = \sum_{\alpha=1}^{N}\frac{1}{2}m_\alpha \dot{\mathbf{r}}_\alpha^2, \tag{10.47}$$

and V the total potential energy of the interaction between the particles,

$$V = \frac{1}{2} \sum_{\substack{\alpha,\beta=1 \\ \beta \neq \alpha}}^{N} V_{\alpha\beta}(r_{\alpha\beta}). \tag{10.48}$$

If the external forces are conservative, i.e. if

$$\mathbf{F}_\alpha(\mathbf{r}) = -\boldsymbol{\nabla} U_\alpha(\mathbf{r}), \tag{10.49}$$

then

$$\sum_{\alpha=1}^{N} \dot{\mathbf{r}}_\alpha \mathbf{F}_\alpha = -\frac{d}{dt} \sum_{\alpha=1}^{N} U_\alpha(\mathbf{r}_\alpha),$$

and the *total energy of the system is conserved*:

$$T + V + \sum_{\alpha=1}^{N} U_\alpha = E_{tot} = const. \tag{10.50}$$

The interaction potentials $V_{\alpha\beta}$, and consequently V, depend only on the relative coordinates:

$$V_{\alpha\beta}(r_{\alpha\beta}) = V_{\alpha\beta}(r'_{\alpha\beta}), \qquad r'_{\alpha\beta} = |\mathbf{r}'_\alpha - \mathbf{r}'_\beta| = r_{\alpha\beta}.$$

In contrast, the total kinetic energy T consists of two contributions:

$$
\begin{aligned}
T &= \frac{1}{2} \sum_{\alpha=1}^{N} m_\alpha \dot{\mathbf{r}}_\alpha^2 = \frac{1}{2} \sum_{\alpha=1}^{N} m_\alpha (\dot{\mathbf{R}} + \dot{\mathbf{r}}'_\alpha)^2 \\
&= \frac{1}{2} M \dot{\mathbf{R}}^2 + \dot{\mathbf{R}} \sum_{\alpha=1}^{N} m_\alpha \dot{\mathbf{r}}'_\alpha + \frac{1}{2} \sum_{\alpha=1}^{N} m_\alpha \dot{\mathbf{r}}'^2_\alpha \\
&= \frac{1}{2} M \dot{\mathbf{R}}^2 + \frac{1}{2} \sum_{\alpha=1}^{N} m_\alpha \dot{\mathbf{r}}'^2_\alpha =: T_S + T_r.
\end{aligned}
\tag{10.51}
$$

(Recall Eq. (10.18), $\sum_{\alpha=1}^{N} m_\alpha \dot{\mathbf{r}}'_\alpha = \mathbf{0}$.) Here,

$$T_S = M\dot{\mathbf{R}}^2/2 \tag{10.52}$$

is the **kinetic energy of the center of mass**, and

$$T_r = \frac{1}{2} \sum_{\alpha=1}^{N} m_\alpha \dot{\mathbf{r}}'^2_\alpha \tag{10.53}$$

is the **kinetic energy of the relative motion** of all the particles about the center of mass, i.e. the kinetic energy in the center of mass frame. The sum of the *potentials of the external forces* generally depends on the coordinate of the center of mass and on the relative coordinates,

$$\sum_{\alpha=1}^{N} U_\alpha(\mathbf{r}_\alpha) = \sum_{\alpha=1}^{N} U_\alpha(\mathbf{R} + \mathbf{r}'_\alpha) = U(\mathbf{R}; \mathbf{r}'_1, \ldots, \mathbf{r}'_N). \qquad (10.54)$$

Hence, in general, the center of mass motion and the relative motion are coupled through the total external potential $U(\mathbf{R}; \mathbf{r}'_1, \ldots, \mathbf{r}'_N)$.

For the <u>gravitational force</u> near the earth's surface, Eq. (10.15) implies that the right hand side of Eq. (10.46) can be written as

$$\sum_{\alpha=1}^{N} \dot{\mathbf{r}}_\alpha \mathbf{F}_\alpha = \mathbf{g} \sum_{\alpha=1}^{N} m_\alpha \dot{\mathbf{r}}_\alpha = M \mathbf{g} \dot{\mathbf{R}}$$

or, alternatively (cf. Eqs. (10.12) and (10.17)),

$$U = -\sum_{\alpha=1}^{N} m_\alpha(\mathbf{R} + \mathbf{r}'_\alpha)\mathbf{g} = -M\mathbf{g}\mathbf{R}. \qquad (10.55)$$

Multiplying Eq. (10.25) by $\dot{\mathbf{R}}$ yields the *conservation of the energy of the center of mass*:

$$E_S = T_S - M\mathbf{R}\mathbf{g} = const. \qquad (10.56)$$

Hence, by virtue of conservation of the total energy, Eq. (10.50), the *energy of the relative motion* (i.e. the *internal energy* of the N particles),

$$E = T_r + V = const, \qquad (10.57)$$

is conserved separately. In the light of general properties derived above – the total energy, momentum and angular momentum each split into two contributions, one for each type of motion – we find that *in a system of N point masses in the homogeneous gravitational field near the earth's surface, the center of mass motion is independent of the relative motion.*

In a <u>closed system</u> ($\mathbf{F}_\alpha = \mathbf{0}$ in Eq. (10.46) or $U_\alpha = 0$ in Eq. (10.50)), the *total energy is conserved*:

$$T + V = T_S + T_r + V = E_{tot} = const. \qquad (10.58)$$

The kinetic energy T_S of the center of mass, Eq. (10.52), is conserved separately because of conservation of center of mass momentum, Eq. (10.27). T_S contains the system's translational degrees of freedom. The kinetic energy T_r of the relative motion contains, for instance, the energy of rotation about an axis through the center of mass (cf. also Subsection 11.3.2). *In a closed system, the kinetic energy T_S of the center of mass motion, and the energy E of the relative motion,* where

$$E = T_r + V = const, \tag{10.59}$$

are conserved separately.

Summary of the conservation laws in an N-particle system.

In a <u>*constant gravitational field,*</u> the motion of the center of mass decouples from the relative motion of the point masses. The center of mass motion is always integrable. The six constants of motion are $\mathbf{R}_0$ and $\mathbf{V}_0$ (cf. Eq. (10.26)). For the relative motion (the 'internal' degrees of freedom), the energy E and the angular momentum $\mathbf{L}$ are conserved, just as they are in the closed N-particle system – so conclusions listed immediately below about relative motion in a closed system hold here, too.

In a <u>*closed system* of N *point masses*</u> (whose interaction forces are conservative), for $N \geq 2$, there are at least ten *independent constants of the motion or nine independent conserved quantities,* respectively:

center of mass momentum ($\mathbf{P}$)	3 quantities
center of mass motion ($\mathbf{R}_0$)	3 quantities
energy of relative motion (E)	1 quantity
angular momentum of the relative motion ($\mathbf{L}$)	3 quantities

The center of mass motion, with three degrees of freedom and five conserved quantities – thus the center of mass motion is always integrable – separates from the remaining $3(N-1)$ degrees of freedom of the relative motion. For the relative motion exist four conserved quantities, namely, the energy and the angular momentum. A subset of three of these, say $(E, \mathbf{L}^2, L_z)$, can be taken that satisfies the integrability conditions in Section 14.1. Therefore, the two-body system ($N = 2$) with three rotational degrees of freedom is always integrable. In the Kepler problem, as we have seen, another independent conserved quantity can be derived from the Runge-Lenz vector, i.e. the vector from the focus to the perihelion. For $N > 2$, the four conserved quantities of the relative motion do not suffice to solve the problem algebraically. To determine the relative motion in a three-particle system, one would need at least six conserved quantities. Special cases of closed three-particle systems will be treated in Chapter 14.

10.3 The Lagrangian of a system of N particles

In analogy with the Lagrangian of *one* mass point (see Subsection 9.3.3), the *Lagrangian* of a system of N point masses subjected to conservative external forces consists of the kinetic energy (10.47),

$$T = \sum_{\alpha=1}^{N} \frac{1}{2} m_\alpha \dot{\mathbf{r}}_\alpha^2,$$

and the potential energy[7] (10.48),

$$V = \frac{1}{2} \sum_{\alpha=1}^{N} \sum_{\beta=1}^{N} V_{\alpha\beta} \left(|\mathbf{r}_\alpha - \mathbf{r}_\beta| \right) + \sum_{\alpha=1}^{N} U_\alpha(\mathbf{r}_\alpha), \qquad \alpha, \beta = 1, \ldots, N,$$

so that we have for the N-particle Lagrangian

$$L\left(\{\mathbf{r}\}, \{\dot{\mathbf{r}}\}, t\right) = T - V. \tag{10.60}$$

The equations of motion (10.10) are now Lagrange's equations,

$$\frac{d}{dt} \frac{\partial L}{\partial \dot{\mathbf{r}}_\alpha} - \frac{\partial L}{\partial \mathbf{r}_\alpha} = 0, \qquad \alpha = 1, \ldots, N,$$

for the extremum of the action integral $S = \int_{t_1}^{t_2} L\left(\{\mathbf{r}\}, \{\dot{\mathbf{r}}\}, t\right) dt$.

We start from the Lagrangian for N point masses in Cartesian coordinates, which we denote explicitly as $L_x = L\left(\{x\}, \{\dot{x}\}, t\right)$, and Lagrange's equations

$$\frac{d}{dt} \frac{\partial L_x}{\partial \dot{x}_j} - \frac{\partial L_x}{\partial x_j} = 0, \qquad j = 1, \ldots, 3N.$$

---

[7]Note that unless stated otherwise, V includes the potential of the external forces, $\sum_{\alpha=1}^{N} U_\alpha$.

In a particular problem, when one transforms to coordinates more appropriate to the symmetry at hand, or one eliminates coordinates with the help of constraints, there appear **generalized coordinates**[8] q_k, $k = 1, \ldots, f$, $(f \leq 3N)$. The Cartesian coordinates x_j are then functions of the f independent generalized coordinates:

$$x_j = x_j(q_1, \ldots, q_f, t), \tag{10.61}$$

where $j = 1, \ldots, 3N$ and $f \leq 3N$. This transformation also determines the velocities $\dot{x}_j$ as functions of the generalized coordinates q_k and their time derivatives $\dot{q}_k$:

$$\dot{x}_j = \sum_{k=1}^{f} \frac{\partial x_j}{\partial q_k} \dot{q}_k + \frac{\partial x_j}{\partial t} = \dot{x}_j \left(\{q\}, \{\dot{q}\}, t\right). \tag{10.62}$$

Inserting Eqs. (10.61) and (10.62) into L_x yields the Lagrangian,

$$L_x \left(\{x(\{q\})\}, \{\dot{x}(\{q\}, \{\dot{q}\}, t)\}, t\right) = L_q \left(\{q\}, \{\dot{q}\}, t\right) =: L \left(\{q\}, \{\dot{q}\}, t\right),$$

(from now on we omit the indices q and x) in the f generalized coordinates q_k:

$$L \left(\{q\}, \{\dot{q}\}, t\right) = T \left(\{q\}, \{\dot{q}\}, t\right) - V \left(\{q\}, t\right). \tag{10.63}$$

The equations of motion in the generalized coordinates are given by Lagrange's equations,

$$\frac{d}{dt} \frac{\partial L}{\partial \dot{q}_k} - \frac{\partial L}{\partial q_k} = 0, \qquad k = 1, \ldots, f, \tag{10.64}$$

which follow from Hamilton's principle of least action (cf. Eq. (9.75)),

$$\delta S = 0, \qquad S = \int_{t_1}^{t_2} L(\{q\}, \{\dot{q}\}, t) dt. \tag{10.65}$$

Recall that if a coordinate q_k is cyclic, i.e. $\partial L / \partial q_k = 0$, then the conjugate momentum is a conserved quantity,

$$p_k = \frac{\partial L}{\partial \dot{q}_k} = const. \tag{10.66}$$

---

[8]Explicitly this has been exemplified for the spherical pendulum in Subsection 9.3.5. There, instead of the former three Cartesian coordinates, one has then only the two angles ϑ and φ.

Conservation of energy

If the Lagrangian does not depend explicitly on time, *viz.* $L = L(\{q\}, \{\dot{q}\})$ and consequently $\partial L/\partial t = 0$, the **Jacobian integral**[9], defined by

$$I_J := \sum_{k=1}^{f} \frac{\partial L}{\partial \dot{q}_k} \dot{q}_k - L = \sum_{k=1}^{f} p_k \dot{q}_k - L = const, \qquad (10.67)$$

is a conserved quantity. The proof is similar to the derivation of the Eq. (9.32). Using Lagrange's equations, one can immediately see that an explicitly time-dependent Lagrangian satisfies

$$\frac{d}{dt} \left(\sum_{k=1}^{f} \frac{\partial L}{\partial \dot{q}_k} \dot{q}_k - L \right) = -\frac{\partial L}{\partial t}.$$

Suppose the Lagrangian in Eq. (10.60) does not depend on time explicitly. Since $\mathbf{p}_\alpha = \partial L/\partial \dot{\mathbf{r}}_\alpha = m \dot{\mathbf{r}}_\alpha$, the Jacobian integral is

$$m \dot{\mathbf{r}}_\alpha^2 - \frac{m}{2} \dot{\mathbf{r}}_\alpha^2 + V = T + V = E.$$

Hence, conservation of energy in the Cartesian coordinate system is tantamount to conservation of the Jacobian integral. If now the transformation from Cartesian to generalized coordinates, Eq. (10.61), does *not explicitly depend on time* (e.g. the constraints are time independent (scleronomic)),

$$x_j = x_j(q_1, \dots, q_f), \qquad (10.68)$$

then

$$\dot{\mathbf{r}}_\alpha = \sum_{k=1}^{f} \frac{\partial \mathbf{r}_\alpha}{\partial q_k} \dot{q}_k, \qquad \alpha = 1, \dots, N,$$

and the kinetic energy

$$T = \frac{1}{2} \sum_{\alpha=1}^{N} m_\alpha \sum_{k,l=1}^{f} \frac{\partial \mathbf{r}_\alpha}{\partial q_k} \frac{\partial \mathbf{r}_\alpha}{\partial q_l} \dot{q}_k \dot{q}_l =: \sum_{k,l=1}^{f} a_{kl}(q_1, \dots, q_f) \dot{q}_k \dot{q}_l \qquad (10.69)$$

---

[9]Carl Gustav Jacob JACOBI (1804-1851), German mathematician.

is a quadratic form, with symmetric coefficients a_{kl}, $a_{kl} = a_{lk}$. Since

$$\sum_{k=1}^{f} \frac{\partial L}{\partial \dot{q}_k} \dot{q}_k = \sum_{k=1}^{f} \frac{\partial T}{\partial \dot{q}_k} \dot{q}_k = 2T,$$

the Jacobian integral, Eq. (10.67), is equal to the total energy. Indeed, we have

$$I_J = 2T - L = T + V = E = const. \tag{10.70}$$

If the Lagrangian L does not depend explicitly on time t, then the Jacobian integral is a conserved quantity. Moreover, if the kinetic energy is a quadratic form in the generalized velocities, the Jacobian integral is equal to the total energy.

10.4   Infinitesimal transformations and conservation laws

We consider infinitesimal transformations of the (generalized) coordinates and time. We show that the behavior of the Lagrangian under these transformations is closely related to the existence of conserved quantities.

10.4.1   Infinitesimal translations of time

If the Lagrangian is invariant under **infinitesimal translation in time**, i.e.

$$L\left(\{q\}_{t'}, \{\dot{q}\}_{t'}, t'\right) = L\left(\{q\}_t, \{\dot{q}\}_t, t\right),$$

with

$$t' = t + \varepsilon, \tag{10.71}$$

then expanding $L\left(\{q\}_{t'}, \{\dot{q}\}_{t'}, t'\right)$ in the explicit time dependence,

$$L\left(\{q\}_{t'}, \{\dot{q}\}_{t'}, t'\right) = L\left(\{q\}_{t'}, \{\dot{q}\}_{t'}, t\right) + \varepsilon \frac{\partial L}{\partial t} + \dots,$$

we see that[10]

$$\partial L / \partial t = 0.$$

---

[10]Of course, the transformation changes the time dependence of the dynamical variables, too. However, the *dependence* of L on the dynamical variables does not change.

Consequently L has no explicit time dependence:

$$L\left(\{q\},\{\dot{q}\},t\right) = L\left(\{q\},\{\dot{q}\}\right).$$

As we have seen, this means that the Jacobian integral, Eq. (10.67), is conserved. (In appropriate cases, the energy is too.)

10.4.2 Infinitesimal coordinate transformations

Noether's theorem

An **infinitesimal transformation of the coordinates** has the form

$$q'_k = q_k + \varepsilon s_k\left(q_1,\ldots,q_f,t\right), \tag{10.72}$$

where ε is an infinitesimal parameter, and the functions s_k depend on a certain number r of *parameters* $(\alpha_1,\ldots,\alpha_r)$ (see below). If L is invariant under this transformation, i.e.

$$L\left(\{q'\},\{\dot{q}'\},t\right) = L\left(\{q\},\{\dot{q}\},t\right), \tag{10.73}$$

where the *same* function L appears on each side, then Lagrange's equations – and consequently the equations of motion – are invariant under this transformation, too.

Now, taking a Taylor expansion of the left hand side of Eq. (10.73) with respect to ε,

$$L\left(\{q'\},\{\dot{q}'\},t\right) = L\left(\{q\},\{\dot{q}\},t\right) + \varepsilon\sum_{k=1}^{f}\left(\frac{\partial L}{\partial q_k}s_k + \frac{\partial L}{\partial \dot{q}_k}\dot{s}_k\right) + \ldots,$$

we see from Eq. (10.73) that to order $\mathcal{O}(\varepsilon)$

$$\sum_{k=1}^{f}\left(\frac{\partial L}{\partial q_k}s_k + \frac{\partial L}{\partial \dot{q}_k}\dot{s}_k\right) = 0. \tag{10.74}$$

Using Lagrange's equations (10.64), it follows that

$$\sum_{k=1}^{f}\left(\left(\frac{d}{dt}\frac{\partial L}{\partial \dot{q}_k}\right)s_k + \frac{\partial L}{\partial \dot{q}_k}\dot{s}_k\right) = \frac{d}{dt}\left(\sum_{k=1}^{f}\frac{\partial L}{\partial \dot{q}_k}s_k\right) = 0.$$

Hence, the expression in the brackets on the right hand side is constant in time,

$$\sum_{k=1}^{f} \frac{\partial L}{\partial \dot{q}_k} s_k = const, \qquad (10.75)$$

and since the s_k depend on r parameters $(\alpha_1, \ldots, \alpha_r)$ one obtains r *constants of the motion*. This close connection between the invariance of a system under transformations and the existence of constants of the motion (conserved quantities) is expressed in

Noether's theorem[11] (1918):

> If a Lagrangian L remains invariant under an r-parametric infinitesimal transformation, then there exist r constants of the motion for Lagrange's equations, $\frac{d}{dt}\frac{\partial L}{\partial \dot{q}_k} - \frac{\partial L}{\partial q_k} = 0$.

Imposing the invariance on L may be too restrictive. What really has to be invariant are the equations of motion, i.e. Lagrange's equations. These are invariant if one allows the Lagrangian to transform as follows:

$$L'\left(\{q'\},\{\dot{q}'\},t\right) := L\left(\{q'\},\{\dot{q}'\},t\right) + \frac{dF}{dt} = L\left(\{q\},\{\dot{q}\},t\right), \quad (10.76)$$

where dF/dt is the total time derivative of a function $F(\{q'\},t)$ that depends on the *coordinates and time*. To show the invariance, we observe that from

$$\frac{dF}{dt} = \sum_{k=1}^{f} \frac{\partial F}{\partial q_k'}\dot{q}_k' + \frac{\partial F}{\partial t},$$

it follows that

$$\frac{\partial}{\partial \dot{q}_k'}\frac{dF}{dt} = \frac{\partial F}{\partial q_k'},$$

and therefore the contribution to Lagrange's equations vanishes:

$$\frac{d}{dt}\frac{\partial}{\partial \dot{q}_k'}\frac{dF}{dt} - \frac{\partial}{\partial q_k'}\frac{dF}{dt} = \frac{d}{dt}\frac{\partial F}{\partial q_k'} - \frac{\partial}{\partial q_k'}\frac{dF}{dt} = 0.$$

---

[11]Emmy NOETHER (1882-1935), influential German mathematician. She worked after her dissertation at the Mathematical Institute of Erlangen from 1907 to 1915 without pay. At the time, women were largely excluded from academic positions.

Lagrangians differing only by a total time derivative of a function of the coordinates lead to the same equations of motion. Since, for $\varepsilon = 0$ in Eq. (10.72) we have the identity transformation, the function F is at least proportional to ε. Therefore F can be written as

$$F = \varepsilon G \left(\{q'\}, t\right).$$ (10.77)

Instead of the more restrictive condition in Eq. (10.74), for an infinitesimal transformation, we now have (to order $\mathcal{O}(\varepsilon)$) the invariance condition

$$\sum_{k=1}^{f} \left(\frac{\partial L}{\partial q_k} s_k + \frac{\partial L}{\partial \dot{q}_k} \dot{s}_k\right) + \frac{dG}{dt} = 0;$$ (10.78)

whence, using Lagrange's equations, instead of Eq. (10.75), the quantity

$$\sum_{k=1}^{f} \frac{\partial L}{\partial \dot{q}_k} s_k + G|_{\varepsilon=0} = const$$ (10.79)

is a *constant of motion*. In general, the infinitesimal transformation of the coordinates given in Eq. (10.72) may also include a *gauge transformation*, Eq. (10.76), of L to L'. L and L' are called *equivalent*, and L is said to be invariant up to a **gauge function** F.

We demonstrate Noether's theorem – the connection between invariance and conservation laws – in the following example. Consider the motion of a particle in the potential $V = V(\rho)$. The Lagrangian in cylindrical coordinates is given in Eq. (9.79). An infinitesimal transformation

$$\varphi' = \varphi + \varepsilon$$

does not change L. Now, in this case, Eq. (10.75) takes the form

$$\frac{\partial L}{\partial \dot{\varphi}} = m\rho^2 \dot{\varphi} = const.$$

This is the conservation law for the z-component of the angular momentum, Eq. (9.83).

The close relation between a symmetry of a system and a constant of motion can be derived without much fuss from the Lagrangian. This is a further advantage of Lagrange's formulation of mechanics.

10.4.3   Galilean transformations and constants of motion

We return to the Lagrangian of an N-particle system in the variables $\mathbf{r}_\alpha$ and $\dot{\mathbf{r}}_\alpha$:

$$L = L\left(\{\mathbf{r}\},\{\dot{\mathbf{r}}\},t\right) = \sum_{\alpha=1}^{N} \frac{1}{2} m_\alpha \dot{\mathbf{r}}_\alpha^2 - V\left(\{r_{\alpha\beta}\}\right). \qquad (10.80)$$

Let the system be *closed*. The total potential V is then invariant under both translations and rotations of the coordinates; i.e., V is of the form shown in Eq. (10.48). We now investigate the consequences of the invariance of the Lagrangian,

$$L\left(\{\mathbf{r}\},\{\dot{\mathbf{r}}\},t\right) = L\left(\{\mathbf{r}'\},\{\dot{\mathbf{r}}'\},t\right), \qquad (10.81)$$

under infinitesimal transformations of the coordinates of the form

$$\mathbf{r}'_\alpha = \mathbf{r}_\alpha + \varepsilon \mathbf{s}_\alpha\left(\{\mathbf{r}\},t\right) \qquad (10.82)$$

that belong to the Galilean group (see Section 8.4): namely, the time-independent translations, rotations, and velocity transformations (boosts).

Translations

An infinitesimal time-independent translation of the particles' coordinates in the direction $\mathbf{s}$ is given by

$$\mathbf{r}'_\alpha = \mathbf{r}_\alpha + \varepsilon \mathbf{s}, \qquad \mathbf{s} = const. \qquad (10.83)$$

The functions $\mathbf{s}_\alpha$ in Eq. (10.82) reduce to $\mathbf{s}_\alpha\left(\{\mathbf{r}\}\right) = \mathbf{s}$. The three independent components of $\mathbf{s}$ are the parameters of the translations. Since T and V are both invariant under such a transformation, Eq. (10.81) holds. From Eq. (10.75), it follows that

$$\mathbf{s}\sum_{\alpha=1}^{N} \frac{\partial L}{\partial \dot{\mathbf{r}}_\alpha} = const.$$

Since, using (10.80),

$$\frac{\partial L}{\partial \dot{\mathbf{r}}_\alpha} = m_\alpha \dot{\mathbf{r}}_\alpha, \qquad (10.84)$$

we have

$$\mathbf{s}\sum_{\alpha=1}^{N} m_\alpha \dot{\mathbf{r}}_\alpha = \mathbf{s}\mathbf{P} = const. \qquad (10.85)$$

The component of the total momentum along **s** is conserved. Since **s** is an arbitrary vector, it must be true that

$$\mathbf{P} = const. \tag{10.86}$$

If L is translationally invariant, the total momentum of a closed system of N particles is conserved.

Each of the three components of **s** corresponds to a component of the conserved momentum **P**.

Rotations

An infinitesimal rotation, by angle φ about the axis **n** (with $\boldsymbol{\varphi} = \varphi \mathbf{n}$), of all the coordinates, is given by (see Eq. (8.20))

$$\mathbf{r}'_\alpha = \mathbf{r}_\alpha + \varepsilon(\boldsymbol{\varphi} \times \mathbf{r}_\alpha).$$

So $\mathbf{s}_\alpha(\{\mathbf{r}\}) = \boldsymbol{\varphi} \times \mathbf{r}_\alpha$, and the three components of $\boldsymbol{\varphi}$ are the parameters of the transformation. Again T and V are invariant under this transformation, and according to (10.75), from the invariance of L (Eq. (10.80)) under time-independent rotations, it follows that

$$\sum_{\alpha=1}^{N} \frac{\partial L}{\partial \dot{\mathbf{r}}_\alpha}(\boldsymbol{\varphi} \times \mathbf{r}_\alpha) = const.$$

Inserting Eq. (10.84) here, yields

$$\boldsymbol{\varphi} \sum_{\alpha=1}^{N} m_\alpha(\mathbf{r}_\alpha \times \dot{\mathbf{r}}_\alpha) = \boldsymbol{\varphi}\mathbf{L}_{tot} = const. \tag{10.87}$$

The component of the angular momentum parallel to the vector $\boldsymbol{\varphi}$ is constant. Since $\boldsymbol{\varphi}$ is arbitrary, it must hold that

$$\mathbf{L}_{tot} = const. \tag{10.88}$$

Invariance of the Lagrangian under rotations implies conservation of angular momentum.

Again, each of the three parameters of $\boldsymbol{\varphi}$ correspond to a conserved component of $\mathbf{L}_{tot}$.

Velocity transformations

Finally, we apply an infinitesimal transformation with time-independent vector parameters $\mathbf{w}$,

$$\mathbf{r}'_\alpha = \mathbf{r}_\alpha + \varepsilon \mathbf{w} t, \qquad \rightarrow \qquad \dot{\mathbf{r}}'_\alpha = \dot{\mathbf{r}}_\alpha + \varepsilon \mathbf{w}, \qquad (10.89)$$

to the Lagrangian $L(\{\mathbf{r}_\alpha\}, \{\dot{\mathbf{r}}_\alpha\}, t)$. From Eq. (10.80), it follows that

$$
\begin{aligned}
L\left(\{\mathbf{r}\}, \{\dot{\mathbf{r}}\}, t\right) &= \sum_{\alpha=1}^{N} \frac{1}{2} m_\alpha (\dot{\mathbf{r}}'_\alpha - \varepsilon \mathbf{w})^2 - V\left(\{r'_{\alpha\beta}\}\right) \\
&= \sum_{\alpha=1}^{N} \frac{1}{2} m_\alpha \dot{\mathbf{r}}'^2_\alpha - V\left(\{r'_{\alpha\beta}\}\right) - \varepsilon \mathbf{w} \sum_{\alpha=1}^{N} m_\alpha \dot{\mathbf{r}}'_\alpha + \mathcal{O}(\varepsilon^2) \\
&= L\left(\{\mathbf{r}'\}, \{\dot{\mathbf{r}}'\}, t\right) - \varepsilon \mathbf{w} \sum_{\alpha=1}^{N} m_\alpha \dot{\mathbf{r}}'_\alpha + \mathcal{O}(\varepsilon^2). \quad (10.90)
\end{aligned}
$$

Here we have used the invariance of the potential V with respect to transformations (10.89),

$$V\left(\{|\mathbf{r}'_\alpha - \mathbf{r}'_\beta|\}\right) = V\left(\{|\mathbf{r}_\alpha - \mathbf{r}_\beta|\}\right).$$

Because

$$\mathbf{w} \sum_{\alpha=1}^{N} m_\alpha \dot{\mathbf{r}}'_\alpha = \frac{d}{dt} M \mathbf{w} \mathbf{R},$$

the relation between the two Lagrangians, to order $\mathcal{O}(\varepsilon)$, is

$$L\left(\{\mathbf{r}\}, \{\dot{\mathbf{r}}\}, t\right) = L\left(\{\mathbf{r}'\}, \{\dot{\mathbf{r}}'\}, t\right) + \varepsilon \frac{dG}{dt},$$

so that for the velocity transformation, the Lagrangians differ by a gauge function $F = \varepsilon G$,

$$G = -M \mathbf{w} \mathbf{R} + \mathcal{O}(\varepsilon). \qquad (10.91)$$

Application of Eqs. (10.77) and (10.79) leads to

$$\sum_\alpha \frac{\partial L}{\partial \dot{\mathbf{r}}_\alpha} \mathbf{w} t - \mathbf{w} \mathbf{R} M = \mathbf{w}(\mathbf{P}t - \mathbf{R}M) = const.$$

Since $\mathbf{w}$ is arbitrary, we have the following constant of the motion:

$$\mathbf{P}t - M\mathbf{R} = const. \qquad (10.92)$$

Gauge invariance of the Lagrangian L under velocity trans-
formations (10.89) implies free center of mass motion.

The three parameters $\mathbf{w}$ are related to the three constants of motion in Eq. (10.92).

Unlike translations and rotations, a velocity transformation changes the total energy, by the amount of kinetic energy T expended in changing the velocities of all point masses by $\mathbf{w}$.

Summary

The symmetry of the Lagrangian, Eq. (10.80), with respect to transformations of the Galilean group, which contains ten parameters (see Section 8.4), generates ten constants of the motion:

> the total energy (one)
> the total momentum (three)
> the total angular momentum (three)
> the free motion of the center of mass (three).

Problems and examples

1. Starting with the Lagrangian L for

   a) The spherical pendulum;

   b) The two-dimensional isotropic oscillator;

   c) Motion in a homogeneous magnetic field (cf. Subsection 9.3.3),

   apply coordinate transformations that leave L invariant, and determine the corresponding conserved quantities.

2. A method for **finding constants of the motion** (see [Saletan/Cromer]) consists in applying an infinitesimal coordinate transformation

$$q'_k = q'_k(\{q\}, t; \varepsilon) \tag{10.93}$$

   to a Lagrangian $L(\{q\}, \{\dot{q}\}, t)$. (The transformation (10.93) is supposed to be invertible and differentiable with respect to ε.) The transformation results in a Lagrangian

$$L_\varepsilon(\{q'\}, \{\dot{q}'\}, t) = L\left(\{q(\{q'\}, t; \varepsilon)\}, \left\{\frac{d}{dt} q(\{q'\}, t; \varepsilon)\right\}, t\right),$$

where

$$q_k = q_k \left(\{q'\}, t; \varepsilon \right) \tag{10.94}$$

is the inverse of the transformation (10.93). The ε-dependence of L_ε is entirely due to the ε-dependence of the transformation (10.94). Therefore (show this!)

$$\frac{dL_\varepsilon}{d\varepsilon} = -\left[\frac{d}{dt}\frac{\partial L}{\partial \dot{q}_k} - \frac{\partial L}{\partial q_k} \right]\frac{\partial q_k}{\partial \varepsilon} + \frac{d}{dt}\left[\frac{\partial L}{\partial \dot{q}_k}\frac{\partial q_k}{\partial \varepsilon} \right], \tag{10.95}$$

where the coordinates $\{q\}$ are considered as functions of the 'new' coordinates $\{q'\}$, in accordance with Eq. (10.94). Now, if the set of 'old' coordinates $\{q\} = \{q'\}|_{\varepsilon=0}$ is a solution of Lagrange's equations, then for $\varepsilon = 0$, the first term on the right hand side of Eq. (10.95) vanishes. Since

$$\frac{dL_\varepsilon}{d\varepsilon} = \frac{d}{dt}\left[\frac{\partial L}{\partial \dot{q}_k}\frac{\partial q_k}{\partial \varepsilon} \right], \quad \varepsilon = 0,$$

$\left. \dfrac{dL_\varepsilon}{d\varepsilon} \right|_{\varepsilon=0}$ is the total time derivative of a function G:

$$\left. \frac{dL_\varepsilon}{d\varepsilon} \right|_{\varepsilon=0} =: \frac{d}{dt}G\left(\{q\}, \{\dot{q}\}, t \right).$$

Hence

$$\frac{d}{dt}\left[\left. \frac{\partial L}{\partial \dot{q}_k}\frac{\partial q_k}{\partial \varepsilon} \right|_{\varepsilon=0} - G \right] = 0. \tag{10.96}$$

In other words: $\left. \dfrac{\partial L}{\partial \dot{q}_k}\dfrac{\partial q_k}{\partial \varepsilon} \right|_{\varepsilon=0} - G$ is a *constant of the motion*.

Application to the *harmonic oscillator*

$$L = \frac{m}{2}\dot{q}^2 - \frac{m\omega^2}{2}q^2.$$

Determine the constant of motion for the infinitesimal transformation

$$q' = q + \varepsilon \sin\omega t,$$

and convince yourself that it is really constant.

11

The rigid body

In this chapter, we consider the mechanics of a rigid body. We imagine a rigid body as an ensemble of point masses that are fixed relative to each other by their mutual forces. This is the limiting case of the more realistic situation, in which the N point masses at positions $\mathbf{r}_\alpha$, $\alpha = 1, ..., N$, interact only via two-particle interactions, so that the total potential is given by (cf. Eq. (10.48)) $V(\{r_{\alpha\beta}\}) = \sum_{\alpha=1}^{N} \sum_{\beta=\alpha+1}^{N}$ $V_{\alpha\beta}(r_{\alpha\beta})$, where $r_{\alpha\beta} = |\mathbf{r}_\alpha - \mathbf{r}_\beta|$. The values $\mathbf{r}_\gamma = \mathbf{a}_\gamma$, $\gamma = 1, ..., N$, for which all internal forces vanish,

$$\mathbf{F}_\alpha = -\partial V/\partial \mathbf{r}_\alpha|_{\{\mathbf{r}_\gamma = \mathbf{a}_\gamma, \gamma = 1, ..., N\}} = 0, \qquad \alpha = 1, ..., N,$$

are the *equilibrium positions* $\mathbf{a}_\alpha$. Whether such positions exist depends on the potential; we assume that equilibrium positions exist.

Analogous to the procedure introduced in Section 3.2 for a one-dimensional system, one can expand the potential about these positions. This will be investigated more closely in the next chapter. For a stable equilibrium, the positions $\mathbf{a}_\alpha$ must be minima of the total potential. In particular, the eigenvalues ω_k^2 of the matrix of coefficients $\partial^2 V/\partial \mathbf{r}_\alpha \partial \mathbf{r}_\beta|_{\{\mathbf{r}_\gamma = \mathbf{a}_\gamma\}}$ in the Taylor expansion of V have to be positive. (Note that for fixed indices α and β, this is already a 9×9-matrix!) If the ω_k^2 are very large[1], a large *displacement from equilibrium* requires more energy than supplied. If, for some energy range, each point mass practically[2] stays in its equilibrium position, then this energy range is

---

[1] We desist from the vanishing eigenvalues of translation and rotation of the system of point masses as a whole (cf. Chapter 12).

[2] We ignore small oscillations of the particles about their equilibrium positions.

the **rigid body** region; only higher energies can lead to deformations of the body. In this chapter, we start off with some basics of statics. Our main concern, however, is the dynamics of rigid bodies.

11.1   Degrees of freedom of a rigid body

In the ideal rigid body, the distances between any pair of the N point masses is fixed,

$$r_{\alpha\beta} = const, \quad r_{\alpha\alpha} = 0, \qquad \alpha, \beta = 1, \dots, N.$$

Since $r_{\alpha\beta} = r_{\beta\alpha}$, we have $(N^2 - N)/2$ conditions. However, these conditions cannot all be independent, since for large enough N they would outnumber the $3N$ degrees of freedom. Furthermore, we know that the rigid body has are at most six degrees of freedom. How does this happen? We note that to fix the position of each particle relative to the others, it is *not necessary to specify all the distances* from each particle to every other particle. All we need is the following. First select three point masses that do not lie on a straight line. The positions of these three point masses, given as their nine coordinates, define a plane in space. Then a fourth point mass is uniquely fixed by its distances to the three point masses selected first[3]. This applies also to the remaining point masses. Hence, all the point masses except the three selected first, are uniquely fixed by their distances to the three reference point masses, giving $3 \times (N - 3)$ conditions for the $N - 3$ point masses. Thus, the number of degrees of freedom is reduced to $3N - 3(N - 3) = 9$. Indeed, the motion of the $N - 3$ particles follows immediately from the motion of the three reference particles. Finally, the degrees of freedom of the three reference particles – originally nine – are diminished by the three independent conditions of fixed distance between the reference particles. Therefore the *total number of the degrees of freedom of a rigid body is six*. This number is independent of N, and therefore also holds in the continuum limit $N \to \infty$. It comprises of three degrees of freedom corresponding to the motion of the center of mass, and three corresponding to the orientational degrees of freedom of the reference plane (i.e. of the body as a whole) in space. In other words, they correspond to the three translational and three rotational degrees of freedom of the rigid body.

---

[3]The position of the fourth mass would not be fixed uniquely if the three point masses lay on a line.

11.2 Some basics of statics

Statics of (rigid) bodies is the science of the equilibrium of bodies. Important in practical applications, statics used to be the central subject in mechanics[4], whereas nowadays mechanics has become more or less identified with dynamics. Statics, as an important discipline in technical applications, such as constructions of buildings, bridges, etc., is now a field in its own.

Because we consider only absolutely rigid bodies here we completely neglect a field important in technical statics: the question of *stability of bodies* (bending of beams, fracture of beams, etc.). This field was initiated by Galileo in his last book "Discorsi e dimostrazioni matematiche, intorno à due nuove scienze" (Discourses and Mathematical Demonstrations Relating to Two New Sciences; Leiden 1638). It contains not only his well-known investigations on motion (free fall, pendulum) but in the beginning it presents also Galileo's rather ignored discoveries on the strength of materials.

After a brief historical survey, we present a few of the basics of statics, which – since this is the nature of statics – are related to the concept of force.

11.2.1 Historical survey

Statics is at least as old as the theory of motion. The work of ARCHIMEDES (287-212 BC) on the laws of levers ('Give me a place to stand on, and I will move the earth') has become urban myth[5]. But equally important is his work on the center of gravity[6]. The famous constructor

---

[4]Lagrange devoted nearly half of his book, "Analytical mechanics", to this extensive subject.

[5]For a detailed account consult Degas's "A History of Mechanics" (see Bibliography).

[6]Archimedes introduced first the concept of 'center of mass' in the form of the 'center of gravity'. He worked with simplified assumptions about gravity that amount to a uniform field, thus arriving at the mathematical properties of what we now call the center of mass. He showed that the center of gravity of several weights placed on the beam is exactly the fulcrum for which the beam is horizontal, i.e. is in equilibrium. In work on floating bodies he demonstrated that the orientation of a floating object is the one that makes its center of mass as low as possible. He developed mathematical techniques for finding the centers of mass of objects of uniform density of various well-defined shapes.

of automates HERO of Alexandria (about 10-70 AD) studied simple machines (see below) in his "Mechanica". The next important step was taken in the 13th century by JORDANUS de Nemore (first half of 13th century) in his studies of the equilibrium of forces, the method of virtual work appears for the first time. About two centuries later, LEONARDO da Vinci (1452-1519) also addressed the problems of statics. Continuous pursuit and development of statics began in the 16th century. We only quote Simon STEVIN (Dutch mathematician and engineer, 1548-1620) using his 'clootcrans' (wreath of spheres) to demonstrate the composition of forces, Galileo mentioning machines to handle great loads in his lectures at Padua (1594), Descartes's principle of virtual work[7], and, in particular, Gilles Personne de ROBERVAL (French mathematician, 1602-1675), who studied equilibria in systems of wheels, ropes, and weights. He was aware of the parallelogram law of forces. 1687 Newton referred shortly to the use and the advantage of 'mechanical powers' in his "Principia" (Scholium of the "Axioms, or Laws of Motion"). Varignon treated the simple machines using the composition of forces in his work "Projet d'une nouvelle mécanique" (1687) and in the more extensive "Nouvelle mécanique, ou Statique" (1725, posthumous) .A version of the 'principle of virtual work' was communicated 1717 by Johann Bernoulli to P.Varignon.

11.2.2 The basic physical principles

In a system of point masses, if the total momentum as well as the total angular momentum are constant in time, then static equilibrium is achieved, if not all the (external[8]) forces vanish. The **equilibrium**

Among later mathematicians who developed the theory of the center of mass are Pappus of Alexandria, Guido Ubaldi, Simon Stevin, Paul Guldin, John Wallis, Pierre Varignon, and Alexis Clairaut. (Wikipedia)

Besides the center of mass, the mass distribution (characterized by its moments, see Section 6.4.2) is of interest in certain problems. We will encounter a particular moment when discussing the dynamics of rigid bodies: the inertia tensor.

[7]In modern mathematical language the *principle of virtual work* is

$$\sum \mathbf{F}_i \delta \mathbf{r}_i = 0.$$

This states that the total work done by the forces $\mathbf{F}_i$ along the *virtual displacements* $\delta \mathbf{r}_i$ in the directions of the applied tools, e.g. towing ropes, must vanish.

[8]In the following, all forces are external, since the internal interaction forces are replaced by constraints of fixed distances between the constituent particles.

conditions are the following:

Equilibrium of forces ($\rightarrow$ no translations of the body):

$$\sum_{\alpha=1}^{N} \mathbf{F}_\alpha = 0; \tag{11.1}$$

and

Equilibrium of the moments ($\rightarrow$ no rotations of the body):

$$\sum_{\alpha=1}^{N} \mathbf{r}_\alpha \times \mathbf{F}_\alpha = \sum_{\alpha=1}^{N} \mathbf{N}_\alpha = 0. \tag{11.2}$$

These are the **basic equations of statics**. The first equilibrium condition requires that the sum of the forces vanishes regardless of their respective contact points. The second equilibrium condition requires that the sum of the moments of the forces with respect to a common, but arbitrary, point must vanish. We prove that this point is arbitrary, using the fact that the momentum does not change if the point is shifted, as follows. Suppose the reference point is shifted, $\mathbf{r}'_\alpha = \mathbf{r}_\alpha + \mathbf{a}$. Since $\mathbf{F}'_\alpha = \mathbf{F}_\alpha$, we have

$$\sum_{\alpha=1}^{N} \mathbf{F}'_\alpha = \sum_{\alpha=1}^{N} \mathbf{F}_\alpha$$

and

$$\sum_{\alpha=1}^{N} \mathbf{N}'_\alpha = \sum_{\alpha=1}^{N} (\mathbf{r}_\alpha + \mathbf{a}) \times \mathbf{F}_\alpha = \sum_{\alpha=1}^{N} \mathbf{r}_\alpha \times \mathbf{F}_\alpha + \mathbf{a} \times \sum_{\alpha=1}^{N} \mathbf{F}_\alpha$$

$$= \sum_{\alpha=1}^{N} \mathbf{N}_\alpha + \mathbf{a} \times \sum_{\alpha=1}^{N} \mathbf{F}_\alpha.$$

If the forces are in equilibrium (Eq. (11.1)), we have

$$\sum_{\alpha=1}^{N} \mathbf{N}'_\alpha = \sum_{\alpha=1}^{N} \mathbf{N}_\alpha.$$

For rigid bodies, the main task in statics is the following. Given a rigid body for which the forces or moments are not in equilibrium, find the force or moment that establishes equilibrium. There are **two basic tasks**:

i) If $\sum_{\alpha=1}^{N} \mathbf{F}_\alpha \neq 0$ but $\sum_{\alpha=1}^{N} \mathbf{N}_\alpha = 0$, it suffices to add the force $\mathbf{F} = -\sum_{\alpha=1}^{N} \mathbf{F}_\alpha$ acting at the common point of reference of the moments (i.e. the pivot point, which is usually situated within the rigid body). The total moment of the forces about the pivot point remains unchanged, and the forces now balance. In practice, forces cannot act at every point of the rigid body (i.e. at every point mass comprising the rigid body). In this case, the fact may sometimes be of use to note that a force can be moved along its line of application, since the moment created by it is not changed by the shift.

ii) If $\sum_{\alpha=1}^{N} \mathbf{F}_\alpha = 0$ but $\sum_{\alpha=1}^{N} \mathbf{N}_\alpha \neq 0$, a single force or a shift in the pivot cannot make the total torque vanish. Only a **couple of forces**, a *basic element of statics*, is helpful here. A couple consists of two forces of equal magnitude acting in opposite directions, applied at two different suitably chosen points. The couple does not contribute to the total force, but the total torque is changed. The couple can be chosen in such a way that the total moment of the forces vanishes.

11.2.3   Simple machines

Starting in antiquity, the most simple equilibrium situations arise for the so-called **simple machines**. They all provide the 'mechanical advantage', i.e. their use makes work easier. The number of these machines varies with time and author. Archimedes studied already three machines: lever, pulley, and screw. Heron investigated five machines: wheel and axle (windlass), lever, pulley, wedge, and screw. In Renaissance times, with the revival of the science of statics, the inclined plane was added[9].

In the following, we will demonstrate the basic principles only for the **lever** and the **pulley**. The lever is a mechanical device consisting of a rigid bar that pivots about a fulcrum point. It is used to transmit – and enhance – the effort force to a load[10]; allowing, for example, heavy

---

[9]These machines are not independent ones. For example the lever plays a role with the pulley (see below), from the inclined plane one can also understand the screw and the wedge.

[10]One distinguishes three different classes of lever, according to the order of load, fulcrum and effort.

bodies to be raised. The wheel (and axle) is mainly used to redirect forces applied by pulling ropes or belts. Finally, the pulley is a machine consisting of a rope, belt or chain wrapped around a grooved wheel. Like levers, a system of pulleys allows heavy loads to be lifted with reduced effort. In all these machines, the equilibrium state is a two-fold balance: between the applied forces, and between the moments of the forces.

We now present three examples of statics problems for rigid bodies, in which the basic tasks of statics are illustrated.

Example 1: Law of the lever

Forces are applied at three points P_1, P_2, P_3 of a rigid rod (i.e. a lever) as shown in the sketch. The force $\mathbf{F}_1$ is known. What is the magnitude of the forces $\mathbf{F}_2$ and $\mathbf{F}_3$ when the rod is in equilibrium? (All forces are applied perpendicular to the lever.)

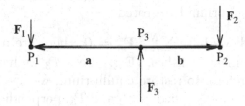

Solution. The equilibrium conditions immediately yield the solutions for $\mathbf{F}_2$ and $\mathbf{F}_3$:

- From equilibrium of forces,

$$\mathbf{F}_3 = -\left(\mathbf{F}_1 + \mathbf{F}_2\right);$$

- From equilibrium of the moments, taking P_3 as fulcrum (lever of type 1),

$$\mathbf{a} \times \mathbf{F}_1 + \mathbf{b} \times \mathbf{F}_2 = \mathbf{0}, \qquad \text{i.e.} \qquad aF_1 = bF_2;$$

or,

$$F_2 = \frac{a}{b} F_1.$$

(This is known as the **law of the lever**.)

Remark: If $\mathbf{F}_1$ and $\mathbf{F}_2$ are gravity acting on masses m_1 and m_2, i.e.
$\mathbf{F}_\alpha = m_\alpha \mathbf{g}$, then $a/b = m_2/m_1$. The supporting force $\mathbf{F}_3$ acts at the
center of gravity $m_1\mathbf{a} + m_2\mathbf{b} = \mathbf{0}$. If P_3 is the fulcrum, only equilibrium
of the moments need be taken into account. The pivot P_3 absorbs the
forces $\mathbf{F}_1$ and $\mathbf{F}_2$. This is the case in a beam balance, or 'steelyard'.

Example 2:

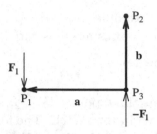

Consider a rectangular rigid rod as shown in the sketch (**b**
is always perpendicular to **a**). Force $\mathbf{F}_1$ acts at P_1 and force
$-\mathbf{F}_1$ acts at P_3. The problem is: apply additional force(s)
such that equilibrium is restored.

Solution. Although we have $\sum_{\alpha=1}^{N} \mathbf{F}_\alpha = \mathbf{0}$, there are no points about
which the total moment vanishes. (E.g. about P_3 we have $\sum_{\alpha=1}^{N} \mathbf{N}_\alpha =$
$\mathbf{a} \times \mathbf{F}_1 \neq \mathbf{0}$.) Therefore, to restore equilibrium, we need a couple. We
try the couple $(\mathbf{F}_2, -\mathbf{F}_2)$ applied at P_2 and P_3, perpendicular to **b**. The
balance of forces $\sum_{\alpha=1}^{N} \mathbf{F}_\alpha = \mathbf{0}$ is unaffected. We must determine $\mathbf{F}_2$
such that the total torque vanishes:

$$\mathbf{a} \times \mathbf{F}_1 + \mathbf{b} \times \mathbf{F}_2 = \mathbf{0}.$$

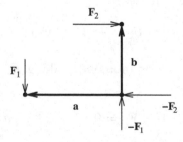

Since the forces $\mathbf{F}_1$ and $\mathbf{F}_2$ are perpendicular to **a** and **b**, respectively,
we have $aF_1 = bF_2$, or

$$F_2 = \frac{a}{b}F_1.$$

A less academic example of such a device is shown in Fig. a) below: a pocket letter balance. The dark part is the rigid body with an additional weight on the backside fixed to it (see Fig. b)).

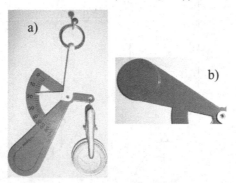

a) A pocket letter balance (size about 8 cm) , b) the circular weight on the backside.

The light L-shaped part is the pointer and there is a clip to hold the item whose weight should be determined. Attaching two coins to the clip, after reaching equilibrium, the pointer shows the weight. (**E**: Explain the situation in more detail.)

Example 3: The pulley

The equilibrium state of a *pulley* – a machine that consists of a grooved wheel, with a rope or chain attached to it, rotating about an axle – can also be discussed in terms of the equilibrium of the lever. A system of pulleys in combination is often used to further reduce the effort in raising loads. A simple system of pulleys is shown in Fig. 11.1. A rope is wrapped around two wheels (each of radius a), with one end of the rope fixed as shown in the figure.

> The force $\mathbf{F}_1$ acting at the center of the left wheel is exerted
> by the load. Find the force $\mathbf{F}_2$ that balances the load.

Solution. We replace the system of pulleys by a system of levers, as shown in Fig. 11.2. Instead of the wheels, we use two levers of length $2a$, whose fulcrum points are D_1 and D_2. The levers are connected by ropes of suitable length so that the levers are parallel (cf. Fig. 11.2). Applying the equilibrium condition for the moments, the left lever (wheel) satisfies

$$aF_1 = 2aF_3,$$

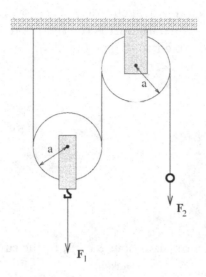

Figure 11.1: A simple system of pulleys.

and the right lever (wheel) satisfies

$$aF_2 = aF_3,$$

so that equilibrium exists if

$$F_2 = F_1/2.$$

Providing *half the force* exerted by the load suffices to maintain equilibrium.

The above examples are also simple as they are only two-dimensional. For a more thorough discussion of statics, we refer the reader to the technical literature.

11.3   Dynamics of the rigid body

11.3.1   Historical landmarks

The field was created from scratch in 1765 by Euler[11] with his epoch-making 'second' mechanics, the "Theoria motus corporum solidorum

---

[11]Already in his 'first' mechanics, the analytical theory of motion (see Bibliography), he presented in paragraph 98 his plan to study the motion of bodies of increasing

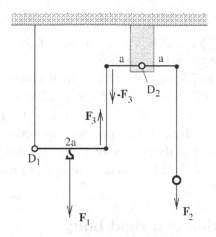

Figure 11.2: An alternative to the system of pulleys in the previous figure.

seu rigidorum" (Theory of the motion of rigid bodies; see Bibliography). Fifteen years ago he already communicated a new 'principle' (equation) of mechanics to the Academy in Berlin[12]. This principle was the equation for a rotating rigid body. With his "Analytical mechanics" Lagrange contributed substantially to the theory of rotating bodies, in particular with his solution of the problem of a freely rotating body of arbitrary shape[13,14]. These two scientists established the theoretical

complexity (cited not literally): 1.) We consider infinitesimally small bodies, which can be considered as points; 2.) we progress to rigid bodies of finite size; 3.) we consider flexible bodies; 4.) ductile ones; 5.) we consider the motion of several loosely connected bodies; 6.) the motion of liquid bodies.

Euler published books only on the first two steps. There are several communications on topics of the remaining ones.

[12] *Decouverte d'un nouveau principe de Mecanique*, Mémoires de l'académie des sciences de Berlin 6, 1752, pp. 185-217 (presented to the Academy September 3, 1750).

[13] Already communicated 1773 as *Nouvelle solution du problème du mouvement de rotation d'un corps de figure quelconque qui n'est animé par aucune force accélératrice* to the Académie royale de Berlin.

[14] In the second edition of the "Theory of the motion of rigid bodies", which appeared posthumously in 1790 with additions by Euler's son Johann Albrecht, Euler refers to Lagrange's solution being more direct than his own. He regrets that he could not grasp the deep thoughts and could not follow the intricate calculations. He

treatment of the dynamics of rigid bodies. Poinsot[15] introduced the ellipsoid of inertia (cf. section 11.3.3) in his "Théorie nouvelle de la rotation des corps" (Paris 1834). The motion of the ellipsoid unifies the motion of rigid body of various shapes.

A kind of answer to Lagrange' s "Analytical Mechanics" is Poisson's[16] "Traité de Mécanique" (A treatise of mechanics; Paris 1811; see Bibliography) according to P. Duhem, who refers to Poisson's book as 'Mécanique physique' (Physical mechanics) opposed to analytical mechanics[17]. In addition to the dynamics of point masses in Poisson's work statics and dynamics of rigid bodies are treated comprehensively for the first time.

11.3.2   The motion of a rigid body

The discussion in the previous chapter of conservation laws in many particle systems also applies to rigid bodies. The difference here is that the interaction reduce to constraints (which decrease the number of degrees of freedom from $3N$ to six). Of the six degrees of freedom of a rigid body, three are translational and three are rotational. Viewing the motion of the rigid body from an *inertial frame* of reference, the center of mass moves according to

$$M\ddot{\mathbf{R}} = \mathbf{F}\left(\mathbf{R}; \mathbf{r}'_1, \ldots, \mathbf{r}'_N\right),$$

where $\mathbf{F}\left(\mathbf{R}; \mathbf{r}'_1, \ldots, \mathbf{r}'_N\right) = \sum_{\alpha=1}^{N} \mathbf{F}_\alpha\left(\mathbf{R} + \mathbf{r}'_\alpha\right)$ is the sum of the external forces $\mathbf{F}_\alpha$ (cf. Eq. (10.20)). Each $\mathbf{F}_\alpha$ acts at the particles' positions $\mathbf{r}'_\alpha$, with the $\mathbf{r}'_\alpha$ fixed relative to each other. The **translational energy** of the body as a whole is

$$T_S = M\dot{\mathbf{R}}^2/2. \tag{11.3}$$

In this inertial frame, the rotation of the body about the center of mass is determined by the equation for the angular momentum $\mathbf{L}$ of the relative motion of the point masses (Eq. (10.39)),

$$\frac{d}{dt}\mathbf{L} = \frac{d}{dt}\left(\sum_{\alpha=1}^{N} m_\alpha \mathbf{r}'_\alpha \times \dot{\mathbf{r}}'_\alpha\right) = \sum_{\alpha=1}^{N} \mathbf{r}'_\alpha \times \mathbf{F}_\alpha = \mathbf{N}. \tag{11.4}$$

points out his admiration for Lagrange's analytical manipulations.

[15]Louis POINSOT (1777-1859), French mathematician and physicist.

[16]Siméon Denis POISSON (1781-1840), French mathematician and physicist.

[17]P. Duhem, "L'Évolution de la Mécanique", Paris 1903; see Bibliography

A body's **rotational energy** is the kinetic energy of the relative motion of the N point masses (Eq. (10.53)),

$$T_r = \frac{1}{2} \sum_{\alpha=1}^{N} m_\alpha \dot{\mathbf{r}}_\alpha'^{\,2}; \tag{11.5}$$

in the case of a rigid body, due to the fixed relative distances, all the radius vectors $\mathbf{r}_\alpha'$ can only rotate simultaneously with (instantaneous) angular velocity $\boldsymbol{\omega}$ about the center of mass, so that (cf. Eq. (8.56))

$$\frac{d\mathbf{r}_\alpha'}{dt} = \dot{\mathbf{r}}_\alpha' = \boldsymbol{\omega} \times \mathbf{r}_\alpha'. \tag{11.6}$$

If the external forces are conservative, $\mathbf{F}_\alpha(\mathbf{r}) = -\boldsymbol{\nabla} U_\alpha(\mathbf{r})$, the Lagrangian of the rigid body is

$$L = T_S + T_r - U, \tag{11.7}$$

since there is no internal potential V (in fact, it is constant, cf. Eq. (10.48)) and U is the sum of the potentials of the external forces (cf. Eq. (10.54)),

$$U = \sum_{\alpha=1}^{N} U_\alpha(\mathbf{R} + \mathbf{r}_\alpha') = U(\mathbf{R}; \mathbf{r}_1', \dots, \mathbf{r}_N')$$

Therefore, in general, the position $\mathbf{R}$ of the center of mass is coupled to the relative coordinates $\{\mathbf{r}_\alpha'\}$ by U. The two motions – translation of the center of mass and rotation of the body about the center of mass – are not independent, even though the relative distances between the constituent point particles are fixed. However, as we saw in the previous chapter, in a homogeneous gravitational field, $U = -M\mathbf{g}\mathbf{R}$ (cf. Eq. (10.55)), as well as for $U = 0$ (no external force), the rotational and the translational degrees of freedom are decoupled. (This is discussed in more detail below.)

In Eqs. (11.4), (11.5), and (11.7), the relative positions $\{\mathbf{r}_\alpha'\}$ (as well as the velocities $\{\dot{\mathbf{r}}_\alpha'\}$) are not independent. The constraints imposed by the body's rigidity reduce the number of independent dynamical variables for the relative motion to the three independent rotational variables for the body as a whole and their time-derivative. This has to be taken into account in the Lagrangian before writing down Lagrange's equations, since the variables of the equations must be independent. We must therefore express $\{r'_\alpha\}$ in terms of three independent variables (e.g. Euler's angles), then determine the equations of rotational motion for the rigid body. When the body is rotating under the influence of is the gravitational force, the body is called a **gyroscope**, or **spinning top**.

The only rotational variable needed to represent the rotational motion of the body is the angular velocity $\boldsymbol{\omega}$. In Eqs. (11.4), (11.5), and (11.7), the angular velocity takes care of all the time dependence. All other features are unchanging (time-independent) in a reference frame rotating with the body, the **body-fixed frame** of reference. In this frame, the center of mass is the origin[18], while the orientation of the axes is still arbitrary. r'_α is the distance of particle α to the origin. Since this distance is fixed, its time derivative in the body-fixed frame vanishes (using the notation of Eq. (8.54)),

$$\left(\frac{d\mathbf{r}'_\alpha}{dt}\right)' = \mathbf{0}. \tag{11.8}$$

In the inertial system, $\mathbf{r}'_\alpha$ is, of course, time dependent, and the time derivative is given by Eq. (11.6), where $\boldsymbol{\omega}$ is the angular velocity of the body-fixed frame. Hence, in the inertial frame the angular momentum is

$$\mathbf{L} = \sum_{\alpha=1}^{N} m_\alpha \mathbf{r}'_\alpha \times \dot{\mathbf{r}}'_\alpha = \sum_{\alpha=1}^{N} m_\alpha \mathbf{r}'_\alpha \times (\boldsymbol{\omega} \times \mathbf{r}'_\alpha), \tag{11.9}$$

and for the rotational energy, Eq. (11.5), we find

$$\begin{aligned} T_r &= \frac{1}{2} \sum_{\alpha=1}^{N} m_\alpha \dot{\mathbf{r}}'^2_\alpha = \frac{1}{2} \sum_{\alpha=1}^{N} m_\alpha (\boldsymbol{\omega} \times \mathbf{r}'_\alpha)(\boldsymbol{\omega} \times \mathbf{r}'_\alpha) \\ &= \frac{1}{2} \sum m_\alpha \boldsymbol{\omega} (\mathbf{r}'_\alpha \times (\boldsymbol{\omega} \times \mathbf{r}'_\alpha)) = \frac{1}{2} \boldsymbol{\omega} \mathbf{L}. \end{aligned} \tag{11.10}$$

Expanding the cross products $\mathbf{r}'_\alpha \times (\boldsymbol{\omega} \times \mathbf{r}'_\alpha)$ in Eq. (11.9), we have

$$\mathbf{L} = \sum_{\alpha=1}^{N} m_\alpha \left(\boldsymbol{\omega} (\mathbf{r}'_\alpha)^2 - \mathbf{r}'_\alpha (\boldsymbol{\omega} \mathbf{r}'_\alpha)\right) =: \mathsf{I}\boldsymbol{\omega}. \tag{11.11}$$

---

[18]The case of a body rotating about a point other than the center of mass will be discussed in Subsection 11.3.5.

(Compare the form of this equation with the relation between (linear) momentum and velocity, $\mathbf{p} = m\mathbf{v}$.) The quantity[19]

$$\mathsf{I} = \sum_{\alpha=1}^{N} m_\alpha \left(1(\mathbf{r}'_\alpha)^2 - \mathbf{r}'_\alpha \otimes \mathbf{r}'_\alpha\right) \tag{11.12}$$

is the **inertia tensor**. Like the *quadrupole moment* in Subsection 6.4.2, the inertia tensor I contains second moments of the mass density of a body (see below) and characterizes the mass distribution in a rigid body with respect to the center of mass[20]. In the body-fixed frame, I is time independent. Expressed in terms of I, the rotational energy of the body has the following form:

$$T_r = \frac{1}{2}\omega\mathsf{I}\omega \tag{11.13}$$

(Again, note the formal similarity of this equation with $T = \frac{1}{2}mv^2$.)

For the **free motion**, $U = 0$, the kinetic energy T_S of the center of mass and the rotational energy T_r are conserved separately (cf. Subsection 10.2.3), so that besides the free motion of the center of mass, $M\ddot{\mathbf{R}} = \mathbf{0}$, we have

$$T_r = \frac{1}{2}\omega\mathsf{I}\omega = const. \tag{11.14}$$

Since $\mathbf{N} = \mathbf{0}$, the equation of motion (11.4) implies that the angular momentum is constant too,

$$\mathbf{L} = \mathsf{I}\omega = const. \tag{11.15}$$

Furthermore, the projection of ω $(= \omega(t))$ on $\mathbf{L}$ is also time independent, since

$$\omega\mathbf{L} = 2T_r = const. \tag{11.16}$$

These three conservation laws determine the temporal development of ω and hence the rotation of the body (see the next section).

As already mentioned above, for **arbitrary external forces**, the total potential $U(\mathbf{R}, \{\mathbf{r}'_\alpha\})$ in the Lagrangian

$$L = M\dot{\mathbf{R}}^2/2 + \frac{1}{2}\omega\mathsf{I}\omega - U(\mathbf{R}, \{\mathbf{r}'_\alpha\})$$

---

[19]The definition of the direct product $\mathbf{a} \otimes \mathbf{b}$ of two vectors is given in Appendix A.

[20]The inertia tensor I differs from the tensor of the quadrupole moment, Eq. (6.62), only by the value of the trace: $I_{kl} + Q_{kl} = \frac{2}{3}\delta_{kl} \operatorname{Tr}(\mathsf{I})$.

in general mixes center of mass motion and rotation about the center of mass[21]. Moreover, in this form, the Lagrangian is not useful for setting up Lagrange's equations, since the (generalized) coordinates parametrizing the orientation of the body do not appear explicitly. They are hidden in the angular velocity ω and in the set $\{r'_\alpha\}$. The Lagrangian formalism can only be applied if ω and $\{r'_\alpha\}$ are represented in terms of, say, Euler's angles (cf. Subsection 11.3.6). The same problem is encountered with the equation of motion (11.4) for the angular momentum,

$$\frac{d}{dt}\mathbf{L} = \sum_{\alpha=1}^{N} \mathbf{r}'_\alpha \times \mathbf{F}_\alpha = \mathbf{N},$$

since $\mathbf{L}$ contains ω (cf. Eq. (11.11)) and the torque $\mathbf{N}$ depends on the relative positions $\{\mathbf{r}'_\alpha\}$ via $\mathbf{F}$ (cf. Eq. (10.23)). Only if $\mathbf{N}$ is independent of $\{\mathbf{r}'_\alpha\}$ can the equation of motion be solved directly for ω (see Subsection 11.3.4).

If $U(\mathbf{R}, \{\mathbf{r}'_\alpha\})$ separates into a part containing only the center of mass motion and another containing only the rotation (relative motion), i.e.

$$U(\mathbf{R}, \{\mathbf{r}'_\alpha\}) = U_1(\mathbf{R}) + U_2(\{\mathbf{r}'_\alpha\}), \tag{11.17}$$

then the motions also separate. This is the case for **gravity in the homogeneous approximation**, Eq. (10.12), in which the total potential (cf. Eq. (10.55)),

$$U = -M\mathbf{R}\mathbf{g} = U(\mathbf{R}), \tag{11.18}$$

has the form of Eq. (11.17). Now, gravity affects only the center of mass motion and not the rotation of the body (cf. also Subsection 10.2.3), the center of mass motion is decoupled from the rotation of the body about the center of mass, and consequently the Lagrangian, Eq. (11.7), splits in two independent parts: the rigid body's center of mass motion,

$$L_{cm} = M\dot{\mathbf{R}}^2/2 + M\mathbf{g}\mathbf{R},$$

and the rigid body's rotation,

$$L_r = T_r = \frac{1}{2}\omega\mathbf{I}\omega. \tag{11.19}$$

---

[21]An example is the external gravitational potential, $U_\alpha \propto m_\alpha/r$. Coupling between the center of mass motion and the rotation is important, for instance, for explaining 'irregularities' in the motion of the Saturn's moon Hyperion. A simplified model predicts chaotic behavior for Hyperion (cf. [Scheck]).

For a homogeneous gravitational field, in the equation of motion (11.4), we have $\mathbf{N} = \mathbf{0}$ (cf. Eq. (10.40)), and consequently we have also

$$\mathbf{L} = \mathsf{I}\boldsymbol{\omega} = const.$$

Also in the center of mass frame $(\mathbf{R} = \dot{\mathbf{R}} = \mathbf{0})$, the body rotates freely about the center of mass (origin), so that we have the same conservation laws as for the free motion, i.e. Eqs. (11.14), (11.15), and (11.16). Nevertheless, if L_r is to be of any use, we must still express $\boldsymbol{\omega}$ in terms of the orientational variables. Before we can evaluate either Lagrange's equations or the equation of motion for $\mathbf{L}$, we must have a closer look at the inertia tensor.

11.3.3 The inertia tensor

The ellipsoid of inertia

Just as the radius vector $\mathbf{r}'_\alpha$ can be decomposed into components in the body-fixed coordinate system with basis vectors[22] $\mathbf{e}'_k$,

$$\mathbf{r}'_\alpha = \sum_{i=1}^{3} \left(x'_k \right)_\alpha \mathbf{e}'_k = x'_\alpha \mathbf{e}'_1 + y'_\alpha \mathbf{e}'_2 + z'_\alpha \mathbf{e}'_3,$$

the inertia tensor I can also be expressed in terms of the base vectors $\mathbf{e}'_k$:

$$\mathsf{I} = \sum_{i,j}^{3} I'_{ij} \mathbf{e}'_i \otimes \mathbf{e}'_j. \tag{11.20}$$

Just as with vectors, one has to distinguish between the tensor I and its components I'_{ij}. The (matrix) representation is often taken as tensor I'; we will also use this notation if there is no ambiguity. The components I'_{ij} of I (cf. Eq. (11.12));

$$I'_{ij} = \sum_{\alpha=1}^{N} m_\alpha \left[\mathbf{r}'^2_\alpha \delta_{ij} - \left(x'_i \right)_\alpha \left(x'_j \right)_\alpha \right], \tag{11.21}$$

---

[22]Since according to our convention, Greek indices label the particles, the i'th component of the radius vector $\mathbf{r}_\alpha$ of particle α is $(x_i)_\alpha$. We shall sometimes use the notation x_α, y_α, z_α.

depend on the choice of the basis. In Cartesian coordinates, the representation of the inertia tensor is the matrix

$$
\mathsf{I}' = \begin{pmatrix}
\sum_{\alpha=1}^{N} m_\alpha \left(y_\alpha'^2 + z_\alpha'^2\right) & -\sum_{\alpha=1}^{N} m_\alpha x_\alpha' y_\alpha' & -\sum_{\alpha=1}^{N} m_\alpha x_\alpha' z_\alpha' \\
-\sum_{\alpha=1}^{N} m_\alpha y_\alpha' x_\alpha' & \sum_{\alpha=1}^{N} m_\alpha \left(x_\alpha'^2 + z_\alpha'^2\right) & -\sum_{\alpha=1}^{N} m_\alpha y_\alpha' z_\alpha' \\
-\sum_{\alpha=1}^{N} m_\alpha z_\alpha' x_\alpha' & -\sum_{\alpha=1}^{N} m_\alpha z_\alpha' y_\alpha' & \sum_{\alpha=1}^{N} m_\alpha \left(x_\alpha'^2 + y_\alpha'^2\right)
\end{pmatrix}.
$$

$$(11.22)$$

The dash indicates that I is referred to the basis $\{e_i'\}$. The diagonal elements are the **moments of inertia** and the off-diagonal elements are the **products of inertia**. Introducing the *mass density* ρ,

$$
\rho(\mathbf{r}') = \sum_{\alpha=1}^{N} m_\alpha \delta(\mathbf{r}' - \mathbf{r}_\alpha'),
\tag{11.23}
$$

with $\int \rho(\mathbf{r}') d^3 x' = \sum_{\alpha=1}^{N} m_\alpha = M$, the components of the tensor can be written

$$
I_{ij}' = \int d^3 x' \rho(\mathbf{r}') \left(\mathbf{r}'^2 \delta_{ij} - x_i' x_j'\right),
\tag{11.24}
$$

as in the procedure in Section 6.4. This is the definition of the inertia tensor for an arbitrary mass density $\rho(\mathbf{r}')$. In the case of a homogeneous mass distribution with density $\rho = M/V$ (V is the volume of the body), it follows that

$$
I_{ij}' = \frac{M}{V} \int d^3 x' \left(\mathbf{r}'^2 \delta_{ij} - x_i' x_j'\right).
\tag{11.25}
$$

Since the matrix I' with components I_{ij}' is symmetric, it can be transformed by a similarity transformation with an orthogonal matrix D to a diagonal matrix I,

$$
\mathsf{I} = \mathsf{D}^T \mathsf{I}' \mathsf{D},
\tag{11.26}
$$

with $\mathsf{D}^T \mathsf{D} = 1$. D is a representation of a rotation of the coordinate system (cf. Appendix B) and rotates the basis vectors $\{e_i'\}$ into vectors[23] $\{e_j\}$ (in the co-rotating, body-fixed frame, this rotation is time independent):

$$
e_j = D_{ij} e_i', \qquad i, j = 1, 2, 3,
\tag{11.27}
$$

---

[23]The basis vectors $\{e_i\}$ in the rotating frame should not be confused with basis vectors of the inertial frame, which do not appear in this chapter.

where we have used the summation convention. The trace of I',

$$\text{Tr } I' = 2 \sum_{\alpha=1}^{N} m_\alpha \mathbf{r}_\alpha'^2, \tag{11.28}$$

is invariant under rotations[24]

$$\text{Tr } I = \text{Tr } \left(D^T I' D \right) = \text{Tr } \left(D D^T I' \right) = \text{Tr } I'.$$

In the new coordinate system with basis vectors $\{\mathbf{e}_i\}$, the radius vector $\mathbf{r}'$ has the decomposition $\mathbf{r}' = x_i \mathbf{e}_i = x\mathbf{e}_1 + y\mathbf{e}_2 + z\mathbf{e}_3$, and the matrix I is diagonal:

$$I = \begin{pmatrix} \int d^3x \rho(\mathbf{r}) \left(y^2 + z^2 \right) & 0 & 0 \\ 0 & \int d^3x \rho(\mathbf{r}) \left(x^2 + z^2 \right) & 0 \\ 0 & 0 & \int d^3x \rho(\mathbf{r}) \left(x^2 + y^2 \right) \end{pmatrix}$$

$$=: \begin{pmatrix} I_1 & 0 & 0 \\ 0 & I_2 & 0 \\ 0 & 0 & I_3 \end{pmatrix}. \tag{11.29}$$

The original matrix representation of the inertia tensor has been transformed to the a representation in terms of the **principal axes** $\mathbf{e}_i$ of the inertia tensor. The three **principal moments of inertia**,

$$I_i = \int d^3x \rho(\mathbf{r}) \left(\mathbf{r}^2 - x_i^2 \right), \tag{11.30}$$

determine the behavior of the rigid body under rotations. From Eq. (11.30), the following conditions for the moments I_i are easily derived:

$$I_i \geq 0 \quad \text{and} \quad |I_i - I_j| \leq I_k \leq I_i + I_j, \quad i \neq j \neq k, \quad i, j, k = 1, 2, 3.$$

---

[24]This can be immediately seen from Eq. (11.28), since the length $\mathbf{r}_\alpha'^2$ is also invariant.

As for vectors, from the representation of the tensor I in the two coordinate systems

$$I = I_{ij}' \mathbf{e}_i' \otimes \mathbf{e}_j' = I_{ij} \mathbf{e}_i \otimes \mathbf{e}_j,$$

one can with the help of (11.27) and the relation

$$I_{ij}' D_{ik} D_{jl} \mathbf{e}_k \otimes \mathbf{e}_l = (D^T)_{ki} I_{ij}' D_{jl} \mathbf{e}_k \otimes \mathbf{e}_l,$$

derive the similarity transformation (11.26) for the matrix representation:

$$I_{kl} = (D^T)_{ki} I_{ij}' D_{jl}.$$

One distinguishes the following **cases of rigid bodies**:

$I_1 \neq I_2 \neq I_3$, **nonsymmetric body**;

$I_1 = I_2 \neq I_3$, **symmetric body**;

$I_1 = I_2 = I_3$, **spherical body**.

For a rigid sphere (radius R) with homogeneous mass density, due to isotropy, no distinct choice of body-fixed axes exists. Hence, the inertia tensor is proportional to the unit tensor, and the matrix I' is therefore always diagonal. From Eq. (11.25), one finds in spherical coordinates (**E**)

$$I'_{ij} = \frac{2}{5}MR^2\delta_{ij}. \tag{11.31}$$

Also simple to derive are the inertia tensors for various ellipsoid-shaped rigid bodies (**E**).

One can assign a scalar quantity $I = I(\mathbf{n})$ to the inertia tensor by multiplying I on both sides by an arbitrary unit vector $\mathbf{n}$:

$$I = I(\mathbf{n}) = \mathbf{n}\mathsf{I}\mathbf{n}. \tag{11.32}$$

I is the **moment of inertia with respect to the axis n**. Introducing the vector $\boldsymbol{\rho} = \mathbf{n}\big/\sqrt{I}$ one has

$$\boldsymbol{\rho}\mathsf{I}\boldsymbol{\rho} = 1.$$

This is the *normal form* of an ellipsoid (a symmetric, positive semidefinite, quadratic form), the so-called **ellipsoid of inertia**. The principal axes of the ellipsoid are the principal axes of the inertia tensor (**E**). The principal moments of inertia are the inverse squares of the semiaxes of the ellipsoid.

The motion of the ellipsoid of inertia

The free rotational motion ($U = 0$ or $U = -M\mathbf{g}\mathbf{R}$) of a rigid body can be illustrated by means of its ellipsoid of inertia. Introducing

$$\boldsymbol{\rho} = \frac{\boldsymbol{\omega}}{\sqrt{2T_r}}, \tag{11.33}$$

– observe that ρ lies along the instantaneous axis of revolution – we recast the rotational energy $T_r = \frac{1}{2}\omega|\omega$ (cf. Eq. (11.13)) into the form

$$\rho|\rho = 1. \tag{11.34}$$

Comparing Eqs. (11.11) and (11.13), one notes that

$$\mathbf{L} = \nabla_\omega T_r \quad (= \partial T_r/\partial\omega), \tag{11.35}$$

and since $\nabla_\omega T_r \propto \nabla_\rho(\rho|\rho)$ is the vector normal to the surface of the ellipsoid of inertia, Eq. (11.34), at the point $\rho = \omega/\sqrt{2T_r}$,

> the *angular momentum* **L** *is perpendicular to the plane tangential to the surface of the ellipsoid*

at the point ρ. The three constants of the motion, Eqs. (11.14), (11.15), and (11.16), determine the rotation of the body: therefore $\omega(t)$ is limited in 3 ways (**Poinsot's construction**). Firstly, Eq. (11.14) – $\omega|\omega$ equals a constant – also defines an ellipsoid, whose size is related to $\rho|\rho = 1$ (Eq. (11.34)) through a constant of proportionality. Secondly, the angular velocity ω changes its direction with time, tracing out a path $\rho(t)\ (= \omega(t)/\sqrt{2T_r})$ on the surface of the latter ellipsoid. This path is determined by the condition that the projection of ω onto the constant angular momentum **L** must remain constant (because $\omega\mathbf{L} = 2T_r = const$). Thirdly, since – due to Eq. (11.35) – **L** is the vector normal to the tangential plane at the point $\rho = \omega/\sqrt{2T_r}$, the tangential plane is fixed in the inertial frame; it is called the **invariable plane**. In other words,

> for **free motion** *the ellipsoid of inertia rolls on the invariable plane* in the **inertial frame**

(cf. Fig. 11.3). The curve of the contact points ρ on the ellipsoid is the **polhode** and the curve of the contact points of the revolving ellipsoid on the invariable plane is called **herpolhode**.

The energy of rotation in the body-fixed frame

In the body-fixed system of principal axes, the rotational energy Eq. (11.13) has a simple form, namely

$$T_r = \frac{1}{2}\left(I_1\omega_1^2 + I_2\omega_2^2 + I_3\omega_3^2\right), \tag{11.36}$$

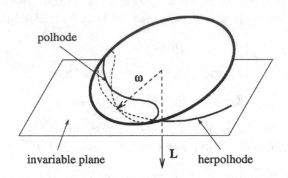

Figure 11.3: The motion of the ellipsoid of inertia.

where
$$\boldsymbol{\omega} = \omega_1 \mathbf{e}_1 + \omega_2 \mathbf{e}_2 + \omega_3 \mathbf{e}_3. \tag{11.37}$$

If $\boldsymbol{\omega}$ is expressed in terms of Euler's angles (see Eq. (8.52), where $\boldsymbol{\omega}' \to \boldsymbol{\omega}$ (cf. Footnote 23)), we have

$$
\begin{aligned}
T_r &= \frac{I_1}{2} \left(\dot{\varphi} \sin\psi \sin\vartheta + \dot{\vartheta} \cos\psi \right)^2 + \frac{I_2}{2} \left(\dot{\varphi} \cos\psi \sin\vartheta - \dot{\vartheta} \sin\psi \right)^2 \\
&\quad + \frac{I_3}{2} \left(\dot{\varphi} \cos\vartheta + \dot{\psi} \right)^2 \\
&= \frac{1}{2} \dot{\varphi}^2 \left\{ (I_1 \sin^2\psi + I_2 \cos^2\psi) \sin^2\vartheta + I_3 \cos^2\vartheta \right\} \\
&\quad + \frac{1}{2} \dot{\vartheta}^2 (I_1 \cos^2\psi + I_2 \sin^2\psi) + \frac{1}{2} \dot{\psi}^2 I_3 \\
&\quad + \dot{\varphi} \dot{\vartheta} (I_1 - I_2) \sin\vartheta \sin\psi \cos\psi + \dot{\varphi} \dot{\psi} I_3 \cos\vartheta. \tag{11.38}
\end{aligned}
$$

Even if the rotation of the body is decoupled from the translation of its center of mass, the rotational energy – and therefore the Lagrangian – has a complex form. For the purpose of writing Eq. (11.38) somewhat more compactly, we introduce the quantities

$$\dot{\boldsymbol{\phi}} = (\dot{\varphi}, \dot{\vartheta}, \dot{\psi}) \tag{11.39}$$

and

$$
\Lambda = \begin{pmatrix} \sin\psi \sin\vartheta & \cos\psi & 0 \\ \cos\psi \sin\vartheta & -\sin\psi & 0 \\ \cos\vartheta & 0 & 1 \end{pmatrix}. \tag{11.40}
$$

Now the energy of rotation reads

$$T_r = \frac{1}{2}\dot{\phi}\Lambda^T |\Lambda\dot{\phi} \tag{11.41}$$

($|$ is the diagonal matrix $| = diag\,(I_1, I_2, I_3)$).

If a body rolls on a surface about a principal axes of inertia, e.g. $\mathbf{e}_3$ ($\omega_1 = \omega_2 = 0$) with $\omega_3 = \dot{\varphi}$, the expression in Eq. (11.36) simplifies to

$$T_r = \frac{1}{2}I_3\dot{\varphi}^2,$$

and the constraint of rolling on the surface relates the motion of the center of mass at position $\mathbf{R}$ to the angular velocity $\dot{\varphi}$ ($\mathbf{E}$). This constraint has to be taken into account either in the Lagrangian or in Lagrange's equations (either by the elimination of variables or introducing a Lagrange multiplier, cf. Subsection 9.2.3).

11.3.4 Euler's equations of motion

We observe the time derivative of the angular momentum (Eq. (11.4)) in two reference frames: the inertial frame and the body-fixed frame. The relation between these two views of the same quantity is (cf. Eq. (8.55))

$$\frac{d\mathbf{L}}{dt} = \left(\frac{d\mathbf{L}}{dt}\right)' + \boldsymbol{\omega} \times \mathbf{L}. \tag{11.42}$$

Therefore, the equation of motion (10.39) in the co-rotating frame is

$$\left(\frac{d\mathbf{L}}{dt}\right)' + \boldsymbol{\omega} \times \mathbf{L} = \mathbf{N}, \tag{11.43}$$

with $\mathbf{L} = |\boldsymbol{\omega}$ or $L_i' = \sum_{j=1}^{3} I_{ij}'\omega_j'$ (cf. Eq. (11.11)). Since in the following we are concerned only with motion in the body-fixed frame, we denote $(d/dt)'$ by d/dt. (That is, $d\mathbf{L}/dt$ is the three-tuple of the time derivatives of the components L_i', where $\mathbf{L} = L_i'\mathbf{e}_i'$, in the body-fixed coordinate system.)

Rotating the coordinate axes into the principal axes of inertia $\mathbf{e}_i$, the matrix $|$ becomes diagonal,

$$| = \begin{pmatrix} I_1 & 0 & 0 \\ 0 & I_2 & 0 \\ 0 & 0 & I_3 \end{pmatrix},$$

and the angular momentum $\mathbf{L}$ is simply

$$\mathbf{L} = \sum_{i=1}^{3} L_i \mathbf{e}_i, \qquad L_i = I_i \omega_i. \tag{11.44}$$

Equation (11.43) then reads, since $dI_j/dt = 0$,

$$I_i \frac{d\omega_i}{dt} + \sum_{j,k}^{3} \varepsilon_{ijk} \omega_j \omega_k I_k = N_i \tag{11.45}$$

(The N_i are the components of $\mathbf{N}$ in the basis $\{\mathbf{e}_i\}$: $\mathbf{N} = \sum_{i=1}^{3} N_i \mathbf{e}_i$.) We write out the three equations (11.45) explicitly. These are known as **Euler's equations of motion** of a rigid body (L. Euler, Theoria motus corporum solidorum ..., Rostock 1765):

$$\begin{aligned}
I_1 \dot{\omega}_1 - \omega_2 \omega_3 (I_2 - I_3) &= N_1 \\
I_2 \dot{\omega}_2 - \omega_3 \omega_1 (I_3 - I_1) &= N_2 \\
I_3 \dot{\omega}_3 - \omega_1 \omega_2 (I_1 - I_2) &= N_3.
\end{aligned} \tag{11.46}$$

Since **the equations are valid in the body-fixed frame**, one must keep in mind that the moment of the force in these seemingly simple, though nonlinear, equations is obtained by a time-dependent transformation from the inertial system. Therefore, the moment of force may have a complicated, time-dependent form, even if, when viewed in the inertial frame, it looks simple and is independent of time. Note that all quantities appearing in the system of equations (11.46) are taken relative to the center of mass, which is the origin of the coordinate system.

In the case of **homogeneous gravity**, $\mathbf{F}_\alpha = m_\alpha \mathbf{g}$, the total moment about the center of mass vanishes, $\mathbf{N} = \mathbf{0}$ (cf. Eq. (10.40)), and we again find that the **motion** of the body **about the center of mass** is a **free rotation**.

For $\mathbf{N} = \mathbf{0}$, Euler's equations of motion (11.46) are a nonlinear, autonomous system in the three variables ω_1, ω_2, and ω_3. Therefore the dynamics of the system may be chaotic if the number of (isolating) integrals is insufficient (cf. Section 4.4).

We now consider two applications.

The free symmetric rigid body

If a rigid body has an axis of rotational symmetry (for example, a disk, or the spinning top shown in Fig. 11.6), two of the principal moments of inertia are equal. We then choose $I_1 = I_2 \neq I_3$. Euler's equations in this case are

$$
\begin{aligned}
I_1\dot{\omega}_1 &= -(I_3 - I_1)\omega_2\omega_3 \\
I_1\dot{\omega}_2 &= (I_3 - I_1)\omega_1\omega_3 \\
I_3\dot{\omega}_3 &= 0.
\end{aligned}
\tag{11.47}
$$

The last equation is easily integrated, giving

$$
\omega_3 = const
\tag{11.48}
$$

as an integral. The two remaining equations can now be written as

$$
\begin{aligned}
\dot{\omega}_1 &= -\Omega\omega_2 \\
\dot{\omega}_2 &= \Omega\omega_1,
\end{aligned}
\tag{11.49}
$$

where

$$
\Omega = \frac{I_3 - I_1}{I_1}\omega_3.
$$

Multiplying the first equation in (11.49) by ω_1 and the second by ω_2, then adding the resulting equations and integrating, yields a further integral,

$$
\omega_1^2 + \omega_2^2 = const.
\tag{11.50}
$$

The existence of the two obviously independent integrals, (11.48) and (11.50), guarantees the integrability of the system of equations (11.47). To find the time dependence $\omega_i = \omega_i(t)$, one still needs another constant of the motion (e.g. from Eq. (11.49), one finds $\dot{\omega}_1^2 + \dot{\omega}_2^2 = const$). But it is easy to solve the system of equations (11.49) directly.

Taking a further time derivative of the two equations (11.49), and combining the equations in the appropriate manner, one finds the set of two equations

$$
\ddot{\omega}_i + \Omega^2\omega_i = 0, \qquad i = 1, 2,
$$

whose solutions are

$$
\begin{aligned}
\omega_1 &= A\sin\Omega t \\
\omega_2 &= A\cos\Omega t.
\end{aligned}
\tag{11.51}
$$

That is, $\boldsymbol{\omega}$ rotates with constant angular velocity Ω ($\omega_3 = const!$) about the z-axis, which is the symmetry axis of the body. Thus,

> in general, in the **body-fixed system** the **axis of rotation**,
> given by $\boldsymbol{\omega}$, undergoes **precession** with frequency Ω, tracing
> out a circular cone

(see the cone about the axis of symmetry in Fig. 11.4). Only if the initial conditions are such that $A = 0$, *viz.* the rotation axis coincides with the symmetry axis (and the angular momentum), there is no precession. Such a situation arises for instance when along the axis of symmetry small rods protrude from the top on both sides and the rotation is generated by means of these rods.

As an example consider the Earth, which is a sphere flattened a little at the poles. For the Earth, $I_3 > I_1$, $(I_3 - I_1)/I_1 = 0.00327$ and hence $\Omega = 0.00327 \times \omega_3 \simeq \omega_3/306$. Assume that the motion of the Earth is nearly free (i.e. assume $U_2(\{\mathbf{r}'_\alpha\}) \simeq 0$ in Eq. (11.17)), then with $\omega_3 = 1/T_{day} = 1$, the parameter Ω has the value $1/306$, so that the period of precession of the axis of the Earth, according to Euler's equations, is 306 days. The actual period of precession is 420 days. Indeed, in reality, the axis of the Earth's angular velocity traces an irregular curve about the north pole. Presumably the difference is partly due to the fact that the Earth is not a rigid body.

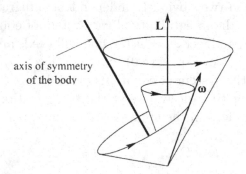

Figure 11.4: Poinsot motion of a free symmetric body in the inertial frame for $I_1 < I_3$.

In the **inertial frame**, $\boldsymbol{\omega}$ rotates about the constant ($\mathbf{N} = \mathbf{0}!$) angular momentum $\mathbf{L}$, the axis of rotation tracing out a circular cone because $\boldsymbol{\omega}\mathbf{L} = 2T_r = const$ (cf. Eq. (11.16)). Therefore, in the inertial frame, we have the situation shown in Fig. 11.4: Both, the axis of rotation and the axis of symmetry perform processions. In order to determine the motion in the inertial coordinate system, one must introduce the representation (8.52) for $\boldsymbol{\omega}$ into the solutions (11.51) and (11.48); *viz.*:

$$
\begin{aligned}
\omega_1 &= \dot{\varphi} \sin\psi \sin\vartheta + \dot{\vartheta}\cos\psi = A\sin\Omega t \\
\omega_2 &= \dot{\varphi}\cos\psi \sin\vartheta - \dot{\vartheta}\sin\psi = A\cos\Omega t \\
\omega_3 &= \dot{\varphi}\cos\vartheta + \dot{\psi} = const.
\end{aligned} \tag{11.52}
$$

If one now chooses the axes in the inertial system such that $\mathbf{L} = (0, 0, L_z)$, the components of $\mathbf{L}$ in the rotating frame are given by (cf. Eq. (8.17)) $\mathbf{DL} = (L_z \sin\psi \ \sin\vartheta, L_z \cos\psi \ \sin\vartheta, L_z \cos\vartheta)$. They must be equal to the components of $\mathbf{l}\boldsymbol{\omega}$, so that by Eq. (11.44),

$$
\begin{aligned}
L_z \sin\psi \ \sin\vartheta &= I_1\omega_1 = I_1\left(\dot{\varphi}\sin\psi \sin\vartheta + \dot{\vartheta}\cos\psi\right) \\
L_z \cos\psi \ \sin\vartheta &= I_1\omega_2 = I_1\left(\dot{\varphi}\cos\psi\sin\vartheta - \dot{\vartheta}\sin\psi\right) \\
L_z \cos\vartheta &= I_3\omega_3 = I_3\left(\dot{\varphi}\cos\vartheta + \dot{\psi}\right).
\end{aligned}
$$

Since L_z, I_1, and I_3 are constant, these equations are compatible only if

$$
\vartheta = const =: a \quad \text{and} \quad \dot{\varphi} = const =: b.
$$

Then one only has to integrate the reduced Eqs. (11.52),

$$
\begin{aligned}
b\sin\psi \sin a &= A\sin\Omega t \\
b\cos\psi \sin a &= A\cos\Omega t \\
b\cos a + \dot{\psi} &= \omega_3.
\end{aligned}
$$

We leave this as an exercise for the reader (**E**; cf. e.g. [Becker]).

For $A = 0$, i.e. rotation about the axis of symmetry, we have $b = 0$ and consequently

$$
\dot{\psi} = \omega_3;
$$

of course, also in the inertial frame the top rotates without precession.

More practically, let us consider a children's top. If we can ensure that the axis of the top is vertical to a horizontal surface (without friction), the body fixed z'-axes and the z-axes of the inertial frame coincide. The center of mass is situated vertically above of the point of support and thus gravity is counterbalanced by the surface; gravity – taken to be homogeneous – has no effect. If the top is rotated about its axis of symmetry (z' -axis; $A = 0$) the top stays rotating vertically to the surface.

The free (nonsymmetric) rigid body

The free motion of a body with no particular symmetry obeys the non-linear system of equations

$$
\begin{aligned}
\dot{\omega}_1 &= \frac{I_2 - I_3}{I_1}\omega_2\omega_3 =: K_1\omega_2\omega_3 \\
\dot{\omega}_2 &= -\frac{I_1 - I_3}{I_2}\omega_1\omega_3 =: -K_2\omega_1\omega_3 \qquad (11.53) \\
\dot{\omega}_3 &= \frac{I_1 - I_2}{I_3}\omega_1\omega_2 =: K_3\omega_1\omega_2.
\end{aligned}
$$

We assume (without loss of generality) that the moments of inertia I_i are ordered such that

$$
I_1 \geq I_2 \geq I_3 \geq 0, \qquad (11.54)
$$

i.e. all K_i are positive. Now the question arises whether the dynamical system defined by this system of equations can manifest chaotic behavior. To answer this, we look to see how many integrals of the motion we can find. Two integrals of the motion are obtained easily. Multiplying each equation in this system by ω_1, ω_2, and ω_3, respectively, one obtains

$$
\begin{aligned}
\omega_1\dot{\omega}_1 &= K_1\omega_1\omega_2\omega_3 \\
\omega_2\dot{\omega}_2 &= -K_2\omega_1\omega_2\omega_3 \\
\omega_3\dot{\omega}_3 &= K_3\omega_1\omega_2\omega_3.
\end{aligned}
$$

Forming appropriate linear combinations yields

$$
\begin{aligned}
\frac{d}{dt}\left(K_2\omega_1^2 + K_1\omega_2^2\right) &= 0 \\
\frac{d}{dt}\left(K_3\omega_1^2 - K_1\omega_3^2\right) &= 0 \\
\frac{d}{dt}\left(K_3\omega_2^2 + K_2\omega_3^2\right) &= 0.
\end{aligned}
$$

Two integrals of these equations are independent. For instance, integrating the first and the third equation gives

$$
\begin{aligned}
C_1 &= K_2\omega_1^2 + K_1\omega_2^2 = const \\
C_3 &= K_3\omega_2^2 + K_2\omega_3^2 = const.
\end{aligned}
\tag{11.55}
$$

The remaining integral is simply the difference of these two integrals: $K_2C_2 = K_3C_1 - K_1C_3$. The system of equations (11.53) is therefore integrable and the solutions are not chaotic. To find the solutions, we express ω_1 and ω_3 in terms of ω_2 using the conservation laws (11.55), and from the second equation (11.53) we obtain

$$
\dot{\omega}_2 = -\sqrt{C_1 C_3}\sqrt{\left(1 - \frac{K_1}{C_1}\omega_2^2\right)\left(1 - \frac{K_3}{C_3}\omega_2^2\right)}.
\tag{11.56}
$$

The solution $t = t(\omega_2)$ is an elliptic integral of the first kind. Inverting $t(\omega_2)$ to get $\omega_2(t)$, and inserting $\omega_2(t)$ into Eq. (11.55), ω_1 and ω_3 can be obtained without much ado.

The **stationary solutions** of the system of equations (11.53) are the solutions of $\dot{\omega}_i = 0$, $i = 1, 2, 3$, i.e.

$$
\omega_2\omega_3 = \omega_1\omega_3 = \omega_1\omega_2 = 0.
$$

Three independent solutions of these equations are

$$
\begin{aligned}
\boldsymbol{\omega}^{(1)} &= (\omega, 0, 0) \\
\boldsymbol{\omega}^{(2)} &= (0, \omega, 0) \\
\boldsymbol{\omega}^{(3)} &= (0, 0, \omega),
\end{aligned}
\tag{11.57}
$$

where ω in each case is arbitrary. The stationary solutions are just the rotations with constant angular velocity about the principal axes.

The question of whether the motion about each axis is stable can be answered to some extent from *linear stability analysis* (cf. Section 3.5). For small displacements from any stationary solution $\boldsymbol{\omega}^{(i)}$,

$$
\boldsymbol{\omega} = \boldsymbol{\omega}^{(i)} + \delta\boldsymbol{\omega}, \qquad i = 1, 2, 3,
\tag{11.58}
$$

we find

$$
\begin{aligned}
\delta\dot{\omega}_1 &= K_1\left(\omega_2^{(i)}\delta\omega_3 + \omega_3^{(i)}\delta\omega_2\right) \\
\delta\dot{\omega}_2 &= -K_2\left(\omega_1^{(i)}\delta\omega_3 + \omega_3^{(i)}\delta\omega_1\right) \\
\delta\dot{\omega}_3 &= K_3\left(\omega_1^{(i)}\delta\omega_2 + \omega_2^{(i)}\delta\omega_1\right).
\end{aligned}
$$

In matrix notation, we have

$$\begin{pmatrix} \delta\dot\omega_1 \\ \delta\dot\omega_2 \\ \delta\dot\omega_3 \end{pmatrix} = \begin{pmatrix} 0 & K_1\omega_3^{(i)} & K_1\omega_2^{(i)} \\ -K_2\omega_3^{(i)} & 0 & -K_2\omega_1^{(i)} \\ K_3\omega_2^{(i)} & K_3\omega_1^{(i)} & 0 \end{pmatrix} \begin{pmatrix} \delta\omega_1 \\ \delta\omega_2 \\ \delta\omega_3 \end{pmatrix}, \qquad (11.59)$$

or more concisely,

$$\delta\dot{\boldsymbol\omega} = \mathsf{M}(\boldsymbol\omega^{(i)})\delta\boldsymbol\omega.$$

The stability of a solution

$$\delta\boldsymbol\omega = \exp(\mathsf{M}t)\delta\boldsymbol\omega_0 \qquad (11.60)$$

can be judged from the eigenvalues λ of the matrix $\mathsf{M}(\boldsymbol\omega^{(i)})$. The eigenvalues are obtained from

$$\det(\mathsf{M}-\lambda 1) = -\lambda^3 + \lambda\left(K_1K_3\left(\omega_2^{(i)}\right)^2 - K_2K_3\left(\omega_1^{(i)}\right)^2 - K_1K_2\left(\omega_3^{(i)}\right)^2\right)$$

$$= 0,$$

or

$$\lambda\left(\lambda^2 - K_1K_3\left(\omega_2^{(i)}\right)^2 + K_2K_3\left(\omega_1^{(i)}\right)^2 + K_1K_2\left(\omega_3^{(i)}\right)^2\right) = 0. \quad (11.61)$$

One solution of Eq. (11.61), the eigenvalue

$$\lambda = 0, \qquad (11.62)$$

is common to all three stationary solutions $\boldsymbol\omega^{(i)}$. In each case the *eigenvector is proportional to the corresponding stationary solution*. Since $\lambda = 0$ is independent from ω, the angular velocity can be arbitrary. Any change of ω in the stationary solution does not influence the stability.

For the two remaining eigenvalues of each of the three stationary solutions $\boldsymbol\omega^{(1)}, \boldsymbol\omega^{(2)}$, and $\boldsymbol\omega^{(3)}$, the *eigenvectors are orthogonal to the stationary solutions*, which means that they correspond to changes in orientation of the respective rotation axes (**E**). The eigenvalues of the three solutions are

i) $\boldsymbol\omega^{(1)} = (\omega, 0, 0)$,

$$\lambda^{(1)} = \pm i\omega\sqrt{K_2K_3} = \pm i\omega\sqrt{(I_1 - I_3)(I_1 - I_2)/I_2I_3}, \quad (11.63)$$

The solution is marginally stable ($|\delta\boldsymbol\omega(t)|$ stays bounded);

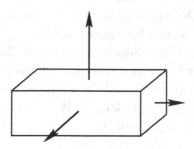

Figure 11.5: The principal inertial moments of a brick: The size of the arrows is according to the magnitude of the inertial moments.

ii) $\boldsymbol{\omega}^{(2)} = (0, \omega, 0)$,

$$\lambda^{(2)} = \pm \omega \sqrt{K_1 K_3} = \pm \omega \sqrt{(I_1 - I_2)(I_2 - I_3)/I_1 I_3}, \qquad (11.64)$$

The solution is unstable ($|\delta\boldsymbol{\omega}(t)|$ increases);

iii) $\boldsymbol{\omega}^{(3)} = (0, 0, \omega)$,

$$\lambda^{(3)} = \pm i\omega \sqrt{K_1 K_2} = \pm i\omega \sqrt{(I_1 - I_3)(I_2 - I_3)/I_1 I_2}, \qquad (11.65)$$

The solution is marginally stable.

The free rotation of a nonsymmetric rigid body about the middle axis (I_2-axis) is unstable in any case. To judge the stability of the two other solutions definitely further investigations are needed (see e.g. [Scheck]).

This behavior is often demonstrated by the (free) rotation of a body with homogeneous mass distribution whose shape is a right parallelepiped, like a brick (cf. Fig. 11.5). The rotations about the axes of the largest and the smallest moment of inertia are stable. If the brick is rotated about the middle axis (and dropped), it starts to wobble immediately, since it is nearly impossible to ensure that the rotation is exactly about the middle axis.

11.3.5 The motion of a spinning top

In all the previous discussion, the position of the center of mass was the reference point for the dynamical quantities T_r, $\mathbf{L}$, and $\mathbf{N}$. If one

changes the point of reference, then – by definition – the angular momentum, the torque and also the inertia tensor change, since these depend
on the reference point. Let $\mathbf{A}$ be the new reference point, which is given
by some constraint. For example, in the case of the spinning top, the
natural reference point is the point of contact with a rough surface, so
that this point is fixed in space, and the rotation axis $\boldsymbol{\omega}$ passes through
the reference point $\mathbf{A}$. Instead of $\mathbf{r}_\alpha = \mathbf{R} + \mathbf{r}'_\alpha$ (Eq. (10.16)), we use
now relative coordinates $\mathbf{r}''_\alpha$, given by

$$\mathbf{r}_\alpha = \mathbf{A} + \mathbf{r}''_\alpha, \tag{11.66}$$

so that $\mathbf{r}'_\alpha$ and $\mathbf{r}''_\alpha$ are related by

$$\mathbf{r}''_\alpha = \mathbf{r}'_\alpha + (\mathbf{R} - \mathbf{A}) =: \mathbf{r}'_\alpha + \mathbf{h}. \tag{11.67}$$

Therefore we have now the condition

$$\sum_{\alpha=1}^{N} m_\alpha \mathbf{r}''_\alpha = M(\mathbf{R} - \mathbf{A}) = M\mathbf{h} \tag{11.68}$$

for the coordinates $\mathbf{r}''_\alpha$ (recall that $\sum_{\alpha=1}^{N} m_\alpha \mathbf{r}'_\alpha = 0$). $\mathbf{h}$ is the vector
between the center of mass position $\mathbf{R}$ and the new reference point $\mathbf{A}$.
The relation between the inertia tensor (denoted $\mathsf{I}^{(\mathbf{A})}$) taken relative
to $\mathbf{A}$, and the inertia tensor (denoted $\mathsf{I}^{(\mathbf{R})}$, previously denoted I) taken
relative to $\mathbf{R}$, can be found by substituting Eq. (11.67) into $\mathsf{I}^{(\mathbf{R})}$:

$$
\begin{aligned}
I_{ij}^{(\mathbf{R})} \;=\; I'_{ij} &= \sum_{\alpha=1}^{N} m_\alpha \left[\mathbf{r}'^2_\alpha \delta_{ij} - (x'_i)_\alpha (x'_j)_\alpha \right] \\
&= \sum_{\alpha=1}^{N} m_\alpha \left[(\mathbf{r}''_\alpha - \mathbf{h})^2 \delta_{ij} - ((x''_i)_\alpha - h_i)((x''_j)_\alpha - h_j) \right] \\
&= \sum_{\alpha=1}^{N} m_\alpha \left[\mathbf{h}^2 \delta_{ij} - h_i h_j \right] + \sum_{\alpha=1}^{N} m_\alpha \left[\mathbf{r}''^2_\alpha \delta_{ij} - (x''_i)_\alpha (x''_j)_\alpha \right] \\
&\quad -2\mathbf{h}\delta_{ij} \sum_{\alpha=1}^{N} m_\alpha \mathbf{r}''_\alpha + h_i \sum_{\alpha=1}^{N} m_\alpha (x''_j)_\alpha + h_j \sum_{\alpha=1}^{N} m_\alpha (x''_i)_\alpha.
\end{aligned}
$$

Because of Eq. (11.68), the terms linear in $\mathbf{r}''_\alpha$ (or $\left(x''_j\right)_\alpha$ respectively),
together with the first term, give $-M \left(\mathbf{h}^2 \delta_{ij} - h_i h_j \right)$. The second term

is the inertia tensor with respect to **A**,

$$\sum_{\alpha=1}^{N} m_\alpha \left[\mathbf{r}''^2_\alpha \delta_{ij} - \left(x''_i\right)_\alpha \left(x''_j\right)_\alpha \right] = I''_{ij} = I^{(\mathbf{A})}_{ij}, \tag{11.69}$$

so that finally we have

$$I^{(\mathbf{R})}_{ij} = I^{(\mathbf{A})}_{ij} - M \left(\mathbf{h}^2 \delta_{ij} - h_i h_j \right), \qquad \mathbf{h} = \mathbf{R} - \mathbf{A}, \tag{11.70}$$

This result is known as **Steiner's theorem**[25].

Not only the moments of the mass distribution, but also the angular momentum and the torque, are now taken with respect to the point **A**. In the most simple case, **A** remains at rest in the inertial frame:

$$\dot{\mathbf{A}} = 0. \tag{11.71}$$

Replacing $\mathbf{r}_\alpha$ in the equations of motion (10.10) by Eq. (11.66), and taking the cross product equation of motion for particle α with $\mathbf{r}''_\alpha$, we have $\sum_{\alpha,\beta=1}^{N} \mathbf{r}''_\alpha \times \mathbf{F}_{\alpha\beta} = 0$, where $\mathbf{F}_{\alpha\beta}$ are the two-particle interaction forces (see Eq. (10.7); recall the derivation in the previous chapter, in particular Footnote 5.) Given that the pivot point remains fixed in the inertial frame (i.e. Eq. (11.71) holds), the angular momentum $\mathbf{L}^{(\mathbf{A})}$ and the moment of the external forces $\mathbf{N}^{(\mathbf{A})}$ are found to be

$$\mathbf{L}^{(\mathbf{A})} = \sum_{\alpha=1}^{N} m_\alpha \mathbf{r}''_\alpha \times \dot{\mathbf{r}}''_\alpha \quad \text{and} \quad \mathbf{N}^{(\mathbf{A})} = \sum_{\alpha=1}^{N} \mathbf{r}''_\alpha \times \mathbf{F}_\alpha.$$

The equation of motion in the inertial frame is derived in analogy with Eq. (11.4) (E),

$$\frac{d}{dt} \mathbf{L}^{(\mathbf{A})} = \mathbf{N}^{(\mathbf{A})}. \tag{11.72}$$

Again, for the time derivative of $\mathbf{r}''_\alpha$ in the rotating frame (cf. Eq. (11.6)), we have

$$\dot{\mathbf{r}}''_\alpha = \boldsymbol{\omega} \times \mathbf{r}''_\alpha,$$

and hence (cf. Eq. (11.11)), the angular momentum relative to the pivot point is

$$\mathbf{L}^{(\mathbf{A})} = \sum_{\alpha=1}^{N} m_\alpha \mathbf{r}''_\alpha \times \left(\boldsymbol{\omega} \times \mathbf{r}''_\alpha \right) = \mathsf{I}^{(\mathbf{A})} \boldsymbol{\omega}. \tag{11.73}$$

---

[25] Jakob STEINER (1796-1863), Swiss mathematician.

Note that it is still true that $d\mathbf{L}^{(\mathbf{A})}/dt + \boldsymbol{\omega} \times \mathbf{L}^{(\mathbf{A})} = \mathbf{N}^{(\mathbf{A})}$ in the co-rotating frame (cf. Eq. (11.43)); hence Euler's equations have the same form as in Eq. (11.46), except that the quantities are now taken relative to $\mathbf{A}$:

$$I_1^{(\mathbf{A})}\dot{\omega}_1 - \omega_2\omega_3(I_2^{(\mathbf{A})} - I_3^{(\mathbf{A})}) = N_1^{(\mathbf{A})}, \quad \text{etc.} \tag{11.74}$$

If the gravitational force $m_\alpha\mathbf{g}$ acts on each constituent particle, then according to Eq. (11.68), the moment of the total force about $\mathbf{A}$ is given by

$$\mathbf{N}^{(\mathbf{A})} = \sum_{\alpha=1}^{N} m_\alpha \mathbf{r}''_\alpha \times \mathbf{g} = M\mathbf{h} \times \mathbf{g}. \tag{11.75}$$

In the co-rotating frame, $\mathbf{h}$ is a constant vector, but the components of the gravitational force are time dependent. The acceleration $\mathbf{g}$ is constant in the inertial frame. In the rotating, body-fixed frame the time dependence $\mathbf{g}$, according to Eq. (8.55), follows from

$$\dot{\mathbf{g}} := \left(\frac{d\mathbf{g}}{dt}\right)' = -\boldsymbol{\omega} \times \mathbf{g}. \tag{11.76}$$

Of course, the magnitude of $\mathbf{g}$ is independent from the frame of reference,

$$|\mathbf{g}| = \sqrt{g_1^2 + g_2^2 + g_3^2} = g. \tag{11.77}$$

The two Eqs. (11.72) and (11.76) determine **the motion of a spinning top in the body-fixed frame** given that the **pivot point $\mathbf{A}$ is at rest**. (From now on we omit the superscripts $(\mathbf{A})$ and $(\mathbf{R})$, if there is no ambiguity.) Inserting the torque, Eq. (11.75), into Euler's equations (11.74), we have

$$\begin{aligned}
I_1\dot{\omega}_1 &= \omega_2\omega_3(I_2 - I_3) + M(h_2g_3 - h_3g_2) \\
I_2\dot{\omega}_2 &= \omega_3\omega_1(I_3 - I_1) + M(h_3g_1 - h_1g_3) \\
I_3\dot{\omega}_3 &= \omega_1\omega_2(I_1 - I_2) + M(h_1g_2 - h_2g_1).
\end{aligned} \tag{11.78}$$

In these equations, the time dependence of $\mathbf{g}$ (cf. Eq. (11.76)) is to be determined from

$$\begin{aligned}
\dot{g}_1 &= \omega_3g_2 - \omega_2g_3 \\
\dot{g}_2 &= \omega_1g_3 - \omega_3g_1 \\
\dot{g}_3 &= \omega_2g_1 - \omega_1g_2.
\end{aligned} \tag{11.79}$$

The six nonlinear first order differential equations, Eqs. (11.78) and (11.79), for the *variables* ω_i and g_i with *parameters* I_i $\left(= I_i^{(A)}\right)$ and h_i, are called the **Euler-Poisson equations**.

The integrability of this system of differential equations depends on the number of integrals. **For arbitrary** values of the **parameters there are three integrals**, namely:

1.) The **total energy** $T_r + U = \frac{1}{2}\omega|\omega - Mgh$, i.e.

$$\frac{1}{2}\sum_{i=1}^{3} I_i\omega_i^2 - M\sum_{i=1}^{3} g_ih_i = const. \qquad (11.80)$$

(Proof: Multiply the i'th equation in (11.78) by ω_i and sum over i.)

2.) The component **Lg** of the **angular momentum parallel to the gravitational acceleration**. Since it follows from Eq. (11.75) that **Ng** $= 0$, and because of Eq. (11.72), it holds that

$$\mathbf{Lg} = \sum_{i=1}^{3} I_i\omega_i g_i = const. \qquad (11.81)$$

(Proof: Multiply the i'th equation of (11.78) by g_i and sum over i.)

3.) Condition (11.77) for **g**,

$$\sum_{i=1}^{3} g_i^2 = g^2. \qquad (11.82)$$

Using these three integrals, the Euler-Poisson system can be reduced to three nonlinear first order ordinary differential equations (say for the unknowns ω_1, g_1, and g_2 only). According to the remark on Page 95 in Section 4.4, chaotic behavior can not be excluded; only if there are four independent integrals of the motion the reduced system of equations is integrable a priori. For arbitrary values of I_i and h_i, there is no other integral. In the following **four special cases**, the system of ordinary differential equations defined by Eqs. (11.78) and (11.79) has a **fourth integral**:

i) **Euler's case**: $h = 0$ ($R = A$). The spinning top is supported at its center of mass. The magnitude of the angular momentum L^2 is the fourth conservation law

$$\sum I_i^2 \omega_i^2 = const.$$ (11.83)

(One can see this by multiplying the i'th equation of the system (11.78) by $I_i \omega_i$, for all i, and adding the resulting equations.) Since for $h = 0$ we also have $U = 0$, it follows that $E = T_r$ ($= \omega | \omega / 2$) $= L\omega/2$, so that conservation of energy (11.80) implies conservation of a further component of L. Since now $|L|$, Lg, and $L\omega$ are fixed, we have $L = const$. This also follows from Eq. (11.72) because $N = 0$. The solution was already discussed in the previous section (see Page 314).

ii) **Lagrange's case** (symmetric top): $I_1 = I_2$, $h_1 = h_2 = 0$. The center of mass is situated on the symmetry axis of the ellipsoid of inertia. From the third of Eqs. (11.78), one finds the fourth conservation law

$$\omega_3 = const.$$ (11.84)

iii) The **total symmetric case**: $I_1 = I_2 = I_3$ (spherical top). The system (11.78) becomes linear. The fourth conservation law is

$$\sum_{i=1}^{3} \omega_i h_i (= \omega h) = const.$$ (11.85)

(One can see this by multiplying the i'th equation of the system (11.78) by h_i, for all i, and subsequently adding the results.)

iv) **Kovalevskaya's case**[26]: $I_1 = I_2 = 2I_3$, $h_3 = 0$. The ellipsoid of inertia is rotational symmetric like in the second case, but the

---

[26]Sofia Kovalevskaya (1850-1891), first major Russian female mathematician. In order to escape the authority of her father and to get access to a study of mathematics she had a fictitious marriage in 1868. Afterwards she attended lectures at the University of Heidelberg, though as a woman she could not register regularly as a student. For a extensive description of her interesting life see the biography (in German) by W. Tuschmann and P. Hartwig, Sofia Kowalewskaja, Birkhäuser, Basel 1993

mass distribution is such that the center of mass lies in the (x, y)-plane of the body-fixed coordinate system (but not necessarily on the symmetry axis, the z-axis). The fourth conservation law is[27]

$$\left[\omega_1^2 - \omega_2^2 + \frac{M}{I_3}(h_1 g_1 - h_2 g_2)\right]^2 + \left[2\omega_1\omega_2 + \frac{M}{I_3}(h_1 g_2 + h_2 g_1)\right]^2 = const.$$
(11.86)

There is a further interesting though restricted case: the **Goryachev-Chaplygin top** with $I_1 = I_2 = 4I_3$ and the center of mass in the horizontal plane through the fixed point **A**. In general this system is not integrable, but if the angular momentum is perpendicular to the direction of gravity, i.e. **Lg** $= 0$, there exists a fourth constant of the motion[28].

Euler's equations (11.46) or (11.74) are a consequence of the equations of motion for the angular momentum viewed in the body-fixed frame. The velocity variables in these equations are not conjugate to the position variables (e.g. the components ω_i are not conjugate to Euler's angles). Hence, Euler's equations cannot be taken directly as Lagrange's equations (for the extremum of the action integral of some Lagrangian). However, we may want to determine the Lagrangian for the motion of a spinning top pivoted in the supporting point **A** in the body-fixed frame. To this end, we introduce the body-fixed coordinates with the help of Eq. (11.66) into the expression for the kinetic energy:

$$T = \frac{1}{2}\sum_{\alpha=1}^{N} m \dot{\mathbf{r}}_\alpha^2 = \frac{1}{2} M \dot{\mathbf{A}}^2 + M \dot{\mathbf{A}}(\boldsymbol{\omega} \times (\mathbf{R} - \mathbf{A})) + \frac{1}{2}\sum_{\alpha=1}^{N} m_\alpha(\boldsymbol{\omega} \times \mathbf{r}_\alpha'')^2.$$

The last term is the rotational energy $T_r^{(\mathbf{A})}$ with respect to **A**. Proceeding analogously to Subsection 11.3.2, and as above for the angular

---

[27]This case was discovered by S. Kovalevskaya in 1888, while she was investigating the analytic properties of the Euler-Poisson equations in the complex t-plane. This method is used in the so-called *Painlevé test* (cf. [Tabor]).

[28]D. Goryachev introduced the system (*On the motion of a rigid material body about a fixed point in the case $A = B = 4C$*. Mat. Sb. **21**, 3, 1900). S.A. Chaplygin integrated the equations in terms of hyperelliptic integrals (*A new case of rotation of a rigid body, supported at one point*. Collected works, Vol. I, Gostekhizdat 1948, pp. 118-124).

momentum $\mathbf{L}^{(\mathbf{A})}$, we can express this last term via the inertia tensor $\mathsf{I}^{(\mathbf{A})}$:

$$T_r^{(\mathbf{A})} = \frac{1}{2} \sum_{\alpha=1}^{N} m_\alpha (\boldsymbol{\omega} \times \mathbf{r}_\alpha'')^2 = \frac{1}{2} \boldsymbol{\omega} \mathsf{I}^{(\mathbf{A})} \boldsymbol{\omega} = \frac{1}{2} \mathbf{L}^{(\mathbf{A})} \boldsymbol{\omega}.$$

If $\mathbf{A}$ is fixed (implying there is no extra time dependence in U), the Lagrangian is

$$L = \frac{1}{2} \boldsymbol{\omega} \mathsf{I}^{(\mathbf{A})} \boldsymbol{\omega} - U(\mathbf{A}, \{\mathbf{r}_\alpha''\}).$$

For the gravitational force, the potential energy is given by Eq. (11.18), $U = -M\mathbf{g}\mathbf{R} = -M\mathbf{g}\mathbf{h} - M\mathbf{g}\mathbf{A}$, and the Lagrangian of the motion with respect to $\mathbf{A}$ is

$$L = \frac{1}{2} \boldsymbol{\omega} \mathsf{I}^{(\mathbf{A})} \boldsymbol{\omega} + M\mathbf{g}\mathbf{h}. \tag{11.87}$$

Since there are only three rotational degrees of freedom, all variables in Eq. (11.87) must be expressible in terms of Euler's angles φ, ϑ, and ψ. These angles parametrize the transformation from the inertial coordinate system, in which $\mathbf{g} = (0, 0, -g)$, to the body-fixed system, in which the inertia tensor is diagonal, $\mathsf{I}^{(\mathbf{A})} = \mathrm{diag}(I_1, I_2, I_3)$ and where $\mathbf{h} = (h_1, h_2, h_3)$ is the position of the center of mass. Using Eq. (8.52) to express $\boldsymbol{\omega}$ in terms of Euler's angles, one obtains for the rotational energy as a function of Euler's angles φ, ϑ, ψ and their time derivatives $\dot{\varphi}$, $\dot{\vartheta}$, $\dot{\psi}$, a rather large expression. And in the potential energy, we have to transform $(0, 0, -g)$ by a rotation D (Eq. (8.17)) to the representation of $\mathbf{g}$ in the body-fixed system, which results in,

$$\mathbf{g} = -g \begin{pmatrix} \sin\psi \sin\vartheta \\ \cos\psi \sin\vartheta \\ \cos\vartheta \end{pmatrix}. \tag{11.88}$$

$\mathbf{g}$ fulfills Eqs. (11.77) and (11.76) (**E**), so that the Lagrangian, Eq. (11.87), in which I, $\boldsymbol{\omega}$, and $\mathbf{g}$ are represented in the body-fixed frame, is tantamount to the system of equations (11.78). We now discuss Lagrange's case – the symmetric spinning top.

11.3.6   The symmetric spinning top

A **symmetric top** spins about its symmetry axis in the gravitational field of the earth, represented by the constant acceleration vector $\mathbf{g}$ $= (0, 0, -g)$. The spinning top is supported at the point $\mathbf{A}$ lying on the

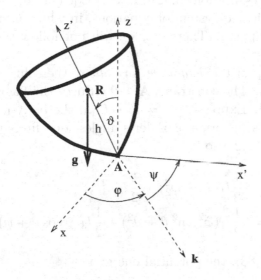

Figure 11.6: The symmetric top.

rotation axis (see Fig. 11.6), by a surface with perfect friction: $\dot{\mathbf{A}} = 0$.
The principal moments of inertia I_i are taken relative to the pivot $\mathbf{A}$.
The relation of the principal moments I_i to the principal moments I_i'
taken with respect to the center of mass position $\mathbf{R}$ is established by
Steiner's theorem, Eq. (11.70), in which, in the body-fixed coordinate
system, we have (cf. Fig. 11.6)

$$\mathbf{h} = \mathbf{R} - \mathbf{A} = (0, 0, h).$$

Since the top is symmetric, the principal moments are given by

$$I_1(= I_2) = I_1' + Mh^2, \ I_3 = I_3'.$$

If one wants to determine the motion of the top from *Euler's equations* (11.78), instead of solving Eqs. (11.79) simultaneously, one could
insert into Euler's equations the moment $\mathbf{N}$ of the gravitational force
with respect to the pivot $\mathbf{A}$ (cf. Eq. (11.75)), where the components of
$\mathbf{g}$ are taken to be in the body-fixed frame, Eq. (11.88),

$$\mathbf{N} = Mgh \begin{pmatrix} \cos\psi\sin\vartheta \\ -\sin\psi\sin\vartheta \\ 0 \end{pmatrix}. \tag{11.89}$$

Utilizing the representation (8.52) for $\boldsymbol{\omega}$ in Eq. (11.78), one produces a somewhat complicated system of equations, in which numerous contributions from $\dot{\boldsymbol{\omega}}$ appear. Therefore, we will not follow the solution via Euler's equations[29].

With the help of the *Lagrange formalism*, the solution is obtained much more easily. The constraint $\dot{\mathbf{A}} = 0$ reduces the Lagrangian to the form Eq. (11.87). Expressing $L = T_r - U$ in the body-fixed system of the principal axes in terms of Euler's angles, one finds the rotational energy (cf. Eq. (8.52)) to be

$$
\begin{aligned}
T_r &= \frac{1}{2}I_1(\omega_1^2 + \omega_2^2) + \frac{1}{2}I_3\omega_3^2 \\
&= \frac{1}{2}I_1(\dot{\varphi}^2 \sin^2\vartheta + \dot{\vartheta}^2) + \frac{1}{2}I_3(\dot{\varphi}\cos\vartheta + \dot{\psi})^2.
\end{aligned}
\qquad (11.90)
$$

Through Eq. (11.88), the potential energy is

$$
U = Mgh\cos\vartheta.
\qquad (11.91)
$$

So the **Lagrangian in terms of Euler's angles** reads

$$
L = \frac{1}{2}I_1(\dot{\varphi}^2 \sin^2\vartheta + \dot{\vartheta}^2) + \frac{1}{2}I_3(\dot{\varphi}\cos\vartheta + \dot{\psi})^2 - Mgh\cos\vartheta.
\qquad (11.92)
$$

One immediately sees that the angles φ and ψ are cyclic variables. Consequently, the respective conjugate momenta p_φ and p_ψ, representing the components of the angular momentum in the direction of the respective rotation axes, are constant. This can also be seen in Fig. 11.6. The torque of the gravitational force is directed along the line denoted by $\mathbf{k}$ in the figure. But the rotation axes of the angles φ and ψ are the 'old' and 'new' z-axis, respectively; that is, they are perpendicular to

---

[29] A clever way to solve the equations in this case is presented in [Becker]. There, the motion of the top is considered in an intermediate coordinate system in which the top rotates only with the angular velocity $\dot{\psi}$ about the z-axis. Euler's equations are transformed to this system in which

$$
\omega = \begin{pmatrix} \dot{\vartheta} \\ \dot{\varphi}\sin\vartheta \\ \dot{\varphi}\cos\vartheta \end{pmatrix} \qquad \text{and} \qquad \mathbf{N} = Mgh \begin{pmatrix} \sin\vartheta \\ 0 \\ 0 \end{pmatrix}.
$$

Now the equations can be solved rather easily.

the vector $\mathbf{k}$, and therefore the corresponding components of the moment vanish, while the two corresponding components of the angular momentum stay constant. For the latter, we find

$$p_\varphi = \frac{\partial L}{\partial \dot\varphi} = (I_1 \sin^2 \vartheta + I_3 \cos^2 \vartheta)\dot\varphi + I_3 \cos\vartheta\dot\psi = const =: I_1 b \quad (11.93)$$

$$p_\psi = \frac{\partial L}{\partial \dot\psi} = I_3(\dot\varphi \cos\vartheta + \dot\psi) = I_3\omega_3 = const =: I_1 a. \quad (11.94)$$

Furthermore, using Eq. (11.94), the integral in Eq. (11.84) implies that

$$\omega_3 = \dot\varphi \cos\vartheta + \dot\psi = const. \quad (11.95)$$

Since the Lagrangian does not depend explicitly on time, the energy

$$E = T + U = \frac{1}{2}I_1(\dot\varphi^2 \sin^2 \vartheta + \dot\vartheta^2) + \frac{1}{2}I_3(\dot\varphi \cos\vartheta + \dot\psi)^2 + Mgh\cos\vartheta \quad (11.96)$$

is also constant.

Altogether we have three conserved quantities for three degrees of freedom – the system is integrable.

Instead of solving the remaining Lagrange's equation for the variable ϑ, we can determine the solution from the conserved quantities. But before doing this, it is advantageous to subtract from E the constant rotational energy about the z-axis; i.e. to consider the conserved quantity

$$E' = E - \frac{p_\psi^2}{2I_3} = \frac{1}{2}I_1(\dot\varphi^2 \sin^2 \vartheta + \dot\vartheta^2) + Mgh\cos\vartheta, \quad (11.97)$$

which is easier to handle. (The kinetic part of E' is the conserved quantity (11.50) of the free symmetric rigid body.)

The three conserved quantities given in Eqs. (11.93), (11.94), and (11.97) contain the angular velocities $\dot\varphi$, $\dot\psi$, and $\dot\vartheta$ in a simple algebraic way. We therefore express $\dot\varphi$, $\dot\psi$, and $\dot\vartheta$ as functions of ϑ. The subsequent integration of $\dot\vartheta = \dot\vartheta(\vartheta)$ allows to calculate $\vartheta(t)$. The solutions $\varphi(t)$ and $\psi(t)$ then follow. The details are as follows. Inserting Eq. (11.94) into Eq. (11.93) yields $I_1\dot\varphi\sin^2\vartheta + I_1 a\cos\vartheta = I_1 b$, so that

$$\dot\varphi = \frac{b - a\cos\vartheta}{\sin^2 \vartheta}. \quad (11.98)$$

Hence Eq. (11.94) reduces to

$$\dot{\psi} = \frac{I_1 a}{I_3} - \cos\vartheta \frac{b - a\cos\vartheta}{\sin^2\vartheta}. \tag{11.99}$$

From Eqs. (11.98) and (11.99), one can calculate $\varphi(t)$ and $\psi(t)$ after $\vartheta = \vartheta(t)$ has been determined.

We obtain the equation for $\vartheta(t)$ from conservation of the energy E', Eq. (11.97). Inserting Eq. (11.98) into Eq. (11.97) leads to

$$E' = \frac{I_1\dot{\vartheta}^2}{2} + \frac{I_1}{2}\left(\frac{b - a\cos\vartheta}{\sin\vartheta}\right)^2 + Mgh\cos\vartheta, \tag{11.100}$$

and substituting $u = \cos\vartheta$, we obtain

$$E'(1 - u^2) = \frac{1}{2}I_1\dot{u}^2 + \frac{1}{2}I_1(b - au)^2 + Mghu(1 - u^2).$$

Introducing the function

$$f(u) = (1 - u^2)(\alpha - \beta u) - (b - au)^2, \tag{11.101}$$

where

$$\alpha = 2E'/I_1, \qquad \beta = 2Mgh/I_1,$$

(α and β are positive parameters!) Eq. (11.100) can be transformed into the equation

$$\frac{du}{dt} = [f(u)]^{1/2}. \tag{11.102}$$

Integrating this equation,

$$t = \int_{u(0)}^{u(t)} \frac{du}{\sqrt{f(u)}},$$

results in an elliptic integral for $t = t(\vartheta)$, from which $\vartheta(t)$ cannot be determined analytically. We therefore present a qualitative view.

Since $f(u \to -\infty) = -\infty$ and $f(u \to +\infty) = +\infty$, there exists at least one real zero of the function f. The only question is whether the zero lies within the physically relevant interval $-1 \le u \le 1$ ($\pi \ge \vartheta \ge 0$). At $u = \pm 1$, the function is $f(\pm 1) = -(b \mp a)^2$, i.e. negative (the case $b = a$ will be discussed later). Only real values of $\dot{u}$ are physically

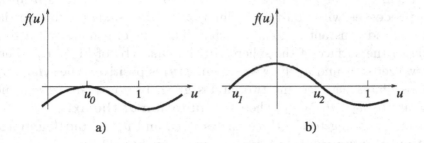

Figure 11.7: The function $f(u)$ (cf. text).

interesting; therefore, in some subinterval of $[-1, 1]$, the function f has to be positive. Hence there are two zeros in that interval, and these may also coincide (cf. Fig. 11.7). These zeros, ϑ_1 and ϑ_2, are the turning points of the ϑ-motion. The time dependence of ϑ between these two values, from Eqs. (11.98) and (11.99), determines the time dependence of φ and ψ.

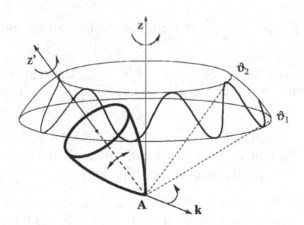

Figure 11.8: The motion of a symmetric top.

The motion of the spinning top is illustrated by letting the symmetry axis trace a figure on the surface of a sphere about the point **A** (Fig. 11.8). In the **case (a)** of Fig. 11.7, $f(u)$ has a twofold zero at u_0. There, since $\dot{u} = 0$, it follows that $\dot{\vartheta} = 0$, i.e. ϑ is equal to a constant

ϑ_0. Then, since we also have $\dot\psi = const$ and $\dot\varphi = const$, **the spinning top precesses** with constant inclination to the z-axis (of the inertial system) and constant angular velocity. The axis of symmetry traces a circle on the surface of the sphere. In the **case (b)** of Fig. 11.7, there are two zeros u_1 and u_2; between them, $f(u)$ is positive. The symmetry axis oscillates between the two circles given by $\vartheta_1 = \arccos(u_1)$ and $\vartheta_2 = \arccos(u_2)$, $\vartheta_1 \geq \vartheta_2$. There is a **nutation of the axis**.

Corresponding to the three angles ψ, φ, and ϑ, the **most general motion of a symmetric top** consists of (cf. Fig. 11.8):

1.) **Rotation about the axis of symmetry** ($=$ principal axis I_3) with angular velocity $\dot\psi$;

2.) **Precession** ($=$ rotation of the axis of symmetry about the z-axis of the inertial system, with **A** lying on the spin axis) with angular velocity $\dot\varphi$;

3.) **Nutation** ($=$ 'bobbing up and down' of the axis of symmetry between the angles ϑ_1 and ϑ_2) with angular velocity $\dot\vartheta$ about the vector **k** (cf. Fig. 11.8), which in turn rotates with angular velocity $\dot\varphi$.

Equation (11.98), $\dot\varphi = \dfrac{b - au}{1 - u^2}$, determines the kind of the nutation that occurs:

i) If $b - a\cos\vartheta_2 = b - au_2 > 0$, i.e. $b/a > u_2$, the sign of $\dot\varphi$ does not change and the motion of the symmetry axis is as shown in Fig. 11.9 (a).

ii) If $b - au_2 < 0$, i.e. $b/a < u_2$ and $u_1 < b/a$, the sign of $\dot\varphi$ changes between ϑ_1 and ϑ_2 at $\vartheta = \arccos(b/a)$ and the spin axis traces loops on the sphere, as shown in Fig. 11.9 (b).

iii) If $b = a\cos\vartheta_2 = au_2$, i.e. $b/a = u_2$, $\dot\varphi$ vanishes just at the turning point ϑ_2. This kind of motion is shown in Fig. 11.9 (c).

A **special case** is $a = b$, i.e. $\vartheta_2 = 0$. The function $f(u)$ has a zero at the boundary of the relevant interval in $u = \cos\vartheta = 1$. The axis of the spinning top is vertical ($\vartheta = 0$)[30]. The question of whether this

---

[30]For $a = b$ and $\vartheta = \dot\vartheta = 0$, we have $E' = Mgh$, i.e. $\alpha = \beta$. $\vartheta = 0$ ($u = 1$) is a zero of $f(u)$ with multiplicity two. The generalized momenta p_ψ and p_φ are degenerate: $p_\psi = p_\varphi = I_3(\dot\varphi + \dot\psi) = const$ (this case is called the 'sleeping top').

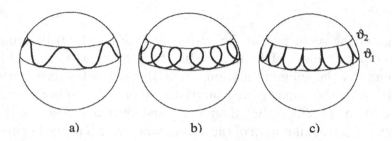

Figure 11.9: Various kinds of nutation.

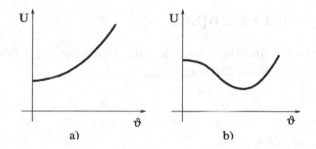

Figure 11.10: The potential $U(\vartheta)$: a) for $\omega_3 > \omega_0$ and b) for $\omega_3 < \omega_0$.

axis position is stable, must be investigated in more detail. To this end, we return to the expression in Eq. (11.100) for the energy E', with $I_1 b = I_1 a = I_3 \omega_3$:

$$E' = \frac{I_1 \dot{\vartheta}^2}{2} + \frac{(I_3 \omega_3)^2}{2 I_1} \left(\frac{1 - \cos \vartheta}{\sin \vartheta} \right)^2 + Mgh \cos \vartheta.$$

E' is the energy of the ϑ-component of the motion in the effective potential,

$$U(\vartheta) = \frac{(I_3 \omega_3)^2}{2 I_1} \tan^2(\vartheta/2) + Mgh \cos \vartheta.$$

For small values of ϑ, a Taylor expansion of $U(\vartheta)$ gives

$$U(\vartheta) = \frac{(I_3 \omega_3)^2}{2 I_1} \frac{\vartheta^2}{4} + Mgh - Mgh \frac{\vartheta^2}{2} = Mgh + \frac{I_3^2}{8 I_1} \left(\omega_3^2 - \omega_0^2 \right) \vartheta^2,$$

where
$$\omega_0^2 = 4I_1 Mgh/I_3^2.$$
The sign of the coefficient of the ϑ^2-term (cf. Fig. 11.10) determines the kind of extremum of U at $\vartheta = 0$. If $\omega_3^2 > \omega_0^2$, then $\vartheta = 0$ is a minimum and the spinning motion about the symmetry axis – which coincides with the z-axis of the inertial system – is stable as long as this condition is fulfilled. But if $\omega_3^2 < \omega_0^2$, the motion is unstable (local maximum of U) and nutation of the axis occurs. Even if at the beginning we have $\omega_3 > \omega_0$, in reality, due to friction, the angular velocity ω_3 of the rotation about the axis will become smaller than the critical value ω_0, and the nutation of the top will set in.

Problems and examples

1. Calculate the inertia tensor for the following rigid bodies with homogeneous mass distribution:

   a) sphere

   b) cube

   c) right parallelepiped

   d) ellipsoid of revolution

   e) circular disk (compare with (d))

2. Two cylinders are rolling down an inclined plane. The one cylinder is full and the other is hollow with an infinitely thin wall. Both cylinders have the same homogeneously distributed total mass. Which cylinder rolls to the base of the inclined plane sooner?

3. **Foucault gyrocompass.** This device indicates the direction towards the actual (kinematic) poles of the earth and not the magnetic poles. The compass consists of a symmetric top spinning with angular velocity $\mathbf{\Omega}$, $|\mathbf{\Omega}| = const$, whose axes of rotation is allowed to precess only in the horizontal plane, i.e. $\mathbf{\Omega} = (\Omega_1, \Omega_2, 0)$. Show that the moment of the Coriolis force exerted on the spinning top in the body-fixed system (cf. Subsection 8.5.3; there is no moment of the gravitational force (why?)) is

$$\mathbf{N}_C = -2 \sum_\alpha m_\alpha \left(\mathbf{r}_\alpha \times \mathbf{\Omega} \right) \left(\boldsymbol{\omega}_e \mathbf{r}_\alpha \right).$$

Calculate $\mathbf{N}_C$ for a circular disk of mass M and radius R rotating about its natural axis (the thickness of the disk does not matter (why?). Answer: $\mathbf{N}_C = -\dfrac{MR^2}{2}(\boldsymbol{\omega}_e \times \boldsymbol{\Omega})$). Only the component of the earth's angular velocity $\boldsymbol{\omega}_e$ in the plane of the compass, $\bar{\boldsymbol{\omega}}_e$, is relevant for the compass. The effective torque on the gyroscope axis is therefore given by $\bar{\mathbf{N}}_C = -\dfrac{MR^2}{2}(\bar{\boldsymbol{\omega}}_e \times \boldsymbol{\Omega})$. Explain how the compass works.

4. A infinitely thin, rectangular sheet (mass M) rotates with constant angular velocity ω about one of its diagonals. Determine the torque that causes such a rotation. What is the result for a square sheet? Explain the difference.

5. From the solution given in Eq. (11.51) for body-fixed frame, by introducing Euler's angles (cf. Eq. (11.52)), determine the motion of a free symmetric body in the inertial frame. Choose the orientation of the axes in the inertial frame such that $\mathbf{L} = (0, 0, L_z)$.

   (The solution is: $\varphi = \dfrac{A}{\sin a}t + \varphi_0$, $\vartheta = a = \arctan\left(\dfrac{AI_1}{\omega_3 I_3}\right)$, $\psi = \Omega t$)

6. Perform in detail the linear stability analysis for the free rotation of a nonsymmetric body (see Eqs. (11.53)).

7. Deduce Eq. (11.72) for the angular momentum with respect to $\mathbf{A}$.

8. Show for Kovalevskaya's case that the integral (11.86) is constant.

9. Determine the Lagrangian and equations of motion of a (plane) **double pendulum**. This device consists of a pendulum (length l_1, mass m_1) where the mass also serves as pivot for a second pendulum (length l_2, mass m_2) that swings in the same plane as the first pendulum (cf. Fig. 12.1). For small displacements from the equilibrium position, find and solve the equations of motion.

10. **The compound pendulum.** Consider a rigid body that can rotate freely about a point $\mathbf{A}$ (a point inside the body that is not the center of mass position $\mathbf{R}$). Suppose the body oscillates about the equilibrium position under the influence of the gravitational force. In the simplest case, the pivot $\mathbf{A}$, at a distance h, $\mathbf{h} =$

R − A, to the center of mass, allows only for planar motion in a plane (i.e. the axis of rotation is fixed) that lies perpendicular to a principal axis and which includes the center of mass. Show that the Lagrangian is

$$L = \frac{1}{2}\left(I_1' + Mh^2\right)\dot{\varphi}^2 - Mgh\cos\varphi,$$

where I_1' is the principal moment with respect to the center of mass.

- Compare with the plane pendulum in Section 3.3.2.
- Compare with the double pendulum above.

11. **The falling rod problem.** (Solution by John Wojdylo.) One end of a uniform rigid rod of length L and mass m is held at angle θ_0 to the vertical, while the other rests on rough, flat ground (assume no slipping). Also, a ball is held level with the free end of the rod. The two are released simultaneously. Which hits the ground first, the ball or the rod? Is there an initial angle for which they both hit the ground at the same time? Does this angle depend on L, or the fact that the experiment is done on earth?

Solution. For the center of mass motion of the rod, conservation of energy implies

$$\dot{\theta}^2 = \frac{mgL}{I}(\cos\theta_0 - \cos\theta).$$

The time $t(\theta_0, \theta) > 0$ taken to get from initial angle θ_0 to final angle θ can be expressed as the integral

$$t(\theta_0, \theta) = \sqrt{\frac{I}{mgL}} \int_{\theta_0}^{\theta} \frac{d\xi}{\sqrt{\cos\theta_0 - \cos\xi}},$$

which cannot be evaluated exactly in closed form. However, a good analytical approximation can be found by noting that the integrand has a singularity at θ_0, so by far the biggest contribution comes from the neighborhood of $\xi = \theta_0$.[31] We expand $\cos\theta_0 -$

---

[31]This is Watson's Lemma. See, for example, A H. Nayfeh, Introduction to Perturbation Techniques (Bibliography).

$\cos \xi$ in a Taylor series about $\xi = \theta_0$ to second order, obtaining $\cos \theta_0 - \cos \xi = a + b\xi + b\xi^2$, where $a = \frac{1}{2}(\theta_0 \cos \theta_0 - 2\sin \theta_0)\theta_0$, $b = \sin \theta_0 - \theta_0 \cos \theta_0$, and $c = \frac{1}{2}\cos \theta_0$. We now have a standard integral:

$$t(\theta_0, \theta) \sim \sqrt{\frac{I}{mgL}} \frac{1}{\sqrt{c}} \int_{\theta_0}^{\theta} \frac{d\xi}{\sqrt{\frac{a}{c} + \frac{b}{c}\xi + \xi^2}}$$

$$= \sqrt{\frac{I}{mgL}} \sqrt{\frac{2}{\cos \theta_0}} \ln \frac{\Delta\theta + \tan \theta_0 + \sqrt{\Delta\theta(\Delta\theta + 2\tan \theta_0)}}{\tan \theta_0},$$

where $\Delta\theta = \theta - \theta_0$. For this rod, $I_{rod} = mL^2/3$. Setting $\theta = \pi/2$ and inserting L and g, we can plot the time $t_{rod}(\theta_0)$ taken for the rod to hit the ground, given $0 \le \theta_0 \le \pi/2$. As expected, at $\theta_0 = 0$, the rod balances on its end and never falls.

Now, the time taken for the ball to hit the ground is $t_{ball}(\theta_0) = \sqrt{2L\cos \theta_0/g}$. To find the starting angle for which the ball and rod hit the ground at the same time, we set $t_{rod}(\theta_0) = t_{ball}(\theta_0)$. This can be solved graphically (see plot below) to give $\theta^* \approx 41°$. (Numerical integration gives about 42°.) The result is independent of L and g.

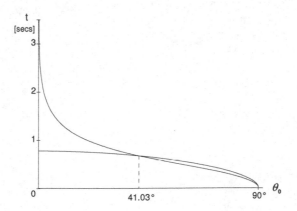

t_{rod} and t_{ball} versus initial rod angle θ_0 (to the vertical). The rod length has been arbitrarily set at $L = 3$m. The cross-over in this approximation occurs at $\theta_0 \approx 41°$ (independent of L and g).

Comment. The rod dynamics is surprisingly complex. The parts closer to the pivot always 'want' to fall at larger angular acceleration than the more distant parts, but are slowed down by them: the distant parts are 'being asked' to travel a much larger distance per unit time than gravity and rigid rotation allow, and they resist, due to inertia. This causes an internal torque in the opposite direction about the free end, which dominates the early motion for small θ_0. But as the rod becomes more horizontal, the torque about the pivot begins to dominate; as the vertical acceleration of the centre of mass approaches g, the more distant points are forced to have vertical acceleration greater than g, and the rod falls 'faster than gravity'. If the rod is brittle enough (e.g. a long chimney), it will break up because of the opposing torques. We leave finding the most probable angle as an exercise!

12

Small oscillations

In the last chapter we have used a system of N particles with fixed distances as a model for a rigid body, having only six degrees of freedom. However, considering the actual motions of the particles about their mean (equilibrium) positions, still using classical concepts[1], we are immediately confronted with a system possessing $3N$ coupled degrees of freedom.

In this chapter motion in a system of N interacting particles is studied. We assume that the interaction is such that there exist equilibrium positions for the particles; for small energies the particles move only in the vicinity of their equilibrium positions. However, even in the simple approach of the harmonic approximation, one has a coupled system of linear equations. It turns out, though, that one regains independent equations of motion if one transforms to so-called normal coordinates. The normal coordinates describe collective (i.e. coherent) displacements – the 'phonons' – of the particles from their equilibrium positions.

12.1 The double pendulum

The simplest non-rigid system consists of three point masses m_1, m_2, and m_3, in which only two distances – distance l_2 between m_1 and m_2, and distance l_1 between m_1 and m_3 – are fixed (see sketch below). If particles 2 and 3 do not interact, then their mutual distance can assume

---

[1]At low temperatures, quantum mechanical concepts are usually appropriate for the treatment of a body consisting of N atoms.

any value between $|l_1 - l_2|$ and $l_1 + l_2$ (otherwise an equilibrium distance may exist between them, and the rigid connections may hang at some angle relative to each other at equilibrium). The system has seven degrees of freedom: three translational and three rotational degrees for the system as a whole, and one internal degree of freedom, the angle between l_1 and l_2. In the following, we assume that there is no interaction between particle 2 and particle 3.

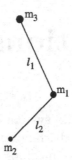

We remove three degrees of freedom by fixing the position of particle 3 such that it serves only as pivot point[2]. We allow gravity to act on the remaining point masses m_1 and m_2. Furthermore, we demand that the two particles and the pivot remain always in a fixed plane. Thus we arrive at the plane **double pendulum** (see Fig. 12.1). The only degrees of freedom left are the two angles ϕ_1 and ϕ_2, as shown in the figure. The only conservation law – that of the total energy – is not enough to guarantee integrability.

To discuss the motion of the (plane) double pendulum, we choose the orientation of the x and y-axes such that the x-axis points in the direction of the gravitational acceleration $\mathbf{g}$. Hence, the forces on the point masses m_1 and m_2 are given by $(m_i g, 0)$, $i = 1, 2$. Expressing the coordinates in terms of the angles ϕ_1 and ϕ_2 (cf. Fig. 12.1), we have

$$\begin{aligned} x_1 &= l_1 \cos \phi_1 \\ y_1 &= l_1 \sin \phi_1, \end{aligned} \qquad (12.1)$$

and

$$\begin{aligned} x_2 &= l_1 \cos \phi_1 + l_2 \cos \phi_2 \\ y_2 &= l_1 \sin \phi_1 + l_2 \sin \phi_2. \end{aligned} \qquad (12.2)$$

---

[2]Note that the center of mass is not fixed by this condition.

In terms of these variables, the potential energy

$$V = -(m_1 g x_1 + m_2 g x_2)$$

is

$$V = -(m_1 + m_2)g l_1 \cos \phi_1 - m_2 g l_2 \cos \phi_2. \qquad (12.3)$$

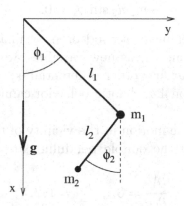

Figure 12.1: The double pendulum.

The x and y components of the velocities of the two point masses are

$$\dot{x}_1 = -l_1 \dot{\phi}_1 \sin \phi_1$$
$$\dot{y}_1 = l_1 \dot{\phi}_1 \cos \phi_1,$$

and

$$\dot{x}_2 = -l_1 \dot{\phi}_1 \sin \phi_1 - l_2 \dot{\phi}_2 \sin \phi_2$$
$$\dot{y}_2 = l_1 \dot{\phi}_1 \cos \phi_1 + l_2 \dot{\phi}_2 \cos \phi_2,$$

respectively. Therefore the kinetic energy is given by

$$T = \frac{1}{2}m_1 l_1^2 \dot{\phi}_1^2 + \frac{1}{2}m_2 \left(l_1^2 \dot{\phi}_1^2 + l_2^2 \dot{\phi}_2^2 + 2l_1 l_2 \dot{\phi}_1 \dot{\phi}_2 (\sin \phi_1 \sin \phi_2 + \cos \phi_1 \cos \phi_2) \right)$$

$$= \frac{1}{2}(m_1 + m_2) l_1^2 \dot{\phi}_1^2 + \frac{1}{2}m_2 l_2^2 \dot{\phi}_2^2 + m_2 l_1 l_2 \dot{\phi}_1 \dot{\phi}_2 \cos(\phi_1 - \phi_2) \qquad (12.4)$$

From the Lagrangian $L = T - V$, we obtain the equations of motion,

$$(m_1 + m_2)l_1^2\ddot{\phi}_1 + m_2l_1l_2\ddot{\phi}_2\cos(\phi_1 - \phi_2) + m_2l_1l_2\dot{\phi}_2^2\sin(\phi_1 - \phi_2)$$

$$+(m_1 + m_2)gl_1\sin\phi_1 = 0, \tag{12.5}$$

$$m_2l_2^2\ddot{\phi}_2 + m_2l_1l_2\ddot{\phi}_1\cos(\phi_1 - \phi_2) - m_2l_1l_2\dot{\phi}_1^2\sin(\phi_1 - \phi_2)$$

$$+m_2gl_2\sin\phi_2 = 0. \tag{12.6}$$

This is a system of nonlinear, second order ordinary differential equations. Introducing $\dot{\phi}_1$ and $\dot{\phi}_2$ as new variables, we get an autonomous system of four nonlinear first order differential equations. Hence, as we have observed in Section 4.4, chaotic behavior cannot be excluded, and, indeed, appears.

We consider only the motion in the vicinity of the equilibrium. The stationary positions of the double pendulum are obtained by setting $\ddot{\phi}_i = \dot{\phi}_i = 0$; i.e.,

$$\frac{\partial V}{\partial \phi_i} = 0, \qquad i = 1, 2.$$

Equation (12.3) implies the conditions

$$\sin\phi_1 = 0 \qquad \text{and} \qquad \sin\phi_2 = 0$$

for the stationary points, which have four solutions: $(\phi_1, \phi_2) = (0, 0)$, $(0, \pi)$, $(\pi, 0)$, (π, π). Only the first is a stable equilibrium, which is intuitively clear. The stable equilibrium corresponds to the absolute minimum of the potential,

$$V_{\min} = -(m_1 + m_2)gl_1 - m_2gl_2.$$

The angles ϕ_1 and ϕ_2 are the displacement angles from the equilibrium position $(0, 0)$. We therefore expand the Lagrangian to second order in ϕ_1 and ϕ_2 and their time derivatives. Omitting the constant $V_{\min}$, we find

$$L = \frac{1}{2}(m_1 + m_2)l_1^2\dot{\phi}_1^2 + \frac{1}{2}m_2l_2^2\dot{\phi}_2^2 + m_2l_1l_2\dot{\phi}_1\dot{\phi}_2$$

$$-\frac{1}{2}(m_1 + m_2)gl_1\phi_1^2 - \frac{1}{2}m_2gl_2\phi_2^2. \tag{12.7}$$

The corresponding equations of motion,

$$(m_1 + m_2)l_1\ddot{\phi}_1 + m_2l_2\ddot{\phi}_2 + (m_1 + m_2)g\phi_1 = 0$$
$$m_2l_2\ddot{\phi}_2 + m_2l_1\ddot{\phi}_1 + m_2g\phi_2 = 0, \quad (12.8)$$

are two coupled linear ordinary differential equations with constant co-efficients. The usual ansatz,

$$\phi_i(t) = \phi_i^0 \exp(-i\omega t), \quad (12.9)$$

with ϕ allowed to be complex – only the real part being physically relevant – leads to an algebraic system of equations that can be written in matrix notation as

$$\left[\begin{pmatrix} (m_1 + m_2)gl_1 & 0 \\ 0 & m_2gl_2 \end{pmatrix} - \omega^2 \begin{pmatrix} (m_1 + m_2)l_1^2 & m_2l_1l_2 \\ m_2l_1l_2 & m_2l_2^2 \end{pmatrix} \right] \begin{pmatrix} \phi_1^0 \\ \phi_2^0 \end{pmatrix} = 0.$$
$$(12.10)$$

Before solving this system, we discuss general properties of this *harmonic approximation*.

12.2 The harmonic approximation

12.2.1 The general theory

The preceding treatment of the double pendulum is an example of a transformation from $3N$ constrained Cartesian coordinates to $f \leq 3N$ generalized independent coordinates that represent the remaining degrees of freedom (cf. Section 10.3). For the double pendulum, four Cartesian coordinates x_1, x_2, y_1, y_2 in the initial description have been reduced the two independent angles ϕ_1 and ϕ_2. These two variables reflect the two degrees of freedom of the plane double pendulum.

Replacing the $3N$ Cartesian coordinates in the Lagrangian of an N particle system by f independent generalized coordinates q_i using the transformation

$$\mathbf{r}_\alpha = \mathbf{r}_\alpha(q_1, q_2, \dots, q_f), \quad \alpha = 1, \dots, 3N, \quad f \leq 3N,$$

we obtain the Lagrangian as function of the generalized coordinates and their time derivatives,

$$L(\{q\}, \{\dot{q}\}) = \frac{1}{2}m_{kl}\dot{q}_k\dot{q}_l - V(\{q\}), \quad k, l = 1, \dots, f; \quad (12.11)$$

here, the summation convention for the indices k, l, numbering the degrees of freedom, is used. Since

$$\dot{\mathbf{r}}_\alpha = \frac{\partial \mathbf{r}_\alpha}{\partial q_k} \dot{q}_k,$$

the coefficients m_{kl} in the kinetic energy

$$m_{kl} = m_{kl}\left(\{q\}\right) = \sum_{\alpha=1}^{N} m_\alpha \frac{\partial \mathbf{r}_\alpha}{\partial q_k} \frac{\partial \mathbf{r}_\alpha}{\partial q_l} \tag{12.12}$$

are functions of the generalized coordinates. They are symmetric with respect to the indices k, l. Since $2T = \sum_{\alpha=1}^{N} m_\alpha (\dot{\mathbf{r}}_\alpha)^2$ is a positive definite quadratic form, this also holds for $m_{kl} \dot{q}_k \dot{q}_l$.

The system of interacting point masses is in *equilibrium* if all the **generalized forces**,

$$Q_k = -\frac{\partial V}{\partial q_k}, \tag{12.13}$$

vanish for certain values of the coordinates q_k, known as the *equilibrium positions* q_k^0:

$$Q_k|_0 = -\left(\frac{\partial V}{\partial q_k}\right)_0 = 0. \tag{12.14}$$

(The index 0 indicates that one has to set $\{q\} = \{q^0\}$.) Expanding the potential in the *displacements* u_k from the equilibrium positions, where

$$u_k = q_k - q_k^0, \tag{12.15}$$

the equilibrium condition (12.14) implies that the linear term in the expansion vanishes at the equilibrium position:

$$V(\{q\}) = V(\{q^0\}) + \frac{1}{2}\left(\frac{\partial^2 V}{\partial q_k \partial q_l}\right)_0 u_k u_l + \dots . \tag{12.16}$$

Whether the system is really in equilibrium is determined by the quadratic form,

$$V = \frac{1}{2} P_{kl} u_k u_l, \qquad P_{kl} := \left(\frac{\partial^2 V}{\partial q_k \partial q_l}\right)_0; \tag{12.17}$$

here, the unimportant constant $V(\{q^0\})$ has been omitted. The necessary condition for a stable equilibrium is that the quadratic form in

Eq. (12.17) is positive definite. If it is only positive semidefinite (at least one eigenvalue of the matrix P_{kl} vanishes), the extremum must be investigated further.

Let us consider now the kinetic energy,

$$T = \frac{1}{2} m_{kl} \dot{q}_k \dot{q}_l = \frac{1}{2} m_{kl} \dot{u}_k \dot{u}_l. \tag{12.18}$$

Expanding the coefficients m_{kl} about the equilibrium positions $\{q^0\}$, we obtain

$$m_{kl}(\{q\}) = M_{kl} + \left(\frac{\partial m_{kl}}{\partial q_n}\right)_0 u_n + \cdots ,$$

with M_{kl} being the symmetric **mass matrix**,

$$M_{kl} = m_{kl}(\{q^0\}). \tag{12.19}$$

Since T is already quadratic in the velocities $\dot{u}_k$, we keep only the first term of the expansion. Thus, in this *harmonic approximation* – only the quadratic terms (12.17) and (12.19) are kept in the expansions – the Lagrangian (12.11) reduces to

$$L(\{u\}, \{\dot{u}\}) = \frac{1}{2}(M_{kl} \dot{u}_k \dot{u}_l - P_{kl} u_k u_l); \tag{12.20}$$

or, in matrix notation,

$$L = \frac{1}{2}(\dot{\mathbf{u}} \mathbf{M} \dot{\mathbf{u}} - \mathbf{u} \mathbf{P} \mathbf{u}), \tag{12.21}$$

with $\mathbf{u} = (u_1, \ldots, u_f)$. The variation of L with respect to the displacements u_k and $\dot{u}_k$ yields the equations of motion,

$$M_{kl} \ddot{u}_l + P_{kl} u_l = 0, \qquad k, l = 1, \ldots, f, \tag{12.22}$$

or,

$$\mathbf{M}\ddot{\mathbf{u}} + \mathbf{P}\mathbf{u} = \mathbf{0}.$$

We solve this system of coupled linear ordinary differential equations by the standard ansatz,

$$u_k(t) = u_k^0 \exp(-i\omega t), \tag{12.23}$$

where, again, the u_k and u_k^0 are now complex. An algebraic system of equations results, *viz.*:

$$(-\omega^2 M_{kl} + P_{kl}) u_l^0 = 0; \tag{12.24}$$

or, in matrix notation,

$$(\mathsf{P} - \omega^2 \mathsf{M})\mathbf{u}^0 = 0. \tag{12.25}$$

A solvability condition for the system is satisfied when the determinant of the matrix vanishes:

$$\det(\mathsf{P} - \omega^2 \mathsf{M}) = 0. \tag{12.26}$$

The *eigenvalues* ω_r^2, $r = 1,\ldots,f$, are solutions of the *characteristic equation* of degree f in the variable ω^2. The corresponding f *eigenvectors* $\mathbf{s}^{(r)}$ (with components $s_k^{(r)}$, $k,r = 1,\ldots,f$) are obtained from the equations,

$$\mathsf{P}\mathbf{s}^{(r)} = \omega_r^2 \mathsf{M}\mathbf{s}^{(r)}. \tag{12.27}$$

A particular eigenvector $\mathbf{s}^{(r)}$ represents a collective displacement of all the particles; the individual displacement of each particle is given by the value of the respective component of $\mathbf{s}^{(r)}$.

Since P as well as M are symmetric matrices, and M is positive definite, it follows that

i) The eigenvalues ω_r^2 are *real* and *non-negative* (Proof: multiply (12.27) by $(\mathbf{s}^{(r)})^*$; E),

$$\omega_r^2 \geq 0; \tag{12.28}$$

ii) The eigenvectors $\mathbf{s}^{(r)}$ can be chosen to be *real* (Proof: consider the difference of the complex conjugate eigenvalue equations; E);

iii) The eigenvectors $\mathbf{s}^{(r)}$ – after an appropriate choice of normalization – satisfy the following *orthonormality condition* (E),

$$\mathbf{s}^{(r)}\mathsf{M}\mathbf{s}^{(s)} = \delta_{rs}. \tag{12.29}$$

(This is also true if the eigenvalues are degenerate.)

From Eqs. (12.27) and (12.29), we find that

$$\mathbf{s}^{(r)}\mathsf{P}\mathbf{s}^{(s)} = \omega_r^2 \delta_{rs}. \tag{12.30}$$

A consequence of Eqs. (12.29) and (12.30) is that both matrices P and M can be diagonalized by a similarity transformation with the same $(f \times f)$-matrix S, where the columns of S are the vectors $\mathbf{s}^{(r)}$:

$$\mathsf{S} = \left(\mathbf{s}^1\, \mathbf{s}^2\, \ldots\, \mathbf{s}^f\right). \tag{12.31}$$

So we have

$$S^T M S = 1 \quad \text{and} \quad S^T P S = \text{diag}(\omega_1^2, \ldots, \omega_f^2). \tag{12.32}$$

Due to Eq. (12.27), the time dependent vectors

$$s_{\omega_r}^{(r)}(t) := s^{(r)} \exp(i\omega_r t) \quad \text{and} \quad s_{-\omega_r}^{(r)}(t) := s^{(r)} \exp(-i\omega_r t), \tag{12.33}$$

$r = 1, \ldots, f$, are solutions of the equation of motion (12.22). Each of these solutions persists in time, oscillating with **eigenfrequency** ω_r. Each such solution is therefore also called an **eigenmode** of the system of particles. The set $\left\{ s_{\omega_r}^{(r)}(t) \right\}$ of the time-dependent, linearly independent vectors form a basis for the solutions of the equation of motion; i.e., an arbitrary solution u can be expanded in terms of these solutions,

$$u(t) = \sum_{r=1}^{f} s^{(r)} \{ c_r^+ \exp(i\omega_r t) + c_r^- \exp(-i\omega_r t) \}. \tag{12.34}$$

The $2f$ complex conjugate coefficients c_r^+ and c_r^- (u is real!),

$$c_r^- = (c_r^+)^* =: c_r,$$

are equivalent to $2f$ real constants, which allow u and $\dot{u}$ to be adjusted to the initial values at time $t = 0$.

Introducing into Eq. (12.34) the **normal coordinate**, $\xi_r(t)$ – i.e. the amplitude of the eigenmode $s^{(r)}$ –

$$\xi_r(t) = c_r^* \exp(i\omega_r t) + c_r \exp(-i\omega_r t) = 2Re\left(c_r \exp(-i\omega_r t) \right), \tag{12.35}$$

we have

$$u(t) = \sum_{r=1}^{f} s^{(r)} \xi_r(t), \tag{12.36}$$

or, conversely, because of Eq. (12.29),

$$\xi_p(t) = s^{(p)} M u(t). \tag{12.37}$$

Taking now the normal coordinates ξ_r as new dynamical variables, the kinetic and the potential energy read (cf. (12.36) and (12.29) or (12.30)),

$$T = \frac{1}{2} \dot{u} M \dot{u} = \frac{1}{2} \sum_{r=1}^{f} \dot{\xi}_r^2, \quad V = \frac{1}{2} u P u = \frac{1}{2} \sum_{r=1}^{f} \omega_r^2 \xi_r^2 \tag{12.38}$$

and hence in terms of the ξ_r, the Lagrangian is

$$L = \frac{1}{2}\sum_{r=1}^{f}\left(\dot{\xi}_r^2 - \omega_r^2\xi_r^2\right).$$

(12.39)

Consequently, Lagrange's equations consist of f *independent* oscillator equations

$$\ddot{\xi}_p + \omega_p^2\xi_p = 0, \qquad p = 1,\dots,f$$

(12.40)

with the initial conditions

$$\xi_p(0) = s^{(p)}\mathbf{M}u(0) \qquad \text{and} \qquad \dot{\xi}_p(0) = s^{(p)}\mathbf{M}\dot{u}(0).$$

(12.41)

12.2.2   The double pendulum (again)

We apply this general procedure to the double pendulum. In harmonic approximation, the Lagrangian (12.7) is

$$L = \frac{1}{2}(\dot{\phi}\mathbf{M}\dot{\phi} - \phi\mathbf{P}\phi).$$

(12.42)

Substituting $\phi_i = \phi_i^0 e^{-i\omega t}$ into the equations of motion, the eigenvalues and eigenvectors are obtained from (recall Eqs. (12.9) and (12.10))

$$(\mathbf{P} - \omega^2\mathbf{M})\phi^0 = 0,$$

where (with $M = m_1 + m_2$)

$$\mathbf{P} = \begin{pmatrix} Mgl_1 & 0 \\ 0 & m_2gl_2 \end{pmatrix} \qquad \text{and} \qquad \mathbf{M} = \begin{pmatrix} Ml_1^2 & m_2l_1l_2 \\ m_2l_1l_2 & m_2l_2^2 \end{pmatrix}.$$

This linear system only has a nontrivial solution if the determinant of the matrix

$$\mathbf{P} - \omega^2\mathbf{M} = \begin{pmatrix} Ml_1(g - \omega^2l_1) & -\omega^2m_2l_1l_2 \\ -\omega^2m_2l_1l_2 & m_2l_2(g - \omega^2l_2) \end{pmatrix}$$

(12.43)

vanishes (cf. (12.26)); i.e., if

$$m_1l_1l_2\omega^4 - Mg(l_1 + l_2)\omega^2 + Mg^2 = 0.$$

The solutions of this equation are the eigenvalues

$$\omega_{1,2}^2 = \frac{Mg}{2m_1}\frac{1}{l_1l_2}\left(l_1 + l_2 \pm \sqrt{(l_1 + l_2)^2 - 4l_1l_2m_1/M}\right).$$

(12.44)

The corresponding eigenvectors $\mathbf{s}^{(r)}$, $r = 1, 2$, are obtained from the equation (see Eqs. (12.27) and (12.43))

$$M(g - \omega_r^2 l_1)s_1^{(r)} - \omega_r^2 m_2 l_2 s_2^{(r)} = 0,$$

so that the ratio of the components is given by

$$\frac{s_1^{(r)}}{s_2^{(r)}} = \frac{\omega_r^2 m_2 l_2}{M(g - \omega_r^2 l_1)}. \tag{12.45}$$

The components still have to be normalized according to Eq. (12.29). For arbitrary parameters, this is a rather tedious procedure. We complete only the special case in which

$$l_1 = l_2 = l.$$

Setting, in Eq. (12.44),

$$\mu = \frac{m_2}{M} \quad \text{and} \quad \omega_0^2 = \frac{g}{l}, \tag{12.46}$$

we have

$$\omega_{1,2}^2 = \omega_0^2 \frac{1}{1 \mp \sqrt{\mu}}, \tag{12.47}$$

in particular $\omega_1^2 \geq \omega_2^2$. The components of the eigenvectors are found from Eq. (12.45),

$$\frac{s_1^{(r)}}{s_2^{(r)}} = \frac{\mu}{\omega_0^2/\omega_r^2 - 1} = (-1)^r \sqrt{\mu};$$

i.e.,

$$\mathbf{s}^{(1)} = c_1 \begin{pmatrix} \sqrt{\mu} \\ -1 \end{pmatrix}, \qquad \mathbf{s}^{(2)} = c_2 \begin{pmatrix} \sqrt{\mu} \\ 1 \end{pmatrix}, \tag{12.48}$$

where the normalization constants c_r are to be determined from condition (12.29). In terms of the definitions (12.46), the matrices P and M read

$$\mathsf{P} = Ml^2 \begin{pmatrix} \omega_0^2 & 0 \\ 0 & \mu\omega_0^2 \end{pmatrix} \quad \text{and} \quad \mathsf{M} = Ml^2 \begin{pmatrix} 1 & \mu \\ \mu & \mu \end{pmatrix}.$$

Using the latter in the orthonormality relation (12.29), we obtain from

$$\left(\mathbf{s}^{(p)}\mathbf{M}\mathbf{s}^{(r)}\right)_{r,p=1,2} = 2M\mu l^2 \left(\begin{array}{cc} \left(1-\sqrt{\mu}\right) c_1^2 & 0 \\ 0 & \left(1+\sqrt{\mu}\right) c_2^2 \end{array} \right)$$

$$= \left(\begin{array}{cc} 1 & 0 \\ 0 & 1 \end{array} \right)$$

the normalization factors

$$c_1^2 = 1/\left(2M\mu l^2(1-\sqrt{\mu})\right)$$
$$c_2^2 = 1/\left(2M\mu l^2(1+\sqrt{\mu})\right).$$

The eigenvectors are now completely determined.

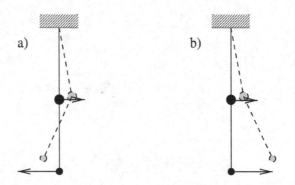

Figure 12.2: The normal modes of the double pendulum.

The eigenvector $\mathbf{s}^{(1)}$ describes motion of the two point masses in opposite directions (Fig. 12.2 (a)), whereas $\mathbf{s}^{(2)}$ describes motion of both particles in the same direction (Fig. 12.2 (b)). The higher eigenfrequency ω_1 is associated with $\mathbf{s}^{(1)}$. This motion is therefore harder to excite than the motion given by $\mathbf{s}^{(2)}$. In the limit $\mu \to 1$, that is $m_1 \to 0$, the frequency $\omega_1 \to \infty$. In this limit, only m_2 oscillates at fixed distance $2l$ from the pivot, with frequency $\omega_2 = \omega_0/\sqrt{2}$. For $\mu \to 0$ ($m_2 \to 0$), we have $\omega_1 = \omega_2 = \omega_0$, and the (massless) rod l_2 ($= l$) oscillates with mass m_1. In both limits we recover the (simple) plane pendulum.

The general solution for the equations of motion (12.8) is a linear combination of the two eigenmodes (12.48),

$$\boldsymbol{\phi}(t) = \xi_1(t)\mathbf{s}^1 + \xi_2(t)\mathbf{s}^2.$$

Since (cf. Eq. (12.30))

$$\left(s^{(p)}Ps^{(r)}\right)_{r,p=1,2} = \begin{pmatrix} \omega_1^2 & 0 \\ 0 & \omega_2^2 \end{pmatrix},$$

the Lagrangian as function of the normal coordinates is

$$L = \frac{1}{2}\left(\dot{\xi}_1^2 + \dot{\xi}_2^2\right) - \frac{1}{2}\left(\omega_1^2\xi_1^2 + \omega_2^2\xi_2^2\right).$$

Hence Lagrange's equations are uncoupled oscillator equations for the normal coordinates, with frequencies ω_r (cf. Eq. (12.47))

$$\ddot{\xi}_r + \omega_r^2\xi_r = 0, \qquad r = 1, 2.$$

Expressing the normal coordinates via Eq. (12.37), $\xi_r = s^{(r)}M\phi$, in terms of ϕ, we find

$$\xi_1 = \sqrt{\frac{M}{2}}l\sqrt{1 - \sqrt{\mu}}\,(\phi_1 - \sqrt{\mu}\phi_2)$$

$$\xi_2 = \sqrt{\frac{M}{2}}l\sqrt{1 + \sqrt{\mu}}\,(\phi_1 + \sqrt{\mu}\phi_2).$$

These relations, taken at $t = 0$, yield the initial values of ξ_1 and ξ_2 as functions of ϕ_1^0 and ϕ_2^0.

12.2.3 Vibrations of a triatomic molecule

As a further example for the application of the harmonic approximation, we consider a *triatomic 'molecule'* of identical interacting classical particles ('atoms'). We assume that the particles are located at positions $x_1, x_2,$ and x_3; and that their two-body interaction force depends only on the distance between them, so there exists a potential

$$V_2(r_{\alpha\beta}) = V_2(r_{\alpha\beta}), \qquad r_{\alpha\beta} = x_\alpha - x_\beta$$

for the interaction between each pair of particles α and β. We assume further that the nature of the potential V_2 is such that there is an equilibrium position for any pair of particles; then, in equilibrium, the particles form an equilateral triangle whose side has length a. At equilibrium, the total potential

$$V = \sum_{\substack{\alpha\neq\beta \\ \alpha,\beta=1,2,3}} V_2(r_{\alpha\beta})$$

is minimal. In our choice of the coordinates, the equilibrium positions $\mathbf{x}_\alpha^0$, $\alpha = 1, 2, 3$, of the atoms lie in the plane $z = 0$, and their (x, y)-coordinates are given by (cf. Fig. 12.3)

$$\mathbf{x}_1^0 = \frac{a}{2}\left(-1, -\frac{\sqrt{3}}{3}\right), \qquad \mathbf{x}_2^0 = \frac{a}{2}\left(1, -\frac{\sqrt{3}}{3}\right), \qquad \mathbf{x}_3^0 = \frac{a}{2}\left(0, \frac{2\sqrt{3}}{3}\right),$$

so that *the center of mass is at the origin.*

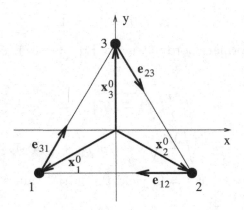

Figure 12.3: The triatomic molecule.

Defining the displacement of the particle α from the equilibrium position $\mathbf{x}_\alpha^0$ to be

$$\mathbf{u}_\alpha = \mathbf{x}_\alpha - \mathbf{x}_\alpha^0, \qquad \alpha = 1, 2, 3, \tag{12.49}$$

we have

$$\mathbf{r}_{\alpha\beta} = \mathbf{u}_\alpha - \mathbf{u}_\beta + \mathbf{x}_\alpha^0 - \mathbf{x}_\beta^0 =: \mathbf{u}_{\alpha\beta} + \mathbf{x}_{\alpha\beta}^0. \tag{12.50}$$

The expansion of $V_2\left(r_{\alpha\beta}\right) = V_2\left(\left\|\mathbf{u}_{\alpha\beta} + \mathbf{x}_{\alpha\beta}^0\right\|\right)$ for small values of $\mathbf{u}_{\alpha\beta}$ about the equilibrium distance $\mathbf{x}_{\alpha\beta}^0$ (for the moment we omit the indices $\alpha\beta$ and denote the components of $\mathbf{r}$ by x_i) is then

$$V_2\left(r\right) = V_2\left(r_0\right) + u_i \left.\frac{\partial V_2}{\partial x_i}\right|_{\mathbf{r}=\mathbf{x}^0} + \frac{1}{2} u_i u_k \left.\frac{\partial^2 V_2}{\partial x_i \partial x_k}\right|_{\mathbf{r}=\mathbf{x}^0} + \dots .$$

This can be simplified by observing that

$$\frac{\partial V_2}{\partial x_i} = \frac{x_i}{r}\frac{dV_2}{dr}, \qquad \frac{\partial^2 V_2}{\partial x_i \partial x_k} = \delta_{ik}\frac{1}{r}\frac{dV_2}{dr} - x_i x_k \frac{1}{r^3}\frac{dV_2}{dr} + x_i x_k \frac{1}{r^2}\frac{d^2 V_2}{dr^2},$$

and that for the equilibrium distance $\mathbf{x}^0$

$$\left.\frac{dV_2}{dr}\right|_{\mathbf{r}=\mathbf{x}^0} = 0.$$

Therefore, in harmonic approximation, we have

$$V_2(r) = V_2(r_0) + \frac{1}{2}u_i u_k x_i^0 x_k^0 \frac{1}{r^2}\frac{d^2 V_2}{dr^2}\bigg|_{\mathbf{r}=\mathbf{x}^0} = V_2(r_0) + \frac{1}{2}\left(\mathbf{u}\mathbf{x}^0\right)^2 \frac{k}{a^2},$$

where the coefficient k is defined by

$$k = a^2 \frac{1}{r^2}\frac{d^2 V_2}{dr^2}\bigg|_{\mathbf{r}=\mathbf{x}^0}. \tag{12.51}$$

Hence, omitting the constant term, the total potential in harmonic approximation is given by

$$V = \frac{k}{a^2}\left[\left(\mathbf{u}_{12}\mathbf{x}_{12}^0\right)^2 + \left(\mathbf{u}_{23}\mathbf{x}_{23}^0\right)^2 + \left(\mathbf{u}_{13}\mathbf{x}_{13}^0\right)^2\right]. \tag{12.52}$$

Defining the unit vectors

$$\mathbf{e}_{\alpha\beta} = (\mathbf{x}_\alpha^0 - \mathbf{x}_\beta^0)/a,$$

i.e.

$$\mathbf{e}_{12} = (-1,0), \qquad \mathbf{e}_{23} = \frac{1}{2}(1,-\sqrt{3}), \qquad \mathbf{e}_{31} = \frac{1}{2}(1,\sqrt{3}), \tag{12.53}$$

which are parallel to the line segments connecting the equilibrium positions of the atoms (see Fig. 12.3), we obtain

$$V = \frac{k}{2}\left[\left(\mathbf{e}_{12}(\mathbf{u}_1 - \mathbf{u}_2)\right)^2 + \left(\mathbf{e}_{23}(\mathbf{u}_2 - \mathbf{u}_3)\right)^2 + \left(\mathbf{e}_{31}(\mathbf{u}_3 - \mathbf{u}_1)\right)^2\right]$$

$$= \frac{k}{2}\left[(\mathbf{e}_{12}\mathbf{u}_1)^2 + (\mathbf{e}_{31}\mathbf{u}_1)^2 + (\mathbf{e}_{12}\mathbf{u}_2)^2 + (\mathbf{e}_{23}\mathbf{u}_2)^2 + (\mathbf{e}_{23}\mathbf{u}_3)^2 + (\mathbf{e}_{31}\mathbf{u}_3)^2 \right.$$
$$\left. - 2\left((\mathbf{e}_{12}\mathbf{u}_1)(\mathbf{e}_{12}\mathbf{u}_2) + (\mathbf{e}_{23}\mathbf{u}_2)(\mathbf{e}_{23}\mathbf{u}_3) + (\mathbf{e}_{31}\mathbf{u}_1)(\mathbf{e}_{31}\mathbf{u}_3)\right)\right]. \tag{12.54}$$

Numbering the components of all the $\mathbf{u}_\alpha$ consecutively (for the moment, we consider only displacements $\mathbf{u}_\alpha$ in the plane of the triangle), we introduce the six-component vector

$$\underline{u} = (\mathbf{u}_1, \mathbf{u}_2, \mathbf{u}_3). \tag{12.55}$$

In this notation we have

$$V = \frac{1}{2}\underline{u}P\underline{u}, \tag{12.56}$$

where

$$P = k \begin{pmatrix} \mathbf{e}_{12} \otimes \mathbf{e}_{12} + \mathbf{e}_{31} \otimes \mathbf{e}_{31} & -\mathbf{e}_{12} \otimes \mathbf{e}_{12} & -\mathbf{e}_{31} \otimes \mathbf{e}_{31} \\ -\mathbf{e}_{12} \otimes \mathbf{e}_{12} & \mathbf{e}_{12} \otimes \mathbf{e}_{12} + \mathbf{e}_{23} \otimes \mathbf{e}_{23} & -\mathbf{e}_{23} \otimes \mathbf{e}_{23} \\ -\mathbf{e}_{31} \otimes \mathbf{e}_{31} & -\mathbf{e}_{23} \otimes \mathbf{e}_{23} & \mathbf{e}_{23} \otimes \mathbf{e}_{23} + \mathbf{e}_{31} \otimes \mathbf{e}_{31} \end{pmatrix}$$

$$= \frac{k}{4} \begin{pmatrix} 5 & \sqrt{3} & -4 & 0 & -1 & -\sqrt{3} \\ \sqrt{3} & 3 & 0 & 0 & -\sqrt{3} & -3 \\ -4 & 0 & 5 & -\sqrt{3} & -1 & \sqrt{3} \\ 0 & 0 & -\sqrt{3} & 3 & \sqrt{3} & -3 \\ -1 & -\sqrt{3} & -1 & \sqrt{3} & 2 & 0 \\ -\sqrt{3} & -3 & \sqrt{3} & -3 & 0 & 6 \end{pmatrix} \tag{12.57}$$

$$(\mathbf{e}_{12}\otimes \mathbf{e}_{12} = \begin{pmatrix} 1 & 0 \\ 0 & 0 \end{pmatrix}, \quad \mathbf{e}_{31}\otimes \mathbf{e}_{31} = \frac{1}{4}\begin{pmatrix} 1 & \sqrt{3} \\ \sqrt{3} & 3 \end{pmatrix} \text{ etc.; cf. Eq.}$$

(A.8)).

Since all three atoms are of equal mass, the mass matrix in the kinetic energy

$$T = \frac{1}{2}\sum_{\alpha=1}^{3} m\dot{\mathbf{x}}_\alpha^2 = \frac{1}{2}\sum_{\alpha=1}^{3} m\dot{\mathbf{u}}_\alpha^2 = \frac{1}{2}\dot{\underline{u}}M\dot{\underline{u}} \tag{12.58}$$

is proportional to the unit matrix: $M = m1$. Therefore the characteristic equation (12.26) reduces to

$$\det(P - \omega^2 m1) = 0,$$

and the squares of the eigenfrequencies, ω_r^2, $r = 1, \ldots, 6$, are the eigenvalues of P/m. Calculating the eigenvalues yields

$$\mathrm{eig}(P)/m = \frac{k}{m}(0, 0, 0, 3/2, 3/2, 3),$$

so that the nonvanishing frequencies ω_r are

$$(\omega_1, \omega_2, \omega_3) = \sqrt{\frac{k}{m}} \left(\sqrt{3/2}, \sqrt{3/2}, \sqrt{3} \right). \qquad (12.59)$$

The three vanishing eigenvalues of the matrix P are easy to understand: two of them correspond to the (two-dimensional) translations of the center of mass and the third corresponds to a rotation of the molecule about its center of mass. The Lagrangian

$$L = \frac{1}{2} \underline{\dot{u}} \mathsf{M} \underline{\dot{u}} - \frac{1}{2} \underline{u} \mathsf{P} \underline{u}$$

is invariant under these operations. From the matrix P, the translational invariance can be recognized from the fact that the sum of the first, third, and fifth – as well as the sum of the second, fourth, and sixth – elements in a row vanish. Therefore, any vector

$$\underline{s} \propto (\mathbf{u}, \mathbf{u}, \mathbf{u}),$$

i.e. a pure translation of *all* atoms (and the center of mass), is an eigenvector of P with eigenvalue zero. For example, translations of each atom along the x-axis and along the y-axis,

$$\mathbf{u}_{t,x} = (1,0) \quad \text{and} \quad \mathbf{u}_{t,y} = (0,1),$$

respectively, can be chosen as a basis for a general translation (cf. Figs. 12.4 (a), (b)); i.e. we have the two linearly independent eigenvectors,

$$\underline{s}_{t1} = c_1(1,0,1,0,1,0) \quad \text{and} \quad \underline{s}_{t2} = c_2(0,1,0,1,0,1). \qquad (12.60)$$

A pure rotation of the molecule is represented by displacement vectors $\mathbf{u}_{r,\alpha}$ that are orthogonal to the vectors $\mathbf{x}_\alpha^0$. This means that $\mathbf{u}_{r,\alpha} = \varphi(\mathbf{n} \times \mathbf{x}_\alpha^0)$, in which the rotation axis $\mathbf{n}$ is perpendicular to the plane. Hence, from the two-dimensional representation

$$\mathbf{u}_{r,\alpha} = \varphi \begin{pmatrix} 0 & -1 \\ 1 & 0 \end{pmatrix} \mathbf{x}_\alpha^0.$$

we find the eigenvector $\underline{s}_r \propto (\mathbf{u}_{r,1}, \mathbf{u}_{r,2}, \mathbf{u}_{r,3})$, i.e.

$$\underline{s}_r = c_3(1, -\sqrt{3}, 1, \sqrt{3}, -2, 0), \qquad (12.61)$$

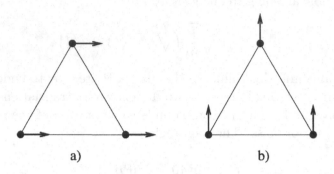

Figure 12.4: Translations of the molecule.

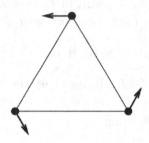

Figure 12.5: A rotation of the molecule.

for the rotation of the molecule (as shown in Fig. 12.5). As one can easily convince oneself, the eigenvalue of this eigenvector of the matrix P is also zero. We leave the calculation of the normalization factors c_i in Eqs. (12.60) and (12.61) for later.

Another easily calculated eigenmode of the molecule is the mode shown in Fig. 12.6, in which for each particle $\mathbf{u}_\alpha \propto \mathbf{x}_\alpha^0$, so that

$$\underline{s}_{o3} = c_6(-\sqrt{3}, -1, \sqrt{3}, -1, 0, 2).\tag{12.62}$$

The molecule 'pulsates'. This eigenvector is orthogonal to the eigenvectors derived above. It is associated with the eigenfrequency

$$\omega_3 = \sqrt{3k/m}$$

in Eq. (12.59). All these modes are easy to visualize.

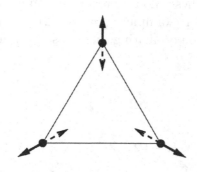

Figure 12.6: The 'pulsating' mode.

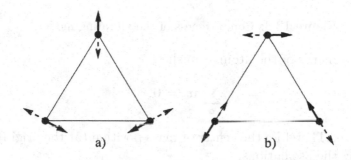

Figure 12.7: Vibrating modes of the molecule.

The remaining two independent eigenvectors are associated with the degenerate eigenfrequencies

$$\omega_1 = \omega_2 = \sqrt{3k/2m}$$

in Eq. (12.59). Therefore, if one has computed two eigenvectors, in general, they are not orthogonal to each other; they must then be orthogonalized. Two possible solutions are (**E**)

$$
\begin{aligned}
\underline{s}_{o1} &= c_4(\sqrt{3}, -1, -\sqrt{3}, -1, 0, 2), \\
\underline{s}_{o2} &= c_5(-1, -\sqrt{3}, -1, \sqrt{3}, 2, 0).
\end{aligned}
$$

(12.63)

These vibrating modes are shown in Fig. 12.7 (a) and (b). The symmetry of the molecule implies, of course, that modes rotated by 120° are also solutions[3]. But these modes are linear combinations of the eigen-modes, Eq. (12.63) (for example: $\underline{s}'_1 = -(1/2)\underline{s}_{o1} + (\sqrt{3}/2)\underline{s}_{o2}$, cf. Fig. 12.8). Note that for the oscillating modes, $\underline{s}_{o1}, \underline{s}_{o2}$, and $\underline{s}_{o3}$, the sum of

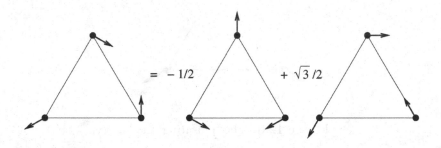

Figure 12.8: Completeness of the vibrating modes.

the displacements of the atoms vanishes,

$$\sum_\alpha \mathbf{u}_\alpha = \mathbf{0},$$

in each case. Therefore the center of mass position (at the origin) is not affected by the oscillations.

We still need to determine the normalization factors. We obtain the c_i from $\underline{s}M\underline{s} = 1$ (cf. Eq. (12.29)):

$$c_1 = c_2 = \sqrt{\frac{1}{3m}}, \quad c_3 = c_4 = c_5 = c_6 = \frac{1}{2}\sqrt{\frac{1}{3m}}. \tag{12.64}$$

A general *vibrational mode* $\underline{s}$ of the molecule is given by a linear combination of the eigenmodes (12.62) and (12.63)

$$\underline{s} = \xi_1(t)\underline{s}_{o1} + \xi_2(t)\underline{s}_{o2} + \xi_3(t)\underline{s}_{o3}, \tag{12.65}$$

where the normal coordinates $\xi_r(t)$ are associated with the eigenfrequencies ω_r (cf. Eq. (12.59)).

---

[3]Discrete symmetries are not related to any conservation law. Only symmetries under transformations which are continuously connected to the identity transformation are associated with a constant of the motion.

The foregoing vibrational modes also appear when the molecule is allowed to move freely in three dimensions. In this case, the number of degrees of freedom $f = 9$, but the number of modes with eigenvalue zero increases from three to six: three degrees of translations of the center of mass and three of rotations about the center of mass. Therefore, again, only three degrees of freedom remain for vibrations, and we have found the three eigenmodes for vibrations already. Because three points always lie on a plane, there are no new vibration modes in three dimensions.

12.3 From the linear chain to the vibrating string

We now directly go over to the case of a body comprising of a large number of interacting point masses. Starting from the harmonic approximation, if we let the inter-particle distances vanish accompanied by an appropriate increase in the particle number – which eventually goes to infinity – we arrive at a field theory for the elastic properties of a continuous body. In this section, we demonstrate the procedure in the simple case of a **linear chain** of harmonically coupled point masses.

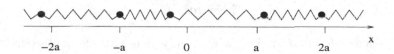

Figure 12.9: The linear chain.

Infinitely many point masses on a (straight) line, each of mass m, at positions x_n, $n \in (-\infty, \infty)$, are coupled in pairs harmonically with strength k so that their equilibrium positions are equidistant (length a) along the x-axis. To begin with, it is assumed that the point masses can only move along the x-axis – they can be displaced only *longitudinal*, as indicated in Fig. 12.9. The displacement of the n'th point mass from its equilibrium position, $x_n^0 = na$, is

$$u_n = x_n - na. \tag{12.66}$$

One readily derives the following expressions for the kinetic and the potential energy:

$$T = \sum_{n=-\infty}^{\infty} \frac{1}{2} m \dot{u}_n^2, \qquad V = \sum_{n=-\infty}^{\infty} \frac{1}{2} k (u_n - u_{n+1})^2. \qquad (12.67)$$

If one consider only *transverse* displacements (in y direction), u_n is the y-coordinate of the n'th particle, $y_n = u_n$. One obtains the same expressions as in Eq. (12.67) for the kinetic and the potential energy. Hence the following applies to both longitudinal and transverse cases.

The Lagrangian

$$L = \sum_{n=-\infty}^{\infty} \left[\frac{1}{2} m \dot{u}_n^2 - \frac{1}{2} k (u_n - u_{n+1})^2 \right] \qquad (12.68)$$

yields the equations of motion,

$$m \ddot{u}_n + 2 k u_n - k (u_{n+1} + u_{n-1}) = 0, \qquad n = -\infty, \ldots, \infty. \qquad (12.69)$$

$\mathbf{u} := (\ldots, u_{n-1}, u_n, u_{n+1}, \ldots)$ is now an infinite dimensional vector; correspondingly, the square matrices M and P are $(\infty \times \infty)$-matrices. M is proportional to the unit matrix,

$$\mathsf{M} = m \mathbf{1},$$

and P is a *banded matrix*, i.e. whose only non-zero elements occur on the main diagonal as well as on either side next to the diagonal. The diagonal entries are all $2k$, and all the other non-vanishing entries are $-k$:

$$\mathsf{P} = \begin{pmatrix} & \cdots & \cdots & \cdots & & & & & \\ & & \cdots & \cdots & \cdots & & & & \\ \cdots & 0 & 0 & -k & 2k & -k & 0 & 0 & \cdots \\ \cdots & 0 & 0 & -k & 2k & -k & 0 & 0 & \cdots \\ \cdots & 0 & 0 & -k & 2k & -k & 0 & 0 & \cdots \\ & & & \cdots & \cdots & \cdots & & & \\ & & & & \cdots & \cdots & \cdots & & \end{pmatrix}.$$

In terms of these matrices, the system of equations of motion can be written as

$$\mathsf{M} \ddot{\mathbf{u}} + \mathsf{P} \mathbf{u} = 0. \qquad (12.70)$$

Because of the infinite dimensional vectors and matrices, the procedure of Subsection 12.2.1 needs to be modified. (This is because, for instance, infinite dimensional vectors cannot be normalized like finite dimensional vectors.) Nevertheless, the ansatz $\mathbf{u}(t) = \mathbf{s}\exp(-i\omega t)$ can be employed, upon which one obtains the equation for the eigenvectors $\mathbf{s}$ and the eigenvalues ω

$$\left(\mathsf{P} - \omega^2 \mathsf{M}\right)\mathbf{s} = 0.$$

As before, a general solution $\mathbf{u}$ is a linear superposition of the eigenmodes

$$\mathbf{u} = \sum_p \xi(p,t)\mathbf{s}(p), \qquad (12.71)$$

where parameter p counts the eigenmodes, and $\xi(p,t)$ is a normal coordinate.

Since M is proportional to the unit matrix – so that any vector is an eigenvector – one only has to find the eigenvectors of P. One recognizes that the vectors $\mathbf{s}\,(p)$, with components

$$s_n(p) = e^{ipna}, \qquad (12.72)$$

where p is an arbitrary real parameter, are eigenvectors, by considering the n'th component of $\mathsf{P}\mathbf{s}$:

$$\begin{aligned}
(\mathsf{P}\mathbf{s})_n &= 2ks_n - k\,(s_{n-1} + s_{n+1}) \\
&= 2k\left(1 - \frac{1}{2}\left(e^{-ipa} + e^{ipa}\right)\right)e^{ipna} \\
&= 2k\,(1 - \cos pa)\,s_n. \qquad (12.73)
\end{aligned}$$

Hence the eigenvalue of P with respect to $\mathbf{s}$ is $2k\,(1 - \cos pa)$. Inserting this into the equation

$$(\mathsf{P} - \omega^2\mathsf{M})\mathbf{s}(p) = \left(2k(1 - \cos pa) - m\omega^2\right)\mathbf{s}(p) = 0$$

yields, for ω, the condition

$$\omega^2 = \frac{2k}{m}(1 - \cos pa) = \frac{4k}{m}\sin^2(pa/2).$$

Therefore the eigenvalue is given by

$$\omega = \pm 2\omega_0 \sin(pa/2), \qquad \omega_0 = \sqrt{k/m}. \qquad (12.74)$$

Let us have a closer look at the eigenmodes $\mathbf{s}(p)$. The real part of the n'th component,

$$\mathrm{Re}\, s_n = \cos(pna), \tag{12.75}$$

is the displacement of particle n. The displacements of the particles in the chain form a wave-like state with wave length

$$\lambda = 2\pi/p =: ra,$$

where ra is the period of the function $\cos px$, e.g. the distance between two successive maxima. The minimal wavelength $\lambda_{\min}$ (minimal value of $r = 2\pi/pa$) is given by the condition that $\lambda_{\min}$ is twice the (equilibrium) distance between two particles,

$$\lambda_{\min} = 2a.$$

Smaller values of λ are meaningless, since there would exist many values of p that describe the same mode with given displacements of the atoms in the chain (i.e. the representation of the modes would not be unique). So $\lambda_{\min}$ fixes the maximal value of p:

$$p_{\max} = \pi/a.$$

There is no physical argument against negative values of p in the displacement (12.75) (in this case λ is given by $|p|$); therefore, the values of the mode indices p lie in the interval

$$p \in [-\pi/a, \pi/a]. \tag{12.76}$$

Since p may take any value in this interval, we replace the sum in Eq. (12.71) by an integral. The displacements $\mathbf{u}$ are then the following linear combinations of the eigenmodes $\mathbf{s}(p)$, with the normal coordinates ξ as coefficients:

$$\mathbf{u} = \int_{-\pi/a}^{\pi/a} dp\, \xi(p,t)\mathbf{s}(p). \tag{12.77}$$

The normal coordinates are the amplitudes of the wave-like eigenmodes $\mathbf{s}(p)$ of the linear chain. The frequencies of the normal coordinates are given by Eq. (12.74), where p lies in the interval given in (12.76).

12.3.1   The vibrating string

In the continuum limit, we let the distance a go to zero while fixing the mean mass density m/a. This results is a model for a vibrating string. Substituting

$$
\begin{aligned}
a &= : \Delta x \\
na &= : x \\
m/a &= : \mu \\
ak &= : Y
\end{aligned}
$$

into the Lagrangian (12.68), and replacing u_n by $u(x)$, we obtain

$$
L = \sum_x \Delta x \left[\frac{1}{2} \mu \dot{u}^2(x) - \frac{1}{2} Y \left\{ (u(x + \Delta x) - u(x))/\Delta x \right\}^2 \right].
$$

The equation of motion (12.69) then has the form

$$
\mu \Delta x \ddot{u}(x) - \frac{Y}{\Delta x} \left[u(x + \Delta x) - 2u(x) + u(x - \Delta x) \right] = 0.
$$

If now $\Delta x \to 0$ with fixed values of μ and Y, we find

$$
L = \int dx \mathcal{L},
$$

where

$$
\mathcal{L} = \frac{1}{2} \left[\mu \dot{u}^2 - Y \left(\frac{\partial u}{\partial x} \right)^2 \right] \tag{12.78}
$$

is the **Lagrangian density** $\mathcal{L}$ for the **displacement field** $u(x, t)$. As equation of motion we obtain in this limit (**E**)

$$
\frac{\partial^2 u}{\partial t^2} - v^2 \frac{\partial^2 u}{\partial x^2} = 0, \qquad v = \sqrt{Y/\mu}. \tag{12.79}
$$

This one-dimensional **wave equation** describes a **vibrating string** (with small displacements). Its general solution is

$$
u(x, t) = f(x - vt) + g(x + vt), \tag{12.80}
$$

where f and g are two arbitrary (at least twice differentiable) functions. $f(x - vt)$ and $g(x + vt)$, respectively, are deformations of the string that propagate at velocity v and $-v$ to $x = \infty$ and to $x = -\infty$, respectively.

In reality, a string has finite length l, and may be fixed on both ends, so that there are the boundary conditions

$$u(0,t) = u(l,t) = 0, \quad -\infty < t < \infty.$$

One may also prescribe an initial deformation of the string by

$$u(x,0) = \sigma(x) \quad \text{with} \quad \sigma(0) = \sigma(l) = 0,$$

and

$$\frac{\partial u}{\partial t}(x,0) = \tau(x) \quad \text{with} \quad \tau(0) = \tau(l) = 0.$$

Apart from the boundary conditions, the functions $\sigma(x)$ and $\tau(x)$ are arbitrary in the interval $(0,l)$. One can easily see that

$$u(x,t) = \frac{1}{2}\left[\sigma(x+vt) + \sigma(x-vt)\right] + \frac{1}{2v}\int_{x-vt}^{x+vt}\tau(s)\,ds \qquad (12.81)$$

is a solution of the wave equation (12.79) of the form (12.80). An example of the functions $\sigma(x)$ and $\tau(x)$ for the initial deformation of the string is the following. If the string is plucked at the point x_0, we have

$$\sigma(x) = \begin{cases} \dfrac{c}{x_0}x & 0 \le x \le x_0 \\[2mm] \dfrac{c}{l-x_0}(l-x) & x_0 \le x \le l \end{cases},$$

$$\tau(x) = 0, \qquad 0 \le x \le l.$$

Inserting this into Eq. (12.81), one immediately obtains the solution $u(x,t)$.

From the Lagrangian density (12.78) – the more general form is $\mathcal{L} = \mathcal{L}(u, \frac{\partial u}{\partial x}, \frac{\partial u}{\partial t}, x, t)$ – one obtains the equations of motion (12.79) of the displacements $u(x,t)$ via the prescription ([Goldstein]; [Fetter and Walecka])

$$\frac{\partial}{\partial t}\frac{\partial \mathcal{L}}{\partial(\partial u/\partial t)} + \frac{\partial}{\partial x}\frac{\partial \mathcal{L}}{\partial(\partial u/\partial x)} - \frac{\partial \mathcal{L}}{\partial u} = 0,$$

which is the generalization of Lagrange's equations for a Lagrangian $\mathcal{L}$. In our case, $\mathcal{L}$ does not depend on u, therefore the last term here vanishes and one recovers the wave equation.

The above presentation of the transition to **continuum mechanics** can only serve as an elementary introduction to the theories of elasticity and hydrodynamics. In both theories, the basic equations are often more complicated and usually nonlinear. Consequently, chaotic behavior enters into these theories, too.

Problems and examples

1. Show, for the characteristic equation (12.27), that:

    i) The frequencies ω_r are real;

    ii) The eigenvectors $\mathbf{s}^{(r)}$ can be chosen to be real;

    iii) The eigenvectors $\mathbf{s}^{(r)}$ satisfy the conditions $\mathbf{s}^{(r)}\mathbf{M}\mathbf{s}^{(s)} = \delta_{rs}$.

2. A point mass m_1 is coupled by a massless spring of strength k_1 to a fixed point. The mass m_1 is coupled to the point mass m_2 by a massless spring of strength k_2 (see sketch below). The equilibrium positions are located at $x_1 = l_1$ and $x_2 = l_1 + l_2$, respectively.

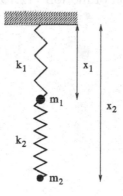

Now suppose gravity is turned on. Derive the Lagrangian for the displacements parallel to the gravitational acceleration $\mathbf{g}$ (x-axis). Determine the new equilibrium positions of the point masses, and the equations of motion for the displacements. Discuss the result for $m_1 = m_2$ and $k_1 = k_2$.

3. In a triatomic molecule, A_2B, the atoms are arranged linearly in their equilibrium positions (cf. sketch).

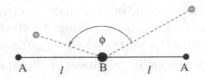

The potential energy of the molecule depends only on the distances AB and BA, as well as on the angle ϕ. The harmonic forces are of different strength along the axis and perpendicular to the axis of the molecule, k_1 and k_2, respectively. Determine

- the potential energy and the Lagrangian in the plane of the atoms;

- the eigenfrequencies and the normal coordinates.

4. Show that the vectors given by Eq. (12.63) are eigenvectors, and deduce the decomposition of the vibrating eigenmode shown in Fig. 12.8.

5. Perform the continuum limit for the one-dimensional chain, and deduce the equation of motion (12.79).

13

Hamiltonian mechanics

In Lagrangian mechanics, the equations of motions are Lagrange's equations for the action integral S (cf. Chapter 9). For the actual orbit, the action integral is extremal,

$$S = \int L dt = Extremum.$$

Starting from this principle (see Section 9.3.4), W.R. Hamilton developed the most unified and elegant account of classical dynamics. Hamilton's approach to classical mechanics was an essential prerequisite for the development of quantum mechanics – the theory of 'atomic mechanics' – that started to grow at the beginning of the 20th century.

13.1 Hamilton's equations of motion in phase space

The Lagrangian of a point mass with a single degree of freedom is a function of the *independent* variables q, $\dot{q}$, and t:

$$L = L(q, \dot{q}, t).$$

Emphasizing the conjugate momentum (9.77) in view of its of its importance to the existence of conserved quantities, one could summarize Lagrangian theory by the equations:

$$p = \frac{\partial L}{\partial \dot{q}}, \qquad \frac{d}{dt} p = \frac{\partial L}{\partial q}.$$

But this form is not particularly useful; one has only added a variable, namely p. Rather, it may be favorable to introduce q and p, instead of q and $\dot{q}$, as independent variables.

This change is achieved by applying a **Legendre transformation** to the Lagrangian L. In this transformation, the derivative of the function in question (L) with respect to one of its arguments $(\dot{q})$ is introduced as new variable (p); in the process, a new function $H(q, p, t)$ is constructed in place of $L(q, \dot{q}, t)$. To find H, we use $p = \partial L/\partial \dot{q}$ to rewrite the total (i.e. exact) differential of L,

$$dL = \frac{\partial L}{\partial q}dq + \frac{\partial L}{\partial \dot{q}}d\dot{q} + \frac{\partial L}{\partial t}dt, \tag{13.1}$$

in the form

$$dL = \frac{\partial L}{\partial q}dq + pd\dot{q} + \frac{\partial L}{\partial t}dt = \frac{\partial L}{\partial q}dq + d(p\dot{q}) - \dot{q}dp + \frac{\partial L}{\partial t}dt.$$

Collecting the total differentials appearing on both sides, we obtain

$$dH := d(p\dot{q} - L) = -\frac{\partial L}{\partial q}dq + \dot{q}dp - \frac{\partial L}{\partial t}dt, \tag{13.2}$$

dH is the total differential of the function $H = p\dot{q} - L$ (The sign is chosen by convention.). Eq. (13.2) shows that the independent variables[1] are q, p, and t. Comparing Eq. (13.2) to total differential of H,

$$dH = \frac{\partial H}{\partial q}dq + \frac{\partial H}{\partial p}dp + \frac{\partial H}{\partial t}dt, \tag{13.3}$$

yields the relations

$$\frac{\partial H}{\partial q} = -\frac{\partial L}{\partial q}, \qquad \dot{q} = \frac{\partial H}{\partial p}, \qquad \frac{\partial H}{\partial t} = -\frac{\partial L}{\partial t}. \tag{13.4}$$

These three equations are in fact generally true for any two functions related by a Legendre transformation. The first and the last relations are typical for all variables not affected by the transformation (here, these are q and t; the sign results from the choice of sign for H). Conversely, the Lagrangian L is the Legendre transform of H, in which one switches from independent variables q and p to the independent variables q and

---

[1] H is in fact independent of $\dot{q}$, since $\partial H/\partial \dot{q} = p - \partial L/\partial \dot{q} = 0$.

$\dot{q}$. The second equation in (13.4) plays the same role as $p = \partial L/\partial \dot{q}$ in the transformation from L to H.

The **Hamiltonian** H,

$$H(q,p,t) = p\dot{q}(q,p,t) - L(q,\dot{q}(q,p,t)t),\qquad (13.5)$$

is the *Legendre transform* of the *Lagrangian* L. Inserting Lagrange's equation, $\dfrac{\partial L}{\partial q} = \dfrac{d}{dt}\dfrac{\partial L}{\partial \dot{q}} = \dot{p}$, into the first condition in (13.4) gives *Hamilton's canonical equations*, or, briefly, **Hamilton's equations**:

$$\dot{q} = \frac{\partial H}{\partial p}$$

$$\dot{p} = -\frac{\partial H}{\partial q}\qquad (13.6)$$

and

$$\frac{\partial H}{\partial t} = -\frac{\partial L}{\partial t}.\qquad (13.7)$$

If the Lagrangian is a function of several dynamical variables q_i and $\dot{q}_i$, $i = 1,\ldots,f$, one proceeds in a quite analogous manner. An f-fold Legendre transformation – i.e. one that replaces the f (generalized) velocities $\dot{q}_i$ by the f (canonical) momenta $p_i = \partial L/\partial \dot{q}_i$ – causes the Lagrangian $L(\{q\},\{\dot{q}\},t)$ to be replaced by the Hamiltonian

$$H(\{q\},\{p\},t) = p_i\dot{q}_i - L(\{q\},\{\dot{q}\},t).\qquad (13.8)$$

(Note that the summation convention has been used here.) The velocities $\dot{q}_i$ in the $p_i\dot{q}_i$ term are now considered as functions of the coordinates q_i and the momenta p_j, $\dot{q}_i = \dot{q}_i(\{q\},\{p\},t)$. Hamilton's equations,

$$\dot{q}_i = \frac{\partial H}{\partial p_i}$$

$$\dot{p}_i = -\frac{\partial H}{\partial q_i}\qquad (13.9)$$

$$\frac{\partial H}{\partial t} = -\frac{\partial L}{\partial t},$$

$i = 1,\ldots,f$, control the temporal development of a classical system, quite analogous to Newton's equations or Lagrange's equations. In this form, the equations of motion are a system of first order partial differential equations for the $2f$ independent variables $\{q_i(t)\}$ and $\{p_i(t)\}$.

Hamilton's equations directly describe the dynamics in phase space. As in Lagrangian theory, a coordinate q_i is *cyclic*, if $\partial H/\partial q_i = 0$. From the second equation in (13.9), it then follows that $\dot{p}_i = 0$; i.e., the conjugate momentum is a constant of motion: $p_i = const.$

If the Lagrangian L does not depend explicitly on time t, since $\partial H/\partial t = -\partial L/\partial t$, we immediately have one constant of motion; namely, the *Jacobian integral* given in Eq. (10.67),

$$I_J = p_i\dot{q}_i - L(\{q\},\{\dot{q}\}) = H(\{q\},\{p\}). \qquad (13.10)$$

In this case, the Hamiltonian does not depend on time explicitly and is itself a first integral. If, in addition, the kinetic energy in generalized coordinates is a (positive definite) quadratic form in the velocities $\{\dot{q}\}$ (cf. Eq. (10.69)), i.e. the Lagrangian has the form

$$L = a_{ik}(\{q\})\,\dot{q}_i\dot{q}_k - V(\{q\}) = T - V, \quad a_{ik} = a_{ki},$$

then we have $p_k = 2a_{ik}(\{q\})\dot{q}_i$, and therefore

$$H = a_{ik}(\{q\})\,p_ip_k + V(\{q\}) = T + V = E.$$

H is equal to the total energy E of the system. One example is the three-dimensional motion of a point mass in a potential $V(\mathbf{r})$. The Lagrangian is

$$L(\mathbf{r},\dot{\mathbf{r}}) = m\dot{\mathbf{r}}^2/2 - V(\mathbf{r}),$$

and $\mathbf{p} = \partial L/\partial\dot{\mathbf{r}} = m\dot{\mathbf{r}}$. It then follows from Eq. (13.8) that the Hamiltonian of the point mass is given by

$$H = \mathbf{p}^2/2m + V(\mathbf{r}) = T + V = E. \qquad (13.11)$$

Below, two more examples illustrate the transition from the Lagrangian to the Hamiltonian view of classical mechanics.

13.1.1   A particle in a central force field

For a particle in a central force field $V(r)$, the motion is restricted to a fixed plane ($z = 0$). According to Eq. (9.79) the Lagrangian is (we denote ρ by r):

$$L = \frac{m}{2}\left(\dot{r}^2 + r^2\dot{\varphi}^2\right) - V(r).$$

The momenta conjugate to r and φ are

$$p_r = \frac{\partial L}{\partial \dot{r}} = m\dot{r} \quad \text{and} \quad p_\varphi = \frac{\partial L}{\partial \dot{\varphi}} = mr^2\dot{\varphi}. \qquad (13.12)$$

For these coordinates, the Hamiltonian has the general form $H = p_r\dot{r} + p_\varphi\dot{\varphi} - L$, which leads to

$$H = \frac{p_r^2}{2m} + \frac{p_\varphi^2}{2mr^2} + V(r). \qquad (13.13)$$

H is independent of the angle φ; φ is a cyclic variable. The canonical equations

$$\begin{aligned}
\dot{r} &= \frac{\partial H}{\partial p_r} = \frac{p_r}{m} \\[2mm]
\dot{\varphi} &= \frac{\partial H}{\partial p_\varphi} = \frac{p_\varphi}{mr^2} \\[2mm]
\dot{p}_r &= -\frac{\partial H}{\partial r} = -\frac{\partial}{\partial r}\left(\frac{p_\varphi^2}{2mr^2} + V\right) \\[2mm]
\dot{p}_\varphi &= -\frac{\partial H}{\partial \varphi} = 0 \quad \rightarrow \quad p_\varphi = const
\end{aligned} \qquad (13.14)$$

yield the conservation of the z-component of the angular momentum p_φ, as well as the familiar equations of motion in a central force field:

$$m\ddot{r} + \frac{\partial}{\partial r}\left(\frac{p_\varphi^2}{2mr^2} + V\right) = m\ddot{r} + \frac{\partial}{\partial r}V_{eff} = 0$$

$$mr^2\dot{\varphi} = p_\varphi = const.$$

13.1.2   The rigid body

Starting from the Lagrangian (cf. Eq. (11.41)

$$L = \frac{1}{2}\dot{\boldsymbol{\phi}}^T \mathsf{I}\Lambda\dot{\boldsymbol{\phi}} - V(\boldsymbol{\phi}),$$

where $\dot{\boldsymbol{\phi}}$ denotes the three-tuple of time derivatives of Euler's angles, $\boldsymbol{\phi} = (\varphi, \vartheta, \psi)$, I is the inertia tensor given in Eq. (11.29), Λ the matrix defined in Eq. (11.40), and $V(\phi)$ is the potential in terms of Euler's angles. The canonical momentum $\mathbf{p}_\phi = (p_\varphi, p_\vartheta, p_\psi)$ is

$$\mathbf{p}_\phi = \frac{\partial L}{\partial \dot{\phi}} = \Lambda^T I \Lambda \dot{\phi}. \tag{13.15}$$

In order to obtain the Hamiltonian, we need the inverse relation,

$$\dot{\phi} = \left(\Lambda^T I \Lambda\right)^{-1} \mathbf{p}_\phi.$$

It then follows from $H = \mathbf{p}_\phi \dot{\phi} - L$ that

$$
\begin{aligned}
H &= \mathbf{p}_\phi \left(\Lambda^T I \Lambda\right)^{-1} \mathbf{p}_\phi - \frac{1}{2}\mathbf{p}_\phi \left(\Lambda^T I^T \Lambda\right)^{-1} \left(\Lambda^T I \Lambda\right) \left(\Lambda^T I \Lambda\right)^{-1} \mathbf{p}_\phi \\
&\quad + V(\phi) \\
&= \mathbf{p}_\phi \left(\Lambda^T I \Lambda\right)^{-1} \mathbf{p}_\phi - \frac{1}{2}\mathbf{p}_\phi \left(\Lambda^T I^T \Lambda\right)^{-1} \mathbf{p}_\phi + V(\phi) \\
&= \frac{1}{2}\mathbf{p}_\phi \left(\Lambda^T I \Lambda\right)^{-1} \mathbf{p}_\phi + V(\phi).
\end{aligned}
$$

(I is diagonal, whence $I^T = I$.) Finally, we obtain the **Hamiltonian of a rigid body**,

$$H = \frac{1}{2}\mathbf{p}_\phi \Lambda^{-1} I^{-1} \left(\Lambda^T\right)^{-1} \mathbf{p}_\phi + V(\phi) \tag{13.16}$$

where

$$\Lambda^{-1} = \frac{1}{\sin\vartheta} \begin{pmatrix} \sin\psi & \cos\psi & 0 \\ \cos\psi\sin\vartheta & -\sin\psi\sin\vartheta & 0 \\ -\sin\psi\cos\vartheta & -\cos\psi\cos\vartheta & \sin\vartheta \end{pmatrix},$$

and

$$I^{-1} = \begin{pmatrix} 1/I_1 & 0 & 0 \\ 0 & 1/I_2 & 0 \\ 0 & 0 & 1/I_3 \end{pmatrix}.$$

For a free *symmetric body* ($I_1 = I_2$, $V = 0$), the Hamiltonian, Eq. (13.16), reduces to (E)

$$H = \frac{(p_\varphi - p_\psi \cos\vartheta)^2}{2I_1 \sin^2\vartheta} + \frac{1}{2I_1}p_\vartheta^2 + \frac{1}{2I_3}p_\psi^2. \tag{13.17}$$

The angles φ and ψ are cyclic variables. Therefore $\dot{p}_\varphi = -\partial H/\partial\varphi = 0$ and $\dot{p}_\psi = -\partial H/\partial\psi = 0$, imply that the momenta conjugate to φ and ψ are constant:

$$\begin{aligned} p_\varphi &= const \\ p_\psi &= const. \end{aligned} \tag{13.18}$$

The remaining canonical equations are

$$\begin{aligned} \dot{\varphi} &= \frac{\partial H}{\partial p_\varphi} = \frac{p_\varphi - p_\psi \cos\vartheta}{I_1 \sin^2\vartheta} \\ \dot{\vartheta} &= \frac{\partial H}{\partial p_\vartheta} = \frac{p_\vartheta}{I_1} \\ \dot{\psi} &= \frac{\partial H}{\partial p_\psi} = -\frac{(p_\varphi - p_\psi \cos\vartheta)\cos\vartheta}{I_1 \sin^2\vartheta} + \frac{p_\psi}{I_3} \\ \dot{p}_\vartheta &= -\frac{\partial H}{\partial\vartheta} = -\frac{(p_\varphi - p_\psi \cos\vartheta)(p_\psi - p_\varphi \cos\vartheta)}{I_1 \sin^3\vartheta}. \end{aligned} \tag{13.19}$$

If we choose the z-axis in the inertial frame to be parallel to the conserved angular momentum, the projection of the angular momentum onto the vector $\mathbf{k}$ (which is orthogonal to the z-axis, cf. Fig. 8.1) vanishes, i.e. $p_\vartheta = 0$. Thus, from the second equation above (for ϑ), we obtain

$$\vartheta = const.$$

Together with Eq. (13.18), we find from the first and the third canonical equations,

$$\begin{aligned} \dot{\varphi} &= const \\ \dot{\psi} &= const. \end{aligned}$$

With this choice of the z-axis, $p_\varphi = L_z \; (= |\mathbf{L}|)$, and p_ψ is the projection of $\mathbf{L}$ onto the symmetry axis in the inertial frame. The body then rotates with constant angular velocity $\dot{\psi}$ about its symmetry axis, and this axis, in turn, rotates about the z-axis with constant angular velocity $\dot{\varphi}$. This is the result obtained in Subsection 11.3.4.

13.1.3 Motion in a central force field and a homogeneous magnetic field

Taking the Lagrangian (9.69) for the motion of a charged particle in a potential and a homogeneous magnetic field, together with the canonical momentum (9.70), one may write down the Hamiltonian[2]:

$$H = \frac{\mathbf{p}^2}{2m} + \frac{q}{2mc}\mathbf{p}\left(\mathbf{r} \times \mathbf{B}\right) + \frac{q^2}{8mc^2}\left(\mathbf{r} \times \mathbf{B}\right)^2 + V(r). \tag{13.20}$$

Using cylindrical coordinates (with $\mathbf{B} = B\mathbf{e}_z$), we find for $\mathbf{p}$, expressed in the momenta p_ρ, p_φ, and p_z conjugate to ρ, φ, and z, respectively (E),

$$\mathbf{p} = p_\rho\mathbf{e}_\rho + \frac{p_\varphi}{\rho}\mathbf{e}_\varphi + p_z\mathbf{e}_z. \tag{13.21}$$

Therefore, setting $\omega_Z = qB/mc$, the Hamiltonian is given by

$$H = \frac{1}{2m}\left(p_\rho^2 + \frac{p_\varphi^2}{\rho^2}\right) + \frac{m}{2}\left(\frac{\omega_Z}{2}\right)^2\rho^2 - \frac{1}{2}\omega_Z p_\varphi + \frac{1}{2m}p_z^2 + V\left(\sqrt{\rho^2 + z^2}\right). \tag{13.22}$$

Since φ is a cyclic coordinate, the conjugate momentum p_φ is constant. Calculating p_φ yields (recall Eq. (9.73); E)

$$p_\varphi = I_{LB}. \tag{13.23}$$

In general, the total energy E – this follows from $\partial H/\partial t = 0$ (see also Subsection 13.4) – and p_φ are the only conserved quantities. This is suggested by numerical solutions of the canonical equations for $V \propto 1/r$ (M. Robnik, J. Phys. **A14**, 3195(1981)), as discussed below.

Let us first have a look at the case where no potential is present, $V(r) \equiv 0$. The Hamiltonian separates into two parts:

$$H = \frac{1}{2m}p_z^2 + H_\rho$$

where

$$H_\rho = \frac{1}{2m}\left(p_\rho^2 + \frac{p_\varphi^2}{\rho^2}\right) + \frac{m}{2}\left(\frac{\omega_Z}{2}\right)^2\rho^2 - \frac{1}{2}\omega_Z p_\varphi. \tag{13.24}$$

Since the canonical equations for z and ρ are not coupled either, these motions are independent (cf. also Section 5.5). In the z-direction we have free motion. The Hamiltonian H_ρ for the motion in the (x, y)-plane

---

[2]In comparing H with $E = m\mathbf{v}^2/2 + V(r)$ (cf. Eq. (5.62)), one must keep in mind that $\mathbf{p}$ is related to $\mathbf{v}$ via Eq. (9.70).

is of the same form as the Hamiltonian for a two-dimensional isotropic harmonic oscillator, with $p_\varphi \to L_z$ and $\omega_Z/2 \to \omega$ (E). Hence, the motion is integrable. But there is a difference between this system and the harmonic oscillator. In the case of the harmonic force (oscillator), the orbits are ellipses whose center is the center of force (cf. Section 4.1). In the case of the magnetic field, the orbits are circles whose centers have finite distances from the arbitrarily chosen origin (cf. Subsection 5.5.1). The difference in the orbits is due to the differing conserved quantities (in particular $\mathbf{u}_\rho$; see Section 5.5).

Consider now the case where a central force is superposed onto the magnetic field, i.e. we have now $V(r) \neq 0$. We observe the motion from the frame of reference rotating about the origin (the center of force) at the *Larmor frequency* (Eq. (8.65)), $\boldsymbol{\omega}_L = -(q/2mc)\,\mathbf{B}$. Using the Lagrangian (9.102), we find the Hamiltonian in the rotating frame to be

$$H' = \frac{\mathbf{p}'^2}{2m} + \frac{q^2}{8mc^2}\,(\mathbf{r} \times \mathbf{B})^2 + V(r). \tag{13.25}$$

Note that – compared to Eq. (13.20) – there is now only *one* term in addition to the potential in the Hamiltonian. Moreover, this term depends only on $\mathbf{r}$, thus merely changing the potential. In the plane perpendicular to $\mathbf{B}$ passing through the center of the potential, this term takes the form of a harmonic potential (E).

Due to the conservation of $p_\varphi = I_{LB}$ (cf. Eq. (13.23)), the system has only two degrees of freedom: in cylindrical coordinates, these correspond to ρ and z. So, for an attractive Coulomb potential $V(r) = -k/r$, $k > 0$, the Hamiltonian in the rotating frame is given by

$$H' = \frac{1}{2m}\left(p_\rho^2 + p_z^2\right) + \frac{I_{LB}^2}{2m\rho^2} + \frac{m}{8}\omega_Z^2\rho^2 - \frac{k}{\sqrt{\rho^2 + z^2}}. \tag{13.26}$$

Choosing the units such that $m = 1$, we have the following canonical equations:

$$\dot{\rho} = \frac{\partial H'}{\partial p_\rho} = p_\rho$$

$$\dot{z} = \frac{\partial H'}{\partial p_z} = p_z$$

$$\dot{p}_\rho = -\frac{\partial H'}{\partial \rho} = \frac{I_{LB}^2}{\rho^3} - \frac{1}{4}\omega_Z^2\rho - \frac{k\rho}{\sqrt{\rho^2 + z^2}^3} \tag{13.27}$$

$$\dot{p}_z = -\frac{\partial H'}{\partial z} = -\frac{kz}{\sqrt{\rho^2 + z^2}^3}.$$

Apart from the energy,

$$E = \frac{1}{2}\left(p_\rho^2 + p_z^2\right) + \frac{I_{LB}^2}{2\rho^2} + \frac{1}{8}\omega_Z^2\rho^2 - \frac{k}{\sqrt{\rho^2 + z^2}}, \qquad (13.28)$$

no further integral of the motion can be found. Chaotic motion can therefore not be excluded[3] (cf. Section 14.1). We study these equations for the same set of parameters used in Robnik's paper (mentioned in Subsection 5.5.2 and above),

$$I_{LB} = \omega_Z = k = 1. \qquad (13.29)$$

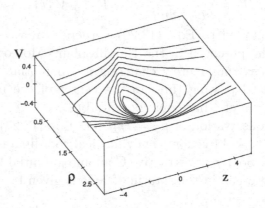

Figure 13.1: The effective potential $V_{eff}(\rho, z) = \frac{1}{2\rho^2} + \frac{1}{8}\rho^2 - \frac{1}{\sqrt{\rho^2+z^2}}$.

The effective potential,

$$V_{eff}(\rho, z) = \frac{1}{2\rho^2} + \frac{1}{8}\rho^2 - \frac{1}{\sqrt{\rho^2 + z^2}}, \qquad (13.30)$$

---

[3]In Quantum Mechanics, the Hamiltonian (13.20) is also used to describe a hydrogen atom in a constant magnetic field. There, for medium values of the magnetic field, the energy levels are distributed in a very strange way (H. Friedrich and D. Wintgen, Phys. Rep. **183**, 37 (1989)).

has an absolute minimum at $z = 0$ and $\rho = .86198 =: \rho_{min}$,

$$V_{eff}(\rho_{min}, 0) = -.39430.$$

Up to energy $E = 0.5$ (in the units fixed by the choice in Eq. (13.29) and $m = 1$), the motion remains bounded in the z-direction[4]. For $E \geq 0.5$, the particle can escape to $z = \infty$. Figure 13.1 shows $V_{eff}(\rho, z)$ in the range where bounded orbits exist, $-.39430 < V < 0.5$.

The dynamics in the four-dimensional phase space (ρ, z, p_ρ, p_z) can be readily represented by a Poincaré map – i.e., by the sequences of intersection points of phase space trajectories with a suitably chosen Poincaré plane (see Chapter 4). Using Eq. (13.28), one can eliminate the coordinate p_z in phase space,

$$p_z = p_z(E, \rho, p_\rho, z).$$

In the remaining independent coordinates (ρ, z, p_ρ) of reduced phase space, we choose the plane

$$z = 0$$

as the *Poincaré surface*, so that the intersection points of the trajectories at positions (ρ, p_ρ) are uniquely fixed by the requirement that $p_z = p_z(E, \rho, p_\rho, z = 0) > 0$. It follows from Eq. (13.28) that the values of ρ and p_ρ lie in the Poincaré plane in the region bounded by the curve

$$p_\rho^2 = 2E - \frac{1}{\rho^2} - \frac{1}{4}\rho^2 + \frac{2}{\rho}. \tag{13.31}$$

Results of the numerical integration for increasing values of the energy E are presented in Figs. 13.2 and 13.3. As in the Hénon-Heiles

---

[4]In order for such motion to occur at all, we must have $E \geq V(\rho, z)$. Since, for all values of z,

$$V(\rho, \infty) = \frac{1}{2\rho^2} + \frac{1}{8}\rho^2 \geq V(\rho, z),$$

and since $V(\rho, \infty)$ has a minimum at $\rho = \sqrt{2}$ with value

$$V\left(\rho = \sqrt{2}, \infty\right) = \frac{1}{2},$$

the energy E required for unbounded motion satisfies

$$E \geq \frac{1}{2} =: E_{esc}.$$

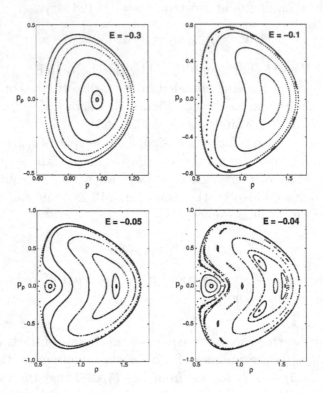

Figure 13.2: Numerically calculated Poincaré sections for the motion in the effective potential.

system (Section 4.2), for small energies, $E < 0.04$, the system seems to be integrable. But at $E = 0.04$, one can clearly recognize the onset of chaotic behavior (see Fig. 13.2). A typical sign is the formation of 'islands' and the disintegration of curves on which the intersection points seem to lie. If the energy is increased further, the chaotic region grows (Fig. 13.3; see also Robnik's results, *loc. cit.*). The return points of *one* trajectory are distributed over an ever increasing part of the region bounded by the curve given by Eq. (13.31), in a completely irregular manner.

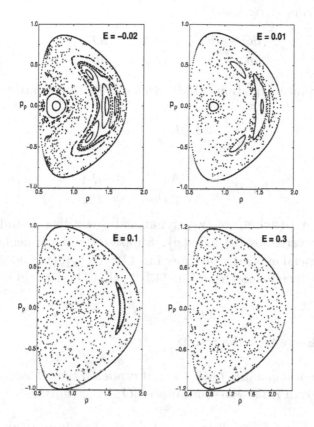

Figure 13.3: At higher energies, the Poincaré sections show the increasing chaotic behavior.

13.2 Poisson brackets

Except for a difference in sign in the equations of motion (13.9), the coordinates and the momenta are on a completely equal footing. Therefore, these are often simply referred to as **canonically conjugate variables** (or briefly, canonical variables, or conjugate variables). In the following formulation of the canonical equations, the variables are completely interchangeable.

Let $O = O(\{q\}, \{p\}, t)$ denote a **dynamical quantity** – that is, a function of the dynamical variables $\{q\}$ and $\{p\}$ with some particular physical meaning (e.g. the energy (Hamiltonian), angular momentum,...). The time derivative of O is

$$\frac{dO}{dt} = \frac{\partial O}{\partial q_i}\dot{q}_i + \frac{\partial O}{\partial p_i}\dot{p}_i + \frac{\partial O}{\partial t}.$$

Using the canonical equations (13.9), this can be recast into the form

$$\frac{dO}{dt} = [O, H]_{q,p} + \frac{\partial O}{\partial t}, \qquad (13.32)$$

where

$$[A, B]_{q,p} := \frac{\partial A}{\partial q_i}\frac{\partial B}{\partial p_i} - \frac{\partial B}{\partial q_i}\frac{\partial A}{\partial p_i} \qquad (13.33)$$

is the **Poisson bracket**[5] of two dynamical quantities A and B with respect to the variables $\{q\}$ and $\{p\}$. Suppose the dynamical quantity O does not depend explicitly on time; i.e., $O = O(\{q\}, \{p\})$, and we have $\partial O/\partial t = 0$. It then follows from Eq. (13.32) that O is time independent, $dO/dt = 0$, if

$$[O, H]_{q,p} = 0. \qquad (13.34)$$

In other words, we have the result:

> *A dynamical quantity O that is not explicitly time-dependent is conserved if the Poisson bracket $[O, H]_{q,p}$ vanishes.*

For the canonical variables q_i and p_i, the **fundamental Poisson brackets** are

$$[q_i, q_j]_{q,p} = [p_i, p_j]_{q,p} = 0, \qquad [q_i, p_j]_{q,p} = \delta_{ij}. \qquad (13.35)$$

---

[5]Some useful relations and properties of the Poisson brackets are given in the following. (We omit the indices q and p, because the relations hold independent of the conjugate variables considered.)

$$
\begin{aligned}
[A, A] &= 0 \\
[A, B] &= -[B, A] \\
[A, B + C] &= [A, B] + [A, C] \\
[A, BC] &= [A, B]C + B[A, C].
\end{aligned}
$$

In addition, **Jacobi's identity** states that

$$[A, [B, C]] + [B, [C, A]] + [C, [A, B]] = 0.$$

For any dynamical quantity O, Eq. (13.33) implies that

$$[q_i, O]_{q,p} = \frac{\partial O}{\partial p_i} \quad \text{and} \quad [p_i, O]_{q,p} = -\frac{\partial O}{\partial q_i}.$$

Setting $O = H$ in these relations, we obtain another form of Hamilton's equations (13.9); the equations are expressed in terms of Poisson brackets; namely,

$$\dot{q}_i = [q_i, H]_{q,p} \quad \text{and} \quad \dot{p}_i = [p_i, H]_{q,p}. \tag{13.36}$$

These equations are completely symmetric with respect to $\{q\}$ and $\{p\}$.

A particle in a central force field

From Eq. (13.13), the Hamiltonian is

$$H = \frac{p_r^2}{2m} + \frac{p_\varphi^2}{2mr^2} + V(r).$$

In terms of the canonical variables r, φ, p_r, and p_φ, the Poisson bracket of two quantities A and B is given by

$$[A, B] = \frac{\partial A}{\partial r}\frac{\partial B}{\partial p_r} - \frac{\partial B}{\partial r}\frac{\partial A}{\partial p_r} + \frac{\partial A}{\partial \varphi}\frac{\partial B}{\partial p_\varphi} - \frac{\partial B}{\partial \varphi}\frac{\partial A}{\partial p_\varphi}.$$

Hence, expressed in terms of Poisson brackets, the canonical equations (13.9) and (13.36), respectively, are

$$\begin{aligned}
\dot{r} &= [r, H] = \frac{\partial H}{\partial p_r} = \frac{1}{m}p_r, \\
\dot{\varphi} &= [\varphi, H] = \frac{\partial H}{\partial p_\varphi} = \frac{p_\varphi}{mr^2},
\end{aligned} \tag{13.37}$$

$$\begin{aligned}
\dot{p}_r &= [p_r, H] = -\frac{\partial H}{\partial r} = \frac{p_\varphi^2}{mr^3} - \frac{\partial V}{\partial r}, \\
\dot{p}_\varphi &= [p_\varphi, H] = -\frac{\partial H}{\partial \varphi} = 0.
\end{aligned} \tag{13.38}$$

The second equation (13.38) reflects the fact that φ is a cyclic coordinate. Together with the second equation (13.37), we obtain the law of conservation of the z-component of the angular momentum ($L_z = mr^2\dot{\varphi}$):

$$L_z = p_\varphi = const.$$

For the $1/r$ potential, the angle $\varphi_K = I_K$ of the *Runge-Lenz vector* (cf. Section 5.4) is another conserved quantity. Expressed in terms of the canonical variables, I_K reads

$$I_K\left(r,\varphi,p_r,p_\varphi\right) = \varphi - \arctan\frac{rp_r}{p_\varphi - m^2kr/L}. \tag{13.39}$$

It is easy to see that

$$[L_z, I_K] = [p_\varphi, \varphi] = -1. \tag{13.40}$$

13.3   Canonical transformations

In a system with Lagrangian $L\left(\{q\},\{\dot{q}\},t\right)$ or Hamiltonian $H\left(\{q\},\{p\},t\right)$, it can often be preferable for various reasons (e.g. symmetry) to introduce new coordinates Q_i,

$$Q_i = Q_i\left(\{q\},t\right), \qquad i = 1,...,f . \tag{13.41}$$

For the Lagrangian case, the coordinate transformation gives rise to a transformation of the Lagrangian from $L\left(\{q\},\{\dot{q}\},t\right)$ to $\hat{L}\left(\{Q\},\{\dot{Q}\},t\right)$. This, in turn, induces a transformation of the momenta p_i to the new momenta P_i,

$$P_i = \partial\hat{L}/\partial\dot{Q}_i.$$

One can now determine the new Hamiltonian[6] $\hat{H}(\{Q\},\{P\},t) = P_i\dot{Q}_i - \hat{L}$.

More generally, we consider transformations of the old coordinates and momenta to

$$\begin{aligned} Q_i &= Q_i(\{q\},\{p\},t) \\ P_i &= P_i(\{q\},\{p\},t). \end{aligned} \tag{13.42}$$

---

[6]Note, that applying the transformation (13.41) directly to $H(\{q\},\{p\},t)$ ($= p_i\dot{q}_i - L\left(\{q\},\{\dot{q}\},t\right)$, with $p_i = \partial L/\partial\dot{q}_i$), gives a function not necessarily equal to $\hat{H}(\{Q\},\{P\},t)$.

It is by no means obvious that the transformed Hamiltonian $\tilde{H}(\{Q\}, \{P\}, t)$, which is obtained from $H(\{q\}, \{p\}, t)$ by inserting $q_i = q_i(\{Q\}, \{P\}, t)$ and $p_i = p_i(\{Q\}, \{P\}, t)$, plays the role of a Hamiltonian in the new the coordinates $\{Q\}$ and $\{P\}$. This leads to the following definition. *A transformation (13.42) of generalized coordinates and momenta is called* **canonical** *if the new coordinates $\{Q\}$ and $\{P\}$ also obey the canonical equations;* i.e. if there exists a function $\hat{H}(\{Q\}, \{P\}, t)$ so that

$$\dot{Q}_i = \frac{\partial \hat{H}}{\partial P_i} = \left[Q_i, \hat{H} \right]_{Q,P}$$

$$\dot{P}_i = -\frac{\partial \hat{H}}{\partial Q_i} = \left[P_i, \hat{H} \right]_{Q,P}. \tag{13.43}$$

The conditions for a transformation to be canonical will be deduced presently.

13.3.1 The generating function of a transformation

The *action integral* is defined as (see Eq. (10.65)) $S = \int_{t_1}^{t_2} dt\, L$. Expressing the Lagrangian in terms of the Hamiltonian, the action integral takes the form

$$S = \int_{t_1}^{t_2} (\dot{q}_i p_i - H(\{q\}, \{p\}, t))\, dt. \tag{13.44}$$

Applying the **principle of least action**, $\delta S = 0$ – varying S with respect to the independent variables $\{q\}$ and $\{p\}$ – one obtains the canonical equations (E). When varying the $\dot{q}_i p_i$-term, one should keep in mind that adding a constant to the action integral does not change the position at which it is extremal. Hence, one can always add a total time derivative of some function $F(\{q\}, \{p\})$ to the integrand. In particular, one can add $d(q_i p_i)/dt$, so that the action integral $\int (-q_i \dot{p}_i - H(\{q\}, \{p\}, t))\, dt$ is equivalent to $\int (\dot{q}_i p_i - H(\{q\}, \{p\}, t))\, dt$. This observation is useful in the variation with respect to p_i.

Under *canonical* transformations, the form of the canonical equations does not change (by definition). Therefore, the least action principle must be invariant under canonical transformations, too. Consequently, after expressing S on the one side in terms of the old variables and on the other in terms of the new variables, one must have

$$\delta \int [p_i dq_i - H(\{q\}, \{p\}, t) \, dt] = \delta \int \left[P_i dQ_i - \hat{H}(\{Q\}, \{P\}, t) \, dt \right].$$
(13.45)

As mentioned above, one can add an exact differential of a function F to the integrands without changing the variational (i.e. the canonical) equations[7]. Thus the integrands are equal up to dF:

$$p_i dq_i - H(\{q\}, \{p\}, t) \, dt = P_i dQ_i - \hat{H}(\{Q\}, \{P\}, t) \, dt + dF. \quad (13.46)$$

The function F may now depend on 'old' *as well as* on 'new' dynamical variables, but only $2f$ of the $4f$ variables $\{q\}$, $\{p\}$, $\{Q\}$, and $\{P\}$ are independent.

Suppose that F depends on $\{q\}$ and $\{Q\}$, $F = F_1(\{q\}, \{Q\}, t)$. Then Eq. (13.46) implies that

$$dF_1 = p_i dq_i - P_i dQ_i - (H - \hat{H}) dt. \quad (13.47)$$

Since dF_1 is an exact differential, we have

$$dF_1 = \frac{\partial F_1}{\partial q_i} dq_i + \frac{\partial F_1}{\partial Q_i} dQ_i + \frac{\partial F_1}{\partial t} dt.$$

Comparing this to Eq. (13.47), we see that

$$p_i \quad = \quad \frac{\partial F_1}{\partial q_i} \quad\quad\quad\quad (13.48)$$

$$P_i \quad = \quad -\frac{\partial F_1}{\partial Q_i} \quad\quad\quad (13.49)$$

$$\hat{H} \quad = \quad H + \frac{\partial F_1}{\partial t}. \quad\quad\quad (13.50)$$

If $F_1 = F_1(\{q\}, \{Q\}, t)$ is given, then these are the equations of the canonical transformation in the form

$$p_i \quad = \quad p_i(\{q\}, \{Q\}, t)$$
$$P_i \quad = \quad P_i(\{q\}, \{Q\}, t).$$

---

[7]In Section 10.4, we showed that the Lagrangian in a given set of coordinates $\{q\}$ is fixed up to a total time derivative of an arbitrary (gauge) function $F(\{q\}, t)$; i.e., $\hat{L} = L + dF/dt$ is also a Lagrangian, and generates the same dynamics as L.

Suppose now that $F = F_2 (\{q\}, \{P\}, t) - Q_i P_i$, i.e. $dF = dF_2 - d(Q_i P_i)$. Then Eq. (13.46) together with the fact that dF_2 is an exact differential, imply that

$$\begin{aligned} dF_2 &= p_i dq_i + Q_i dP_i - (H - \hat{H}) dt \\ &= \frac{\partial F_2}{\partial q_i} dq_i + \frac{\partial F_2}{\partial P_i} dP_i + \frac{\partial F_2}{\partial t} dt, \end{aligned} \qquad (13.51)$$

which by comparison yields the equations

$$p_i = \frac{\partial F_2}{\partial q_i} \qquad (13.52)$$

$$Q_i = \frac{\partial F_2}{\partial P_i} \qquad (13.53)$$

$$\hat{H} = H + \frac{\partial F_2}{\partial t}. \qquad (13.54)$$

Now the canonical transformation has the form

$$\begin{aligned} p_i &= p_i (\{q\}, \{P\}, t) \\ Q_i &= Q_i (\{q\}, \{P\}, t) . \end{aligned}$$

There are two further possible combinations of variables in F. In these cases, one sets in Eq. (13.46) $dF = dF_3 (\{p\}, \{Q\}, t) + d(q_i p_i)$ or $dF = dF_4 (\{p\}, \{P\}, t) + d(q_i p_i) - d(Q_i P_i)$. The respective equations involving $F_3 (\{p\}, \{Q\}, t)$ and $F_4 (\{p\}, \{P\}, t)$ are easy to derive (E).

The functions F_1, F_2, F_3, F_4 are called **generating functions** (of canonical transformations). The standard notation for the functions is

$$\begin{aligned} F_1 &= F_1 (\{q\}, \{Q\}, t) , & F_2 &= F_2 (\{q\}, \{P\}, t) \\ F_3 &= F_3 (\{p\}, \{Q\}, t) , & F_4 &= F_4 (\{p\}, \{P\}, t) . \end{aligned} \qquad (13.55)$$

We will mainly be concerned with generating functions of the kind F_2.

Given a function F in one of the four forms presented in (13.55), the transformation from $\{q\}, \{p\}$ to the new variables $\{Q\}, \{P\}$ is canonical, and the new Hamiltonian $\hat{H} (\{Q\}, \{P\}, t)$ is obtained from the appropriate equation (for example, from Eq. (13.54), in which one must substitute $q_i = q_i(\{Q\}, \{P\}, t)$ and $p_i = p_i(\{Q\}, \{P\}, t)$ into the right hand side). If the generating function F – and therefore the canonical transformation – does not explicitly depend on time, the respective third conditions (e.g. Eqs. (13.50) or (13.54)) reduce to

$$\hat{H} (\{Q\}, \{P\}, t) = H (\{q(\{Q\}, \{P\})\}, \{p(\{Q\}, \{P\})\}, t) . \qquad (13.56)$$

A particular class of canonical transformations is the class of *transformations of the coordinates* (Eq. (13.41)) only, also called **point transformations**. The generating function of a point transformation has the form

$$F_2(\{q\}, \{P\}, t) = \sum_{i=1}^{f} Q_i(\{q\}, t) P_i. \qquad (13.57)$$

By Eq. (13.53), the coordinate transformation is

$$Q_i = Q_i(\{q\}, t), \qquad i = 1, \ldots, f, \qquad (13.58)$$

and Eqs. (13.52), (13.54) give

$$p_i = \sum_{j=1}^{f} P_j \frac{\partial}{\partial q_i} Q_j(\{q\}, t) \qquad (13.59)$$

$$\hat{H} = H + \sum_{j=1}^{f} P_j \frac{\partial}{\partial t} Q_j(\{q\}, t). \qquad (13.60)$$

The f linear relations (13.59) between the momenta p_i and the momenta P_j can be inverted,

$$P_j = \sum_{i=1}^{f} p_i \frac{h_{ij}(\{q\}, t)}{\det[\partial Q/\partial q]},$$

where the h_{ij} are some coefficients, if the determinant of the matrix $[\partial Q_k/\partial q_l]$ does not vanish; i.e., if the coordinate transformation (13.58) is invertible.

We point out two particular generating functions that are of interest. The first is generated by the function

$$F_1 = q_i Q_i. \qquad (13.61)$$

The equations of the transformation are (cf. Eqs. (13.48) – (13.50))

$$p_i = Q_i, \qquad P_i = -q_i, \qquad \text{and} \qquad \hat{H} = H. \qquad (13.62)$$

The canonical transformation,

$$Q_i = p_i$$
$$P_i = -q_i$$

$$\hat{H}\left(\{Q\},\{P\},t\right) = H\left(\{-P\},\{Q\},t\right),$$

is just an *interchange of the coordinates and the momenta*. This transformation shows also that in Hamilton's theory, the conceptual difference between coordinates and momenta is meaningless. However, these names are kept for convenience.

The second transformation is generated by

$$F_2 = q_i P_i. \tag{13.63}$$

Equations (13.52), (13.53), and (13.54) imply that

$$p_i = P_i, \qquad Q_i = q_i, \qquad \text{and} \qquad \hat{H} = H. \tag{13.64}$$

This is the *identity transformation*. (Also $F_3 = -Q_i p_i$ generates the identity transformation.) This generating function is the starting point of infinitesimal canonical transformations, i.e. transformations that differ only slightly from the identity transformation (see below).

Exemplification: The harmonic oscillator

The Hamiltonian of a one-dimensional harmonic oscillator is

$$H = p^2/2m + m\omega^2 x^2/2.$$

If we apply the canonical transformation generated by F_1,

$$F_1(x,Q) = \frac{1}{2}m\omega x^2 \cot Q,$$

the equations of the transformations (obtained from Eqs. (13.48) and (13.49)) are

$$p = \frac{\partial F_1}{\partial x} = m\omega x \cot Q \qquad \text{and} \qquad P = -\frac{\partial F_1}{\partial Q} = \frac{m\omega x^2}{2}\frac{1}{\sin^2 Q}.$$

The inverse transformation is

$$x = \sqrt{\frac{2P}{m\omega}}\sin Q \qquad \text{and} \qquad p = \sqrt{2m\omega P}\cos Q.$$

Inserting this into H, we obtain (since $\partial F_1/\partial t = 0$)

$$\hat{H} = \omega P. \tag{13.65}$$

The new Hamiltonian $\hat{H}$ is a much simpler function of (Q, P) than H in terms of the old variables (q, p). Moreover, Q is a cyclic coordinate; hence $\dot{P}$ $(= -\partial H/\partial Q) = 0$; i.e. $P = const$, and (since $\hat{H} = H = E$)

$$P = E/\omega. \tag{13.66}$$

Furthermore, relation $\dot{Q} = \dfrac{\partial H}{\partial P} = \omega$ implies

$$Q = \omega t + \varphi, \tag{13.67}$$

which yields the solution in terms of x and p:

$$x = \sqrt{\frac{2E}{m\omega^2}} \sin(\omega t + \varphi), \qquad p = \sqrt{2mE} \cos(\omega t + \varphi).$$

In the coordinates (Q, P) the Hamiltonian (13.65) and the solutions of the canonical equations, (13.66) and (13.67), assume particularly simple (standard) forms (cf. Section 14.1).

13.3.2   Canonical invariants

Of particular interest are quantities that are invariant under canonical transformations. The fundamental Poisson brackets, Eq. (13.35), are invariant under canonical transformations. Moreover, one can proof the following theorem (for the proof, see e.g. [Goldstein]):

> *A transformation is canonical, if and only if the fundamental*
> *Poisson brackets are invariant; i.e., if*

$$[Q_i, P_i]_{q,p} = \delta_{ij}, \qquad [Q_i, Q_j]_{q,p} = [P_i, P_j]_{q,p} = 0. \quad (13.68)$$

This theorem implies that the canonical property of a transformation does not depend on the particular Hamiltonian. If a transformation is canonical, it is canonical for every Hamiltonian. A consequence of Eq. (13.68) is the following relation between two dynamical functions A and B (E):

$$[A, B]_{q,p} = [A, B]_{Q,P}, \tag{13.69}$$

where A and B are considered as functions of $(\{q\}, \{p\})$ and $(\{Q\}, \{P\})$ in the respective Poisson bracket. We therefore need not specify the set of dynamical variables in Poisson brackets, and will omit the indices from now on.

Other quantities that are invariant under canonical transformations are the **Poincaré's invariants**. Among these invariants we consider only the *volume element in phase space* $\prod_{i=1}^{f} dq_i dp_i$. Its invariance rests on the following property of the **Jacobian determinant** of a canonical transformation[8]:

$$D := \det \left(\frac{\partial(Q_1, \ldots, Q_f, P_1, \ldots, P_f)}{\partial(q_1, \ldots, q_f, p_1, \ldots, p_f)} \right) = 1. \tag{13.70}$$

Consequently, the volume element is transformed as

$$\prod_{i=1}^{f} dQ_i dP_i = D \prod_{i=1}^{f} dq_i dp_i = \prod_{i=1}^{f} dq_i dp_i. \tag{13.71}$$

Considering, in particular, the evolution of a system in time, the fact that this is also a canonical transformation (see below) implies that the magnitude of a (finite) volume in phase space is conserved (see also Liouville's theorem regarding the volume in phase space, in Subsection 13.5). This statement is relevant for chaotic behavior in Hamiltonian systems.

13.3.3 Infinitesimal canonical transformations

In Section 10.4, we studied infinitesimal transformations in configuration space (recall Eq. (10.72)). They turned out to be useful in establishing the connection between (continuous) symmetry[9] and conservation law. We now extend the concept of infinitesimal transformation to phase space, allowing for a transformation of coordinates and momenta. An infinitesimal transformation has the form

$$\begin{aligned} Q_i &= q_i + \varepsilon f_i(\{q\}, \{p\}) \\ P_i &= p_i + \varepsilon g_i(\{q\}, \{p\}), \end{aligned} \tag{13.72}$$

---

[8] A proof is given in [Landau/Lifshitz]. For an infinitesimal canonical transformation, this is shown in Subsection 13.5.

[9] An example of continuous symmetry is rotational symmetry about an axis. The rotation angle appearing in the transformation can assume any value; in particular, setting the angle to zero, yields the identity transformation.

where the functions f_i and g_i also depend on a set of parameters α_i, $i = 1, \ldots, r$ (as in Eq. (10.72)), and the parameter ε is infinitesimally small. If this transformation is canonical, it will be generated by a function infinitesimally close to the generating function for the identity transformation. We look for a generating function of the kind F_2, since the identity transformation is also induced by a generating function of that kind (see Eq. (13.63)). We therefore try the generating function

$$F_2 = q_i P_i + \varepsilon J(\{q\}, \{P\}), \tag{13.73}$$

where $J(\{q\}, \{P\})$ has to be determined. Using Eqs. (13.52), (13.53) yields

$$p_i = \frac{\partial F_2}{\partial q_i} = P_i + \varepsilon \frac{\partial J}{\partial q_i} \quad \text{and} \quad Q_i = \frac{\partial F_2}{\partial P_i} = q_i + \varepsilon \frac{\partial J}{\partial P_i}.$$

Comparing with Eqs. (13.72) shows that we have the conditions for J

$$g_i = -\frac{\partial J}{\partial q_i} \quad \text{and} \quad f_i = \frac{\partial J}{\partial P_i}. \tag{13.74}$$

If such a function J exists, then the infinitesimal transformation (13.72) is canonical. J is the **generator** *of an infinitesimal canonical transformation.*

On the other hand, if F_2 – in particular J – is given, the corresponding canonical transformation follows from Eqs. (13.52), (13.53). Since P_i and p_i differ only by order $\mathcal{O}(\varepsilon)$, we also have

$$J(\{q\}, \{P\}) \simeq J(\{q\}, \{p\}) \quad \text{and} \quad f_i \simeq \frac{\partial J}{\partial p_i}. \tag{13.75}$$

The change in a dynamical quantity $O(\{q\}, \{p\})$ subject to the infinitesimal canonical transformation (13.72) is

$$\delta O = O(\{q + \varepsilon f\}, \{p + \varepsilon g\}) - O(\{q\}, \{p\}) = \varepsilon \left(\frac{\partial O}{\partial q_i} f_i + \frac{\partial O}{\partial p_i} g_i \right).$$

Inserting Eq. (13.74) into this expression, and observing Eq. (13.75), gives the infinitesimal change in O induced by the generator J:

$$\delta O = \varepsilon [O, J]. \tag{13.76}$$

In the following, we give examples of infinitesimal canonical transformations.

i) Rotations

Consider a *time independent rotation* of the coordinate axes by the infinitesimal angle $\varepsilon\varphi$. Let φ be the vector in the axis of rotation, so that $\varphi = |\varphi|$. The components $\mathbf{r}$ of the radius vector of the particle are transformed to $\mathbf{r}'$ according to (cf. Eq. (8.20)):

$$\mathbf{r}' = \mathbf{r} + \varepsilon(\varphi \times \mathbf{r}). \tag{13.77}$$

(The q_i are now regarded as the Cartesian components of $\mathbf{r}$, and the Q_i the Cartesian components of $\mathbf{r}'$.) Rotation of the coordinate system also affects the conjugate momentum vector[10] $\mathbf{p}$:

$$\mathbf{p}' = \mathbf{p} + \varepsilon(\varphi \times \mathbf{p}) \tag{13.78}$$

According to Eq. (13.74), we have

$$\mathbf{r}' - \mathbf{r} = \varepsilon(\varphi \times \mathbf{r}) = \varepsilon\frac{\partial J}{\partial \mathbf{p}}$$
$$\mathbf{p}' - \mathbf{p} = \varepsilon(\varphi \times \mathbf{p}) = -\varepsilon\frac{\partial J}{\partial \mathbf{r}}. \tag{13.79}$$

The conditions (13.74) read now $\partial J/\partial \mathbf{p} = (\varphi \times \mathbf{r})$ and $\partial J/\partial \mathbf{r} = -(\varphi \times \mathbf{p})$; they can be integrated to give

$$J = -\mathbf{r}(\varphi \times \mathbf{p}) = \varphi(\mathbf{r} \times \mathbf{p}) = \varphi\mathbf{L}. \tag{13.80}$$

Therefore, infinitesimal rotations of the coordinate system are canonical transformations. *The component of the angular momentum parallel to the axis of rotation is the generating function of an infinitesimal rotation.*

ii) Translations

Likewise, one can show for an *infinitesimal translation* proportional to the vector $\mathbf{s}$, $\mathbf{r}' = \mathbf{r} + \varepsilon\mathbf{s}$ and $\mathbf{p}' = \mathbf{p}$, there exists a generator (E),

$$J = \mathbf{sp}. \tag{13.81}$$

Therefore, infinitesimal translations of the coordinate system are canonical transformations. *The component of the momentum parallel to a vector is the generator of an infinitesimal translation in the direction of this vector.*

---

[10]This can also be seen from Eq. (13.77) by multiplying the time derivative of the equation by the mass.

iii) Translations in time

Also the *evolution* of a system *in time* can be considered as a canonical transformation. To show this, we set $J = H(\{q\}, \{p\})$ and $\varepsilon = dt$ in Eq. (13.73). Then Eqs. (13.72) and (13.74) imply that

$$Q_i = q_i + \frac{\partial H}{\partial p_i} dt$$

$$P_i = p_i - \frac{\partial H}{\partial q_i} dt. \tag{13.82}$$

Using the canonical equations (13.9), we get

$$Q_i = q_i + \dot{q}_i dt \simeq q_i(t + dt)$$
$$P_i = p_i + \dot{p}_i dt \simeq p_i(t + dt). \tag{13.83}$$

The Hamiltonian is the generator of infinitesimal translations in time of the coordinates and of the momenta[11].

Since the product of two canonical transformations is again a canonical transformation, the *temporal development of a system* can be considered as product of *infinitesimal canonical transformations,* each of them is *generated by the Hamiltonian.* A sequence of such infinitesimal transformations produces a finite shift in time of the system.

13.4   Symmetries and conservation laws

We now ask how the Hamiltonian H behaves under an infinitesimal canonical transformation generated by a function J. The answer is given by Eq. (13.76). Inserting H into this equation as dynamical quantity $(O = H)$, we find that under a canonical transformation, the change in H is

$$\delta H = \varepsilon [H, J]. \tag{13.84}$$

Consequently, H is invariant, $\delta H = 0$, with respect to the transformation generated by J if

$$[H, J] = 0. \tag{13.85}$$

---

[11] And therefore it translates also any dynamical quantity in time. Thus, the Hamiltonian controls a system's behavior in time.

On the other hand, Eq. (13.32) with $O = J$ implies that the time dependence of $J = J(\{q\}, \{p\})$ obeys

$$\frac{dJ}{dt} = [J, H].$$

(13.86)

Therefore, the invariance of H, Eq. (13.85), is linked to the time-independence of the generator J,

$$\frac{dJ}{dt} = 0.$$

(13.87)

If the Hamiltonian of a system is invariant under an infinitesimal canonical transformation, the generator of the transformation is a conserved quantity.

This is the central statement about the connection between symmetry and conservation law. It corresponds to Noether's theorem (cf. Page 280). Compared to the Lagrangian theory, in Hamiltonian mechanics, the close relation between invariance (symmetry) and conserved quantity is expressed more directly and simply. The invariance of H is coupled to the time dependence of J (cf. Eqs. (13.84) and (13.87)): $\delta H = 0$ implies $J = const.$ We illustrate this general statement with the following concrete examples[12] for an N particle systems.

For an N particle system, let the Hamiltonian H (cf. (10.80)),

$$H(\{\mathbf{r}\}, \{\mathbf{p}\}, t) = \sum_{\alpha=1}^{N} \frac{1}{2m_\alpha} \mathbf{p}_\alpha^2 + V(\{r_{\alpha\beta}\})$$

(13.88)

be invariant under *infinitesimal rotations* (cf. Eqs. (13.77), (13.78)):

$$\mathbf{r}_\alpha' = \mathbf{r}_\alpha + \varepsilon(\boldsymbol{\varphi} \times \mathbf{r}_\alpha)$$
$$\mathbf{p}_\alpha' = \mathbf{p}_\alpha + \varepsilon(\boldsymbol{\varphi} \times \mathbf{p}_\alpha), \qquad \alpha = 1, \ldots, N.$$

Now $\partial J/\partial \mathbf{p}_\alpha = (\boldsymbol{\varphi} \times \mathbf{r}_\alpha)$ and $\partial J/\partial \mathbf{r}_\alpha = -(\boldsymbol{\varphi} \times \mathbf{p}_\alpha)$, so that the generator of this transformation is the component of the total angular momentum along the rotation axis $\mathbf{n}$ (cf. Eq. (13.80)),

$$J = \sum_{\alpha=1}^{N} \boldsymbol{\varphi}(\mathbf{r}_\alpha \times \mathbf{p}_\alpha) = \boldsymbol{\varphi} \mathbf{L}.$$

---

[12]For $\partial H/\partial t = 0$, a trivial example is the Hamiltonian itself. ($J = H$; H is the generator of time translations.)

Hence according to Eq. (13.85),

$$[H, \varphi \mathbf{L}] = 0, \tag{13.89}$$

since H is invariant[13]. But, by Eq. (13.86), this also implies that $\varphi \mathbf{L}$ is a conserved quantity. If H is invariant under rotations with arbitrary parameters φ, then Eq. (13.89) implies

$$[H, \mathbf{L}] = 0. \tag{13.90}$$

But according to Eq. (13.86), this means that the total angular momentum, $\mathbf{L} = \mathbf{L}(\{\mathbf{r}\}, \{\mathbf{p}\})$, is conserved:

$$d\mathbf{L}/dt = 0.$$

Quite analogously, one shows that for *infinitesimal translations*, $\mathbf{r}'_\alpha = \mathbf{r}_\alpha + \varepsilon \mathbf{s}$ and $\mathbf{p}'_\alpha = \mathbf{p}_\alpha$, the generator $J = \mathbf{sP}$, that is the total momentum in the direction of the translation $\mathbf{s}$, is conserved. If $\mathbf{s}$ can be arbitrary chosen, then the total momentum is constant (**E**).

13.5   The flow in phase space

If a system has $2f$ phase space variables, any point in *phase space* is given by $\mathbf{u} = (q_1, \ldots, q_f, p_1, \ldots, p_f) = (\{q\}, \{p\})$. The canonical equations (13.9), are first order differential equations,

$$\dot{\mathbf{u}} = \mathbf{G}(\mathbf{u}), \tag{13.91}$$

with the $2f$-component vector field

$$\mathbf{G}(\mathbf{u}) = \left(\left\{ \frac{\partial H}{\partial p} \right\}, -\left\{ \frac{\partial H}{\partial q} \right\} \right). \tag{13.92}$$

The set of all solution trajectories in phase space – i.e. for all possible initial values of the form $\mathbf{u}_{t_0} = \mathbf{u}(t = t_0)$, having components

$$u_i = u_i(\{u_{t_0}\}, t, t_0), \qquad i = 1, \ldots, 2f$$

– is called the **flow in phase space**. If the flow is known, then for any initial value $\mathbf{u}_{t_0}$, one can specify the point $\mathbf{u}$ at which the system will be

---

[13]Compare this derivation to the one given in Section 10.4 for Eq. (10.87).

at time t; this holds for both directions of time. For a time independent Hamiltonian, $H = H(\mathbf{u})$, the canonical equations are an autonomous system. In this case, they are invariant with respect to translations of time.

The **symplectic** (= braided) structure[14] of the canonical equations has two consequences in particular (see also the remarks at the end of Section 4.4):

i) To solve the autonomous system of $2f$ equations, f integrals are sufficient (otherwise one would need $2f - 1$ integrals) – this is addressed in the next chapter;

ii) The size of a volume in phase space stays constant as it flows (evolves in time).

The explanation for the second point – which is peculiar to Hamiltonian systems – is the property (13.70) of the Jacobi determinant of an arbitrary canonical transformation applied to the temporal evolution of the system. As explained above, this temporal evolutions is a sequence of infinitesimal canonical transformations. If one writes the infinitesimal canonical transformation for the evolution of a system in time (Eqs. (13.82) with $dt = t - t_0$) in terms of the variables $\mathbf{U} = (\{Q\}, \{P\})$ and $\mathbf{u}$, the transformation reads:

$$\mathbf{U} = \mathbf{u} + (t - t_0)\,\mathbf{G},$$

where $\mathbf{G}$ is given by Eq. (13.92). From this expression, one obtains for the derivatives

$$\frac{\partial U_i}{\partial u_j} = \delta_{ij} + (t - t_0)\frac{\partial G_i}{\partial u_j}, \qquad i, j = 1, \ldots, 2f.$$

Now, a matrix $1 + \varepsilon\mathbf{M}$ 'close' to the unit matrix has the property that

$$\det(1 + \varepsilon\mathbf{M}) = 1 + \varepsilon\,\mathrm{Tr}\,\mathbf{M} + \mathcal{O}(\varepsilon^2).$$

In particular,

$$\det\left[\frac{\partial U_i}{\partial u_j}\right] = 1 + (t - t_0)\,\mathrm{Tr}\left[\frac{\partial G_i}{\partial u_j}\right] = 1 + (t - t_0)\frac{\partial G_i}{\partial u_i}.$$

[14]The time derivatives of the coordinates $\{q\}$ (the first f components of $\mathbf{u}$) are equal to the derivatives of H with respect to the momenta $\{p\}$ (the second f components of $\mathbf{u}$); and vice versa.

(Again, the square brackets indicate a matrix whose elements are $\partial U_i/\partial u_j$.) Making use of the definition (13.92) of the vector field $\mathbf{G}$, it follows that

$$\frac{\partial G_i}{\partial u_i} = \frac{\partial^2 H}{\partial q_i \partial p_i} - \frac{\partial^2 H}{\partial p_i \partial q_i} = 0; \tag{13.93}$$

or, in vectorial notation,

$$\boldsymbol{\nabla}_{\mathbf{u}}\mathbf{G} := \frac{\partial G_i}{\partial u_i} = 0. \tag{13.94}$$

Consequently,

$$\det\left[\frac{\partial U_i}{\partial u_j}\right] = 1, \tag{13.95}$$

in accordance with the property given in Eq. (13.70) of the Jacobi determinant of a canonical transformation.

The state of a system at time t is fixed by specifying all coordinates and momenta. This state is associated with a point in the $2f$-dimensional phase space. The motion of this point is determined by Hamilton's equations. If we consider many realizations of the system (each of them starting from different but neighboring initial values), their starting points are mapped onto a cloud of points in phase space, enclosed at time t_0 by a certain volume $V(t_0)$,

$$V(t_0) = \int_{\mathcal{R}} \prod_{i=1}^{2f} du_i,$$

where $\mathcal{R}$ is the domain of integration of the variables $\mathbf{u}$, determined by the (convex) hull of the cloud of points. Since the motion of each point corresponds to canonical transformations to new variables $\mathbf{U}$, at time $t_0 + dt$ the volume occupied by the cloud of points is

$$V(t_0 + dt) = \int_{\mathcal{R}'} \prod_{i=1}^{2f} dU_i.$$

The domain $\mathcal{R}$ is mapped onto $\mathcal{R}'$. According to Eqs. (13.71) or (13.95), respectively, one gets

$$V(t_0 + dt) = \int_{\mathcal{R}'} \prod_i dU_i = \int_{\mathcal{R}} \prod_i du_i \det\left(\frac{\partial U_i}{\partial u_j}\right) = \int_{\mathcal{R}} \prod_i du_i = V(t_0).$$

This is **Liouville's theorem**[15] (for the evolution of a volume in phase space):

> For Hamiltonian systems, the size of a volume in phase space remains constant as it evolves.

The shape of the cloud of points (and also of the enclosing volume) will generally change. The cloud of points moves like an incompressible fluid. Therefore, in a Hamiltonian dynamical system, motion in phase space, even if chaotic, cannot shrink asymptotically to an attractor of lower dimensionality (cf. Section 4.4).

Problems and examples

1. Determine the Hamiltonian for a two-dimensional isotropic oscillator in polar coordinates.

2. Calculate the Hamiltonian for a free symmetric rigid body (cf. Eq. (13.17)).

3. Derive the Hamiltonian and the canonical equations for the

    i) Two-dimensional anharmonic oscillator;
    ii) Motion of a point mass in the $1/r$ potential.

4. Show that Eq. (13.21) can be obtained as the canonical momentum from a Lagrangian in cylindrical coordinates, Eq. (9.79). Hence, deduce the form (13.22) of the Hamiltonian.

5. Consider the planar motion of a charged point mass in Cartesian coordinates, viewed in a rotating reference frame, described by the Hamiltonian

$$H = \frac{1}{2m}\left(p_x^2 + p_y^2\right) + V\left(\sqrt{x^2+y^2}\right) + \frac{q^2 B^2}{8mc^2}\left(x^2+y^2\right).$$

Derive the equations of motion, and discuss the solutions for

$$V\left(\sqrt{x^2+y^2}\right) = \frac{1}{\sqrt{x^2+y^2}}.$$

---

[15] Joseph LIOUVILLE (1809-1882), French mathematician.

6. Show that for a canonical transformation, $[A, B]_{q,p} = [A, B]_{Q,P}$.

7. Convince yourself that the relations (13.39) and (13.40) are true.

8. Use Eq. (13.44) to derive Lagrange's equations.

9. Derive the integrability conditions for generators of the kind F_3 and F_4.

10. Consider the Hamiltonian for the one-dimensional harmonic oscillator

$$H = \frac{1}{2m}p^2 + \frac{m\omega^2}{2}q^2.$$

For the following transformation,

$$Q = \lambda\,(p + im\omega q)\,, \quad P = \lambda\,(p - im\omega q)$$

determine the constant λ such that the transformation from (q, p) to (Q, P) is canonical. What is the generating function $F_2\,(q, P)$? Give $\hat{H}\,(Q, P)$ and the canonical equations for Q and P.

11. For the transformation induced by $F_1(q, Q) = \frac{1}{2}m\omega q^2 \cot Q$, derive the generator of the kind $F_2(q, P)$.

Solution: $F_2(q, P) = \dfrac{m\omega q}{2}\sqrt{\dfrac{2P}{m\omega} - q^2} + P\arcsin\dfrac{q}{\sqrt{2P/m\omega}}.$

12. Determine the generator J for infinitesimal translations in an N particle system.

13. What is the generator J for the infinitesimal transformation giving rise to the constant the center of mass motion?

14. Determine the infinitesimal canonical transformation generated by the Runge-Lenz vector (cf. Section 5.4) (Hint: $J = \boldsymbol{\lambda}\mathbf{K}$, where $\boldsymbol{\lambda}$ is an arbitrary vector.)

14

Hamilton-Jacobi theory

In the previous chapter, we saw that canonical transformations permit a clear presentation of the connection between symmetries of a system and constants of the motion. At least equally importantly, canonical transformations may serve also as a tool for systematically calculating the constants of the motion. In the following, we show how this can be done.

14.1   Integrability

We have seen in section 10.1, that in a system with f degrees of freedom, $2f - 1$ conserved quantities exist in principle. The dynamics of the system are determined when we find all these $2f-1$ conserved quantities. In practice, this can very rarely be achieved. Nevertheless, if the number of conserved quantities known suffices to solve the problem, the system is termed 'integrable'.

If we succeed, for instance, in finding a canonical transformation

$$Q_i = Q_i(\mathbf{q}, \mathbf{p}, t)$$
$$P_i = P_i(\mathbf{q}, \mathbf{p}, t),$$

with[1] $\mathbf{q} = (q_1, ..., q_f) = \{q\}$, $\mathbf{p} = (p_1, ..., p_f) = \{p\}$, that transforms all coordinates (or momenta) into cyclic coordinates (or momenta), then

---

[1] For convenience, from now on we use vector notation for the f coordinates and f momenta.

the transformed system is integrable. If, for instance, in the transformed system, all coordinates Q_i are cyclic, $\partial \hat{H} / \partial Q_i = 0$, then the Hamiltonian $\hat{H}$ depends only on the momenta P_i,

$$\hat{H} = \hat{H}(\mathbf{P}), \qquad \mathbf{P} = (P_1, ..., P_f). \tag{14.1}$$

Since $\dot{P}_i = -\partial \hat{H} / \partial Q_i = 0$, the momenta are constant,

$$P_i = const =: \beta_i, \tag{14.2}$$

and $\hat{H}(\mathbf{P}) = \hat{H}(\boldsymbol{\beta})$. The second set of canonical equations implies that

$$\dot{Q}_i = \frac{\partial \hat{H}}{\partial P_i} = \frac{\partial \hat{H}}{\partial \beta_i} =: \nu_i = const, \tag{14.3}$$

and thus

$$Q_i = \nu_i t + \alpha_i, \qquad i = 1, \ldots, f, \tag{14.4}$$

with α_i being the constants of integration. Together with Eq. (14.2), this is the solution for the dynamical system (cf. the example of the oscillator on Page 385).

For a Hamiltonian system with f degrees of freedom to be integrable already f conserved quantities with particular properties, are enough (see below).

14.1.1 Liouville's theorem on integrability

We assume in the following that the *motion remains bounded*. Then the Q_i must be finite for all time and thus the quantities ν_i must be angular variables; i.e., quantities whose argument is evaluated modulo 2π.

Liouville's theorem[2]

If for a Hamiltonian system with f degrees of freedom (and $2f$ generalized coordinates q_i, p_i) f functions I_i (on a subset U of phase space) with the following properties are known (without loss of generality $I_1 = H$):

---

[2]According to H.S. Dumas (see Bibliography) the concept of integrable systems started 1843 with Jacobi's "Vorlesungen über Dynamik" (see Footnote 4) and was made more explicit by Liouville in *Note sur l'intégration des équations differentielles de la dynamique*, J. des Math. Pures et Appl. **20**, 137 (1855). In more modern form it was presented and proved in the 1930s by H. Mineur and in a form suited to Hamiltonian perturbation theory by V.I. Arnold in the 1960s (see in particular [Arnold]).

i) $[I_i, H] = 0$, $i = 1, \ldots, f$; *i.e., the I_i are conserved quantities*;

ii) $[I_i, I_j] = 0$, $i, j = 1, \ldots, f$; *i.e., the functions I_i are in* **involution**;

iii) *The total differentials* $dI_i = \dfrac{\partial I_i}{\partial q_k} dq_k + \dfrac{\partial I_i}{\partial p_k} dp_k$ *are linearly inde-*

pendent; i.e., the $f \times 2f$ matrix of coefficients $\left[\dfrac{\partial I_i}{\partial q_k}, \dfrac{\partial I_i}{\partial p_l}\right]$, $i, k, l =$
$1, \ldots, f$, has rank f;

then the system is **integrable** *(on U), i.e. the solution of Hamilton's equations can be obtained by quadrature (integration).*

Examples

1.) Quantities not in involution are the components of (conserved) angular momentum **L**. The Poisson brackets are readily calculated:

$$[L_i, L_j] = \varepsilon_{ijk} L_k. \tag{14.5}$$

Out of the three components of **L**, only two conserved quantities can be constructed that are in involution. A standard choice is $\mathbf{L}^2 = \sum_i L_i^2$ and $L_3 = L_z$ (E). This fact must be observed when checking the integrability of a system by counting the conserved quantities.

2.) Consider the motion of a *particle in a $1/r$ potential*. Since the system has three degrees of freedom, according to the general discussion in Section 10.1, $2f - 1 = 5$ independent conserved quantities are necessary to determine the orbit by purely algebraic methods (as the curve of intersection of the surfaces defined by these conserved quantities). In the $1/r$ potential case, five isolating conserved quantities (H, **L**, I_K; cf. Chapter 5) are known, but only three ($= f$) of them are in involution, e.g. H, $\mathbf{L}^2$, and L_z. As we have shown already, I_K derived from the Runge-Lenz vector satisfies (cf. Eq. (13.40); now putting $L_z = L_\varphi$)

$$[L_z, I_K] = -1,$$

and is therefore not in involution with L_z.

3.) Similarly for the *two-dimensional isotropic harmonic oscillator* (cf. Section 4.1) the maximal number of independent conserved quantities is three. A possible choice is

$$H = \frac{1}{2m} \left(p_x{}^2 + p_y{}^2\right) + \frac{m\omega^2}{2} \left(x^2 + y^2\right)$$

$$H_x = \frac{1}{2m}p_x{}^2 + \frac{m\omega^2}{2}x^2, \qquad L_z = xp_y - p_xy.$$

But H_x and L_z are not in involution, since (**E**)

$$[L_z, H_x] = \frac{1}{m}p_xp_y + m\omega^2 xy = mI_4,$$

where I_4 is the conserved quantity (4.16). So there are only two $(= f)$ conserved quantities fulfilling the conditions of Liouville's theorem.

4.) Let us consider finally the *motion of a charged particle in a homogeneous magnetic field* in the plane perpendicular to **B** (cf. Section 5.5). The conserved quantities are (cf. (13.22) and (13.23))

$$H = \frac{p_\rho^2}{2m} + \frac{p_\varphi^2}{2m\rho^2} + \frac{m}{2}\left(\frac{\omega_Z}{2}\right)^2 \rho^2 - \frac{1}{2}\omega_Z p_\varphi$$

$$p_\varphi = I_{LB}, \qquad \varphi_u = \varphi + \arctan\frac{p_\varphi + \frac{1}{2}m\rho^2\omega_Z}{\rho p_\rho}.$$

Indeed we have $[H, p_\varphi] = 0$, but

$$[p_\varphi, \varphi_u] = -1.$$

In all these examples, only f independent integrals of the motion are in involution.

14.1.2   Sketched proof of the theorem

We present only sketch proof of Liouville's theorem, in several steps. (For a more in-depth proof, see, for example, [Arnold], [Thirring].) The differentiable manifold on which the motion takes place is embedded in the $2f$-dimensional phase space. Each integral of the motion I_i, reduces the dimension of the (embedding subspace U of the) manifold by one, since we can eliminate one of the dynamical variables; this implies that for an integrable system, the motion takes place on a manifold $\mathfrak{P}$ in phase space whose dimension is at most f. In the following, we consider only bounded motion, implying that the manifold $\mathfrak{P}$ is compact (it is a closed and finite set). We assume, furthermore, that $\mathfrak{P}$ is simply connected. These two properties make it easy to prove that $\mathfrak{P}$ can be mapped onto a particular 'simpler' manifold[3], namely, the f-dimensional surface of a *hypertorus* (or, simply, an f-dimensional **torus**). We sketch the proof of this last statement here (cf. [Berry]).

Using the integrals of motion I_i, we define the $2f$-component vector fields

$$\mathbf{V}_i = (\boldsymbol{\nabla}_p I_i, -\boldsymbol{\nabla}_q I_i), \qquad i = 1, \ldots, f,$$

($\boldsymbol{\nabla}_q = (\partial/\partial q_1, \ldots, \partial/\partial q_f)$, and analogously $\boldsymbol{\nabla}_p$). The $2f$-dimensional vectors normal to $\mathfrak{P}$, i.e. the vectors normal to the surfaces, $I_i = const$ (remember that $\mathfrak{P}$ is the set formed by the intersections of all these surfaces; see, for example, Fig. (4.3)), are given by

$$\mathbf{n}_j = (\boldsymbol{\nabla}_q I_j, \boldsymbol{\nabla}_p I_j), \qquad j = 1, \ldots, f;$$

and these are perpendicular to the vectors $\mathbf{V}_i$,

$$\mathbf{V}_i \mathbf{n}_j = -[I_i, I_j] = 0 \quad \forall i, j,$$

since – according to statement (ii) in Liouville's theorem – the I_i are in involution. The vectors $\mathbf{V}_i$ are therefore everywhere tangential to the surface of the manifold $\mathfrak{P}$. Since $\mathfrak{P}$ is compact, this implies that no points exist on $\mathfrak{P}$ such that the vector field $\mathbf{V}_i$ is singular – the vector field $\mathbf{V}_i$ can conceivably be 'combed' (through some transformation) without 'vortices' ('crowns') or 'partings' (where $\mathbf{V}_i \mathbf{n}_j$ would be indeterminate). Therefore, $\mathfrak{P}$ is topologically equivalent to a (hyper-)torus. (For a sphere, in contrast, at least one point exists where $\mathbf{V}_i$ is not defined.) Since the size of $\mathfrak{P}$ is determined by the values of the integrals I_i (for bounded motion, the integrals of the motion restrict the maximum values of the dynamical variables, cf. Chapter 4), the dimensions of the torus are also fixed by them: the radii of the torus are functions of the I_i.

---

[3]The mathematically correct terminology for such a mapping is *diffeomorphism*; i.e. an isomorphism exists between the two (differentiable) manifolds.

The dynamics of the system consists in the motion of a point in $2f$-dimensional phase space on an f-dimensional manifold $\mathfrak{P}$. A canonical transformation – which must be found or constructed – can be used to map this trajectory into the f-dimensional surface of a torus in an $(f + 1)$-dimensional space. The motion of a point in phase space is thereby represented by the motion of a point on the torus.

The motion on the torus

Let us now consider the motion on the surface of the torus. The position of any point on the surface is fixed by giving f angles θ_i:

$$\boldsymbol{\theta} = (\theta_1, \ldots, \theta_f). \tag{14.6}$$

The variation of θ_j, with the remaining angles held fixed, describes a circle on the torus, whose radius J_j is fixed by the conserved quantities I_i. Figure 16.2 shows the situation for two degrees of freedom. Since, on the torus, the radii J_j are the conjugate quantities to the angles θ_j, the dynamics is described completely by these variables. Therefore, the transformation of $\mathfrak{P}$ into the torus must map the (constant) f-tuple of integrals of the motion, $\mathbf{I} = (I_1, \ldots, I_f)$, to a likewise constant f-tuple of quantities $\mathbf{J} = (J_1, \ldots, J_f)$. These quantities J_i must be conjugate to the angles θ_i, so that the fundamental Poisson brackets must be

$$[J_i, J_j] = [\theta_i, \theta_j] = 0, \qquad [J_i, \theta_j] = \delta_{ij}. \tag{14.7}$$

Also, the Hamiltonian of the original system is mapped to the Hamiltonian $\hat{H}(\boldsymbol{\theta}, \mathbf{J})$ to give the dynamics on the torus. Since the J_i are constant, the condition $\dot{\mathbf{J}} = -\boldsymbol{\nabla}_\theta \hat{H} = \mathbf{0}$ implies that the angles θ_i are cyclic variables and hence that $\hat{H}(\boldsymbol{\theta}, \mathbf{J}) = \hat{H}(\mathbf{J})$. Therefore, the second set of the canonical equations, $\dot{\boldsymbol{\theta}} = \boldsymbol{\nabla}_J \hat{H} = const$, yields the equations of motion

$$\frac{d}{dt}\boldsymbol{\theta} = \boldsymbol{\omega}, \tag{14.8}$$

where the constant angular velocities $\boldsymbol{\omega} = (\omega_1, \ldots, \omega_f)$ follow from

$$\boldsymbol{\omega} = \boldsymbol{\omega}(\mathbf{J}) := \boldsymbol{\nabla}_J \hat{H}. \tag{14.9}$$

For arbitrary values of the angular velocities ω_i, the motion on the torus is *nearly periodical* (*quasiperiodical*), meaning that with time, the whole torus is covered by the trajectory $\boldsymbol{\theta}(t)$ (cf. Section 16.2). Integrating the equations of motion (14.8), hence obtaining the solution, is easy. The difficult part of the procedure – namely, the explicit construction of the variables $\boldsymbol{\theta}$ and $\mathbf{J}$ parametrizing the motion on the torus – will be demonstrated in the next chapter.

In general, one is left with the problem of finding f conserved quantities with properties (i) – (iii) of Liouville's Theorem. We shall now present a method by which – at least in principle – one can determine the canonical transformation to a Hamiltonian that is cyclic in the new coordinates.

14.2 Time-independent Hamilton-Jacobi theory

14.2.1 The Hamilton-Jacobi equation

We look for a method of finding the generator of a canonical transformation that transforms a given, time-independent Hamiltonian, $H = H(\mathbf{q}, \mathbf{p})$, to a Hamiltonian that is obviously integrable[4]. The new Hamiltonian may, for instance, depend only on the 'new' momenta, as is assumed below. This can be achieved by a time-independent generator of the kind F_2 (cf. Subsection 13.3.1), which, as is customary, we denote by

$$W = W(\mathbf{q}, \mathbf{P}).$$

The generator W is also called **Hamilton's characteristic function**. The relations between the 'old' and the 'new' canonical variables, according to Eqs. (13.52) and (13.53) are given by

$$\mathbf{p} = \boldsymbol{\nabla}_q W \tag{14.10}$$

$$\mathbf{Q} = \boldsymbol{\nabla}_P W, \tag{14.11}$$

and according to (13.54) we have

$$\hat{H}(\mathbf{Q}, \mathbf{P}) = H(\mathbf{q}, \mathbf{p}), \tag{14.12}$$

since W is supposed not to depend on time explicitly.

---

[4]C.G.J. Jacobi, "Vorlesungen über Dynamik" (Lectures on dynamics), Königsberg 1842/43, ed. by A. Clebsch 1866, as well as "Vorlesungen über analytische Mechanik" (Lectures on analytical mechanics), Berlin 1847/48, ed. by H. Pulte 1996.

But now we assume in addition that $\hat{H}$ depends only on the 'new' momenta,

$$\hat{H}(\mathbf{Q}, \mathbf{P}) = \hat{H}(\mathbf{P}), \tag{14.13}$$

so that the new momenta are constant,

$$\mathbf{P} = \boldsymbol{\beta}.$$

(See Eqs. (14.1) and (14.2). For the P_i, we assume that conditions (i) – (iii) of Liouville's theorem on Page 398 are satisfied.) The new Hamiltonian,

$$\hat{H}(\mathbf{P}) = \hat{H}(\boldsymbol{\beta}) =: E(\boldsymbol{\beta}), \tag{14.14}$$

is not an additional conserved quantity, since it is given by the independent, conserved momenta.

The equation for the generator W of such a transformation is obtained from Eq. (14.12), taking into account Eqs. (14.13) and (14.14). Inserting Eq. (14.10) into $H(\mathbf{q}, \mathbf{p})$, we obtain the **Hamilton-Jacobi differential equation** for a time-independent system:

$$H(\mathbf{q}, \boldsymbol{\nabla}_q W) = H\left((q_1, ..., q_f), \left(\frac{\partial W}{\partial q_1}, ..., \frac{\partial W}{\partial q_f}\right)\right) = E. \tag{14.15}$$

This is a first order partial differential equation for the generator W in terms of the f variables q_i. We search for a **complete integral**; i.e., a solution W that depends on some f independent constants β_i: $W = W(\mathbf{q}, \boldsymbol{\beta})$. These constants β_i can be directly taken as the (constant) momenta (integrals of the motion) P_i:

$$W(\mathbf{q}, \boldsymbol{\beta}) = W(\mathbf{q}, \mathbf{P}).$$

The conjugate coordinates[5] Q_i are calculated from Eq. (14.11),

$$Q_i = \frac{\partial W}{\partial P_i} = \frac{\partial W}{\partial \beta_i}.$$

According to Eq. (14.3), the time dependence of the Q_i is given by the constants

$$\nu_i = \frac{\partial \hat{H}(\boldsymbol{\beta})}{\partial \beta_i} = \frac{\partial E(\boldsymbol{\beta})}{\partial \beta_i}. \tag{14.16}$$

---

[5]The conjugate variables $\mathbf{Q}$ are cyclic, since $\hat{H} = \hat{H}(\mathbf{P})$ by assumption (14.13).

The Hamilton-Jacobi equation is a tool that, in principle, enables one to systematically – without intuition or trial and error – transform the Hamiltonian of a system into a manifestly integrable one, whose solution is easily obtained. Despite the attractiveness of this method, in practice, however, one soon learns the 'fundamental law of physics' – the law of conservation of difficulty: the difficulty in finding a solution is merely shifted from one step to another. In other words, solving the Hamilton-Jacobi equation is, in general, equally hard or easy as solving the equations of motion directly. Nevertheless, the Hamilton-Jacobi approach is useful in approximately calculating the conserved quantities (see Section 16.3).

In general, E is a function of all the β_i. In practice, one chooses (without loss of generality) the constants β_i beginning with

$$\beta_1 = E. \tag{14.17}$$

Because $\dot{Q}_1 = \dfrac{\partial \hat{H}}{\partial \beta_1} = \dfrac{\partial E}{\partial \beta_1} = 1$, we then have

$$Q_1 = t + \alpha_1, \tag{14.18}$$

and since the remaining β_i, $i = 2, \ldots, f$, are independent of $E = \beta_1$, it follows that $\dot{Q}_i = \partial \hat{H}/\partial \beta_i = \partial E/\partial \beta_i = \nu_i = 0$; i.e.,

$$Q_i = \alpha_i, \qquad i = 2, \ldots, f. \tag{14.19}$$

These $(f - 1)$ equations fix the trajectory; and Q_1 determines the time dependence of the motion along the trajectory. The $2f$ constants α_i and β_i are fixed by the initial values of the old dynamical variables q_i and p_i.

14.2.2 Separation of variables

In the Hamilton-Jacobi equation (14.15), suppose we set

$$W = \sum_{i=1}^{f} W_i(q_i, \boldsymbol{\beta}), \tag{14.20}$$

where each function W_i depends *only* on q_i. If this leads to uncoupled equations for the W_i,

$$h_i\left(q_i, \frac{\partial W_i}{\partial q_i}, \boldsymbol{\beta}\right) = \beta_i, \tag{14.21}$$

then the **system is separable**. (No summation convention is used in this section.) This technique is called **separation of variables**. The solution $\partial W_i / \partial q_i$ of Eq. (14.21) has the form $\partial W_i / \partial q_i = p_i(q_i, \boldsymbol{\beta})$ (cf. (14.10)); then $W_i(q_i, \boldsymbol{\beta})$ follows after integrating with respect to q_i.

The separability of the Hamilton-Jacobi equation depends strongly on the choice of the coordinates q_i. If the Hamiltonian in some set of coordinates has the form

$$H(\mathbf{q}, \mathbf{p}) = \sum_{i=1}^{f} h_i(q_i, p_i), \qquad (14.22)$$

then the characteristic function W is obviously separable. One example is

The two-dimensional harmonic oscillator

In Section 4.1, we focussed on the anisotropic two-dimensional harmonic oscillator. Its Hamiltonian is given by

$$H = \frac{1}{2m}(p_x^2 + p_y^2) + \frac{m}{2}(\omega_x^2 x^2 + \omega_y^2 y^2). \qquad (14.23)$$

There is no coupling between x and y-motion, so that the ansatz

$$W = W_x(x) + W_y(y)$$

for the characteristic function suggests itself. Indeed, the Hamilton-Jacobi equation

$$\frac{1}{2m}\left(\left(\frac{\partial W_x}{\partial x}\right)^2 + \left(\frac{\partial W_y}{\partial y}\right)^2\right) + \frac{m}{2}(\omega_x^2 x^2 + \omega_y^2 y^2) = E = \beta_1$$
$$(14.24)$$

separates, since after writing it in the form

$$\beta_1 - \frac{1}{2m}\left(\frac{\partial W_x}{\partial x}\right)^2 - \frac{m}{2}\omega_x^2 x^2 = \frac{1}{2m}\left(\frac{\partial W_y}{\partial y}\right)^2 + \frac{m}{2}\omega_y^2 y^2 = \beta_2,$$

we see that each side depends on a different, independent variable; and therefore each side is equal to a constant, β_2 say. The energies E_x and E_y appear as separation constants:

$$\frac{1}{2m}\left(\frac{\partial W_x}{\partial x}\right)^2 + \frac{m}{2}\omega_x^2 x^2 = \beta_1 - \beta_2 = E - E_y = E_x$$

$$\frac{1}{2m}\left(\frac{\partial W_y}{\partial y}\right)^2 + \frac{m}{2}\omega_y^2 y^2 = \beta_2 = E_y. \qquad (14.25)$$

The solutions of these equations are

$$W_x\left(x, E_x\right) = \int^x dx'\sqrt{2mE_x - m^2\omega_x^2 x'^2}$$

$$= \frac{1}{2}\left[x\sqrt{2mE_x - m^2\omega_x^2 x^2} + \frac{2E_x}{\omega_x}\arcsin\left(x\sqrt{\frac{m\omega_x^2}{2E_x}}\right)\right]$$

$$(14.26)$$

$$W_y\left(y, E_y\right) = \int^y dy'\sqrt{2mE_y - m^2\omega_y^2 y'^2}$$

$$= \frac{1}{2}\left[y\sqrt{2mE_y - m^2\omega_y^2 y^2} + \frac{2E_y}{\omega_y}\arcsin\left(y\sqrt{\frac{m\omega_y^2}{2E_y}}\right)\right].$$

$$(14.27)$$

The coordinates Q_i conjugate to the generalized momenta $\beta_i\ (E_i)$ determine the motion. Their time dependence follows from (cf. Eqs. (14.11) and (14.18))

$$Q_1 = \frac{\partial W}{\partial \beta_1} = \frac{\partial W}{\partial E} = \frac{\partial W_x}{\partial E} = t + \alpha_1.$$

Using

$$\frac{\partial W_x}{\partial E} = \frac{1}{\omega_x}\int^x dx'\left(\frac{2(E - E_y)}{m\omega_x^2} - x'^2\right)^{-1/2}$$

$$= \frac{1}{\omega_x}\arcsin\left(x\sqrt{\frac{m\omega_x^2}{2E_x}}\right),$$

we obtain

$$x = \sqrt{\frac{2E_x}{m\omega_x^2}}\sin(\omega_x t + \bar{\alpha}_1), \qquad \bar{\alpha}_1 = \omega_x\alpha_1, \qquad (14.28)$$

and from (cf. Eq. (14.19))

$$Q_2 = \frac{\partial W}{\partial \beta_2} = \frac{\partial W}{\partial E_y} = \alpha_2,$$

the equation of the orbit results[6],

$$\alpha_2 = \frac{1}{\omega_y} \arcsin\left(y\sqrt{\frac{m\omega_y^2}{2E_y}}\right) - \frac{1}{\omega_x} \arcsin\left(x\sqrt{\frac{m\omega_x^2}{2E_x}}\right),$$
(14.29)

which – using Eq. (14.28) – leads to the second part of the solution

$$y = \sqrt{\frac{2E_y}{m\omega_y^2}} \sin(\omega_y t + \overline{\alpha}_2), \qquad \overline{\alpha}_2 = \omega_y(\alpha_1 + \alpha_2). \quad (14.30)$$

The Hamiltonian in generalized momenta is

$$\hat{H} = \beta_1; \tag{14.31}$$

the cyclic coordinates Q_1 and Q_2 do not appear. The solution is more involved than the direct solution of the equation of motion, but it yields the constants of motion and the 'normal form' of H.

Separation of variables is particularly simple if the Hamiltonian is independent of one or several coordinates. If, for instance, q_k is cyclic, then for the characteristic function one can set

$$W\left(\mathbf{q}, \boldsymbol{\beta}\right) = q_k \beta_k + \sum_{i \neq k} W_i\left(q_i, \boldsymbol{\beta}\right), \tag{14.32}$$

---

[6]The solution for the isotropic oscillator, $\omega_x = \omega_y = \omega$,

$$x\sqrt{\frac{m\omega_x^2}{2E_x}} =: x/a = \cos\varphi, \qquad y\sqrt{\frac{m\omega_y^2}{2E_y}} =: y/b = \sin\varphi$$

is obtained if $\alpha_2\omega$ is chosen to be $-\pi/2$. This gives the equation of an ellipse: $(x/a)^2 + (y/b)^2 = 1$.

For arbitrary ratios ω_x/ω_y, the constant α_2 can be expressed from the equation for the orbit in terms of initial values by inserting in the right hand side of Eq. (14.29) the values x_0 and y_0 instead of x and y. With $v_x^2/\omega_x^2 = 2E_x/m\omega_x^2 - x^2$ and the corresponding relations for v_y, v_x^0, and v_y^0, and observing that arcsin $z = \arctan\left(z/\sqrt{1-z^2}\right)$, one can thus derive from Eq. (14.29) the 'useless' integral given in Eq. (4.32) of Section 4.3.

since the conjugate momentum p_k is a constant: $p_k = \partial W / \partial q_k = \beta_k$. This constant is chosen to be the new momentum, $P_k = \beta_k$. We apply this ansatz to the

Motion in a central force

In spherical coordinates, the Hamiltonian reads (E)

$$H = \frac{1}{2m}\left(p_r^2 + \frac{p_\vartheta^2}{r^2} + \frac{p_\varphi^2}{r^2 \sin^2 \vartheta}\right) + V(r). \qquad (14.33)$$

This Hamiltonian turns out to be separable. For the characteristic function W we try the ansatz

$$W = W_r(r) + W_\vartheta(\vartheta) + W_\varphi(\varphi). \qquad (14.34)$$

Since φ is cyclic, in accordance with Eq. (14.32), we set

$$W_\varphi(\varphi) = \beta_\varphi \varphi, \qquad (14.35)$$

where the separation constant β_φ is the z-component of the angular momentum,

$$\beta_\varphi = L_z \qquad (14.36)$$

(because $p_\varphi = \partial W / \partial \varphi = \beta_\varphi$). Inserting now $p_i = \partial W / \partial q_i$, one obtains the Hamilton-Jacobi equation

$$\left(\frac{\partial W_r}{\partial r}\right)^2 + \frac{1}{r^2}\left[\left(\frac{\partial W_\vartheta}{\partial \vartheta}\right)^2 + \frac{\beta_\varphi^2}{\sin^2 \vartheta}\right] + 2mV(r) = 2mE.$$

$$(14.37)$$

The expression in the square bracket is a function of ϑ only. If we now recast the equation into

$$\left[\left(\frac{\partial W_\vartheta}{\partial \vartheta}\right)^2 + \frac{\beta_\varphi^2}{\sin^2 \vartheta}\right] = r^2\left[2m(E - V(r)) - \left(\frac{\partial W_r}{\partial r}\right)^2\right],$$

the left hand side depends only on ϑ, while the right hand side is a function of r only. Consequently, both sides are equal to a constant, which we denote by β_ϑ^2 (with β_ϑ real) since the left hand side is always positive. Because the representation of the angular momentum in spherical coordinates is (E)

$$\mathbf{L}^2 = p_\vartheta^2 + p_\varphi^2 / \sin^2 \vartheta,$$

it follows that

$$\beta_\vartheta = |\mathbf{L}|\,. \tag{14.38}$$

So we have now the equations

$$\left(\frac{\partial W_\vartheta}{\partial \vartheta}\right)^2 = \beta_\vartheta^2 - \frac{\beta_\varphi^2}{\sin^2 \vartheta}$$

$$\left(\frac{\partial W_r}{\partial r}\right)^2 = -\frac{\beta_\vartheta^2}{r^2} + 2m(E - V(r)), \tag{14.39}$$

with solution (cf. Eq. (14.34))

$$W(r, \vartheta, \varphi, E, \beta_\vartheta, \beta_\varphi) = \beta_\varphi \varphi + \int^r dr' \sqrt{2m(E - V) - \beta_\vartheta^2/r'^2}$$

$$+ \int^\vartheta d\vartheta' \sqrt{\beta_\vartheta^2 - \beta_\varphi^2/\sin^2 \vartheta'}. \tag{14.40}$$

The three separation constants E $(=: \beta_1)$, β_ϑ, β_φ are the constant new momenta. The equation of the orbit and the time dependence of the motion are obtained according to Eqs. (14.11), (14.18), and (14.19):

$$t + \alpha_1 = \frac{\partial W}{\partial E} = m \int^r dr' \frac{1}{\sqrt{2m(E - V) - \beta_\vartheta^2/r'^2}} \tag{14.41}$$

$$\alpha_\varphi = \frac{\partial W}{\partial \beta_\varphi} = -\frac{\beta_\varphi}{\beta_\vartheta} \int^\vartheta d\vartheta' \frac{1}{\sin \vartheta' \sqrt{\sin^2 \vartheta' - (\beta_\varphi/\beta_\vartheta)^2}} + \varphi \tag{14.42}$$

$$\alpha_\vartheta = \frac{\partial W}{\partial \beta_\vartheta} = -\beta_\vartheta \int^r dr' \frac{1}{\sqrt{2m(E - V) - \beta_\vartheta^2/r'^2}}$$

$$- \int^\vartheta d\vartheta' \frac{1}{\sqrt{1 - (\beta_\varphi/\beta_\vartheta)^2 \sin^2 \vartheta'}}, \tag{14.43}$$

with α_1, α_φ, α_ϑ being the constants of integration. From Eq. (14.42), one finds the equation of the plane in which the orbit resides. Setting $\cos \psi = \beta_\varphi / \beta_\vartheta = L_z / |\mathbf{L}|$, the angles ϑ and φ in the plane are related by (F. Schweiger, Acta.Phys.Austr.19,138 (1964))

$$\alpha_\varphi = - \arcsin \left(\cot \psi \cot \vartheta \right) + \varphi. \tag{14.44}$$

Together with Eq. (14.43), the orbital curve is fixed. Equation (14.41) describes the time dependence of the motion along the orbit.

We could have taken advantage of angular momentum conservation for the motion in the central force field right from the start. Using polar coordinates r and φ in the plane perpendicular to $\mathbf{L}$, we have

$$W = W_r(r) + W_\varphi(\varphi) = \int^r dr' \sqrt{2m(E - V) - \beta_\varphi^2 / r'^2} + \beta_\varphi \varphi. \tag{14.45}$$

This coincides with Eq. (14.40) if we set $\beta_\vartheta = \beta_\varphi$ and $\vartheta = \pi/2$.

14.3 The problem of two centers of gravity

The motion of a body in the field of two fixed centers of gravity[7] shows rotational symmetry about the axis between the two centers, which we take as x-axis. In cylindrical coordinates the motion around the x-axis can be separated yielding the constant component of the angular moment along this axis ($\mathbf{E}$). The interesting part of the motion occurs in a plane containing the x-axis.

The planar motion of a point mass $m = 1$ at position $\mathbf{r}$ in the gravitational fields of two bodies m_1 and m_2 fixed at $\mathbf{R}_1$ and $\mathbf{R}_2$, respectively, is determined by canonical equations that follow from the Hamiltonian

$$H = \frac{1}{2}(p_x^2 + p_y^2) - \frac{\kappa_1}{r_1} - \frac{\kappa_2}{r_2}, \qquad \kappa_1 = Gm_1, \ \kappa_2 = Gm_2, \tag{14.46}$$

---

[7]This problem is also called **Euler's three-body problem**. Its solution was discussed by Euler in the contribution *Un corps etant attire en raison reciproque quarree des distances vers deux points fixes...*, Memoires de l'academie des sciences de Berlin **16**, 1767, pp. 228-249.

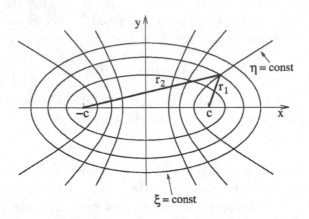

Figure 14.1: Elliptical coordinates.

where
$$r_1 = |\mathbf{r} - \mathbf{R}_1| \qquad \text{and} \qquad r_2 = |\mathbf{r} - \mathbf{R}_2|.$$

In order that this two-dimensional problem be integrable, in addition to the energy $E \ (= H)$, we must find at least one more conserved quantity.

For convenience, we choose the origin and the axes such that

$$r_1 = \sqrt{(x-c)^2 + y^2} \qquad \text{and} \qquad r_2 = \sqrt{(x+c)^2 + y^2}, \qquad (14.47)$$

where $|\mathbf{R}_1 - \mathbf{R}_2| = 2c$. That is, the centers of force are symmetric with respect to the y-axis. This symmetry suggests the use of **elliptic coordinates**[8],

$$
\begin{aligned}
x &= c \cosh \xi \cos \eta \\
y &= c \sinh \xi \sin \eta
\end{aligned}
\qquad (14.48)
$$

where
$$\xi \geq 0 \quad \text{and} \quad 0 \leq \eta < 2\pi. \qquad (14.49)$$

---

[8]More details about these coordinates can be found in, for example, [Born], [Corben/Stehle], or [Thirring]. Elliptic coordinates are a further example of orthogonal curvilinear coordinates (cf. Appendix A; E).

The curves $\xi = const$ are confocal ellipses, and the curves $\eta = const$ confocal hyperbolas (see Fig. 14.1). For both family of curves the foci are the points $(c, 0)$ and $(-c, 0)$. It follows from Eq. (14.48) that the distances to the centers of force are

$$r_1 = c(\cosh \xi - \cos \eta) \quad \text{and} \quad r_2 = c(\cosh \xi + \cos \eta). \quad (14.50)$$

The conjugate momenta in elliptic coordinates are conveniently obtained from the Lagrangian (cf. Eq. (15.12) for $\omega = 0$)

$$L = \frac{1}{2}(\dot{x}^2 + \dot{y}^2) + \frac{\kappa_1}{r_1} + \frac{\kappa_2}{r_2}$$

expressed in elliptic coordinates

$$L = \frac{c^2}{2}(\cosh^2 \xi - \cos^2 \eta)(\dot{\xi}^2 + \dot{\eta}^2) + \frac{\kappa_1/c}{\cosh \xi - \cos \eta} + \frac{\kappa_2/c}{\cosh \xi + \cos \eta}. \quad (14.51)$$

One finds

$$p_\xi = \frac{\partial L}{\partial \dot{\xi}} = c^2(\cosh^2 \xi - \cos^2 \eta)\dot{\xi}$$

$$p_\eta = \frac{\partial L}{\partial \dot{\eta}} = c^2(\cosh^2 \xi - \cos^2 \eta)\dot{\eta}. \quad (14.52)$$

Hence, the Hamiltonian $H = H(\xi, \eta, p_\xi, p_\eta)$ in elliptic coordinates is given by

$$H = \frac{1}{2c^2}\frac{1}{(\cosh^2 \xi - \cos^2 \eta)}(p_\xi^2 + p_\eta^2) - \frac{\kappa_1/c}{\cosh \xi - \cos \eta} - \frac{\kappa_2/c}{\cosh \xi + \cos \eta}. \quad (14.53)$$

This can be rewritten as

$$H = \frac{1}{(\cosh^2 \xi - \cos^2 \eta)}(H_\xi + H_\eta) \quad (14.54)$$

where

$$H_\xi = \frac{1}{2c^2}\left[p_\xi^2 - \frac{\kappa_1 + \kappa_2}{c}\cosh \xi\right] \quad (14.55)$$

$$H_\eta = \frac{1}{2c^2}\left[p_\eta^2 - \frac{\kappa_1 - \kappa_2}{c}\cos \eta\right]. \quad (14.56)$$

The Hamiltonian is an integral of the equations of motion. In order that the system is integrable we need at least a second one. Multiplying (14.54) by $\left(\cosh^2 \xi - \cos^2 \eta\right)$, inserting $H = E$, and separating the variables ξ and η. we get

$$E \cosh^2 \xi - H_\xi = E \cos^2 \eta + H_\eta =: B.$$

Each side of the first equality depends only on ξ and η respectively. Therefore each side must be equal to a constant B: B is the *second conserved quantity*[9]. Inserting for E (14.54) B can be written in a more symmetric form

$$B = \frac{1}{(\cosh^2 \xi - \cos^2 \eta)} \left[H_\xi \cos^2 \eta + H_\eta \cosh^2 \xi\right]. \tag{14.57}$$

Since we have now the Hamiltonian H and the constant of motion B as independent integrals (**E**) *the system of two centers is integrable.*

To utilize the Hamilton-Jacobi method we rewrite the energy conservation $H = E$ with H given by (14.53)

$$p_\xi^2 + p_\eta^2 - 2\kappa_1 c(\cosh \xi + \cos \eta) - 2\kappa_2 c(\cosh \xi - \cos \eta) = 2Ec^2(\cosh^2 \xi - \cos^2 \eta),$$

and get

$$p_\xi^2 - 2c\left(\kappa_1 + \kappa_2\right)\cosh \xi - 2Ec^2 \cosh^2 \xi$$

$$= - \left[p_\eta^2 - 2c\left(\kappa_1 - \kappa_2\right)\cos \eta - 2Ec^2 \cos^2 \eta\right].$$

The obvious separability leads us to the following ansatz for the characteristic function W (cf. Subsection 14.2.2):

$$W\left(\xi, \eta\right) = W_\xi\left(\xi\right) + W_\eta\left(\eta\right).$$

The substitution $p_\xi = \dfrac{\partial W_\xi}{\partial \xi}$ and $p_\eta = \dfrac{\partial W_\eta}{\partial \eta}$ (cf. Eq. (14.10)) yields

$$\left(\frac{\partial W_\xi}{\partial \xi}\right)^2 - 2c(\kappa_1 + \kappa_2)\cosh \xi - 2c^2 E \cosh^2 \xi$$

---

[9]This further integral of motion is related to the so-called super-generalized Runge-Lenz vector (N. Kryukov and E. Oks, International Review of Atomic and Molecular Physics 2(2), 105-108, 2011). See also H. Waalkens, H. R. Dullina, and P. H. Richter, Physica **D 196**, 265, 2004.

$$= -\left(\frac{\partial W_\eta}{\partial \eta}\right)^2 + 2c(\kappa_1 - \kappa_2)\cos\eta - 2c^2 E \cos^2\eta =: -\beta_2.$$

$$(14.58)$$

The separation constant β_2 is nothing but the second conserved quantity $(\beta_2 = 2c^2 B)$. The equations

$$\left(\frac{\partial W_\xi}{\partial \xi}\right)^2 - 2c(\kappa_1 + \kappa_2)\cosh\xi - 2c^2 E \cosh^2\xi = -\beta_2$$

$$\left(\frac{\partial W_\eta}{\partial \eta}\right)^2 - 2c(\kappa_1 - \kappa_2)\cos\eta + 2c^2 E \cos^2\eta = \beta_2$$

have solutions

$$W_\xi = \int^\xi d\xi' \left[2c^2 E \cosh^2\xi' + 2c(\kappa_1 + \kappa_2)\cosh\xi' - \beta_2\right]^{1/2}$$

$$(14.59)$$

$$W_\eta = \int^\eta d\eta' \left[-2c^2 E \cos^2\eta' + 2c(\kappa_1 - \kappa_2)\cos\eta' + \beta_2\right]^{1/2},$$

where $W_\xi = W_\xi(\xi, E, \beta_2)$ and $W_\eta = W_\eta(\eta, E, \beta_2)$ are elliptic integrals. The motion of the point mass follows from (cf. Eqs. (14.11) (with $P_i = \beta_i$), (14.18), and (14.19)):

$$t + \alpha = \frac{\partial W}{\partial E} = \frac{\partial W_\xi}{\partial E} + \frac{\partial W_\eta}{\partial E}$$

$$\alpha_2 = \frac{\partial W}{\partial \beta_2} = \frac{\partial W_\xi}{\partial \beta_2} + \frac{\partial W_\eta}{\partial \beta_2}.$$

The second equation is the equation for the orbit,

$$2\alpha_2 = -\int^\xi \left[2c^2 E \cosh^2\xi' + 2c(\kappa_1 + \kappa_2)\cosh\xi' - \beta_2\right]^{-1/2} d\xi'$$

$$+ \int^\eta \left[-2c^2 E \cos^2\eta' + 2c(\kappa_1 - \kappa_2)\cos\eta' + \beta_2\right]^{-1/2} d\eta', \quad (14.60)$$

as a function of the parameters E, β_2, and α_2, while the first equation yields the time dependence of the motion.

Thus the problem is reduced to standard integrals that can be found in tables or can be evaluated numerically. This solution, with the help of the Hamilton-Jacobi method, is simple and short – according to Corben and Stehle (see the Bibliography), considerably less difficult than integrating Lagrange's equations directly[10].

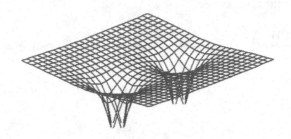

Figure 14.2: The potential V_{2C}; $x \in [-1, 1]$, $y \in [-1, 1]$.

In the limit $c \to 0$, the centers of force coincide. The transition to polar coordinates is established by $\xi \to \infty$, where

$$c \cosh \xi \simeq c \sinh \xi \simeq \frac{c}{2} \exp(\xi) =: r$$

remains finite. Now we only need rename η as φ. Equation (14.60) then implies (recall: $d\xi = dr/r$)

$$2\alpha_2 = -\int^{r} \frac{dr'}{r'^2} \left[2E + 2(\kappa_1 + \kappa_2)/r' - \beta_2/r'^2\right]^{-1/2} + \beta_2^{-1/2}\varphi.$$

The integral yields the expected equation (5.31) for a Keplerian orbit.

Instead of solving the integral in Eq. (14.60), we solve the equations of motion numerically. Taking $\kappa_1 = \kappa_2 = 1/2$ and $c = 1/2$, the Cartesian components of the equation of motion in the gravitational field of the two centers are given by

$$\ddot{x} = -\frac{1}{2} \left(\frac{x - \frac{1}{2}}{\sqrt{\left(x - \frac{1}{2}\right)^2 + y^2}^3} + \frac{x + \frac{1}{2}}{\sqrt{\left(x + \frac{1}{2}\right)^2 + y^2}^3} \right)$$

[10]In his "Analytical Mechanics", Lagrange solves the equations of motion by reducing them to an integral ('quadrature'), which he believes to be 'impossible to solve in general'.

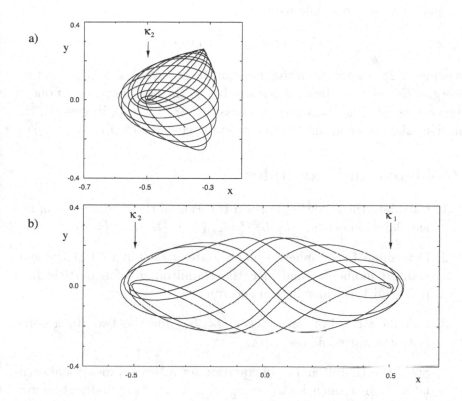

Figure 14.3: Two orbits in the problem of two centers for (a) $E = -2.2$; and (b) $E = -1.8.(\kappa_1 = \kappa_2 = 1/2, c = 1/2)$.

$$\ddot{y} = -\frac{y}{2}\left(\frac{1}{\sqrt{\left(x - \frac{1}{2}\right)^2 + y^2}^3} + \frac{1}{\sqrt{\left(x + \frac{1}{2}\right)^2 + y^2}^3}\right). \quad (14.61)$$

The potential energy,

$$V_{2C}(x, y) = -\frac{1}{2}\left(\frac{1}{\sqrt{\left(x - \frac{1}{2}\right)^2 + y^2}} + \frac{1}{\sqrt{\left(x + \frac{1}{2}\right)^2 + y^2}}\right) \quad (14.62)$$

has two infinitely deep wells at the location of the two centers of force $(-1/2, 0)$ and $(1/2, 0)$. It vanishes for $|x| \to \infty$ as well as for $|y| \to \infty$. Between the wells, there is a saddle point at $S = (0, 0)$. The value of the potential at the saddle point is

$$V_{2C}(S) = V_{2C}(0,0) = -2$$

(see Fig 14.2). Hence, from the shape of the potential, it follows that for energies $E < -2$, the motion is restricted to the neighborhood of one of the centers (cf. Fig. 14.3 (a)), whereas for $-2 < E < 0$, in general, the motion takes place around both attracting centers (cf. Fig. 14.3 (b)).

Problems and examples

1. Calculate the following Poisson brackets of the components of the angular momentum: $[L_i, L_j]$ $(= \varepsilon_{ijk} L_k)$, $[\mathbf{L}^2, L_z]$ $(= 0)$.

2. Determine $\mathbf{L}^2$ in spherical coordinates as a function of the generalized momenta, and give the Hamiltonian of a particle in a potential $V(r)$ in spherical coordinates.

3. Calculate the Poisson bracket $[L_z, H_x]$ for the two-dimensional isotropic harmonic oscillator.

4. Solve the Hamilton-Jacobi equation for a three-dimensional oscillator with frequencies $\omega_x = \omega_y =: \omega \neq \omega_z$ in cylindrical coordinates [Born].

5. Discuss the three-dimensional motion in Euler's three-body problem using cylindrical coordinates.

6. **Elliptic coordinates**: Determine the (local) basis vectors, the Jacobian determinant and the components of the velocity $\dot{\mathbf{r}}$ (cf. Appendix A, Section A.3).

7. Show that the integrals H and B of the problem of two centers of gravitation are independent.

15

Three-body systems

We will now go beyond the two-body problem. Though two-body systems with central field interactions are integrable, in the case of the long-ranged gravitational interaction, they are unrealistic. The gravitational force of further bodies can change the dynamics of the two bodies considered in an essential way. If one takes for instance the Earth and its moon one encounters very quickly irregularities of the moon's motion. Effects which are due to the presence of the sun (at least) appear. Thus the tides cannot be explained by the action of the sun only – as Galileo supposed[1] –, but one has to take into account also the moon. Already Newton had a closer look at three-body systems in Proposition. LXVI, Book I of the "Principia". In Book III he explained thus the *tides* (Prop. XXII) and the *motion of the moon's nodes* [2,3].

The small number of conserved quantities for a system of $N \geq 3$ interacting bodies suggests that looking for an exact analytical solution, in general, will be futile. But in special situations, one can make useful statements about the solutions (see [Thirring] and references in the article by D.G. Saari and Zhihong Xia, Notices of the AMS,**42**,538(1995)).

When more particles are involved, unphysical[4] solutions appear. An

---

[1] *Discorso sul flusso e il reflusso del mare* (Discourse on the Tides), letter to Cardinal Orsini, 1616.

[2] "Of the motion of the moon's nodes" is a separate section inserted between Propositions XXXIII and XXXIV of Book III.

[3] The lunar nodes are the points where the moon's orbit crosses the ecliptic (i.e. the plane of the apparent orbit of the sun).

[4] Regarding the relation between physical reasoning and reality outlined in Chapter

example occurs with $N \geq 4$ particles: solutions exist for which one particle can escape to infinity in finite time ([Thirring]; D.G. Saari and Zhihong Xia, loc.cit.). The cause of such unphysical solutions can be traced back to the assumption of instantaneous interaction (implying the existence of infinite velocities) and the concept of point mass. The first defect is removed in a relativistic field theory of the interaction: the velocity of light is the speed limit for the propagation of the interaction. All these problems and aspects are beyond the scope of this book.

Even the three-body problem is not solvable in general. The general three-body problem has nine degrees of freedom. For a solution, one needs nine integrals of the motion with the properties given in Liouville's theorem (P. 398). The motion of the center of mass with three degrees of freedom is integrable, but the remaining six degrees of freedom for the relative motion generally yield only three conserved quantities: $E, \mathbf{L}^2, L_z$. It would be too ambitious to attempt the solution of the general problem of three gravitationally interacting bodies. We consider only the so-called restricted three body problem.

15.1   The restricted three-body problem

This model for the dynamics of three point masses interacting via the gravitational force is defined by the following assumptions.

i) For one of the point masses, the mass m is so small that the influence on the motion of the two other point masses, m_1 and m_2, can be neglected.

ii) The orbit of the two 'heavy' point masses m_1 and m_2 is circular.

The third assumption is:

iii) The 'light' point mass moves in the plane of the circular orbit. (If the initial position and the initial velocity of the light mass lie in the plane of the circle, then the conservation of angular momentum keeps its orbit in that plane.)

---

one, we may add the following. The images of physical processes can sometimes entail consequences that do not match reality (consequences entailed by nature). The mathematical description may contain information that is not allowed physically. One trivial example occurs every time when an equation has several roots, but physics insists on keeping just some of them (e.g. the positive or real ones).

This model is approximately applicable to the motion of an asteroid[5] in the gravitational fields of the biggest members of our solar system, the sun and Jupiter[6]. Such asteroids moving in the orbital plane of Jupiter actually exist (see later). Poincaré's outstanding work on new methods of celestial mechanics, mentioned already earlier, is primarily concerned with this constellation (see Bibliography).

We first describe briefly how the general three-body problem is reduced to the restricted one. In terms of the coordinates $\mathbf{R}_1$, $\mathbf{R}_2$, $\mathbf{r}_m$ of the three point masses m_1, m_2, and m, respectively, the Lagrangian is given by (E)

$$L_{12m} = \frac{m_1 m_2}{2\mathcal{M}} \left(\dot{\mathbf{R}}_1 - \dot{\mathbf{R}}_2 \right)^2 + \frac{Gm_1 m_2}{|\mathbf{R}_1 - \mathbf{R}_2|} + L_m, \qquad (15.1)$$

where

$$\mathcal{M} := m_1 + m_2 + m \qquad (15.2)$$

is the total mass, and

$$L_m = \qquad\qquad\qquad\qquad\qquad\qquad\qquad\qquad\qquad (15.3)$$
$$m \left[\frac{m_1}{2\mathcal{M}} \left(\dot{\mathbf{r}}_m - \dot{\mathbf{R}}_1 \right)^2 + \frac{m_2}{2\mathcal{M}} \left(\dot{\mathbf{r}}_m - \dot{\mathbf{R}}_2 \right)^2 + \frac{Gm_1}{|\mathbf{r}_m - \mathbf{R}_1|} + \frac{Gm_2}{|\mathbf{r}_m - \mathbf{R}_2|} \right]$$

describes the motion of the light particle m under the influence of the heavy ones, m_1 and m_2. Let the mass m be much smaller than m_1 and m_2. Then the latter two point masses – since they are not (appreciably) influenced by m, *the contribution L_m in total Lagrangian* (15.1) *can be neglected* – move on Keplerian orbits under their mutual interaction. Thus the orbits are ellipses with circles as special case.

Let our frame of reference be the center of mass reference frame. In our case, for the three interacting point masses (without external force), this is an inertial frame. We assume that the origin is situated at the center of mass, i.e.

$$m_1 \mathbf{R}_1 + m_2 \mathbf{R}_2 + m\mathbf{r}_m = \mathbf{0}. \qquad (15.4)$$

---

[5]Asteroids orbit the sun but are not big enough to be considered planets.

[6]The system that first comes to mind – the sun, earth, and moon – does not fit the criteria. Apart from the moon's mass being too large, the moon's orbital plane is inclined with respect to the orbital plane of the earth.

Therefore, we can recast Eq. (15.3) into the form

$$L_m = m \left[\frac{\mathcal{M}+m}{2\mathcal{M}} \dot{\mathbf{r}}_m^2 + \frac{m_1}{2\mathcal{M}} \dot{\mathbf{R}}_1^2 + \frac{m_2}{2\mathcal{M}} \dot{\mathbf{R}}_2^2 + \frac{Gm_1}{|\mathbf{r}_m - \mathbf{R}_1|} + \frac{Gm_2}{|\mathbf{r}_m - \mathbf{R}_2|} \right].$$

We *neglect the light mass* m in the square brackets compared to the sum of the heavier ones,

$$M = m_1 + m_2 \simeq \mathcal{M}, \tag{15.5}$$

since $m/M \ll 1$. Hence, we obtain the Lagrangian,

$$L_m = m \left[\frac{1}{2} \dot{\mathbf{r}}_m^2 + \frac{m_1}{2M} \dot{\mathbf{R}}_1^2 + \frac{m_2}{2M} \dot{\mathbf{R}}_2^2 + \frac{Gm_1}{|\mathbf{r}_m - \mathbf{R}_1|} + \frac{Gm_2}{|\mathbf{r}_m - \mathbf{R}_2|} \right] \tag{15.6}$$

for the motion of the light point mass m under the influence of the two heavier ones, m_1 and m_2, that follow Keplerian orbits $\mathbf{R}_1(t)$ and $\mathbf{R}_2(t)$ about their center of mass – which is situated at the origin, since, in this approximation,

$$\frac{m_1}{M} \mathbf{R}_1 + \frac{m_2}{M} \mathbf{R}_2 = \mathbf{0}. \tag{15.7}$$

$\mathbf{r}_m$ is now the radius vector of m with respect to the center of mass of m_1 and m_2. For L_m, *only* $\mathbf{r}_m$ *and* $\dot{\mathbf{r}}_m$ *are the dynamical variables*; $\mathbf{R}_1$, $\mathbf{R}_2$ *as well as* $\dot{\mathbf{R}}_1$, $\dot{\mathbf{R}}_2$ *are parameters*, whose time dependence is known. This makes L_m explicitly time-dependent, $L_m = L_m(\mathbf{r}_m, \dot{\mathbf{r}}_m, t)$.

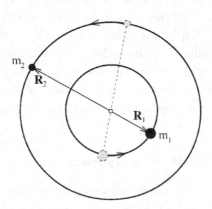

Figure 15.1: The circular orbits of the heavy bodies.

The next simplifying assumption is that the point masses m_1 and m_2 travel along a *circular orbit* about their center of mass. From Eq. (15.7), we see that the radius vectors $\mathbf{R}_1$ and $\mathbf{R}_2$ must satisfy the conditions

$$R_1 =: a = const, \quad R_2 =: b = const \quad \text{with} \quad a/b = m_2/m_1. \tag{15.8}$$

Thus $\mathbf{R}_1$ and $\mathbf{R}_2$ have opposite directions,

$$\mathbf{R}_1 = a\mathbf{n} \quad \text{and} \quad \mathbf{R}_2 = -b\mathbf{n},$$

where $\mathbf{n}$ is a unit vector (cf. Fig. 15.1). Using the condition $a/b = m_2/m_1$, we introduce the parameter λ

$$a/m_2 = b/m_1 =: \lambda . \tag{15.9}$$

In terms of λ, the distance between m_1 and m_2 is

$$|\mathbf{R}_1 - \mathbf{R}_2| = a + b = \lambda M. \tag{15.10}$$

The value of λ depends on the units of length and mass chosen. As shown in Chapter 5 – for motion in the $1/r$ potential – the relative vector $\mathbf{R}_1 - \mathbf{R}_2$, and consequently the vectors $\mathbf{R}_1$, $\mathbf{R}_2$, and $\mathbf{n}$, all rotate with angular velocity

$$\omega = \sqrt{\frac{GM}{(a+b)^3}} = \sqrt{\frac{G}{\lambda^3 M^2}} \tag{15.11}$$

(See Eq. (5.30), and set $k = GM$ and $r_{\min} = |\mathbf{R}_1 - \mathbf{R}_2| = a+b = \lambda M$.)

The time dependence of $\mathbf{R}_1$ and $\mathbf{R}_2$ can be eliminated by transforming to a system rotating with angular velocity ω about the center of mass, with rotation axes parallel to the angular momentum of the pair m_1 and m_2. To determine the Lagrangian in the rotating system, we insert the relation between the velocities in the inertial and the rotating frame – derived in Chapter 8 (cf. (8.56)) –

$$(\dot{\mathbf{r}})_{\text{inertial}} = (\dot{\mathbf{r}})_{\text{rot}} + \boldsymbol{\omega} \times \mathbf{r}$$

into L, Eq. (15.6). Denoting

$$\mathbf{r} := \mathbf{r}_m \quad \text{and} \quad \dot{\mathbf{r}} := (\dot{\mathbf{r}}_m)_{\text{rot}},$$

we obtain

$$L(\mathbf{r}, \dot{\mathbf{r}}, t) = m \left[\frac{1}{2}\dot{\mathbf{r}}^2 + \dot{\mathbf{r}}(\boldsymbol{\omega} \times \mathbf{r}) + \frac{1}{2}(\boldsymbol{\omega} \times \mathbf{r})^2 + \frac{Gm_1}{|\mathbf{r} - \mathbf{R}_1|} + \frac{Gm_2}{|\mathbf{r} - \mathbf{R}_2|} \right],$$

$$(15.12)$$

where the constant contribution from $\dot{\mathbf{R}}_1$ and $\dot{\mathbf{R}}_2$ has been omitted $((\dot{\mathbf{R}}_i)_{\text{rot}} = 0, \ i = 1, 2)$. In order to display the relevant parameters, we express m_1 and m_2 everywhere in terms of the mass M (cf. Eq. (15.5)) and the ratio

$$\mu = m_1/M, \qquad (15.13)$$

so that

$$m_1 = \mu M, \quad m_2 = (1 - \mu) M, \qquad (15.14)$$

and then change the length and time scales. We have

$$\frac{Gm_1}{|\mathbf{r} - \mathbf{R}_1|} = \frac{Gm_1}{|\mathbf{r} - \lambda m_2 \mathbf{n}|} = \frac{GM\mu}{|\mathbf{r} - \lambda M (1 - \mu) \mathbf{n}|} = \frac{G}{\lambda} \frac{\mu}{|\mathbf{r}/\lambda M - (1 - \mu) \mathbf{n}|},$$

and similarly,

$$\frac{Gm_2}{|\mathbf{r} - \mathbf{R}_2|} = \frac{G}{\lambda} \frac{1 - \mu}{|\mathbf{r}/\lambda M + \mathbf{n}|}.$$

The angular velocity $\boldsymbol{\omega}$, Eq. (15.11), is already expressed in terms of these parameters:

$$\boldsymbol{\omega} = \frac{1}{\lambda M} \sqrt{G/\lambda} \, \hat{\boldsymbol{\omega}}, \quad \hat{\boldsymbol{\omega}}^2 = 1, \qquad (15.15)$$

where the unit vector $\hat{\boldsymbol{\omega}}$ is perpendicular to the orbital plane of m_1 and m_2. If one now measures the radius vector $\mathbf{r}$ in units of λM, and time t in units of $\lambda M \sqrt{\lambda/G}$ – by setting[7]

$$\mathbf{r}' = \mathbf{r}/(\lambda M) \quad \text{and} \quad t' = t \Big/ \left(\lambda M \sqrt{\lambda/G} \right), \qquad (15.16)$$

then rescaling the Lagrangian via $L' = \dfrac{\lambda}{G} L$, and omitting dashes for notational convenience – we arrive at the **Lagrangian in the rotating**

---

[7]If $m_1 \ll m_2$, as is the case for the sun and another planet, then $b \simeq 0$, $m_2 \simeq M$ and consequently $\lambda \simeq a/M$. The unit of length λM in this case is just the distance $a + b \simeq a$ between m_1 and m_2. The natural unit of time is then the period T_1 of the body m_1, since $\lambda M \sqrt{\lambda/G} \simeq a^{3/2} \sqrt{1/MG} = T_1/2\pi$ (cf. Eq. (6.42)).

system:

$$L = m \left[\frac{1}{2}\dot{\mathbf{r}}^2 + \dot{\mathbf{r}}(\hat{\boldsymbol{\omega}} \times \mathbf{r}) + \frac{1}{2}(\hat{\boldsymbol{\omega}} \times \mathbf{r})^2 + \frac{\mu}{|\mathbf{r} - (1 - \mu)\,\mathbf{n}|} + \frac{1 - \mu}{|\mathbf{r} + \mu\mathbf{n}|} \right].$$

$$\tag{15.17}$$

The only parameter in this Lagrangian is the mass ratio μ. Therefore the orbits of m depend only on μ. In the 'natural' units, Eq. (15.16), the parameters a and b reduce to

$$a = \lambda m_2 = 1 - \mu \quad \text{and} \quad b = \lambda m_1 = \mu \tag{15.18}$$

and the distance between m_1 and m_2 is

$$|\mathbf{R}_1 - \mathbf{R}_2| = 1. \tag{15.19}$$

Now there is no explicit time dependence in L anymore. Therefore the *Jacobian integral* (cf. Section 10.3),

$$I_J = \frac{\partial L}{\partial \dot{\mathbf{r}}}\dot{\mathbf{r}} - L$$

$$= m \left[\frac{1}{2}\dot{\mathbf{r}}^2 - \frac{1}{2}(\hat{\boldsymbol{\omega}} \times \mathbf{r})^2 - \frac{\mu}{|\mathbf{r} - (1 - \mu)\,\mathbf{n}|} - \frac{1 - \mu}{|\mathbf{r} + \mu\mathbf{n}|} \right] \tag{15.20}$$

is conserved. Expressing $\dot{\mathbf{r}}$ in the Jacobian integral in terms of the canonical momentum,

$$\mathbf{p} = \frac{\partial L}{\partial \dot{\mathbf{r}}} = m(\dot{\mathbf{r}} + \hat{\boldsymbol{\omega}} \times \mathbf{r}), \tag{15.21}$$

the Jacobian integral is equal to the **Hamiltonian in the rotating system:**

$$H = \frac{1}{2m}\mathbf{p}^2 - \hat{\boldsymbol{\omega}}(\mathbf{r} \times \mathbf{p}) - \frac{\mu m}{|\mathbf{r} - (1 - \mu)\,\mathbf{n}|} - \frac{(1 - \mu)\,m}{|\mathbf{r} + \mu\mathbf{n}|}. \tag{15.22}$$

The last restriction in the definition of the model – the requirement that the *motion of m be in the plane* orthogonal to $\hat{\boldsymbol{\omega}}$ – can be fulfilled by appropriate initial values. We may therefore restrict ourselves to a two-dimensional description in the (x, y)-plane (in this case $\hat{\boldsymbol{\omega}} = (0, 0, 1)$). We choose the x-axis to lie along the line passing through bodies m_1 and m_2, so that (cf. Eq. (15.18))

$$\mathbf{R}_1 = (a, 0) = (1 - \mu, 0), \qquad \mathbf{R}_2 = (-b, 0) = (-\mu, 0). \tag{15.23}$$

The distances of the small mass m to m_1 and m_2 are therefore given by

$$
\begin{aligned}
r_1 &= |\mathbf{r} - \mathbf{R}_1| = |\mathbf{r} - (1 - \mu)\,\mathbf{n}| = \sqrt{(x - (1 - \mu))^2 + y^2} \\
r_2 &= |\mathbf{r} - \mathbf{R}_2| = |\mathbf{r} + \mu\mathbf{n}| = \sqrt{(x + \mu)^2 + y^2}. \quad (15.24)
\end{aligned}
$$

Choosing the unit of mass such that $m = 1$, the Hamiltonian is finally

$$
H = \frac{1}{2}(p_x^2 + p_y^2) + (p_x y - p_y x) - \frac{\mu}{r_1} - \frac{1 - \mu}{r_2}. \quad (15.25)
$$

We have reduced the problem of the motion of a 'light' point mass in the gravitational fields of two 'heavy' point masses (with three simplifying assumptions) to its most simple form. The *problem of two centers of gravitation,* investigated in the last chapter, is related to the restricted three-body problem. By insisting the heavy bodies be at rest (in the inertial frame), i.e. setting $\hat{\omega} = 0$ in the Lagrangian (15.17) and in the Hamiltonian (15.22), we arrive at the Hamiltonian (14.46)

If one succeeds in finding two integrals of the motion, the restricted three-body problem would be integrable. The only known integral is the energy in the rotating frame of reference, the Jacobian integral I_J, Eq. (15.20)

$$
I_J = \frac{1}{2}(\dot{x}^2 + \dot{y}^2) - \frac{1}{2}(x^2 + y^2) - \frac{\mu}{r_1} - \frac{1 - \mu}{r_2}. \quad (15.26)
$$

There is no second conserved quantity in sight that would make the problem integrable. Apparently chaotic solutions of the system – obtained numerically – suggest that no other isolating integral of the motion exists.

15.2 Solutions of the restricted three-body problem

From the Hamiltonian (15.25), we obtain the canonical equations ($p_x = \dot{x} =: u$, $p_y = \dot{y} =: v$)

$$
\begin{aligned}
\dot{x} &= u \\
\dot{y} &= v \\
\dot{u} &= 2v + x - \frac{\mu(x - (1 - \mu))}{r_1^3} - \frac{(1 - \mu)(x + \mu)}{r_2^3} \quad (15.27)
\end{aligned}
$$

$$\dot{v} \;=\; -2u + y - \frac{\mu y}{r_1^3} - \frac{(1-\mu)\,y}{r_2^3},$$

where r_1 and r_2 are given by Eq. (15.24). The *stationary solutions* of the equations of motion are obtained from the conditions $\dot{x} = \dot{y} = \dot{u} = \dot{v} = 0$; i.e.,

$$\frac{\mu(x - (1-\mu))}{r_1^3} + \frac{(1-\mu)\,(x+\mu)}{r_2^3} - x \;=\; 0 \qquad (15.28)$$

$$\left(\frac{\mu}{r_1^3} + \frac{(1-\mu)}{r_2^3} - 1 \right) y \;=\; 0. \qquad (15.29)$$

These equations yield also the extrema of the potential

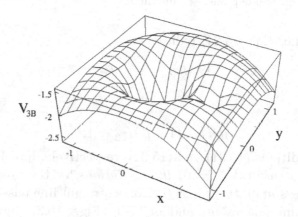

Figure 15.2: The potential $V_{3B}\,(x,y)$ at the mass ratio $\mu = 0.1$.

$$V_{3B}\,(x,y) = -\frac{1}{2}(x^2 + y^2) - \frac{\mu}{r_1} - \frac{1-\mu}{r_2}. \qquad (15.30)$$

Figure 15.2 shows the potential surface for $\mu = 0.1$. The positions of the two heavy masses m_1 and m_2 are at the centers of the two wells. In Fig. 15.3, the equipotential lines of $V_{3B}\,(x,y)$ are shown for three different values of μ. In each plot, three saddle points $(+)$ and two maxima $(\star)$ are clearly visible. These five equilibrium positions are called **Lagrangian points**.

Equation (15.29) implies either

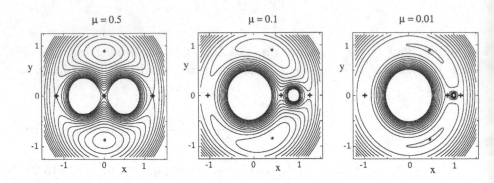

Figure 15.3: The equipotential lines and the extrema of $V_{3B}(x,y)$ for $\mu = 0, 5, 0.1, 0.01$; $+$: saddlepoint, $\bigstar$: maximum.

a) $y = 0$, i.e. Eq. (15.24) reduces to

$$r_1 = |x - (1-\mu)| \quad \text{and} \quad r_2 = |x + \mu|, \tag{15.31}$$

and/or

b)

$$\mu/r_1^3 + (1-\mu)/r_2^3 = 1. \tag{15.32}$$

Inserting conditions (15.31) and (15.32), respectively, into Eq. (15.28), one obtains in case (a) $(y = 0)$ *three solutions* for the position of the light point mass m on the x-axis (i.e. on a straight line passing through the heavy point masses m_1 and m_2). In Fig. 15.3, these positions are the three saddle points $(+)$. Stability analysis shows that these solutions are unstable. No actual configurations of celestial bodies seem to correspond to these collinear solutions.

In case (b), from Eq. (15.28), we find

$$1/r_1^3 - 1/r_2^3 = 0,$$

which, together with Eq. (15.32), yields

$$r_1 = r_2 = 1 = (a + b) = |\mathbf{R}_1 - \mathbf{R}_2|. \tag{15.33}$$

Hence the coordinates of the fixed points are (recall that we are using the natural units defined in Eq. (15.16))

$$x_0 = \frac{1}{2} - \mu, \quad y_0 = \pm\frac{\sqrt{3}}{2}. \tag{15.34}$$

These *two equilibrium positions* for the light point mass m in respect to the heavy bodies at $\mathbf{R}_1 = (1 - \mu, 0)$ and $\mathbf{R}_2 = (-\mu, 0)$ form *equilateral triangles*, regardless of the mass ratio μ. In Figs. 15.2 and 15.3 these positions of the light point mass appear as maxima ($\star$) of the potential surface! Linear stability analysis of Eqs. (15.27) shows, however, that these stationary points are stable solutions as long as $\mu < 0.038$ (**E**). The Coriolis force prevents the point mass m from dropping off the maxima. The particle may move in some neighborhood of the maxima remaining close to them. For decreasing values of the mass ratio μ, the ridges of $V_{3B}(x, y)$ – whose peaks are the maxima – become narrower and longer (cf. Fig. 15.3). The region of stability expands in the same manner. The mass can oscillate along the ridges about the equilibrium positions, whereas there is (nearly) no motion perpendicular to the ridges. Finally, for very small values of μ (e.g. $\mu = 0.001$), the light particle can even go from one stable point to the other; the orbits are horseshoe-shaped, going around both stationary points[8].

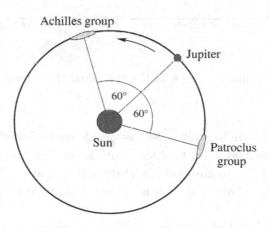

Figure 15.4: The Trojans.

Celestial configurations corresponding to the triangular solutions have been found; they consist of the sun, Jupiter ($\mu \simeq m_{Jupiter}/m_{Sun} \simeq 0.001$), and any of the asteroids in the **Trojan groups**. Both the Trojan groups have the same orbit and velocity as Jupiter. The Trojan groups

---

[8]E.W. Schmid, G. Spitz, and W. Lösch, "Theoretische Physik mit dem Personal Computer", Springer, Berlin 1987.

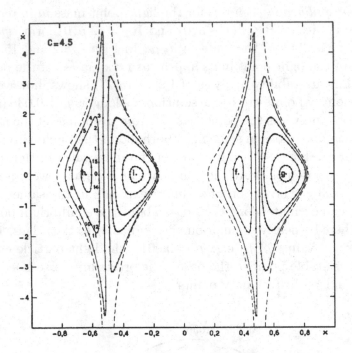

Figure 15.5: Poincaré section of the restricted three body problem for $C = 4.5$ (see text).

comprise of approximately 2,300 asteroids with diameters greater than 15 kilometers. About 1,300 of these are located in the leading **Achilles group**, and 1,000 in the trailing **Patroclus group**. The sun, Jupiter, and each of the Trojan groups form an equilateral triangle as sketched in Fig. 15.4.

M. Hénon[9] performed a detailed numerical investigation of the equations of motion (15.27) for

$$\mu = 1/2.$$

At this value of the mass ratio, the three collinear Lagrangian points – the saddle points (+) in the left plot of Fig. 15.3 – are situated at

$$L_1 = (0,0), \quad L_2 = \left(\sqrt{5}/2, 0\right), \quad L_3 = \left(-\sqrt{5}/2, 0\right). \tag{15.35}$$

---

[9]M. Hénon, "Numerical Exploration of Hamiltonian Systems", in "Chaotic Behavior of Deterministic Systems", Les Houches 1981, North Holland 1983.

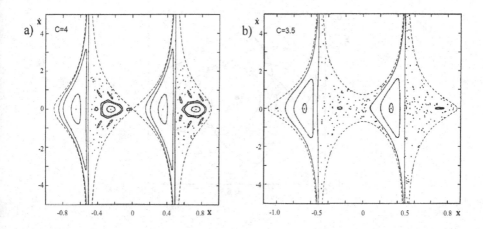

Figure 15.6: Two more Poincaré sections in the restricted three body problem.

The two remaining Lagrangian points ($\star$) – at the corners of the equilateral triangles – have the planar coordinates

$$L_4 = \left(0, \sqrt{3}/2\right), \quad L_5 = \left(0, -\sqrt{3}/2\right). \tag{15.36}$$

For those values of the Jacobian integral (15.26), that are smaller than

$$V_{3B}(L_1) = V_{3B}(0,0) = -2, \tag{15.37}$$

for appropriate initial values of the dynamical variables, the motion of the light mass always remains inside of a single well of the potential shown in Fig. 15.3. If the energy I_J is larger than $V_{3B}(L_1) = -2$ but smaller than

$$V_{3B}(L_2) = V_{3B}(L_3) = -\left(5/8 + \sqrt{5}/2\right) = -1.7430, \tag{15.38}$$

its motion is restricted to a neighborhood of the two heavy bodies. (The values of the potential at the extrema L_4 and L_5 are: $V_{3B}(L_4) = V_{3B}(L_5) = -\frac{11}{8} = -1.375$.)

Hénon used the parameter $C = -2I_J$, so that instead the limit in Eq. (15.37), his limit is

$$C > 4 \tag{15.39}$$

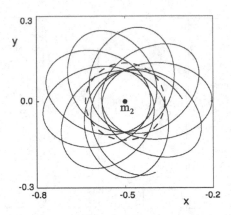

Figure 15.7: A non-periodic orbit around m_2.

for bounded motion of the light mass only about a *single* heavy body; and for bounded motion around both bodies, the limits are

$$\left(5/4 + \sqrt{5}\right) (= 3.\,4861) < C < 4. \tag{15.40}$$

As in Chapter 4, the dynamics of this two-dimensional system can be represented by a Poincaré map – the sequence of intersection points with a suitably chosen Poincaré section in the reduced phase space with coordinates $(x, y, \dot{x})$ ($\dot{y}$ is eliminated with the help of Eq. (15.26)). Choosing $y = 0$ as the Poincaré surface, the dashed curves in the Poincaré maps of Figs. 15.5 and 15.6 border the accessible region of the variables x and $\dot{x}$. These curves are obtained from Eq. (15.26) by setting $y = \dot{y} = 0$.

There are four fixed points in the Poincaré section, corresponding to closed orbits around each of the heavy bodies: h and i for the motion around m_2 and f, g for the motion around m_1. The orbit belonging to the fixed point h is shown as a dashed circle in Fig. 15.7. For sufficiently large values of C (i.e. low values of I_J), the sequences of intersection points seem to indicate that the system is integrable; they appear to lie on closed curves around the fixed points (Fig. 15.5, $C = 4.5$). The points on the curve around the fixed point h are numbered to display the temporal sequence of a trajectory's repeated returns to the Poincaré section. A corresponding orbit is shown in Fig. 15.7. This orbit is not closed; the motion is only nearly periodical. After a full period (that

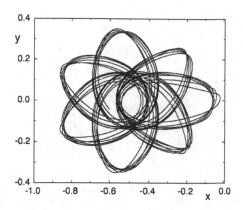

Figure 15.8: A band-like orbit around m_2.

is the time between two successive passes of the pericenter), the orbit is similar to a Keplerian ellipse, but due to the existence of the second heavy body the 'ellipse' rotates. All return sequences situated on curves around h, correspond to orbits having the same sense of rotation around the body m_1. The same applies to return sequences near the fixed point f. The fixed points i and g, as well as the neighboring curves of return sequences, correspond to orbits with a *reversed sense of rotation* around the respective body. (For these values of I_J, the sense of rotation in an orbit does not change.) The asymmetry about the ($x = 0$)-line in Figs. 15.5 and 15.6 results from the sense of rotation of the two bodies m_1 and m_2 on their circular orbits (cf. Fig. 15.1).

As the energy I_J is increased – i.e. the value of C is lowered – sequences of 'islands' belonging to a single orbit appear. So for $C = 4$ in Fig. 15.6 (a), a sequence of six 'islands' appears. The corresponding orbit in configuration space is shown in Fig. 15.8. Over a long enough period of time, the orbit would fill the band that is already taking shape. The appearance of such orbits is a characteristic of chaotic behavior. Such a behavior is clearly present at this value of C. In Fig. 15.6 (a) all the 'scattered' points belong to a *single* orbit.

The chaotic regions grow for decreasing values of C (see, e.g. $C = 3.5$, Fig. 15.6 (b)). Figure 15.9 shows a seemingly completely irregular orbit at[10] $C = 3.5$. The difference to the regular orbits in the problem

---

[10]The value $C = 3$ given by Hénon in his Fig. 29 must be a misprint. For this

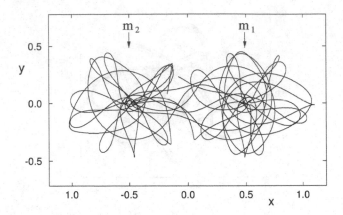

Figure 15.9: An evidently chaotic orbit around both heavy bodies.

of two centers can clearly be seen (cf. Fig. 14.3 (b)). In particular, in the orbit shown in Fig. (15.9), the sense of rotation about the heavy bodies changes in an irregular way.

Some beautiful regular orbits are shown in Fig. 15.10 for several values of μ (H.J. Scholz, Praxis der Naturwissenschaften – Physik 7,10,1987).

15.3   Is our planetary system also chaotic?

Even this simplified model for the motion of three interacting bodies shows chaotic behavior[11]. If the model is applicable to a real configuration of celestial bodies, then their motion cannot be predicted in general even if there current positions and velocities are known. The initial values are never known to perfect accuracy; therefore exact prediction of the light body's orbit is impossible in the long term.

---

value, the motion is already unbounded.

[11] According to A. Moszkowski ("Einstein – Einblicke in seine Gedankenwelt" (Einstein – insights in his ideas), Fontane, Berlin 1922), Einstein was of the opinion that the three body problem can be considered as solved by the mere statement of the equations of motion, since their numerical solution can be determined to any given precision, 'so that in this respect, one hardly can expect undreamt-of conclusions out of future revolutions of science'.

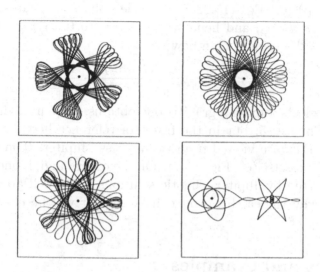

Figure 15.10: Several nonchaotic orbits in the restricted three body problem.

These situation may be worrying if one thinks of the stability and regularity of the planetary motion in our solar system (recall Lagrange's clock!). Why is the simple explanation of the planetary orbits by considering only the interaction of the sun and the respective planet so exact?

To gain some understanding, we briefly investigate the influence of our solar system's biggest planet, Jupiter, on the earth's motion around the sun. The earth's motion seems to be regular and by no means chaotic. Due to the relative mass ratios, the center of mass of the system comprising of the sun, Jupiter, and earth is practically located at the center of the sun, i.e. to a good approximation, we can set $\mathbf{R}_1$ $(= \mathbf{R}_S) = 0$. The Jovian orbit is assumed to be circular, and Jupiter sweeps around with constant angular velocity ω_J. This is exactly the configuration of the restricted three body problem (the orbits of the earth and Jupiter are (nearly) coplanar), with

$$m_1 = m_{S(un)}, \qquad m_2 = m_{J(upiter)} \quad \text{and} \quad m = m_{E(arth)}.$$

We pass to the frame of reference rotating with Jupiter, keeping the sun in the origin. As a result the Hamiltonian acquires an additional Coriolis term $\boldsymbol{\omega}_J(\mathbf{r} \times \mathbf{p})$ (cf. Eq. (15.22)). In plane polar coordinates, we have $\mathbf{p} = p_r\mathbf{e}_r + (p_\varphi/r)\mathbf{e}_\varphi$, and hence (with $\boldsymbol{\omega}_J = (0,0,\omega_J)$) $\boldsymbol{\omega}_J(\mathbf{r} \times \mathbf{p}) = \omega_J p_\varphi$, so that the Hamiltonian now reads

$$H = \frac{1}{2m_E}\left(p_r^2 + p_\varphi^2/r^2\right) - \omega_J p_\varphi - \frac{Gm_S m_E}{r} - \frac{Gm_J m_E}{|\,\mathbf{r} - \mathbf{R}_J\,|}. \qquad (15.41)$$

If one neglects Jupiter's gravity, one obtains, of course, the ellipses we met in Chapter 5; but in the frame of reference here, the *ellipses rotate*, since they are viewed from (a massless) Jupiter. The orbit has the shape of a rosette (cf. Fig. 15.7). On the other hand, if one includes the force exerted by Jupiter, chaotic solutions could well appear. The question of why the earth's orbit can be regular will be elucidated in the next chapter.

Problems and examples

1. Derive the equations of motion (15.27) for the restricted three body problem from the Hamiltonian (15.25). For the triangular solution (15.34), determine the stability limit for the mass ratio μ.

2. A further approach to solutions of the **planar three-body problem** is based on an idea of Lagrange. He wondered which triangles – formed by three bodies – remain similar in the course of the bodies' motion.

   Since the center of mass motion is trivial, it suffices to consider the relative motion only. In the center of mass frame, the coordinates of the three bodies, $\mathbf{r}_j = (x_j, y_j)$, $j = 1,2,3$ satisfy the condition $\sum_j m_j \mathbf{r}_j = \mathbf{0}$. The following ansatz encodes the assumption that the triangle (or, in general, for more than three bodies in a plane, the polygon) formed by the coordinates $\mathbf{r}_j(t)$ always remains similar (cf. C.L. Siegel and J.K. Moser, Lectures on Celestial Mechanics, Springer, Berlin 1995; [Thirring]):

   $$x_j(t) + iy_j(t) = c_j z(t), \qquad c_j \text{ and } z(t) \text{ complex}, \quad j = 1,2,3.$$

   The equation of motion can now be separated into an ordinary differential equation for the time-dependent scaling factor $z(t)$,

   $$\ddot{z} = -\omega^2 \frac{z}{|z|^3}, \qquad (15.42)$$

and three coupled algebraic equations for the constant quantities c_j,

$$-\omega^2 c_j = G \sum_{k \neq j} \frac{c_k - c_j}{|c_k - c_j|^3} m_k \qquad (15.43)$$

(ω^2 is the separation constant). Obviously the c_j satisfy the condition $\sum_j m_j c_j = 0$. The solutions of Eq. (15.43) are the stable configurations of the three bodies. One finds the familiar situation of an equilateral triangle ($|c_k - c_j| = a, \forall j, k \to \omega^2 = G(m_1 + m_2 + m_3)/a^3$) and a linear configuration ($\operatorname{Re} c_j = 0, \forall j$, consequently there are two equations for the two independent 'distances' ($\operatorname{Im} c_1 - \operatorname{Im} c_2$) and ($\operatorname{Im} c_2 - \operatorname{Im} c_3$)). Equation (15.42) is independent of the number of bodies. The solutions describe various changes in the polygon formed by the bodies (rotation, change in size).

The method can be generalized to N bodies in a plane. (For $N = 2$, Eq. (15.43) can always be solved trivially: the line between two bodies can change only into a line when the bodies move.)

16

Approximating
non-integrable systems

Only a few dynamical systems in classical mechanics are solvable exactly, while very many more can be solved only approximately or numerically. Some of these are extensions of an exactly solvable system. In such cases, the exactly solvable system can be taken as the starting point in a perturbative calculation. Even if the perturbative expansion does not converge, it may shed light on the behavior of the system.

16.1 Action-angle variables

16.1.1 Definition and general properties

In Section 14.1, we sketched the proof of Liouville's theorem on integrability of classical dynamical systems. We showed that bounded motion in an integrable system is equivalent to motion on the surface of a torus. Now we want to explicitly map *bounded motion in a completely separable system* onto a torus. *Completely separable* means that in appropriate coordinates q_i, Hamilton's characteristic function is a sum of f functions W_i,

$$
W = \sum_{i=1}^{f} W_i(q_i, \boldsymbol{\beta}),
\tag{16.1}
$$

where each function W_i depends solely on the coordinate q_i. (The quantities β_i are constant parameters; see Section 14.2.) This implies that each of the conjugate momenta p_i,

$$p_i = \frac{\partial W}{\partial q_i} = \frac{dW_i(q_i, \boldsymbol{\beta})}{dq_i}, \qquad i = 1, ..., f,$$

depends *exclusively on the conjugate coordinate* q_i:

$$p_i = p_i(q_i). \qquad (16.2)$$

Examples of such completely separable systems include the two-dimensional harmonic oscillator, and a particle in a central force field, both discussed in the previous chapter.

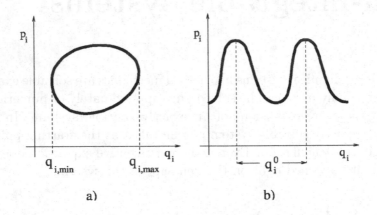

Figure 16.1: Two kinds of motion: a) libration, b) rotation.

In completely separable systems, the coordinates q_i can be classified into two types. Consider, for example, the rosette-like orbits in a central force field as shown in Fig. 5.4. The variable r is periodic, and takes values in the interval $[r_{min}, r_{max}]$, while the angle φ runs through the interval $[0, 2\pi)$. The motion as a whole is aperiodic. In general, in a system with bounded motion, two kinds of time dependence of the generalized coordinates are possible[1]:

i) The coordinate q_i, as well as the conjugate momentum p_i, Eq. (16.2), are periodic with the same period (as illustrated in

---

[1]The following is kind of Gedankenexperiment. In reality, the values taken by the different coordinates do not vary separately.

Fig. 16.1 (a)). The domain of each q_i is the interval $[q_{i,\min}, q_{i,\max}]$. (In the rosette-like orbit above, this is the radial coordinate.) The corresponding motion is called **libration**. In a full cycle, q_i runs from $q_{i,\min}$ to $q_{i,\max}$ and back to $q_{i,\min}$.

ii) The momentum p_k is periodic in coordinate q_k with period q_k^0,

$$p_k(q_k + q_k^0) = p_k(q_k).$$

(In the rosette-like orbit above, this is the polar coordinate φ.) The corresponding motion is called *rotation*. A single cycle of motion occurs after q_k increases by q_k^0 (see Fig. 16.1 (b)).

To each coordinate q_i we now associate an **action variable**[2] J_i, defined by

$$J_i = \frac{1}{2\pi} \oint p_i dq_i, \qquad i = 1, \ldots, f. \tag{16.3}$$

(No summation convention is used in this section!) The integral $\oint$ indicates that the domain of integration is over a full cycle, either by returning to the initial value of q_i or by increasing the value of q_k by q_k^0.

Since we assume that H is time independent, completing a cycle at different times must lead to the same result for J_i. Hence J_i is time independent, and therefore an integral of the motion. So it suggests itself to transform the variables $\mathbf{q}$ and $\mathbf{p}$ to new variables via the canonical transformation generated by a – yet unknown – characteristic function $W(\mathbf{q}, \mathbf{J})$, i.e. with $\mathbf{J} = (J_1, \ldots, J_f)$ as the f new, constant momenta. The variables conjugate to the J_i are the **angle variables** θ_i that are calculated using (recall Eq. (14.11))

$$\theta_i = \frac{\partial W}{\partial J_i}. \tag{16.4}$$

In this transformation, $H(\mathbf{q}, \mathbf{p})$ is replaced by the new Hamiltonian $\hat{H}(\boldsymbol{\theta}, \mathbf{J})$, $\boldsymbol{\theta} = (\theta_1, \ldots, \theta_f)$ (cf. Eq. (14.12)). The definition (16.3) shows that the action variables J_i have dimensions of angular momentum. Consequently, the θ_i are dimensionless (angles).

---

[2]Action variables play a central role in the quantum theory of Bohr, Sommerfeld, and Born.

Since the J_i are constant, the canonical equations

$$-\frac{\partial \hat{H}}{\partial \theta_i} = \dot{J}_i = 0$$

show that $\hat{H}$ is independent of the angle variables θ_i:

$$\hat{H} = \hat{H}(\mathbf{J}) = E(\mathbf{J}). \tag{16.5}$$

The second set of canonical equations

$$\dot{\theta}_i = \frac{\partial \hat{H}}{\partial J_i} =: \omega_i \ (= const)\,, \tag{16.6}$$

implies that the time dependence of the angle variables is given by

$$\theta_i = \omega_i t + \theta_i^0. \tag{16.7}$$

Now we focus on the characteristic function generating the transformation. Since the system is supposed to be completely separable (cf. Eq. (16.1); note that each W_i depends only on q_i), and because of $p_i = dW_i/dq_i$, the definition (16.3) of the action variables can be written as

$$J_i = J_i\,(\boldsymbol{\beta}) = \frac{1}{2\pi} \oint \frac{dW_i}{dq_i} dq_i. \tag{16.8}$$

Inverting these relations for $\boldsymbol{\beta}$, i.e. after finding

$$\boldsymbol{\beta} = \boldsymbol{\beta}\,(\mathbf{J})\,,$$

we have

$$W = \sum_i W_i(q_i, \mathbf{J}). \tag{16.9}$$

Therefore, in general, the angle variables θ_i,

$$\theta_i = \partial W / \partial J_i,$$

still depend on *all* q_i. Because the motion is bounded, each q_i only takes values in a finite interval $[q_{i,\min}, q_{i,\max}]$. In this interval, the function

$$W_i(q_i) = \int p_i dq_i$$

(obtained from $p_i = dW_i/dq_i$) is not unique, since otherwise we would have

$$J_i = \frac{1}{2\pi} \oint \frac{dW_i}{dq_i} dq_i = \frac{1}{2\pi} \oint dW_i = 0.$$

So it must hold that

$$J_i = \frac{1}{2\pi} \Delta W_i, \tag{16.10}$$

where ΔW_i is the increase in W_i when q_i completes one period. If q_i describes librations, then the sign of p_i ($= dW_i/dq_i$) depends on the direction in which the interval $[q_{i,\min}, q_{i,\max}]$ is traversed[3]; and similarly for W_i. The integral over the cycle is then

$$J_i = \frac{1}{2\pi} \oint \frac{dW_i}{dq_i} dq_i = \frac{1}{2\pi} \left(\int_{q_{i,\min}}^{q_{i,\max}} \frac{dW_i}{dq_i} dq_i - \int_{q_{i,\max}}^{q_{i,\min}} \frac{dW_i}{dq_i} dq_i \right)$$

$$= \frac{1}{\pi} \int_{q_{i,\min}}^{q_{i,\max}} \frac{dW_i}{dq_i} dq_i.$$

If the coordinate is an angle, $q_i = \varphi$, the increase is simply defined by

$$J_i = \frac{1}{2\pi} \oint \frac{dW_i}{dq_i} dq_i = \frac{1}{2\pi} \int_0^{2\pi} \frac{dW_i}{d\varphi} d\varphi.$$

In practice, one determines a characteristic function $W(\mathbf{q}, \boldsymbol{\beta})$ as a function of constant momenta β_i. Then, with $p_i = \partial W/\partial q_i$, one calculates the action variables J_i according to Eq. (16.3) as functions of the β_i. The inverse of these relations, $\beta_i = \beta_i(\mathbf{J})$, is then inserted into W in order to obtain the desired characteristic function $W(\mathbf{q}, \mathbf{J})$. These steps are illustrated in the following examples.

16.1.2   Transforming to action and angle variables

The two-dimensional harmonic oscillator

In the previous chapter, we derived from the Hamiltonian (14.23) the derivatives $\partial W_x/\partial x$ and $\partial W_y/\partial y$ of the characteristic function (cf. Eqs.

---

[3]In fact, this can be seen in the solutions. For example, in Eqs. (14.26) and (14.27) actually two signs appear for the root.

(14.25)). Using relation (16.8) for the action variables, we obtain

$$J_1 = \frac{1}{2\pi} \oint \frac{\partial W_x}{\partial x} dx$$

$$= \frac{1}{2\pi} \oint (\pm) \sqrt{2m(\beta_1 - \beta_2) - m^2\omega_x^2 x^2} dx, \quad \beta_1 - \beta_2 > 0, \quad (16.11)$$

$$J_2 = \frac{1}{2\pi} \oint \frac{\partial W_y}{\partial y} dy$$

$$= \frac{1}{2\pi} \oint (\pm) \sqrt{2m\beta_2 - m^2\omega_y^2 y^2} dy, \quad \beta_2 > 0, \quad (16.12)$$

where we have explicitly indicated the two roots for the derivatives of W. The y-integration starts at one zero of the root, say $y_{min} = -\sqrt{2m\beta_2}/m\omega_y$, and extends to the other, $y_{max} = \sqrt{2m\beta_2}/m\omega_y$. Then the integration runs back to y_{min}, with the momentum $p_y = -\sqrt{2m\beta_2 - m^2\omega_y^2 y^2}$ now having opposite sign, so that

$$J_2 = \frac{1}{2\pi} \int_{y_{min}}^{y_{max}} dy \sqrt{2m\beta_2 - m^2\omega_y^2 y^2} - \frac{1}{2\pi} \int_{y_{max}}^{y_{min}} dy \sqrt{2m\beta_2 - m^2\omega_y^2 y^2}$$

$$= \frac{1}{\pi} \int_{y_{min}}^{y_{max}} dy \sqrt{2m\beta_2 - m^2\omega_y^2 y^2}$$

$$= \beta_2/\omega_y. \quad (16.13)$$

Similarly, we obtain for the x-component of the motion,

$$J_1 = (\beta_1 - \beta_2)/\omega_x. \quad (16.14)$$

The inverted relations

$$\beta_2 = \omega_y J_2 \quad \text{and} \quad \beta_1 - \beta_2 = \omega_x J_1 \quad (16.15)$$

imply that

$$\hat{H}(J_1, J_2) = E = \beta_1 = \omega_x J_1 + \omega_y J_2, \quad (16.16)$$

since (cf. Eq. (14.25)) $\beta_1 = E$ and $\beta_2 = E_y$. Expressing now the β_i in the generator $W = W_x + W_y$ (see Eqs. (14.26) and (14.27)) in terms of the action variables J_i, then calculating the angle variables θ_i according to Eq. (16.4), one finds

$$\theta_1 = \arcsin\left(x/\sqrt{2J_1/m\omega_x}\right)$$

$$\theta_2 = \arcsin\left(y/\sqrt{2J_2/m\omega_y}\right). \tag{16.17}$$

Equation (16.6) then yields the frequencies

$$\omega_1 = \omega_x, \qquad \omega_2 = \omega_y. \tag{16.18}$$

The frequencies ω_i are equal to the frequencies of the harmonic oscillator, and are independent of the action variables J_i. This is a peculiarity of the harmonic oscillator (remember, its periods are independent from the displacement and the energy!). The time dependence of the angle variables θ_i is given by Eq. (16.7):

$$\theta_1 = \omega_x t + \theta_1^0 \qquad \text{and} \qquad \theta_2 = \omega_y t + \theta_2^0. \tag{16.19}$$

Inverting relations (16.17) yields the first part of the canonical transformation,

$$x = \sqrt{2J_1/m\omega_x}\,\sin\theta_1$$

$$y = \sqrt{2J_2/m\omega_y}\,\sin\theta_2 \tag{16.20}$$

and by Eqs. (16.19), we have the familiar solution given in Section 4.1. The second part of the transformation, the equations for the conjugate momenta, are obtained from $p_x = \partial W_x/\partial x$ and $p_y = \partial W_y/\partial y$:

$$p_x = \sqrt{2m\omega_x J_1 - m^2\omega_x^2 x^2} = \sqrt{2m\omega_x J_1 - m^2\omega_x^2 \frac{2J_1}{m\omega_x}\sin^2\theta_1}$$

$$= \sqrt{2m\omega_x J_1}\,\cos\theta_1 \tag{16.21}$$

and similarly

$$p_y = \sqrt{2m\omega_y J_2}\,\cos\theta_2. \tag{16.22}$$

The isotropic oscillator in polar coordinates

It is instructive to contrast the foregoing treatment of the oscillator in the case $\omega_x = \omega_y =: \omega$ to a treatment involving polar coordinates.

The Hamiltonian for the isotropic oscillator (cf. Eq. (14.23)) in polar coordinates (ρ, φ) reads

$$H = \frac{1}{2m} \left(p_\rho^2 + \frac{p_\varphi^2}{\rho^2} \right) + \frac{m\omega^2}{2} \rho^2. \tag{16.23}$$

Since φ is cyclic, we have $p_\varphi = \beta_\varphi = const$, and hence we make the following ansatz for W:

$$W(\rho, \varphi) = W_\rho(\rho) + \beta_\varphi \varphi.$$

The Hamilton-Jacobi equation (14.15) then reduces to

$$\frac{\partial W_\rho}{\partial \rho} = (\pm) \sqrt{2mE - \frac{\beta_\varphi^2}{\rho^2} - (m\omega)^2 \rho^2}, \tag{16.24}$$

such that the action variables are

$$J_\rho = \frac{1}{\pi} \int_{\rho_1}^{\rho_2} d\rho \sqrt{2mE - \frac{\beta_\varphi^2}{\rho^2} - (m\omega)^2 \rho^2} \tag{16.25}$$

$$J_\varphi = \frac{1}{2\pi} \int_0^{2\pi} d\varphi \, \beta_\varphi = \beta_\varphi, \tag{16.26}$$

where ρ_1 and ρ_2 are the turning points of the motion. Observe that β_φ is already an action variable. J_ρ is given by the integral

$$\int_{\rho_1}^{\rho_2} d\rho \sqrt{a - \frac{b^2}{\rho^2} - c^2 \rho^2}, \qquad a \geq 0,$$

where, since $\rho \geq 0$,

$$\rho_{1,2} = \sqrt{\frac{1}{2c^2} \left(a \pm \sqrt{a^2 - 4b^2 c^2} \right)}$$

are the positive roots of $(a - b^2/\rho^2 - c^2\rho^2)$. Supposing $\omega > 0$, we get from $a^2 - 4b^2 c^2 \geq 0$ the condition for β_φ,

$$-\frac{E}{\omega} \leq \beta_\varphi \leq \frac{E}{\omega}.$$

The result for the integral is

$$\int_{\rho_1}^{\rho_2} d\rho \sqrt{a - \frac{b^2}{\rho^2} - c^2 \rho^2} = \frac{\pi}{2} \left(\frac{a}{2c} - |b| \right), \qquad (16.27)$$

so that we have for the action variable J_ρ,

$$J_\rho = \frac{1}{2} \left(\frac{E}{\omega} - |J_\varphi| \right).$$

Therefore, the Hamiltonian (respectively the energy) expressed in terms of the action variables reads

$$E = \hat{H} \left(J_\rho, J_\varphi \right) = \omega \left(2J_\rho + |J_\varphi| \right) \qquad (16.28)$$

(compare to Eq. (16.16)), and consequently, the time derivatives of the angle variables $\dot{\theta}_i = \partial \hat{H} / \partial J_i = \omega_i$ are given by

$$\omega_\rho = 2\omega_\varphi = 2\omega. \qquad (16.29)$$

The frequency of the radial motion is twice the frequency of the revolution. In one revolution, in which φ varies from $\varphi = 0$ to $\varphi = 2\pi$, the radius ρ cycles twice from ρ_1 (which is, for instance, the pericenter) to ρ_2 (the apocenter) and back to ρ_1. We recognize this as motion along an elliptical orbit with the origin (center of force) in the center of the ellipse. Calculation of $W \left(\rho, \varphi, J_\rho, J_\varphi \right)$ and the canonical transformation is straightforward (**E**).

Motion in a homogeneous magnetic field

The motion of a charged particle in the plane perpendicular to a homogeneous magnetic field $\mathbf{B} = B e_z$ obeys the canonical equations derived from the Hamiltonian (cf. Subsection 13.1.3, in particular Eq. (13.24))

$$H = \frac{1}{2m} \left(p_\rho^2 + \frac{p_\varphi^2}{\rho^2} \right) + \frac{m}{2} \left(\frac{\omega_Z}{2} \right)^2 \rho^2 - \frac{\omega_Z}{2} p_\varphi. \qquad (16.30)$$

Except for the last term, this Hamiltonian is of the same form as that given by Eq. (16.23). Therefore, we can build on the above calculation (again $p_\varphi = I_{LB} = \beta_\varphi$) by setting in Eq. (16.27)

$$a = 2mE + m\omega_Z \beta_\varphi \qquad \text{and} \qquad c = \frac{m\omega_Z}{2}.$$

The result for J_ρ is

$$J_\rho = \frac{1}{2}\left(\frac{2mE + m\omega_Z \beta_\varphi}{m\omega_Z} - |\beta_\varphi|\right) = \frac{1}{2}\left(\frac{2E}{\omega_Z} + \beta_\varphi - |\beta_\varphi|\right).$$

Hence,

$$E = \hat{H}\left(J_\rho, \beta_\varphi\right) = \omega_Z J_\rho - \frac{\omega_Z}{2}\left(\beta_\varphi - |\beta_\varphi|\right),$$

so that for $\beta_\varphi \geq 0$,

$$\hat{H} = E = \omega_Z J_\rho, \tag{16.31}$$

and for $\beta_\varphi \leq 0$,

$$E = \omega_Z\left(J_\rho + |\beta_\varphi|\right). \tag{16.32}$$

At first glance, this splitting in two cases may seem strange. We recall the circular orbit Eq. (5.84); in polar coordinates, the orbit is given by

$$\rho^2 + 2\rho d \sin\left(\varphi - \varphi_u\right) + d^2 = r_0^2,$$

where for the radius r_0, the distance d of the circle's center from the origin, and the angle φ_u we have (cf. Eqs. (5.83), (5.85), (5.86)):

$$r_0^2 = \frac{2E}{m\omega_Z^2}, \qquad d^2 = \frac{1}{\omega_Z^2}\left(u_1^2 + u_2^2\right) = \frac{2E}{m\omega_Z^2} + \frac{2\beta_\varphi}{m\omega_Z}, \qquad \tan\varphi_u = u_2/u_1.$$

For $p_\varphi = \beta_\varphi \; (= I_{LB}) \geq 0$ (i.e. $d^2 - r_0^2 = 2\beta_\varphi/m\omega_Z \geq 0$), the circle does not include the (arbitrarily chosen) origin. The variable φ cannot perform a cycle from 0 to 2π. If $p_\varphi \leq 0$ (i.e. $d^2 - r_0^2 \leq 0$), the origin is included and φ cycles from 0 to 2π. In the first case φ is *not* an angle variable (in the sense of Section 16.1) anymore. Only a subinterval $[\varphi_1, \varphi_2]$ is realized in the motion, and Eq. (16.26) is not applicable. In this case, φ only takes values between (E)

$$\varphi_{1,2} = \varphi_u - \frac{\pi}{2} \pm \arctan\sqrt{\frac{r_0^2}{d^2 - r_0^2}} = \varphi_u - \frac{\pi}{2} \pm \arctan\sqrt{\frac{E}{\beta_\varphi\omega_Z}}. \tag{16.33}$$

Hence, for $\beta_\varphi \geq 0$, the action variable J_φ vanishes; i.e.,

$$J_\varphi = \frac{1}{2\pi}\left[\int_{\varphi_1}^{\varphi_2} \beta_\varphi d\varphi + \int_{\varphi_2}^{\varphi_1} \beta_\varphi d\varphi\right] = 0$$

and, as already presented in Eq. (16.31), we have

$$\hat{H} = \omega_Z J_\rho.$$

For $\beta_\varphi \leq 0$ the origin is situated within the circular orbit. Now φ is an angle variable, and the action variable J_φ is

$$J_\varphi = \beta_\varphi.$$

Therefore, from (16.32) we have

$$E = \hat{H}(J_\rho, J_\varphi) = \omega_Z (J_\rho + |J_\varphi|). \tag{16.34}$$

The situation becomes completely clear when one transforms to a frame of reference rotating with the Larmor frequency, $\omega_L = -\omega_Z/2$. In the Hamiltonian (16.30) the last term simply vanishes (recall Eq. (9.102)), and

$$H = \frac{1}{2m}\left(p_\rho^2 + \frac{p_\varphi^2}{\rho^2}\right) + \frac{m}{2}\left(\frac{\omega_Z}{2}\right)^2 \rho^2.$$

The Hamiltonian – and hence the orbits – are those of a harmonic oscillator with frequency $\omega_Z/2$. Because the reference frame rotates with angular velocity $\omega_L = -\omega_Z/2$ – i.e. in the opposite direction to the particle's direction of motion – the particle's originally circular orbit, due to the frequency ratio $\omega_Z/\omega_L = 2$, is an ellipse in the rotating system, with the center of the ellipse coinciding with the origin.

Motion in a central force

In the case of the motion in a central force, the action variables are given by (see Subsection 14.2.2)

$$J_r = \frac{1}{2\pi}\oint \frac{\partial W_r}{\partial r}dr = \frac{1}{\pi}\int_{r_{\min}}^{r_{\max}} dr \sqrt{2m(E-V) - \beta_\vartheta^2/r^2} \tag{16.35}$$

$$J_\vartheta = \frac{1}{2\pi}\oint \frac{\partial W_\vartheta}{\partial \vartheta}d\vartheta = \frac{1}{2\pi}\int_0^{2\pi} d\vartheta \sqrt{\beta_\vartheta^2 - \beta_\varphi^2/\sin^2\vartheta} \tag{16.36}$$

$$J_\varphi = \frac{1}{2\pi}\oint \frac{\partial W_\varphi}{\partial \varphi}d\varphi = \frac{1}{2\pi}\int_0^{2\pi} \beta_\varphi d\varphi = \beta_\varphi, \tag{16.37}$$

where β_ϑ $(= |\mathbf{L}|)$ is restricted to $\beta_\vartheta \geq 0$, while $\beta_\varphi (= L_z)$ may take positive and negative values as well; $r_{\min}$, $r_{\max}$ are the two turning points, i.e. the zeros of $\left(2m(E - V)r^2 - \beta_\vartheta^2\right)$. Integration yields for J_ϑ (see e.g. [Saletan/Cromer])

$$J_\vartheta = (\beta_\vartheta - |\beta_\varphi|). \tag{16.38}$$

Hence, from Eq. (16.37), we find

$$\beta_\vartheta = J_\vartheta + |J_\varphi|, \qquad \beta_\varphi = J_\varphi. \tag{16.39}$$

Only J_r depends on the potential $V(r)$.

Let us consider first a perturbed $1/r$ potential of the form

$$V(r) = -\frac{mk}{r} + \frac{c}{r^2}. \tag{16.40}$$

A rather involved calculation yields (cf. e.g. [Saletan/Cromer]; [Goldstein])

$$J_r = -\sqrt{(J_\vartheta + |J_\varphi|)^2 + 2mc} + \frac{mk}{2}\sqrt{\frac{2m}{-E}}. \tag{16.41}$$

(For bounded motion, E must be negative.) From this, we obtain

$$E = \hat{H}\left(J_r, J_\vartheta, J_\varphi\right) = -\frac{m^3 k^2}{2\left(J_r + \sqrt{(J_\vartheta + |J_\varphi|)^2 + 2mc}\right)^2}. \tag{16.42}$$

The frequencies ω_i of the angle variables θ_i are given by $\omega_i = \partial \hat{H}/\partial J_i$ (cf. Eq. (16.6)). Unlike the case of the oscillator (see Eq. (16.18)), the frequencies depend on the values of the action variables. Since $\omega_\varphi = \omega_\vartheta \neq \omega_r$ the orbit for irrational values ω_φ/ω_r is not closed and manifests the rosette-shape mentioned in Section 5.3. Also, neither the trajectory in phase space, nor its mapping onto a torus, is closed after one cycle (this is similar to the situation shown in Fig. 16.2). After sufficiently many revolutions, the trajectory 'covers' the torus, in the sense that it will eventually come arbitrarily close to any point given on the surface.

For a pure $1/r$ potential, $c = 0$, we have

$$E = \hat{H} = -\frac{m^3 k^2}{2\left(J_r + J_\vartheta + |J_\varphi|\right)^2}, \tag{16.43}$$

and hence all frequencies are equal[4]:

$$\omega_r = \omega_\vartheta = \omega_\varphi =: \omega = \frac{m^3 k^2}{(J_r + J_\vartheta + |J_\varphi|)^3}. \tag{16.44}$$

The orbits are closed even after one cycle of any variable (remember the *Runge-Lenz vector!*). The angle variables can be determined from $\theta_i = \partial W(\mathbf{q}, \mathbf{J})/\partial J_i$ (cf. Eq. (16.4); E). Their time dependence is given by $\theta_i = \omega(\mathbf{J})t + \theta_i^0$ (cf. Eq. (16.7)).

If, instead, one considers motion in the central force right from the beginning as a two-dimensional problem (in the plane of the orbit), then we have $J_\vartheta = \beta_\vartheta - |\beta_\varphi| = 0$ above, and the relation between the energy and the action variables reduces to

$$E = \hat{H} = -\frac{m^3 k^2}{2(J_r + |J_\varphi|)^2}. \tag{16.45}$$

The frequencies of the angle variables θ_r and θ_φ are, of course, still equal. If, however, one passes to a system rotating with angular velocity ω_J (cf. Subsection 15.3), then again we have

$$J_\varphi = \frac{1}{2\pi} \oint p_\varphi d\varphi = p_\varphi = L_z, \tag{16.46}$$

and together with

$$p_r = \sqrt{2m_E(E + \omega_J J_\varphi + Gm_S m_E/r) - J_\varphi^2/r^2}$$

obtained from (15.41), one finds (cf. [Berry]; see also Eq. (16.41))

$$J_r = \frac{1}{2\pi} \oint p_r dr = -|J_\varphi| + \frac{Gm_S m_E^2}{\sqrt{-2m(E + \omega_J J_\varphi)}}. \tag{16.47}$$

Hence, for this system,

$$E = -\omega_J J_\varphi - \frac{G^2 m_S^2 m_E^3}{2(J_r + |J_\varphi|)^2}, \tag{16.48}$$

so that the frequencies (for $J_\varphi \geq 0$) are

$$\omega_r = \frac{G^2 m_S^2 m_E^3}{(J_r + J_\varphi)^3} \quad \text{and} \quad \omega_\varphi = -\omega_J + \omega_r. \tag{16.49}$$

That is, *in the rotating frame of reference, the degeneracy is removed!*

---

[4]Expressing the sum of the action variables in terms of the energy, then relating the energy via Eq. (6.39) to the semimajor axis, one obtains Kepler's third law.

16.2   Dynamics on the torus

Even in the completely separable case, the angle variables θ_i depend in general on all coordinates q_j (cf. Eq. (16.4)),

$$\theta_i = \theta_i(\mathbf{q})\,.$$

One can gain an idea of the meaning of the angle variables θ_i by considering the change of θ_i in the course of a complete cycle of the variable q_k at *fixed* values of the remaining coordinates q_j, $j \neq k$:

$$
\begin{aligned}
\Delta_k \theta_i &= \oint \frac{\partial \theta_i}{\partial q_k} dq_k = \oint \frac{\partial^2 W}{\partial q_k \partial J_i} dq_k = \frac{\partial}{\partial J_i} \oint \frac{\partial W}{\partial q_k} dq_k \\
&= \frac{\partial}{\partial J_i} \oint p_k dq_k = 2\pi \frac{\partial J_k}{\partial J_i} = 2\pi \delta_{ik}.
\end{aligned}
$$

In such a fictitious cycle of only the coordinate q_k – one ought to bear in mind that in reality, in the course of one period of the entire system, all coordinate values change – θ_k increases by 2π. So the action variables are left unchanged, whereas each of the angle variables increases by 2π individually (i.e. not at the same rate). This behavior indeed describes the motion on a *torus* of constant dimensions as introduced in Section 14.1. The orbit is produced by the time dependence of the angle variables only. Conversely, if the angle variable θ_i changes by 2π at fixed values of $\mathbf{J}$, then the coordinate $q_i(\boldsymbol{\theta})$ *runs through a period*, and either returns to its initial value or increases by q_i^0; whereas the other coordinates q_k, $k \neq i$, return in general to their initial values *without completing a period*. Therefore, we have for the librational coordinates,

$$q_i(\theta_1 + 2\pi n_1, \ldots, \theta_f + 2\pi n_f, \mathbf{J}) = q_i(\theta_1, \ldots, \theta_f, \mathbf{J}),$$

whereas for rotational coordinates,

$$q_i(\theta_1 + 2\pi n_1, \ldots, \theta_f + 2\pi n_f, \mathbf{J}) = q_i(\theta_1, \ldots, \theta_f, \mathbf{J}) + n_i q_i^0.$$

In these relations the n_i, $i = 1, \ldots f$, are arbitrary integers, and q_i^0 is the period of q_i. In the case of rotations, if one switches to the variable

$$q_i^* = q_i - \frac{1}{2\pi} q_i^0 \theta_i,$$

then $q_i^*(\boldsymbol{\theta}, \mathbf{J})$ satisfies the same relation as a librational coordinate:

$$q_i^*(\boldsymbol{\theta} + 2\pi \mathbf{n}, \mathbf{J}) = q_i^*(\boldsymbol{\theta}, \mathbf{J})\,. \tag{16.50}$$

The q_i and the q_i^* are **multiply periodic functions** in the angle variables. For this reason, the coordinates can be expanded in a Fourier series:

$$q_i(\boldsymbol{\theta}, \mathbf{J}) = \sum_{\mathbf{n}} q_{i,\mathbf{n}}(\mathbf{J}) \exp(i\mathbf{n}\boldsymbol{\theta}), \tag{16.51}$$

where $\mathbf{n} = (n_1, ..., n_f)$, $n_i \in (-\infty, \infty)$, n_i integer. Inserting here the time dependence of $\boldsymbol{\theta}$, $\boldsymbol{\theta} = \boldsymbol{\omega}t + \boldsymbol{\theta}^0$, we have

$$q_i(\boldsymbol{\theta}, \mathbf{J}) = \sum_{\mathbf{n}} q_{i,\mathbf{n}}(\mathbf{J}) \exp\left(i\mathbf{n}\left(\boldsymbol{\omega}t + \boldsymbol{\theta}^0\right)\right).$$

This expression makes it quite clear that the quantity $\mathbf{n}\boldsymbol{\omega}$ determines whether an orbit in configuration space is closed. If a set of integer values n_i^0, $i = 1, \ldots, f$ exists having the property

$$\mathbf{n}^0\boldsymbol{\omega} = 0, \tag{16.52}$$

i.e. if the frequencies are commensurable, then the orbit is closed. In other words, orbits are closed if the ratios between the periods ω_i are rational.

As we saw in Eq. (16.10), the characteristic function W_i increases in the course of one cycle by $2\pi J_i$. This implies that $W_i - \theta_i J_i$ is periodic, so that the 'reduced' charactcristic function,

$$W^* = W(\mathbf{q}, \mathbf{J}) - \boldsymbol{\theta}\mathbf{J}, \tag{16.53}$$

must be a multiply periodic function in the angle variables. Since we have $\boldsymbol{\theta} = \nabla_J W$, we realize that $W^* = W^*(\mathbf{q}, \boldsymbol{\theta})$ is the Legendre transform of W. (Note that $\nabla_J = (\partial/\partial J_1, \ldots, \partial/\partial J_f)$; similarly for ∇_q and ∇_θ.) The relations

$$\mathbf{p} = \nabla_q W^*(\mathbf{q}, \boldsymbol{\theta}) \qquad \text{and} \qquad \mathbf{J} = -\nabla_\theta W^*(\mathbf{q}, \boldsymbol{\theta})$$

are valid. The left relation implies that the momenta p_i, as well as any dynamical function $O(\mathbf{q}, \mathbf{p})$, are multiply periodic functions, too.

In terms of the action-angle variables, different systems with the same number of degrees of freedom are mapped on a (hyper-)torus. So motion on a torus is a universal type of dynamical system – it embraces the dynamics of seemingly different separable systems. In this sense, the systems are equivalent.

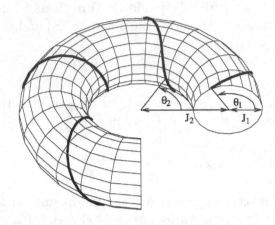

Figure 16.2: The torus and a trajectory of a two-dimensional system.

Let us consider a separable two-dimensional system. Phase space is then four-dimensional, and the trajectories lie on the three-dimensional energy surface given by

$$H = \hat{H}(J_1, J_2) = E.$$

The two-dimensional manifold on which the motion occurs is equivalent (diffeomorphic) to the surface of a torus in a three-dimensional space. At a given energy E, the conserved quantities J_1 and J_2 are the radii of the torus, as shown in Fig. 16.2. Motion along the perimeter of the cross section occurs according to $\theta_1 = \omega_1 t + \theta_1^0$, while about the torus axis, the time dependence is given by $\theta_2 = \omega_2 t + \theta_2^0$. Thus each trajectory is a spiral on the surface of the torus (cf. Fig. 16.2). Depending on the ratio ω_1/ω_2 – which is either rational or irrational – the trajectory is closed, or in the long term it 'covers' the entire surface (in the sense given above); in the latter case, the trajectory is called *nearly periodic* or *quasiperiodic*. If one wishes to represent the dynamics in a Poincaré section, a convenient Poincaré surface is a plane containing the symmetry axis of the torus. Depending on the frequency ratio, either the return points become dense on the curve of the cross-section (the ratio is irrational) or only a finite number of return points exists (the ratio is rational). As an example recall Fig. (4.2) for the two-dimensional oscillator.

Motion on the torus is sometimes represented in a unit square with coordinates $\theta_i/2\pi$. Coordinates increasing with time are taken modulo 1, so their trajectories always remain in the square. Using this square to represent the dynamics has been mentioned earlier in Section 4.4.

For a system with f degrees of freedom, with action-angle variables $\mathbf{J}, \boldsymbol{\theta}$, and the Hamiltonian $H = H(\mathbf{J})$, a graphical representation of the motion on the f-dimensional surface of a torus is not feasible; nevertheless, the reasoning is similar to the two-dimensional case.

16.3 Canonical perturbation theory

Canonical perturbation theory is intended to allow the approximate solution of the canonical equations of a system that cannot be solved exactly, but which differ only 'slightly' from the canonical equations of an exactly solvable one. We present the idea of canonical perturbation theory as an application of action-angle variables.

Suppose, the Hamiltonian $H\,(\mathbf{q}, \mathbf{p})$ of the system to be solved, differs from the Hamiltonian $H_0\,(\mathbf{q}, \mathbf{p})$ of the integrable system by an additional term[5] λH_1:

$$H = H_0 + \lambda H_1. \tag{16.54}$$

The strategy of perturbation theory starts from H_0. Since the system is integrable, after a canonical transformation of the conjugate variables $\mathbf{q}$ and $\mathbf{p}$ to action-angle variables $\mathbf{J}$ and $\boldsymbol{\theta}$, the Hamiltonian H_0 is cyclic in the angle variables $\boldsymbol{\theta}$:

$$H_0 = H_0(\mathbf{J}). \tag{16.55}$$

Consequently, the canonical equations for $\mathbf{J}$ and $\boldsymbol{\theta}$ are:

$$\dot{\mathbf{J}} = -\nabla_\theta H_0(\mathbf{J}) = 0 \tag{16.56}$$

$$\dot{\boldsymbol{\theta}} = \nabla_J H_0(\mathbf{J}) =: \boldsymbol{\omega}_0(\mathbf{J}). \tag{16.57}$$

Applying the transformation from $(\mathbf{q}, \mathbf{p})$ to $(\boldsymbol{\theta}, \mathbf{J})$ to H also – in particular to H_1 – we obtain

$$H(\boldsymbol{\theta}, \mathbf{J}) = H_0(\mathbf{J}) + \lambda H_1(\boldsymbol{\theta}, \mathbf{J}). \tag{16.58}$$

---

[5]Of course, it is desirable to have λ small, but in any case it supports book-keeping of terms in the approximation.

The idea of dividing H in two parts, an exactly soluble one and the remainder, can be found already in the second work of W.R. Hamilton cited in footnote 14 on page 249.

In this system, the canonical equations[6] in terms of $\mathbf{J}$ and $\boldsymbol{\theta}$ are:

$$\dot{\mathbf{J}} = -\nabla_\theta H(\boldsymbol{\theta}, \mathbf{J})$$
$$\dot{\boldsymbol{\theta}} = \nabla_J H(\boldsymbol{\theta}, \mathbf{J}).$$

However, in this system, $\boldsymbol{\theta}$ and $\mathbf{J}$ do not serve as action-angles variables – $\mathbf{J}$ is not time independent anymore, and $\dot{\boldsymbol{\theta}}$ is not constant anymore. Rather, using Eqs. (16.56) and (16.57), the time dependence is to be derived from the equations

$$\dot{\mathbf{J}} = -\lambda \nabla_\theta H_1(\boldsymbol{\theta}, \mathbf{J})$$
$$\dot{\boldsymbol{\theta}} = \boldsymbol{\omega}_0(\mathbf{J}) + \lambda \nabla_J H_1(\boldsymbol{\theta}, \mathbf{J}).$$

The aim now is to make a canonical transformation to new action-angle variables $\mathbf{K}$ and $\boldsymbol{\phi}$, so that the new Hamiltonian $\hat{H}(\boldsymbol{\phi}, \mathbf{K}) = H(\boldsymbol{\theta}, \mathbf{J})$ is cyclic in $\boldsymbol{\phi}$,

$$\hat{H}(\mathbf{K}) = H(\boldsymbol{\theta}, \mathbf{J}), \tag{16.59}$$

and $\mathbf{K}$ and $\boldsymbol{\phi}$ satisfy the canonical equations

$$\dot{\mathbf{K}} = -\nabla_\phi \hat{H}(\mathbf{K}) = 0$$
$$\dot{\boldsymbol{\phi}} = \nabla_K \hat{H}(\mathbf{K}) =: \boldsymbol{\omega}(\mathbf{K}). \tag{16.60}$$

The point is, that the last two equations can easily be solved. The transformation is accomplished by a characteristic function $W(\boldsymbol{\theta}, \mathbf{K})$, a generator of the second kind (see Subsection 13.3.1), since for $\lambda = 0$, we expect the identity transformation. Therefore, in addition to Eq. (16.59), we have the relations (cf. Eqs. (14.10) and (14.11))

$$\mathbf{J} = \nabla_\theta W(\boldsymbol{\theta}, \mathbf{K}) \tag{16.61}$$
$$\boldsymbol{\phi} = \nabla_K W(\boldsymbol{\theta}, \mathbf{K}). \tag{16.62}$$

Using Eq. (16.61), the characteristic function W can be determined from the Hamilton-Jacobi equation (cf. Eq. (16.59)),

$$H(\boldsymbol{\theta}, \nabla_\theta W) = E = \hat{H}(\mathbf{K}). \tag{16.63}$$

Due to the fact that the canonical equations associated with H are not solvable exactly, we must use an approximation method to achieve this transformation.

---

[6]Recall that the property of canonical conjugacy of variables is independent of the Hamiltonian.

Since for $\lambda = 0$, W must reduce to the identity transformation, we set

$$W = W_0 + \lambda W_1 + \lambda^2 W_2 + \ldots, \tag{16.64}$$

where

$$W_0 = \boldsymbol{\theta}\mathbf{K} \tag{16.65}$$

induces the identity transformation (cf. Eq. (13.63)), and the functions W_i are to be calculated from the Hamilton-Jacobi equation (16.63). We require the $W_i(\boldsymbol{\theta},\mathbf{K})$, $i \geq 1$, to be multiply periodic functions in the angle variables $\boldsymbol{\theta}$ (recall the discussion preceding Eq. (16.53)). Also expanding $\hat{H}$ in Eq. (16.63) in a power series in λ, we have the equation[7]

$$H_0(\boldsymbol{\nabla}_\theta W) + \lambda H_1(\boldsymbol{\theta}, \boldsymbol{\nabla}_\theta W) = \hat{H}_0(\mathbf{K}) + \lambda\hat{H}_1(\mathbf{K}) + \ldots \ . \tag{16.66}$$

Replacing now W by its expansion (16.64), both H_0 and H_1 also become power series in λ:

$$\begin{aligned} H_0(\boldsymbol{\nabla}_\theta W) &= H_0(\mathbf{K} + \lambda\boldsymbol{\nabla}_\theta W_1 + \ldots) \\ &= H_0(\mathbf{K}) + \lambda(\boldsymbol{\nabla}_\theta W_1)(\boldsymbol{\nabla}_J H_0(\mathbf{J} = \mathbf{K})) + \ldots \end{aligned} \tag{16.67}$$

and

$$\begin{aligned} H_1(\boldsymbol{\theta}, \boldsymbol{\nabla}_\theta W) &= H_1(\boldsymbol{\theta}, \mathbf{K} + \lambda\boldsymbol{\nabla}_\theta W_1 + \ldots) \\ &= H_1(\boldsymbol{\theta}, \mathbf{K}) + \lambda(\boldsymbol{\nabla}_\theta W_1)(\boldsymbol{\nabla}_J H_1(\boldsymbol{\theta}, \mathbf{J} = \mathbf{K})) + \ldots \ . \end{aligned}$$

Comparing in Eq. (16.66) the coefficients of like powers of λ, we find

$$\lambda^0 : \qquad H_0(\mathbf{K}) = \hat{H}_0(\mathbf{K}) \tag{16.68}$$

$$\lambda^1 : \qquad \boldsymbol{\omega}_0(\mathbf{K})(\boldsymbol{\nabla}_\theta W_1) + H_1(\boldsymbol{\theta}, \mathbf{K}) = \hat{H}_1(\mathbf{K}) \tag{16.69}$$

$$\vdots \ ,$$

where (cf. Eq. (16.57))

$$\boldsymbol{\omega}_0(\mathbf{K}) = \boldsymbol{\nabla}_J H_0(\mathbf{K} = \mathbf{J}). \tag{16.70}$$

Since $\hat{H}_1(\mathbf{K})$ is independent of $\boldsymbol{\theta}$, it can be calculated by averaging Eq. (16.69) over the angles θ_i. Denoting the average of a function

---

[7]Due to the canonical transformation (16.61) and (16.62), higher powers of λ in H may arise.

$F(\boldsymbol{\theta}, \mathbf{K})$ by $\bar{F}(\mathbf{K})$, where

$$\bar{F}(\mathbf{K}) = \frac{1}{(2\pi)^f} \int_0^{2\pi} \dots \int_0^{2\pi} d\theta_1 \dots d\theta_f F(\boldsymbol{\theta}, \mathbf{K}),$$

the multiple periodicity of each of the W_i (except W_0) implies that[8]

$$\frac{\overline{\partial W_1}}{\partial \theta_i} = 0, \qquad i = 1, \dots, f,$$

and hence

$$\hat{H}_1(\mathbf{K}) = \bar{H}_1(\mathbf{K}). \tag{16.71}$$

W_1 can now be determined from the equation

$$\boldsymbol{\omega}_0(\mathbf{K})(\boldsymbol{\nabla}_\theta W_1) = \bar{H}_1(\mathbf{K}) - H_1(\boldsymbol{\theta}, \mathbf{K}). \tag{16.72}$$

The results of the calculation, to first order in λ, are the Hamiltonian (cf. Eqs. (16.67) and (16.71))

$$\hat{H}(\mathbf{K}) = H_0(\mathbf{J})|_{\mathbf{J}=\mathbf{K}} + \lambda \bar{H}_1(\mathbf{K}), \tag{16.73}$$

the frequencies of the angle variables $\boldsymbol{\phi}$

$$\boldsymbol{\omega}(\mathbf{K}) = \boldsymbol{\nabla}_K \hat{H}(\mathbf{K}) = \boldsymbol{\omega}_0(\mathbf{K}) + \lambda \boldsymbol{\nabla}_K \bar{H}_1(\mathbf{K}), \tag{16.74}$$

and the function W_1. From W_1, one obtains the new variables in terms of the old ones via Eqs. (16.61) and (16.62).

We illustrate this procedure in the following.

---

[8]Multiple periodicity has the consequence that $W_1(\mu, \mathbf{K})$ can be expanded in a Fourier series
$$W_1(\mu, \mathbf{K}) = \sum_{\mathbf{m}} W_{1,\mathbf{m}}(\mathbf{K}) \exp[i\mathbf{m}\mu].$$
In the derivatives with respect to θ_i, the $m_i = 0$ term disappears, and the integration in the averaging procedure yields
$$\int_0^{2\pi} d\theta_i \frac{\partial W_1}{\partial \theta_i} = \frac{\overline{\partial W_1}}{\partial \theta_i} = 0.$$

16.3.1   The one-dimensional anharmonic oscillator

We reconsider the solutions for the one-dimensional anharmonic oscillator of Chapter 3, known as *Duffing's oscillator* ($m = 1$, $\mu/4 \to \lambda$), whose Hamiltonian is

$$H = p^2/2 + (\omega_0^2/2)x^2 + \lambda x^4. \tag{16.75}$$

In order to apply the perturbational method, we separate the Hamiltonian in two parts,

$$H_0 = p^2/2 + (\omega_0^2/2)x^2 \quad \text{and} \quad H_1 = x^4, \tag{16.76}$$

and introduce the action-angle variables into H_0. From the former investigation of the harmonic oscillator, we know that (cf. Eq. (16.16))

$$H_0 = \omega_0 J, \tag{16.77}$$

which is achieved by the canonical transformation consisting of Eqs. (16.20) and (16.21). Applying in particular

$$x = \sqrt{2J/\omega_0} \sin \theta \tag{16.78}$$

to H_1, we find

$$H_1(\theta, J) = (2J/\omega_0)^2 \sin^4 \theta, \tag{16.79}$$

so that the total Hamiltonian reads

$$H = \omega_0 J + \lambda(2J/\omega_0)^2 \sin^4 \theta.$$

This is the starting point for the transformation we are really interested in.

We perform the canonical transformation from (J, θ) to (K, ϕ), i.e. to the proper action-angle variables of H, with the generator

$$W(\theta, K) = \theta K + \lambda W_1(\theta, K) + \dots .$$

According to Eq. (16.68), this gives, to zeroth order,

$$\hat{H}_0(K) = H_0(K) = \omega_0 K. \tag{16.80}$$

Averaging $H_1(\theta, J)$ over θ (since $\dfrac{1}{2\pi} \int\limits_0^{2\pi} d\theta \sin^4 \theta = \dfrac{3}{8}$) yields

$$\bar{H}_1(K) = (3/2)(K/\omega_0)^2,$$

and by Eq. (16.71), it follows that

$$\hat{H}_1(K) = (3/2)(K/\omega_0)^2, \tag{16.81}$$

so that $\hat{H} = \hat{H}_0 + \lambda\hat{H}_1$ to this order is given by Eqs. (16.80) and (16.81):

$$\hat{H} = \omega_0 K + \frac{3}{2}\lambda(K/\omega_0)^2.$$

With the frequency (cf. Eq. (16.6))

$$\omega = \frac{\partial\hat{H}}{\partial K} = \omega_0 + 3\lambda K/\omega_0^2, \tag{16.82}$$

the equation of motion for the variable ϕ, $\dot{\phi} = \dfrac{\partial\hat{H}}{\partial K}$, has the simple solution

$$\phi(t) = \omega t + \phi_0 \tag{16.83}$$

(see Eq. (16.7)).

According to Eq. (16.69), W_1 satisfies the equation ($\omega_0(K) = \omega_0$, cf. Eq. (16.77))

$$\omega_0\frac{\partial}{\partial\theta}W_1(\theta, K) + (2K/\omega_0)^2\sin^4\theta = \hat{H}_1(K),$$

which, after inserting Eq. (16.81), takes the form

$$\omega_0\frac{\partial}{\partial\theta}W_1(\theta, K) = (K/\omega_0)^2\left(3/2 - 4\sin^4\theta\right).$$

Integrating leads to[9]

$$W_1(\theta, K) = \frac{K^2}{\omega_0^3}\left(\sin 2\theta - \frac{1}{8}\sin 4\theta\right). \tag{16.84}$$

From the generator $W = \theta K + \lambda W_1$ we can now determine to first order the relations between (K, ϕ) and the old variables (J, θ) (Eqs. (16.61) and (16.62)):

$$\phi = \frac{\partial W}{\partial K} = \theta + 2\lambda K/\omega_0^3\left(\sin 2\theta - \frac{1}{8}\sin 4\theta\right) \tag{16.85}$$

$$J = \frac{\partial W}{\partial\theta} = K + 2\lambda K^2/\omega_0^3\left(\cos 2\theta - \frac{1}{4}\cos 4\theta\right). \tag{16.86}$$

---

[9]An additional function $f(K)$ can be excluded, since W_1 has to be periodic in θ.

Equation (16.85) yields θ as a function of K and ϕ,

$$\theta = \phi - 2\lambda K/\omega_0^3 \left(\sin 2\phi - \frac{1}{8} \sin 4\phi \right), \qquad (16.87)$$

and subsequently from Eq. (16.86), we have

$$J = K + 2\lambda K^2/\omega_0^3 \left(\cos 2\phi - \frac{1}{4} \cos 4\phi \right). \qquad (16.88)$$

Equations (16.87) and (16.88) are the canonical transformation from (K, ϕ) to (J, θ). We can now use these relations in the initial canonical transformation (16.78). With $\sqrt{J}$ from Eq. (16.88) and $\sin \theta$ from Eq. (16.87),

$$\sqrt{J} = \sqrt{K} \left(1 + \lambda K/\omega_0^3 \left(\cos 2\phi - \frac{1}{4} \cos 4\phi \right) \right)$$

$$\sin \theta = \sin \phi - 2\lambda K/\omega_0^3 \left(\sin 2\phi - \frac{1}{8} \sin 4\phi \right) \cos \phi,$$

we obtain

$$x = \sqrt{2K/\omega_0} \sin \phi$$
$$+ \lambda \frac{K}{\omega_0^3} \sqrt{2K/\omega_0} \left(\sin \phi \left(\cos 2\phi - \frac{1}{4} \cos 4\phi \right) \right.$$
$$\left. - 2 \cos \phi \left(\sin 2\phi - \frac{1}{8} \sin 4\phi \right) \right)$$
$$= \sqrt{2K/\omega_0} \sin \phi - \frac{\lambda}{4} \frac{K}{\omega_0^3} \sqrt{2K/\omega_0} \left(6 \sin \phi + \sin 3\phi \right). \quad (16.89)$$

Inserting for $\phi(t)$ the solution (16.83), we get the time dependence of the motion in the original coordinates, to zeroth order:

$$x = \sqrt{2K/\omega_0} \sin(\omega_0 t + \phi_0). \qquad (16.90)$$

Since also $K = J$, we recover, as expected, the solution of the harmonic oscillator.

In $\mathcal{O}(\lambda)$, the frequency determining the temporal development is given by Eq. (16.82). The choice $\phi_0 = \pi/2$ leads – for $t = 0$ in Eq. (16.89) – to

$$x_0 = x(t = 0) = \sqrt{2K/\omega_0} \left(1 - \frac{5\lambda}{4} \frac{K}{\omega_0^3} \right),$$

or, conversely, to

$$\sqrt{2K/\omega_0} = x_0 \left(1 + \frac{5\lambda}{8} \frac{x_0^2}{\omega_0^2}\right),$$

so that the solution, expressed in terms of x_0, reads

$$\begin{aligned} x &= x_0 \left(1 + \frac{5\lambda}{8} \frac{x_0^2}{\omega_0^2}\right) \cos \omega t - \lambda \frac{x_0^3}{8\omega_0^2} \left(6 \cos \omega t - \cos 3\omega t\right) \\ &= x_0 \cos \omega t - \lambda \frac{x_0^3}{8\omega_0^2} \left(\cos \omega t - \cos 3\omega t\right). \end{aligned}$$

This is the solution (3.86) given in Chapter 3, with $\lambda = \mu/4$. Replacing K also in Eq. (16.82) by the initial position x_0, we obtain the same result as in Eq. (3.87), namely:

$$\omega = \omega_0 \left(1 + \frac{3}{2}\lambda x_0^2/\omega_0^2\right). \qquad (16.91)$$

16.3.2   First order corrections

We return now to the general discussion of perturbation theory, in particular to Eq. (16.72). As we have seen, this equation can be solved in the one-dimensional case. Solving it in several dimensions turns out to be much harder.

Since $H_1(\boldsymbol{\theta}, \mathbf{K})$ is a multiply periodic function too, it can be expanded in an f-dimensional Fourier series:

$$H_1(\boldsymbol{\theta}, \mathbf{K}) = \sum_{\mathbf{m}} H_{1\mathbf{m}}(\mathbf{K}) \exp(i\mathbf{m}\boldsymbol{\theta}), \qquad (16.92)$$

$\mathbf{m} = (m_1, \ldots, m_f)$ with $m_i \in (-\infty, \infty)$ integer. In the Fourier expansion of W_1,

$$W_1(\boldsymbol{\theta}, \mathbf{K}) = \sum_{\substack{\mathbf{m} \\ \mathbf{m} \neq 0}} W_{1\mathbf{m}}(\mathbf{K}) \exp(i\mathbf{m}\boldsymbol{\theta}), \qquad (16.93)$$

the $\mathbf{m} = 0$ term does not appear, since according to the premiss, W_1 is periodic. This has the consequence that $\bar{H}_1$ must be equal to the $\mathbf{m} = 0$ contribution in the Fourier series representation of H_1, so that

$$H_1 - \bar{H}_1 = \sum_{\substack{\mathbf{m} \\ \mathbf{m} \neq 0}} H_{1\mathbf{m}}(\mathbf{K}) \exp(i\mathbf{m}\boldsymbol{\theta}). \qquad (16.94)$$

Inserting this into Eq. (16.72), one obtains an equation for the Fourier coefficients $W_{1\mathbf{m}}$,

$$i\mathbf{m}\boldsymbol{\omega}_0(\mathbf{K})W_{1\mathbf{m}} = -H_{1\mathbf{m}}(\mathbf{K}),$$

and for the total correction to first order

$$W_1 = i \sum_{\substack{\mathbf{m} \\ \mathbf{m} \neq 0}} \frac{H_{1\mathbf{m}} \exp(i\mathbf{m}\boldsymbol{\theta})}{\mathbf{m}\boldsymbol{\omega}_0(\mathbf{K})}. \tag{16.95}$$

This expression can be used to demonstrate problems that appear even in the first-order perturbation expansion for $f \geq 2$. The convergence of the sum depends essentially on the denominator $\mathbf{m}\boldsymbol{\omega}_0(\mathbf{K})$. Let us consider a specific set of values for the components of $\boldsymbol{\omega}_0(\mathbf{K})$. If the frequencies $\omega_{0,i}(\mathbf{K})$ are **resonant** (or commensurable), i.e. if there exists a set of integers m_i such that

$$\mathbf{m}\boldsymbol{\omega}_0(\mathbf{K}) = 0, \tag{16.96}$$

then the sum in Eq. (16.95) clearly diverges. One must use *degenerate perturbation theory* (see e.g. [Born]). But even when the $\omega_{0,i}$ are *non-resonant*, that is for *no* choice of the m_i relation (16.96) is satisfied, the value of $\mathbf{m}\boldsymbol{\omega}_0(\mathbf{K})$ can be made arbitrarily small by choosing (large) values for the m_i. Only for sufficiently non-resonant values of the $\omega_{0,i}$ does the sum converge absolutely. This is called the 'problem of small divisors' (cf. also the resonance catastrophe in Chapter 3). Hence the question arises: does the series converge at all for finite values of λ?

16.4 The KAM theorem

What happens to the motion on a torus of an integrable system if a perturbation is switched on? An answer is provided by the so-called Kolmogorov-Arnold-Moser theorem[10] (A.N. Kolmogorov (1954), V.I. Arnold (1963), J. Moser (1962)). The behavior of the motion on the torus depends essentially on the frequencies $\boldsymbol{\omega}_0(\mathbf{K})$ of the unperturbed system (cf. Eq. (16.70)). For a chosen set of values for the conserved quantities K_i, these frequencies determine the motion on the surface of the torus. Loosely-speaking the **KAM theorem** says

---

[10]For details see [Lichtenberg/Liebermann] and [Tabor]. The recent "The KAM Story" by H.S. Dumas illustrates the importance of the KAM theorem and sketches its history extensively (see Bibliography).

If the frequencies $\omega_{0,i}$ on a torus in the unperturbed system H_0 are sufficiently non-resonant, then there exists also in the system $H = H_0 + \lambda H_1$ at small, but finite values of λ a distorted torus, which for $\lambda \to 0$ turns into the torus of the unperturbed system H_0.

Remarks

i) The condition for 'sufficiently non-resonant frequencies' can be stated quantitatively.

ii) Tori for which this condition is not satisfied, are 'destructed'. Their remnants are situated amongst the 'surviving' tori.

Example: The modified Hénon-Heiles system

The Hénon-Heiles system presented in Chapter 4 is not directly suitable for demonstrating this theorem, since even in the unperturbed Hamiltonian, $H_0 = (p_x^2 + p_y^2)/2 + (x^2 + y^2)/2$, the frequencies are most commensurable (they are equal: $\omega_x = \omega_y = 1$). Treatment of the perturbation

$$\lambda H_1 = x^2 y - y^3/3 \qquad (16.97)$$

requires degenerate perturbation theory. Therefore we choose an irrational number as frequency ratio in H_0 – e.g. $\pi/3$ in the coefficient of the y^2-term – so that the basic Hamiltonian H_0 of this **modified Hénon-Heiles system** is

$$H_0 = \frac{1}{2}(p_x^2 + p_y^2) + \frac{1}{2}(x^2 + \frac{\pi}{3}y^2), \qquad (16.98)$$

where now

$$\omega_x = 1 \qquad \text{and} \qquad \omega_y = \sqrt{\pi/3}.$$

The shape of the potential

$$V(x,y) = \frac{1}{2}(x^2 + \frac{\pi}{3}y^2) + x^2 y - y^3/3$$

is hardly different from what it was originally. In the minimum at $(0,0)$ the potential vanishes: $V = 0$. There are also three saddle points at $(0, \pi/3)$, $\left(\pm\frac{1}{2}\sqrt{2\pi/3 + 1}, -\frac{1}{2}\right)$. The potential in the latter two points,

$V = \frac{1}{24}(\pi + 1)$, is lower than in the first one, $V = \frac{1}{6}\pi/3$, so that the maximum energy allowing bounded motion is $(1/6)(\pi + 1)/4 = (1/6) \times 1,0354 = 0,1726$ (instead of $1/6$ in the original Hénon-Heiles system). Since, for small energies, the potential well is similar to that of a harmonic oscillator – and the anharmonic terms gain influence only for larger energies – the energy can be taken as a measure of the perturbational parameter[11] λ.

Expressed in terms of the action variables J_i (Eqs. (16.14) and (16.13)), we have instead of Eq. (16.98) (cf. Eq. (16.16))

$$H_0 = J_1 + \sqrt{\pi/3}J_2,$$

and the perturbation (16.97) in terms of action-angle variables (16.20) is

$$\lambda H_1 = 2\sqrt{\frac{2J_2}{\sqrt{\pi/3}}} \sin\theta_2 \left(J_1 \sin^2\theta_1 - (J_2/\sqrt{3\pi})\sin^2\theta_2 \right). \qquad (16.99)$$

We avoid the perturbation theory analysis, and merely state the numerical results for the Poincaré sections. For small energies, the sequences of points belonging to returning trajectories resemble the ellipses (sections of the tori) of the harmonic oscillator (cf. Fig. 16.3). If the energy increases, the picture changes. Tori disappear, and the chaotic region in phase space grows. Compared to the original Hénon-Heiles system

---

[11] Starting from the energy

$$E = \frac{m}{2}\left(\dot{x}^2 + \dot{y}^2\right) + \frac{1}{2}\left(\omega_x^2 x^2 + \omega_y^2 y^2\right) + \left(x^2 y - \frac{1}{3}y^3\right),$$

changing the length scale by

$$(x, y) \to (x', y') = (x, y)/\alpha,$$

one obtains

$$E = \frac{m}{2}\alpha^2\left(\dot{x}'^2 + \dot{y}'^2\right) + \frac{1}{2}\alpha^2\left(\omega_x^2 x'^2 + \omega_y^2 y'^2\right) + \alpha^3\left(x'^2 y' - \frac{1}{3}y'^3\right).$$

Setting $\alpha = \lambda$, it follows that

$$E/\lambda^2 = \frac{m}{2}\left(\dot{x}'^2 + \dot{y}'^2\right) + \frac{1}{2}\left(\omega_x^2 x'^2 + \omega_y^2 y'^2\right) + \lambda\left(x'^2 y' - \frac{1}{3}y'^3\right).$$

Fixing E/λ^2 shows that $\lambda \sim \sqrt{E}$.

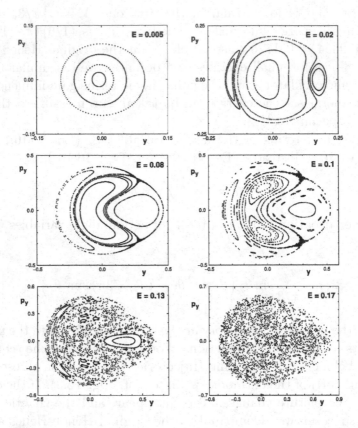

Figure 16.3: Poincaré sections in the modified Hénon-Heiles system for increasing energies.

one has the impression that the tori persist longer for increasing energy. The system is not integrable; besides the energy, no other (global) isolating integral exists. One cannot expect the perturbation expansion to be valid. Nevertheless, in the original Hénon-Heiles system, for moderate energies, the behavior in phase space is rather well reproduced by the perturbational calculation of Gustavson, as demonstrated in his Poincaré sections (cf. Fig. 4.8). Supposedly this is the case also for this modified system.

16.5 Is the solar system stable?

Finally we address ourselves to questions vital for our life on earth: Is the solar system stable? or more specific: Is the orbit of the earth stable? Changes in the semi-major axis (or mean distance from the sun) would be fatal for life, as we know it. Also a change in the (small) eccentricity would have a considerable influence on the life conditions[12].

16.5.1 A few historical landmarks

After Kepler found his laws[13], in particular the elliptic orbits of the planets, it was Newton who explained Kepler's laws by the universal gravitation[14]. However Newton's explanation, as well as Kepler's laws, are strictly valid only for two bodies gravitating towards one another. Therefore, to explain some anomalies of the moon's orbit, Newton studied the influence of a third body, the sun, on the orbit. And it seems, that from these investigations some doubt about the stability of the planetary orbits arose. In the final General Scholium of the "Principia" he expressed his opinion that the 'most beautiful system of the sun, planets, and comets, could only proceed from the counsel and dominion of an intelligent and powerful being'. Also in a short notice in Question 31 at the end of his "Opticks" Newton referred to that unsettled question[15]. This was the beginning of investigations on the stability of the solar system (or system of the world as Newton called it). Also 'to Euler it had seemed doubtful whether forces so numerous, so variable in position, so different in intensity, as those in the solar system, could be capable of maintaining permanently a condition of equilibrium'[16].

---

[12]Moreover the inclination of the rotation axis of the earth with respect to the orbital plane is important for the climate.

[13]In a letter to a friend Kepler introduced the comparison of the solar system with a clockwork (R. Newton, From Clockwork to Crapshoot: A History of Physics, Harvard University Press, 2007).

[14]A very readable account of the history is "Newton's Clock" by I. Petersen (see Bibliography).

[15]'... blind Fate could never make all the Planets move one and the same way in Orbs concentrick, some inconsiderable Irregularities excepted, which may have risen from the mutual Actions of Comets and Planets upon one another, and which will be apt to increase, till this System wants a Reformation.' (I. Newton, Opticks, Dover, New York 1952, p. 402).

[16]F. Cajori, A History of Mathematics, p. 260 (see Bibliography).

'*In 1773 Laplace brought out a paper in which he proved that the mean motions or mean distances of planets are invariable or merely subject to small periodic changes. This was the first and most important step in establishing the stability of the solar system.... This paper was the beginning of a series of profound researches by Lagrange and Laplace on the limits of variation of the various elements of planetary orbits, in which the two great mathematicians alternately surpassed and supplemented each other.*'[17]. Laplace has treated such 'perturbations' extensively in the second book of his "Celestial Mechanics". In chapter VII he finds that the disturbances of the elliptical motion cause a slow time dependence of the elements of the motion (such as the semi-major axes, the eccentricities, and inclinations). Such variations are called *secular equations*. He also showed that there is a small precession of the perihelion due to the other planets.

In the years 1892-99 Poincaré[18] provided the proof for the impossibility to integrate the equations of motion of the three body problem. Moreover he demonstrated that one cannot find an analytic solution representing the planetary motion for an infinite time interval. He also showed that the approximating series expansions used in astronomy are in general divergent. But he conjectured that for incommensurable starting values the series may converge[19].

About fifty years later, in 1954, Kolmogorov considered the convergence properties of the perturbation series and sketched the proof of a

---

[17]F. Cajori, *ibid.*

[18]As mentioned already in the first Chapter, in 1889 Poincaré won a mathematical competition of the Swedish Academy with a contribution on the restricted three-body problem. But a serious error in the paper was detected and Poincaré was asked to submit a revised paper. It appeared in 1890.

His most famous work in celestial mechanics is the three-volume book "Les méthodes nouvelles de la mécanique céleste", mentioned already earlier in Chapter 4 (see Bibliography). Poincaré studied in detail the properties of perturbative solutions (stability, asymptotic behavior, and so on). He introduced the small parameter method, fixed points, integral invariants, variational equations (Wikipedia). The book contains many new ideas used today in the perturbation theoretical approach to dynamical systems.

[19]J. Laskar, Is the solar system stable?, arXiv : 1209.5996v1, 2012. This article provides an insight into the use of computers to answer that question. Computers enable extensive analytical calculations as well as voluminous numerical integrations of the equations of motion.

theorem which was later called KAM theorem[20] Starting from 1961 V. Arnold and J. Moser completed the proof in a series of papers.

More and more activities in numerical studies of the planetary system followed and they last until today. With increasing computing power more demanding questions could be studied. Typical results are: Lyapunov-exponents, planetary motion over several billion years (Gyrs).

16.5.2 On the stability of planetary orbits

Let us now consider the question of stability of planetary orbits[21], raised in Subsection 15.3. What can be inferred from the KAM theorem about the motion of celestial bodies? A problem immediately arises: The unperturbed Hamiltonian of planetary motion depends only on the sum of the action variables, see (16.43), and consequently the frequencies are completely degenerate, Eq. (16.44). But one can overcome this degeneracy to some extent[22].

Earth's orbit

Even the earth's orbit around the sun – taking Jupiter also into account (see the Hamiltonian (15.41)) – could be chaotic. The ratio of the radial frequency to the angular frequency in the motion of the 'unperturbed' earth (i.e. if Jupiter did not exist) – in the system rotating with Jupiter at angular velocity ω_J is (cf. Eq. (16.49)),

$$\omega_r/\omega_\varphi = 1 + \omega_J/\omega_\varphi. \tag{16.100}$$

---

[20]For more details we refer again to the book by [Dumas].

[21]When comparing the planetary model system consisting of point masses with the real planetary system, additional effects have to be taken into account possibly, e.g.:

   i) the finite size of the planets; they have a nonspherical shape, an inhomogeneous mass distribution that is even time dependent (tides, liquid core).

  ii) the loss of mass of the sun

 iii) the mass distribution outside the planetary system

  iv) effects from general relativity (important for the inner planets; in the calculations an effective potential is added)

Some of these effects have been included in systems with fewer bodies (planets). But whether they may change the behavior of a model system including all planets is still an open problem.

[22]J. Laskar, *ibid.*

Applying now the KAM theorem to the Hamiltonian (including the attraction of Jupiter), due to the small mass ratio m_J/m_S the motion still will take place on a torus, except the frequencies ω_r and ω_φ are resonant, i.e. the frequency ratio is rational. Since the period of Jupiter is 11.86223 years, we have $\omega_J/\omega_\varphi = 1/11.86223$, and from Eq. (16.100) we find

$$\omega_r/\omega_\varphi \cong 1.0843.$$

This ratio seems to be sufficiently irrational for both the torus and earth's orbit to remain stable under the influence of Jupiter.

Other manifestations of the KAM theorem

In fact, the destruction of the tori at resonant frequencies can be observed in our solar system (cf. [Berry]). An example is the gaps in the **asteroid belt** between Mars and Jupiter. The gaps occur at rational frequency ratios ω/ω_J (cf. Fig. 16.4). Gaps also occur in the **rings of Saturn**[23] (cf. Fig. 16.5) – which consist of small-sized lumps of matter that move essentially independently, without collisions, on circular orbits. Here, the main attractive force is caused by Saturn, and the main perturbation is caused by Mimas, one of Saturn's inner moons. However, some of the remaining moons cannot be quite neglected. Therefore it is not easy to relate a gap to some rational frequency ratio. Anyhow, also in this case the idea is that, in ancient times, lumps that originally moved on stable orbits were eventually knocked out of their orbit when

---

[23] 1610 Galileo detected the strange shape of Saturn, when scanning the sky with his telescope: '*Saturn is not a single star, but a compound of three that almost touch each other ...*', (Opere X, p. 410, Florence 1900).

1658 Huygens recognized that this shape is due to a surrounding ring (Oeuvres, XV, p. 57, The Hague 1925).

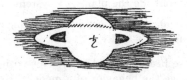

On the basis of Newton's theory of gravitation Laplace treated first the figure and the motion of Saturn's rings in his "Celestial Mechanics".

it became unstable. This caused the gaps that now appear as thin black rings in high-resolution photographs.

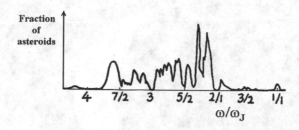

Figure 16.4: The gaps in the asteroid belt (from [Berry]).

Of course, these two examples are only a hint at why the orbits in a system of more than three bodies can be stable. Still, despite enormous recent effort using complex models and numerical techniques the question of the stability of our solar system has not been answered definitively - and presumably it can't. (See, for example, various contributions in "Chaos, resonance and collective dynamical phenomena in the solar system", ed. S. Ferraz-Mello, Kluwer, Dordrecht 1992 as well as the already cited article by J. Laskar). An appropriate conclusion seems to be the one drawn by D. Shiga[24]: *The studies suggest that the solar system's planets will continue to orbit the Sun stably for at least 40 million years. But after that, they show there is a small but not insignificant chance that things could go terribly awry.*

These applications are the fruits of the rather formal Hamilton-Jacobi theory. Even if the mathematical effort is considerable, the insight gained into general features of dynamical systems and their solutions is extremely valuable. Numerical solutions can be achieved given enough time and resources. But general features and connections can only be understood through analytical methods. Here, the Hamilton-Jacobi theory, and in particular action-angle variables, play an important role. A detailed presentation of nonintegrable, chaotic dynamical systems is beyond the scope of the book. We refer the reader to the special literature already cited (see the Bibliography).

---

[24]David Shiga, *Solar system could go haywire before the Sun dies*, NewScientist.com News Service, 2008.

Figure 16.5: The rings of Saturn (Courtesy of NASA/JPL/Caltech).

Concerning the initially raised question on the stability of earth's orbit Laplace's result seems still to apply. So we can hope to enjoy further favorable living conditions.

Problems and examples

1. Determine $W(\rho, \varphi, J_\rho, J_\varphi)$ and the canonical transformation for the isotropic oscillator (see Eq. (16.1.2)). Compare your answers to those obtained for $\omega_x = \omega_y = \omega$ in Cartesian coordinates (cf. Eq. (16.1.2)).

2. Determine the angle variables θ_i for the motion of a particle in the $1/r$ potential, and sketch the motion in these variables.

3. Derive Eq. (16.33). Convince yourself that the orbits in the system rotating with the angular velocity $\omega = -\omega_Z/2$ are really ellipses. What happens when $d = r_0$?

4. The Hamiltonian of a plane pendulum (cf. Section 3.3.2) is

$$H = \frac{p_\varphi^2}{2ml^2} - mgl\cos\varphi.$$

Calculate the action variable $J = \frac{1}{2\pi}\oint p_\varphi d\varphi$ and discuss its dependence on the energy $E\,(\geq mgl)$; i.e. when

- $E = -mgl + \varepsilon,\ \varepsilon \geq 0$;
- $E = mgl - \varepsilon$;
- $E \gg mgl$.

and determine $J(E)$ qualitatively ([Berry]).

5. Apply canonical perturbation theory to the modified Hénon-Heiles system, Eqs. (16.98) and (16.97). Calculate x and y as functions of the action-angle variables to first order, and determine the time dependence of the coordinates.

6. The observation of **adiabatic invariance** is due to P. Ehrenfest (1914). For a single degree of freedom, the statement of adiabatic invariance is as follows. Let the Hamiltonian H depend also on a variable $\alpha\,(t)$ varying only slowly in time (in terms of typical time scales in the system; see below):

$$H\,(q,p;\alpha) = \frac{p^2}{2m} + V\,(q;\alpha).$$

We make a canonical transformation to action-angle variables, so that

$$H = H\,(J;\alpha),$$

with

$$J = \frac{1}{2\pi}\oint p\,dq.$$

For the variation of α to be slow, the condition

$$\frac{1}{\alpha}\frac{d\alpha}{dt} \ll \omega = \frac{\partial H}{\partial J}$$

must hold. Then adiabatic invariance says that J does not change on average over a period $T = 2\pi/\omega$; i.e. that

$$\overline{\frac{dJ}{dt}} := \frac{1}{T} \int_0^T \frac{dJ}{dt} dt = 0.$$

(For a proof see [Landau/Lifschitz] or [Born].) *For a slow change of α, the action does not change* (adiabatic invariance of the action).

In the case of the one-dimensional oscillator with time dependent frequency $\omega(t)$ the adiabatic invariance of

$$J = E/\omega$$

implies that the energy changes in proportion to the frequency $\omega(t)$. We apply this to the plane pendulum for small displacements ($H = \omega J$, $\omega^2 = g/l$) with weakly time-dependent length $l(t)$ of the pendulum (as in raising or lowering as swinging chandelier), and ask:

How does maximal angle of displacement φ_{max} depend on the slowly varying length l?

Answer: Since $E = \frac{1}{2}\omega^2 l^2 \varphi_{max}^2$, we have $J = \frac{\sqrt{g}}{2} l^{3/2} \varphi_{max}^2$, and the maximal angle depends on the length l according to

$$\varphi_{max}(t) \propto [l(t)]^{-3/4}.$$

Also for f degrees of freedom, with Hamiltonian

$$H = H(\mathbf{J}; \alpha),$$

we can observe adiabatic invariance, but there is a difference. Adiabatic invariance only holds for the subset of action variables J_k whose frequencies $\omega_k = \partial H/\partial J_k$ are incommensurate:

$$\sum m_k \theta_k \neq 0 \quad \text{for any set of integers } \{m_k\}.$$

More details can be found in [Born] and [Arnold].

In retrospect

In *Newton's lex secunda,* the interrelation between motion and force in different mechanical systems is ingeniously formulated as a universal law, yielding the basic *equations of motion* in mechanics. With the computing power we have at hand today, one could, without further considerations, solve the equations of motions even for very complex dynamical systems in which analytical methods are hopelessly ineffective for producing the solution. Over the years, the dominance of analytical methods in establishing general properties of the equations of motion and solving them has waned, in favor of computer calculations backing up the argumentation. Nevertheless, the insights obtained from analytical methods provide a more comprehensive view and deeper understanding of dynamical systems than numerical calculations can hope to allow.

The development of *Lagrangian mechanics* was a big extension to Newtonian mechanics. In the Lagrangian scheme, Newton's equations result from an *extremum principle for the Lagrangian function.* The Lagrangian approach allows a more systematic study of mechanical systems; in particular, it makes the connection between symmetries and conserved quantities of a system obvious (especially through Noether's theorem).

Finally, the most clear and elegant formulation of classical mechanics was established by *Hamilton* and *Jacobi.* Newton's equations of motion turn into the *canonical equations* derived from the *Hamiltonian.* The canonical equations determine the flow in *phase space,* a basic concept in understanding the dynamics of a mechanical system. The generators of symmetry operations in a system are conserved quantities of the canonical equations. If there are sufficiently many conserved quantities in the form of integrals of the motion, the system is integrable, and its dynamics – initially represented as a flow in phase space – is equivalent to the flow on an hypertorus. If there are too few integrals of the motion (due to lack of symmetry in phase space), in nonlinear systems, chaotic behavior may appear.

As a consequence of the discovery of chaotic behavior in the 1960s, believe in the deterministic behavior of the mechanical world governed by Newton's equations – as professed by Laplace – was finally smashed. One had to learn that apparently unstable, analytically nonsolvable systems are potential candidates for this newly observed feature of nature – chaotic, effectively unpredictable behavior. After the theory of relativity and quantum mechanics, this represented the third 'scientific revolution*' of the 20th century. From then on, we had to keep in mind the fact that predictions may depend very sensitively on the initial values, and these can never be completely stated exactly. So we know now that a certain amount of chance exists also in classical mechanics. This adds a new flavor to classical mechanics and makes the field vital and challenging again.

---

*The term 'scientific revolution' was created by T.S. Kuhn with his book "The Structure of Scientific Revolutions" (Univ. of Chicago Press 1962). But none of this three 'revolutions' caused a 'change in paradigm' in the sense of Kuhn; i.e. the traditional paradigms were *not replaced* by the new ones; all three revolutions turned out to be extensions of the physical world view.

Appendix A

Coordinate systems and some vector analysis

A.1 The Euclidean space $\mathbb{E}^3$

The Euclidean space $\mathbb{E}^3$ is obtained from the vector space[1] $\mathbb{R}^3$ by defining the *scalar product* $(\mathbf{r}_1, \mathbf{r}_2) = \mathbf{r}_1 \mathbf{r}_2$ of two elements $\mathbf{r}_1, \mathbf{r}_2$. The scalar product is a mapping of two vectors $\mathbf{r}_1$ and $\mathbf{r}_2$ onto the real numbers. It is distributive, commutative, homogeneous $((\alpha \mathbf{r}_1, \mathbf{r}_2) = \alpha (\mathbf{r}_1, \mathbf{r}_2))$, and positive definite $((\mathbf{r}, \mathbf{r}) \geq 0)$. This scalar product allows to define the *length of a vector* $\mathbf{r}$,

$$r = |\mathbf{r}| = \sqrt{(\mathbf{r}, \mathbf{r})} = \sqrt{\mathbf{r}^2} \tag{A.1}$$

and *the angle ϕ between vectors* $\mathbf{r}_1$ and $\mathbf{r}_2$

$$\cos \phi = \frac{1}{r_1 r_2} \sqrt{\mathbf{r}_1 \mathbf{r}_2}. \tag{A.2}$$

---

[1] In a vector space

- there are two operations which always produce again an element of $\mathbb{R}^3$: addition of elements: $\mathbf{r}_1 + \mathbf{r}_2 = \mathbf{r}_3$ and multiplication by a real number α: $\alpha \mathbf{r}_1 = \mathbf{r}_2$.

- addition is associative and commutative; for every element $\mathbf{r}$ there exists a unique vector $\mathbf{0}$ (the origin) such that $\mathbf{r} + \mathbf{0} = \mathbf{r}$ and a unique inverse element, called $-\mathbf{r}$, such that $\mathbf{r} + (-\mathbf{r}) = \mathbf{0}$.

- multiplication by a real number α is associative and distributive and for all elements $\mathbf{r}$: $1\mathbf{r} = \mathbf{r}$.

Two vectors $\mathbf{r}_1$ and $\mathbf{r}_2$ are called *orthogonal* if $\phi = \pi/2$; i.e., if

$$\mathbf{r}_1\mathbf{r}_2 = 0.$$

In $\mathbb{E}^3$, there are at most three linearly independent[2] vectors that are orthogonal to each other. One may therefore select three orthogonal vectors $\mathbf{e}_i$, $i = 1, 2, 3$, of unit length

$$\mathbf{e}_i\mathbf{e}_j = \delta_{ij}, \tag{A.3}$$

where

$$\delta_{ij} = \begin{cases} 1 \\ 0 \end{cases} \quad \text{for} \quad \begin{array}{l} i = j \\ i \neq j \end{array} \tag{A.4}$$

is the Kronecker δ. The *orthonormal* vectors $\mathbf{e}_i$ form a *basis*[3] in $\mathbb{E}^3$; i.e., any vector $\mathbf{a}$ in $\mathbb{E}^3$ can be represented as a linear combination of the three basis vectors:

$$\mathbf{a} = a_1\mathbf{e}_1 + a_2\mathbf{e}_2 + a_3\mathbf{e}_3.$$

The *position* or *radius vector* $\mathbf{r}$ is given the linear combination

$$\mathbf{r} = x\mathbf{e}_1 + y\mathbf{e}_2 + z\mathbf{e}_3 = x_1\mathbf{e}_1 + x_2\mathbf{e}_2 + x_3\mathbf{e}_3.$$

Its change in time can be described by the *velocity vector* $\dot{\mathbf{r}}$ and by the *acceleration vector* $\ddot{\mathbf{r}}$. Expressed in terms of the basis vectors, the vectors $\dot{\mathbf{r}}$ and $\ddot{\mathbf{r}}$ are written as

$$\dot{\mathbf{r}} = \dot{x}_1\mathbf{e}_1 + \dot{x}_2\mathbf{e}_2 + \dot{x}_3\mathbf{e}_3 \quad \text{and} \quad \ddot{\mathbf{r}} = \ddot{x}_1\mathbf{e}_1 + \ddot{x}_2\mathbf{e}_2 + \ddot{x}_3\mathbf{e}_3, \tag{A.5}$$

respectively. (Here, the basis vectors are independent of time.)

---

[2]Three vectors $\mathbf{r}_1, \mathbf{r}_2, \mathbf{r}_3 \neq \mathbf{0}$ are said to be linearly independent, if

$$\alpha\mathbf{r}_1 + \beta\mathbf{r}_2 + \gamma\mathbf{r}_3 = \mathbf{0}$$

implies

$$\alpha = \beta = \gamma = 0.$$

[3]A maximal set of linearly independent vectors is called a basis.

A.2 Cartesian coordinates

In a Cartesian coordinate system, the three standard axes can be thought of as being generated by the orthogonal unit vectors $\mathbf{e}_1, \mathbf{e}_2, \mathbf{e}_3$. The three-tuples (a_1, a_2, a_3) and (x, y, z) (or (x_1, x_2, x_3)) of the *Cartesian components* are obtained from $\mathbf{a}$ and $\mathbf{r}$ via the relations

$$a_i = \mathbf{a}\mathbf{e}_i \quad \text{and} \quad x_i = \mathbf{r}\mathbf{e}_i,$$

respectively. These three-tuples are often used as representation of vectors in a Cartesian coordinate system[4]; *viz.*:

$$\mathbf{a} = (a_1, a_2, a_3), \quad \mathbf{r} = (x, y, z).$$

Moreover we have

$$\dot{\mathbf{r}} = (\dot{x}_1, \dot{x}_2, \dot{x}_3) \quad \text{and} \quad \ddot{\mathbf{r}} = (\ddot{x}_1, \ddot{x}_2, \ddot{x}_3).$$

The basis vectors have the representation

$$\mathbf{e}_1 = (1, 0, 0), \quad \mathbf{e}_2 = (0, 1, 0), \quad \mathbf{e}_3 = (0, 0, 1).$$

The *scalar product* of the two vectors $\mathbf{a}$ and $\mathbf{b}$ is given by

$$\mathbf{a}\mathbf{b} = a_1 b_1 + a_2 b_2 + a_3 b_3 = a_i b_i, \tag{A.6}$$

where the *summation convention* (summation over repeated indices) is used in the second term on the right[5]. The angle θ between $\mathbf{a}$ and $\mathbf{b}$ is given by (cf. (A.2))

$$\cos\theta = \frac{\mathbf{a}\mathbf{b}}{|\mathbf{a}|\,|\mathbf{b}|},$$

---

[4]There are also other coordinates, e.g. curvilinear coordinates (see below), but these representations of vectors do not fulfill the axioms of a linear space.

[5]Writing the scalar product written in terms of the *metric tensor* g_{ik},

$$\mathbf{a}\mathbf{b} = a_i b_k g_{ik},$$

and comparing with

$$a_i b_i = a_i b_k \delta_{ik},$$

one obtains the *Euclidean metric tensor*

$$g_{ik} = \delta_{ik}.$$

Using the metric tensor, one can impose Euclidean geometry on a vector space (with elements (x_1, x_2, x_3)).

where $|\mathbf{a}|$ is the length of $\mathbf{a}$ (cf. (A.1)):

$$|\mathbf{a}| = \sqrt{a_1^2 + a_2^2 + a_3^2}.$$

The *cross* or *vector product* of two vectors $\mathbf{a}$ and $\mathbf{b}$

$$\mathbf{c} = \mathbf{a} \times \mathbf{b}$$

can be expressed with the help of the ε-tensor[6] in terms of the components

$$c_i = (\mathbf{a} \times \mathbf{b})_i = \varepsilon_{ijk} a_j b_k. \tag{A.7}$$

The *direct product* $\mathbf{a} \otimes \mathbf{b}$ of two vectors $\mathbf{a}$ and $\mathbf{b}$ is given in terms of the coordinate representation $\mathbf{a} = (a_1, a_2, a_2)$ and $\mathbf{b} = (b_1, b_2, b_3)$ by the matrix

$$(\mathbf{a} \otimes \mathbf{b})_{ik} = a_i b_k. \tag{A.8}$$

Operators of vector analysis

The vector differential operator ∇, the *del* or *nabla operator*, is defined by

$$\nabla = \sum_i \mathbf{e}_i \frac{\partial}{\partial x_i} =: \sum_i \mathbf{e}_i \partial_i.$$

Under rotations of the coordinate system, $\nabla = (\partial_1, \partial_2, \partial_3)$ is transformed like a vector (cf. (B.2) and (B.14)).

Let $\psi(\mathbf{r})$ be a scalar function (i.e. it is invariant under rotations of the coordinate system) and $\mathbf{a}(\mathbf{r})$ a vector function (i.e. it is transformed

---

[6]The definition of ε_{ijk}, $i, j, k = 1, 2, 3$, is:

- $\varepsilon_{123} = 1$;

- For any even permutation of the indices $\{123\}$, ε_{ijk} equals unity: e.g., $\varepsilon_{231} = \varepsilon_{312} = 1$;

- Interchange of two indices changes the sign: e.g., $\varepsilon_{ijk} = -\varepsilon_{jik}$. Consequently, $\varepsilon_{ijk} = 0$ if any two indices are equal.

A useful relation is

$$\varepsilon_{ijk} \varepsilon_{imn} = \delta_{jm}\delta_{kn} - \delta_{jn}\delta_{km}.$$

like a tensor of rank one under rotations of the coordinate system; see (B.14)). We define the following operators:

$$\text{gradient of a scalar field:} \quad \text{grad } \psi \;=\; \boldsymbol{\nabla}\psi$$
$$\text{divergence of a vector field:} \quad \text{div } \mathbf{a} \;=\; \boldsymbol{\nabla}\mathbf{a}$$
$$\text{curl of a vector field:} \quad \text{curl } \mathbf{a} \;=\; \boldsymbol{\nabla} \times \mathbf{a}$$

The following relations are valid: $\boldsymbol{\nabla} \times \boldsymbol{\nabla}\psi = 0$, $\boldsymbol{\nabla}\left(\boldsymbol{\nabla} \times \mathbf{a}\right) = 0$.

A.3 Orthogonal, curvilinear coordinates

A.3.1 General relations

Given a Cartesian coordinate system (x, y, z) in a Euclidean space $\mathbb{E}^3$, a *curvilinear coordinate system* (u, v, w) is introduced via differentiable, invertible functions

$$
\begin{aligned}
u &= u(x, y, z) \\
v &= v(x, y, z) \\
w &= w(x, y, z),
\end{aligned}
\tag{A.9}
$$

whose inverse is, respectively,

$$
\begin{aligned}
x &= x(u, v, w) \\
y &= y(u, v, w) \\
z &= z(u, v, w).
\end{aligned}
\tag{A.10}
$$

We now find a standard basis in the new coordinate system. A *coordinate surface* is defined by setting any of u, v, or w equal to a constant; i.e., any of $u = const$, $v = const$, or $w = const$ is a coordinate surface. The intersection of any two coordinate surfaces, where the third coordinate is allowed to run free, defines a curve: the intersection of the surfaces $v = const$ and $w = const$ defines the u curve; and similarly for the v and w curves. A set of coordinates is called *orthogonal curvilinear* if at each point (u, v, w), the curves formed from the intersections of coordinate surfaces in pairs are mutually perpendicular. A set of *basis vectors* at the point (u, v, w) consists of the normalized tangent vectors to these curves. More precisely, consider $\mathbf{r} = \mathbf{r}(u, v, w)$. The first derivative $\partial \mathbf{r}/\partial u$ of the radius $\mathbf{r} = x\mathbf{e}_x + y\mathbf{e}_y + z\mathbf{e}_z$ along the u curve, together with the first derivatives of the radius $\mathbf{r}$ along the v and w curves, respectively, gives three orthogonal vectors. Normalizing these, we obtain

$$\mathbf{e}_u = \frac{\partial \mathbf{r}}{\partial u} \bigg/ \left|\frac{\partial \mathbf{r}}{\partial u}\right|, \qquad \mathbf{e}_v = \frac{\partial \mathbf{r}}{\partial v} \bigg/ \left|\frac{\partial \mathbf{r}}{\partial v}\right|, \qquad \mathbf{e}_w = \frac{\partial \mathbf{r}}{\partial w} \bigg/ \left|\frac{\partial \mathbf{r}}{\partial w}\right| \quad (A.11)$$

as a basis at the point (u, v, w). The orientation of these basis vectors generally depends on the point (u, v, w). Defining

$$h_u = \left|\frac{\partial \mathbf{r}}{\partial u}\right|, \quad h_v = \left|\frac{\partial \mathbf{r}}{\partial v}\right|, \quad h_w = \left|\frac{\partial \mathbf{r}}{\partial w}\right|,$$

and writing (A.11)

$$\frac{\partial \mathbf{r}}{\partial u} = h_u \mathbf{e}_u \quad \text{etc.},$$

we have

$$\begin{aligned} d\mathbf{r}(u, v, w) &= \frac{\partial \mathbf{r}}{\partial u} du + \frac{\partial \mathbf{r}}{\partial v} dv + \frac{\partial \mathbf{r}}{\partial w} dw \\ &= h_u du\, \mathbf{e}_u + h_v dv\, \mathbf{e}_v + h_w dw\, \mathbf{e}_w. \end{aligned}$$

The *line element ds* in curvilinear coordinates is then given by

$$\begin{aligned} (ds)^2 &= (d\mathbf{r})^2 = (dx)^2 + (dy)^2 + (dz)^2 \\ &= h_u^2 (du)^2 + h_v^2 (dv)^2 + h_w^2 (dw)^2 \\ &= (ds_u)^2 + (ds_v)^2 + (ds_w)^2, \end{aligned} \quad (A.12)$$

where $ds_u = h_u du$ denotes the line element along the u curve, and so on.

The $\mathbf{r}$-dependence of the basis in curvilinear coordinates has to be taken into account particularly when transforming an expression such as $\boldsymbol{\nabla}\psi$, with ψ a scalar function. The decomposition of $\boldsymbol{\nabla}\psi$ in the curvilinear basis is

$$\boldsymbol{\nabla}\psi = (\boldsymbol{\nabla}\psi)_u\, \mathbf{e}_u + (\boldsymbol{\nabla}\psi)_v\, \mathbf{e}_v + (\boldsymbol{\nabla}\psi)_w\, \mathbf{e}_w.$$

Inserting into the definition of the gradient $\boldsymbol{\nabla}\psi$,

$$d\psi = \boldsymbol{\nabla}\psi\, d\mathbf{r},$$

the total differential of $\psi(u, v, w)$,

$$d\psi = \frac{\partial \psi}{\partial u} du + \frac{\partial \psi}{\partial v} dv + \frac{\partial \psi}{\partial w} dw,$$

we obtain, with $d\mathbf{r} = (ds_u, ds_v, ds_w)$,

$$(\boldsymbol{\nabla}\psi)_u \, h_u = \frac{\partial \psi}{\partial u}, \quad \dots \; .$$

Therefore,

$$(\boldsymbol{\nabla}\psi)_u = \frac{1}{h_u}\frac{\partial \psi}{\partial u}, \quad (\boldsymbol{\nabla}\psi)_v = \frac{1}{h_v}\frac{\partial \psi}{\partial v}, \quad (\boldsymbol{\nabla}\psi)_w = \frac{1}{h_w}\frac{\partial \psi}{\partial w}. \qquad (A.13)$$

The expressions for the divergence of a vector function $\mathbf{A}$, $\boldsymbol{\nabla}\mathbf{A}$; curl of $\mathbf{A}$, $\boldsymbol{\nabla} \times \mathbf{A}$; and the Laplacian of a scalar function ψ, $\boldsymbol{\nabla}^2\psi$; have a much more complicated form in curvilinear coordinates than in Cartesians. The reader is referred to, for example, [Spiegel].

Upon transformation of coordinates (A.9), the *element of volume* $dV = dxdydz$ changes to

$$dV = |J| \, dudvdw,$$

where J is the *Jacobian* of the transformation,

$$J = \det(\partial x_i / \partial u_j) = h_u h_v h_w, \qquad (A.14)$$

with $(x_1, x_2, x_3) = (x, y, z)$ and $(u_1, u_2, u_3) = (u, v, w)$.

For some transformations from Cartesian to curvilinear coordinates, the mapping may be not one-to-one (e.g. the origin is not mapped uniquely in the transformations below). Points were this occurs sometimes have to be treated separately.

A.3.2　Spherical coordinates

The transformation from Cartesian coordinates (x, y, z) to spherical coordinates (r, ϑ, φ) is given by

$$\begin{aligned} x &= r\sin\vartheta\cos\varphi \\ y &= r\sin\vartheta\sin\varphi \\ z &= r\cos\vartheta, \end{aligned} \qquad (A.15)$$

with the inverse transformation

$$
\begin{aligned}
r &= (x^2 + y^2 + z^2)^{1/2} & 0 \le r \le \infty \\
\vartheta &= \arccos\left(z \big/ (x^2 + y^2 + z^2)^{1/2}\right) & 0 \le \vartheta \le \pi \quad \text{(A.16)} \\
\varphi &= \arctan(y/x) & 0 \le \varphi < 2\pi.
\end{aligned}
$$

The surfaces $r = const$ are spherical shells centered at the origin; the surfaces $\vartheta = const$ are cones with apex at the origin; and the surfaces $\varphi = const$ are planes perpendicular to the (x, y) plane containing the z axis. At every point (r, ϑ, φ), these three coordinate surfaces are orthogonal to each other.

The Cartesian components of the orthogonal unit vectors (A.11) in (r, ϑ, φ) are

$$
\begin{aligned}
\mathbf{e}_r &= (\sin\vartheta\cos\varphi, \sin\vartheta\sin\varphi, \cos\vartheta) \\
\mathbf{e}_\vartheta &= (\cos\vartheta\cos\varphi, \cos\vartheta\sin\varphi, -\sin\vartheta) \quad \text{(A.17)} \\
\mathbf{e}_\varphi &= (-\sin\varphi, \cos\varphi, 0).
\end{aligned}
$$

It follows that $\mathbf{e}_r \times \mathbf{e}_\vartheta = \mathbf{e}_\varphi$ (and cyclic permutations). The *Jacobian* (A.14) is easily shown to be

$$
J = r^2 \sin\vartheta,
$$

and hence, the *element of volume* is given by

$$
dV = r^2 dr \sin\vartheta d\vartheta d\varphi. \quad \text{(A.18)}
$$

The *gradient* of a scalar field ψ is

$$
\boldsymbol{\nabla}\psi = \mathbf{e}_r \frac{\partial\psi}{\partial r} + \mathbf{e}_\vartheta \frac{1}{r}\frac{\partial\psi}{\partial\vartheta} + \mathbf{e}_\varphi \frac{1}{r\sin\vartheta}\frac{\partial\psi}{\partial\varphi}. \quad \text{(A.19)}
$$

The *time derivative* of the radius vector $\mathbf{r}$ $(= r\mathbf{e}_r)$ is $\dot{\mathbf{r}} = \dot{r}\mathbf{e}_r + r\dot{\mathbf{e}}_r$; since $\dot{\mathbf{e}}_r = \dot{\vartheta}\mathbf{e}_\vartheta + \dot{\varphi}\sin\vartheta\mathbf{e}_\varphi$, it follows that

$$
\dot{\mathbf{r}} = \dot{r}\mathbf{e}_r + r\dot{\vartheta}\mathbf{e}_\vartheta + r\sin\vartheta\dot{\varphi}\mathbf{e}_\varphi. \quad \text{(A.20)}
$$

For the *second time derivative* one finds

$$
\begin{aligned}
\ddot{\mathbf{r}} = {} & \left(\ddot{r} - r\dot{\vartheta}^2 - r\dot{\varphi}^2\sin^2\vartheta\right)\mathbf{e}_r + \left(2\dot{r}\dot{\vartheta} + r\ddot{\vartheta} - r\dot{\varphi}^2\sin\vartheta\cos\vartheta\right)\mathbf{e}_\vartheta \\
& + \left(2\dot{r}\dot{\varphi}\sin\vartheta + 2r\dot{\vartheta}\dot{\varphi}\cos\vartheta + r\ddot{\varphi}\sin\vartheta\right)\mathbf{e}_\varphi. \quad \text{(A.21)}
\end{aligned}
$$

A.3.3   Cylindrical coordinates

The transformation from Cartesian coordinates (x, y, z) to cylindrical coordinates (ρ, φ, z) is given by

$$
\begin{aligned}
x &= \rho \cos \varphi \\
y &= \rho \sin \varphi \\
z &= z,
\end{aligned}
\tag{A.22}
$$

and the inverse transformation is

$$
\begin{aligned}
\rho &= (x^2 + y^2)^{1/2} & 0 \le \rho \le \infty \\
\varphi &= \arctan(y/x) & 0 \le \varphi < 2\pi \\
z &= z & -\infty \le z \le \infty.
\end{aligned}
\tag{A.23}
$$

Cylindrical coordinates are simply plane polar coordinates together with a z-axis. The surfaces $\rho = const$ are (infinitely long) cylinders whose axis is the z axis; the surfaces $\varphi = const$ are planes perpendicular to the (x, y)-plane containing the z-axis; and the surfaces $z = const$ are planes parallel to the (x, y)-plane.

The Cartesian components of the orthogonal unit vectors (A.11) in (ρ, φ, z)-coordinates can be shown to be

$$
\begin{aligned}
\mathbf{e}_\rho &= (\cos \varphi, \sin \varphi, 0) \\
\mathbf{e}_\varphi &= (-\sin \varphi, \cos \varphi, 0) \\
\mathbf{e}_z &= (0, 0, 1),
\end{aligned}
\tag{A.24}
$$

with orthogonality relations: $\mathbf{e}_\rho \times \mathbf{e}_\varphi = \mathbf{e}_z$ and cyclic permutations. The *Jacobian* is given by

$$
J = \rho,
$$

and the *element of volume* is therefore

$$
dV = \rho d\rho dz d\varphi.
\tag{A.25}
$$

The expression for $\boldsymbol{\nabla} \psi$ in cylindrical coordinates reads

$$
\boldsymbol{\nabla} \psi = \mathbf{e}_\rho \frac{\partial \psi}{\partial \rho} + \mathbf{e}_\varphi \frac{1}{\rho} \frac{\partial \psi}{\partial \varphi} + \mathbf{e}_z \frac{\partial \psi}{\partial z}.
\tag{A.26}
$$

The *time derivative* of the radius vector $\mathbf{r} = \rho \mathbf{e}_\rho + z \mathbf{e}_z$, using $\dot{\mathbf{e}}_\rho = \dot{\varphi} \mathbf{e}_\varphi$, is obtained as

$$
\dot{\mathbf{r}} = \dot{\rho} \mathbf{e}_\rho + \rho \dot{\varphi} \mathbf{e}_\varphi + \dot{z} \mathbf{e}_z,
\tag{A.27}
$$

and the *second time derivative* is then

$$\ddot{\mathbf{r}} = (\ddot{\rho} - \rho\dot{\varphi}^2)\mathbf{e}_\rho + (2\dot{\rho}\dot{\varphi} + \rho\ddot{\varphi})\mathbf{e}_\varphi + \ddot{z}\mathbf{e}_z. \tag{A.28}$$

Appendix B

Rotations and tensors

B.1 Rotations

Imagine a point P in space. Suppose P is viewed in two systems K and K' with orthogonal basis vectors $\{e_i\}$ and $\{e'_i\}$, respectively, whose origins coincide. The radius vectors of the point P in K and K' are $\mathbf{r} = x_j e_j$ and $\mathbf{r}' = x_j e_j$, respectively, where (x_1, x_2, x_3) and (x'_1, x'_2, x'_3) are the respective Cartesian components. Since the origins coincide, we have $\mathbf{r} = \mathbf{r}'$, or

$$x'_j e'_j = x_i e_i. \tag{B.1}$$

The components (x'_1, x'_2, x'_3) are related to the components (x_1, x_2, x_3) by a linear transformation; *viz.*:

$$x'_i = D_{ij} x_j. \tag{B.2}$$

Inserting (B.2) yields $D_{ji} x_i e'_j = x_i e_i$, and since P is arbitrary, the set of coordinates x_i is arbitrary; consequently,

$$D_{ji} e'_j = e_i. \tag{B.3}$$

Multiplying both sides by e'_k and using $e'_j e'_k = \delta_{jk}$ gives

$$D_{ki} = e'_k e_i; \tag{B.4}$$

the D_{ki} are the projections of the basis vectors of K' onto the basis vectors of K. Taking the D_{ji} as elements of a matrix D we have

$$\mathsf{D} = \begin{pmatrix} e'_1 e_1 & e'_1 e_2 & e'_1 e_3 \\ e'_2 e_1 & e'_2 e_2 & e'_2 e_3 \\ e'_3 e_1 & e'_3 e_2 & e'_3 e_3 \end{pmatrix}. \tag{B.5}$$

The nine matrix elements are not independent. Since the e'_i are unit vectors, e'_1 is already fixed by specifying two of the *direction cosines* $e'_1 e_i$ with respect to the basis $\{e_i\}$. Now, because $e'_2 e'_1 = 0$ the second unit vector e'_2 is already fixed by giving just one of its direction cosines. Therefore the matrix elements $e'_k e_i$ can be parametrized by only three independent parameters. If the basis vectors in the two systems are, for instance, both right handed, D is a rotation matrix (see below) and the three free parameters correspond to the three degrees of freedom of rotation.

On the other hand, without referring to a basis, one may directly interpret (B.2) as a transformation between the Cartesian coordinates (x_1, x_2, x_3) and (x'_1, x'_2, x'_3) of a point P considered in two coordinate systems K and K'. If the origins of the two systems coincide, for the distance between P and the origin we have the condition

$$x'_j x'_j = x_i x_i.$$

(In the above view, this relation is a consequence of (B.1).) Inserting (B.2), we obtain

$$D_{ji} D_{jk} x_i x_k = x_i x_i = x_i x_k \delta_{ik}.$$

Because the point P and hence the x_i are arbitrary, this is tantamount to

$$D_{ji} D_{jk} = \delta_{ik}, \tag{B.6}$$

which, in terms of the matrix D, reads

$$D^T D = 1, \tag{B.7}$$

where D^T is the *transpose* of the matrix D and 1 is the *unit*, or *identity*, matrix[1]. Since we have also $DD^T = 1$, Eq. (B.7) implies that the transpose of the matrix D is also the *inverse* matrix of D,

$$D^{-1} = D^T. \tag{B.8}$$

The *rotation matrix* D is an *orthogonal matrix*. If the elements of D^{-1} are denoted by D^{-1}_{ij}, Eq. (B.8) reads

$$D^{-1}_{ij} = D_{ji}. \tag{B.9}$$

---

[1]Of course, the properties of D follow also from the representation (B.5).

Multiplying (B.3) by D_{ik}^{-1} (and summing over the repeated indices), one finds $e_k' = e_i D_{ik}^{-1}$, and finally, from (B.9), one obtains an equation for rotation of the basis vectors:

$$e_k' = D_{ki} e_i. \tag{B.10}$$

The basis *vectors* are transformed like the *components* of the radius vector $\mathbf{r}$ (cf. Eq. (B.2)).

Since $\det \mathsf{D}^T = \det \mathsf{D}$, it follows from (B.7) that

$$\det \mathsf{D}^T \det \mathsf{D} = (\det \mathsf{D})^2 = \det \mathbf{1} = 1;$$

therefore, the determinant of an orthogonal matrix is given by

$$\det \mathsf{D} = \pm 1. \tag{B.11}$$

Transformations with $\det \mathsf{D} = 1$ are continuously linked to the identity transformation $\mathbf{1}$ (cf. (8.13) and (B.13)); these are called *proper rotations*. On the other hand, every rotation with $\det \mathsf{D} = -1$ contains a *reflection* P; i.e. one or all three axes (basis vectors) are reflected. Here are two examples:

$$\mathsf{P} = \begin{pmatrix} -1 & 0 & 0 \\ 0 & -1 & 0 \\ 0 & 0 & -1 \end{pmatrix}, \qquad \mathsf{P} = \begin{pmatrix} -1 & 0 & 0 \\ 0 & 1 & 0 \\ 0 & 0 & 1 \end{pmatrix}. \tag{B.12}$$

Then the transformation D may be decomposed into a reflection and proper rotation; *viz.*: $\mathsf{D} = \mathsf{P}\mathsf{D}'$, where D' is a proper rotation.

Due to the symmetry $i \leftrightarrow k$, (B.6) yields only six independent conditions for the nine elements of the matrix D, so that only three free parameters remain, which are just the quantities needed to specify an arbitrary rotation. Two useful sets of the three parameters are: a) the direction of rotation $\mathbf{n}$ (specified as a unit vector along the rotation axis), together with the rotation angle α; and b) the Euler angles φ, ϑ, and ψ. In the former case, the rotation matrix is denoted by $\mathsf{D}_\mathbf{n}(\alpha)$, and in the latter case, by $\mathsf{D}(\psi, \vartheta, \varphi)$ (cf. Section 8.3.2). $\mathsf{D}_\mathbf{n}(\alpha)$ has the following properties:

i) For $\alpha = 0$, the matrix $\mathsf{D}_\mathbf{n}$ reduces to the unit matrix, i.e. to the identity transformation:

$$\mathsf{D}_\mathbf{n}(\alpha = 0) = \mathbf{1} \qquad \text{or} \qquad D_{ij}(0) = \delta_{ij}.$$

ii) Rotation by $-\alpha$ is the inverse of the rotation by α:

$$D_n(-\alpha) = D_n^{-1}(\alpha).$$

Rotation about a coordinate axis is a simple special case. It is easy to see that a rotation by the angle α about the z-axis, for instance, is represented by the matrix

$$D_z(\alpha) = \begin{pmatrix} \cos\alpha & \sin\alpha & 0 \\ -\sin\alpha & \cos\alpha & 0 \\ 0 & 0 & 1 \end{pmatrix}. \qquad (B.13)$$

The eigenvalues of this matrix are 1, $e^{i\alpha}$, and $e^{-i\alpha}$ (i.e., the eigenvalues of D_z lie on the unit circle in the complex plane).

B.2   Tensors

Quantities $T_{j_1 \dots j_n}$ that transform under rotation like an n-fold product of components $x_{i_1} x_{i_2} \dots x_{i_n}$ (cf. B.2),

$$T'_{i_1 \dots i_n} = D_{i_1 j_1} \dots D_{i_n j_n} T_{j_1 \dots j_n} , \qquad (B.14)$$

are called[2] *tensors of rank n*. A vector is a tensor of rank one. Examples of tensors of rank two are the invariant tensor δ_{ij} and the direct product (A.8). The total antisymmetric tensor ε_{ijk} (cf. Appendix A) is an invariant tensor of rank three.

The rank of a tensor is reduced by the operation of *contraction*, i.e. the summation over repeated indices. For example, the rank four tensor T_{ijkl}, can be contracted to a rank two tensor:

$$T_{ijkk} = T_{ij},$$

as can be seen from (B.14) using Eq. (B.8). A common example of contraction is the trace of a matrix M (here M is a tensor of rank two):

$$\text{Tr } M = M_{ii}.$$

Taking the trace of

$$M'_{ij} = D_{il} D_{jk} M_{lk},$$

---

[2]Here, we define a tensor as an object that has certain behavior under rotations.

Eq. (B.6) implies

$$\begin{aligned}
\text{Tr } \mathsf{M}' &= M_{ii}' = D_{il}D_{ik}M_{lk} = \delta_{lk}M_{lk} \\
&= M_{ll} = \text{Tr } \mathsf{M}.
\end{aligned} \tag{B.15}$$

The trace is therefore an invariant property of a tensor.
Further operations with tensors:

- Addition of tensors (of equal rank)

- Product of two tensors:

    outer product: e.g. $A_{ij}B_{klm} = C_{ijklm}$;

    inner product: e.g. $A_{ij}B_{ilm} = C_{jlm}$.

Appendix C

Green's functions

C.1 The Dirac δ-function

C.1.1 Distributions

The Dirac δ-function (P.A.M. Dirac, 1902-1984, English physcist) is properly called a *distribution*. Distributions are linear, continuous functionals in the space of *test functions*.

Applying a distribution D to a test function $f(x)$ yields a (real) number r: $D(f) = r \in \mathbb{R}$. A distribution can be defined in terms of the limit of a series of functions $\{g_n(x)\}$:

$$D_n(f) = \int_{-\infty}^{\infty} dx g_n(x) f(x) = r_n,$$

which for $n \to \infty$ gives the distribution

$$D(f) = \lim_{n \to \infty} \int_{-\infty}^{\infty} dx g_n(x) f(x) = r, \qquad r = \lim_{n \to \infty} r_n.$$

Often the limiting function $g_\infty(x) = \lim_{n \to \infty} g_n(x)$ is itself called a distribution (cf. the representations of the δ-function below).

493

C.1.2 The δ-function

Representations of the δ-function

There are many representations of the δ-function, we just mention a few of them.

- In terms of the *theta function* $\Theta(t)$ (also called *Heaviside (step) function*), defined as

$$\Theta(t) = \begin{cases} 1 & t > 0 \\ 0 & t < 0 \end{cases}, \qquad (C.1)$$

the δ-function is the limit of the rectangular shaped functions:

$$\delta(t) = \lim_{\tau \to 0} \frac{1}{\tau} \left[\Theta(t) - \Theta(t - \tau) \right].$$

- As the infinite limit of a sequence of other functions, e.g. (cf. Fig. C.1):

$$\delta(t) = \lim_{n \to \infty} \sqrt{\frac{n}{\pi}} e^{-nt^2}.$$

- As the Fourier transform (cf. Section C.2) of the function $f(t) = 1$:

$$\delta(t) = \frac{1}{2\pi} \int_{-\infty}^{\infty} e^{i\omega t} d\omega. \qquad (C.2)$$

Properties of the δ-function

Some useful properties of the δ-function are listed below.

1.) $\delta(t - t_0) = 0$ for $t \neq t_0$ and $\delta(0) = \infty$, where $\int_a^b \delta(t - t_0)dt = 1$ if $a < t_0 < b$;

2.) $f(t)\delta(t - t_0) = f(t_0)\delta(t - t_0)$, in particular:

$t\delta(t) = 0$ (more exactly: $\int_a^b f(t)\delta(t-t_0)dt = f(t_0)$, where $a < t_0 < b$ and f continuous in t_0);

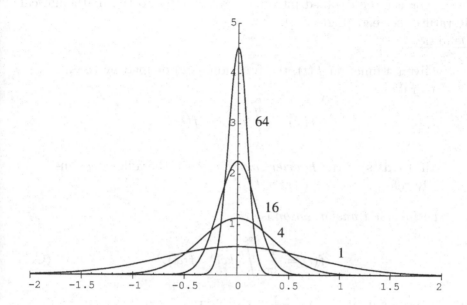

Figure C.1: The sequence of functions $\sqrt{\frac{n}{\pi}}e^{-nt^2}$ for $n = 1, 4, 16, 64$

3.) $\delta(-t) = \delta(t)$;

4.) $\delta(\alpha t) = \dfrac{1}{|\alpha|}\delta(t)$;

5.) Relation to the theta function: $\dfrac{d}{dt}\Theta(t) = \delta(t)$;

6.) Derivative δ' of the δ-function:

$$\int\limits_a^b f(t)\delta'(t - t_0)dt = \int\limits_a^b f(t)\frac{d}{dt}\delta(t - t_0)dt = -\left.\frac{d}{dt}f(t)\right|_{t=t_0}, \quad a < t_0 < b$$

(by partial integration).

C.2 Fourier transforms

We present a very brief outline of the essential properties of Fourier transforms (J.-B. Fourier, 1768-1830, French mathematician and physi-

cist). For a more detailed introduction, we refer to the mathematical
literature (see e.g. [Lighthill]).

Definition:

Given a function $f(t)$, the function $\tilde{f}(\omega)$ defined by the integral

$$\tilde{f}(\omega) = \int\limits_{-\infty}^{\infty} dt e^{-i\omega t} f(t) \qquad\qquad (C.3)$$

(if it exists) is the *Fourier transform*[1] (in the following simply referred to as FT) of $f(t)$.

The *inverse transformation* is

$$f(t) = \frac{1}{2\pi} \int\limits_{-\infty}^{\infty} d\omega e^{i\omega t} \tilde{f}(\omega). \qquad\qquad (C.4)$$

We note the following properties of the FT.

- The FT is linear:

$$\tilde{f}(\omega) + \tilde{g}(\omega) = \int\limits_{-\infty}^{\infty} dt e^{-i\omega t} \left(f(t) + g(t) \right);$$

- The FT of the derivative of $f(t)$ is

$$\int\limits_{-\infty}^{\infty} dt e^{-i\omega t} \frac{d}{dt} f(t) = i\omega \tilde{f}(\omega)$$

('Proof': integrate by parts); the generalization to higher derivatives is obvious;

- *Convolution theorem:*

$$\tilde{f}(\omega)\tilde{g}(\omega) = \int\limits_{-\infty}^{\infty} dt e^{-i\omega t} \int\limits_{-\infty}^{\infty} d\tau f(\tau) g(t - \tau).$$

---

[1]In physics the actually necessary distinction between the *different* functions f and $\tilde{f}$ is usually not made; we also use this convention. One has to keep in mind that $f(t)$ and its Fourier transform $f(\omega)$ are really different functions.

Two frequently occuring examples

- The FT of a Gaussian $f(t) = e^{-t^2/\sigma}$ is again a Gaussian $\tilde{f}(\omega) = \sqrt{\sigma\pi}e^{-\frac{1}{4}\sigma\omega^2}$.

- The FT of an exponential $f(t) = e^{-\alpha|t|}$ is a Lorentzian $\tilde{f}(\omega) = \dfrac{2\alpha}{\omega^2 + \alpha^2}$.

Distributions can also be Fourier transformed (see e.g. [Lighthill]); for example, the δ-function is the FT of the constant function $\tilde{f}(\omega) = 1$ (cf. relation (C.2)).

C.3 Linear differential equations and Green's functions

C.3.1 The Green's function

In a *linear* differential equation

$$L_t x(t) = K(t), \tag{C.5}$$

the differential operator L_t has the general form

$$L_t = a_n(t)\frac{d^n}{dt^n} + \ldots + a_1(t)\frac{d}{dt} + a_0(t). \tag{C.6}$$

A solution $G(t, t')$ of the equation

$$L_t G(t, t') = \delta(t - t') \tag{C.7}$$

is called *Green's function* (G. Green, 1793-1841, British mathematical physicist).

Once a Green's function $G(t, t')$ has been found, a solution $x(t)$ of equation (C.5) can be given for any inhomogeneous term $K(t)$, namely

$$x(t) = \int\limits_{-\infty}^{\infty} G(t, t')K(t')dt'. \tag{C.8}$$

Proof: $L_t x(t) = \int\limits_{-\infty}^{\infty} L_t G(t, t')K(t')dt' = \int\limits_{-\infty}^{\infty} \delta(t - t')K(t')dt' = K(t)$.

The integral operator $\int dt' G(t, t')\ldots$ is the inverse of the differential-operator L_t.

If L_t is invariant with respect to translations $t \to t + a$, i.e.

$$L_{t+a} = L_t,$$

and if, for example, all coefficients $a_i(t)$ are independent of t, $a_i(t) = a_i$, then the Green's function depends only on the difference $t - t'$

$$G(t, t') = G(t - t').$$

C.3.2   The equation of the damped oscillator

We want to calculate a Green's function for the differential operator of the damped oscillator,

$$L_t = \frac{d^2}{dt^2} + 2\gamma \frac{d}{dt} + \omega_0^2. \tag{C.9}$$

That is, we look for a solution of

$$\left(\frac{d^2}{dt^2} + 2\gamma \frac{d}{dt} + \omega_0^2 \right) G(t - t') = \delta(t - t').$$

The Fourier transform of this equation is an algebraic equation for the Fourier transform $G(\omega)$ of the Green's function:

$$\left(-\omega^2 + 2i\gamma\omega + \omega_0^2 \right) G(\omega) = 1.$$

It can be readily solved to give

$$G(\omega) \;=\; \frac{1}{-\omega^2 + 2i\gamma\omega + \omega_0^2} \tag{C.10}$$

$$=\; -\frac{1}{\left(\omega - i\gamma - \sqrt{\omega_0^2 - \gamma^2} \right) \left(\omega - i\gamma + \sqrt{\omega_0^2 - \gamma^2} \right)}.$$

The calculation of the Fourier transform,

$$G(t) = \frac{1}{2\pi} \int\limits_{-\infty}^{\infty} d\omega\, e^{i\omega t} G(\omega) \tag{C.11}$$

requires some mathematical tricks. First, it is advantageous to extend ω to complex plane, $\omega = \operatorname{Re}\omega + i\operatorname{Im}\omega = |\omega| e^{i\varphi}$, and to continue the integrand to the complex ω-plane. Suppose $t > 0$. To the integral (C.11), we add another integral over a semicircular path beginning just above the positive real axis and traversing the upper complex plane anticlockwise around to the negative real axis (cf. Fig. C.2), which, together with the original integral, forms a closed path of integration. Letting the radius of the semicircle go to infinity, we have

$$\int_{-\infty}^{\infty} d\omega\, e^{i\omega t} G(\omega) + \int_{SC} d\omega\, e^{i\omega t} G(\omega) = \oint d\omega\, e^{i\omega t} G(\omega),$$

where the contour integral over SC is along the semicircle with radius $|\omega| \to \infty$ from $\omega = \infty$ to $\omega = -\infty$. The semicircle is chosen such that there is no contribution to the total contour integral. Observing that the term $\exp(-t\,\mathrm{Im}\,\omega)$ in the exponential $e^{i\omega t}$ dominates the integrand as $|\omega| \to \infty$, one recognizes that the contour for $t > 0$ has to be chosen in the upper half-plane ($\mathrm{Im}\,\omega > 0$), and for $t < 0$ in the lower half-plane ($\mathrm{Im}\,\omega < 0$), in order that the contour integral over SC does not contribute, i.e. in order that $\lim_{|\omega| \to \infty} \exp(-t\,\mathrm{Im}\,\omega) \to 0$. In both cases, we have now

$$\int_{-\infty}^{\infty} d\omega\, e^{i\omega t} G(\omega) = \oint d\omega\, e^{i\omega t} G(\omega). \tag{C.12}$$

The closed contour integral can be evaluated using the *theorem of residues*,

$$\oint_{C} f(z)\, dz = 2\pi i \sum_{k} \mathrm{Res}\, f(z)|_{z=z_k}\,,$$

where the closed contour C borders a region in the complex z-plane. The sum of the residues is the sum of the contributions of the isolated singular points z_k of the function $f(z)$ within this region. The poles of the integrand in (C.12) are the two simple poles of the function $G(\omega)$, Eq. (C.10); these lie in the upper half of the ω-plane (cf. Fig. C.2). At the pole $\omega = i\gamma - \bar{\omega}$, $\bar{\omega} = \sqrt{\omega_0^2 - \gamma^2}$, the residue is

$$\mathrm{Res} \left[\frac{e^{i\omega t}}{\omega - i\gamma - \bar{\omega}} \right]_{\omega = i\gamma - \bar{\omega}} = -\frac{e^{-\gamma t} e^{-i\bar{\omega} t}}{2\bar{\omega}};$$

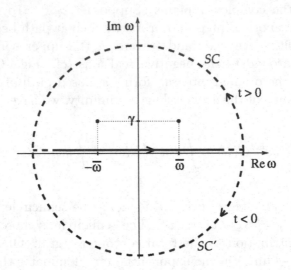

Figure C.2: The poles (•) and the contour of integration consisting of the straight line $-\infty \leq \mathrm{Re}\,\omega \leq \infty$ (shown displaced from its actual position on the real ω-axis) and one of the semicircles SC or SC' which actually are situated at infinity (cf. text).

at the pole $\omega = i\gamma + \overline{\omega}$ the residue is

$$\mathrm{Res}\left[\frac{e^{i\omega t}}{\omega - i\gamma + \overline{\omega}}\right]_{\omega = i\gamma + \overline{\omega}} = \frac{e^{-\gamma t}e^{i\overline{\omega}t}}{2\overline{\omega}}.$$

Hence, the result of the integration is the *causal* Green's function

$$G(t) = \begin{cases} \dfrac{1}{\overline{\omega}}e^{-\gamma t}\sin\overline{\omega}t & t > 0 \\ 0 & t \leq 0 \end{cases}, \qquad \overline{\omega} = \sqrt{\omega_0^2 - \gamma^2}. \qquad (C.13)$$

This Green's function $G(t)$ is recognized as 'causal' by inspecting the solution of $\left(\dfrac{d^2}{dt^2} + 2\gamma\dfrac{d}{dt} + \omega_0^2\right)x(t) = K(t)$

$$x(t) = \int_{-\infty}^{\infty} G(t - t')K(t')dt'$$

$$= \frac{1}{\bar{\omega}} \int_{-\infty}^{t} e^{-\gamma(t-t')} \sin\left[\bar{\omega}\left(t-t'\right)\right] K(t')dt'. \qquad (C.14)$$

The solution $x(t)$ is only influenced by forces $K(t')$ which act at times $t' \leq t$: the effect follows the cause.

The Green's function obtained here is *only one of many* possible Green's functions, because we may add any solution $F(\omega)$ of the homogeneous equation

$$\begin{aligned} 0 &= \left(\omega^2 - 2i\gamma\omega - \omega_0^2\right) F(\omega) \\ &= \left(\omega - i\gamma + \bar{\omega}\right)\left(\omega - i\gamma - \bar{\omega}\right) F(\omega) \end{aligned}$$

to (C.10) and obtain a new Green's function. Using to the second property of the δ-function (Appendix C.1), the general solution $F(\omega)$,

$$F(\omega) = c_1 \delta\left(\omega - i\gamma + \bar{\omega}\right) + c_2 \delta\left(\omega - i\gamma - \bar{\omega}\right), \qquad (C.15)$$

is fixed up to two constants. The Fourier transform of $F(\omega)$ is the general solution of the homogeneous, damped oscillator equation

$$\begin{aligned} F(t) &= \frac{1}{2\pi} \int_{-\infty}^{\infty} d\omega e^{i\omega t} F(\omega) \\ &= \frac{1}{2\pi} \int_{-\infty}^{\infty} d\omega e^{i\omega t} \left[c_1 \delta\left(\omega - i\gamma + \bar{\omega}\right) + c_2 \delta\left(\omega - i\gamma - \bar{\omega}\right)\right] \\ &= c_1 e^{-\gamma t} e^{-i\bar{\omega} t} + c_2 e^{-\gamma t} e^{i\bar{\omega} t}. \qquad (C.16) \end{aligned}$$

Bibliography

Textbooks in Mechanics

"Classical" books

BOLTZMANN L., Vorlesungen über die Prinzipe der Mechanik, J.A. Barth, Leipzig 1922; I found no English translation

EULER L., Mechanica sive motus scientia analytice exposita (Mechanics or the science of motion analytically presented), 2 Volumes, St. Petersburg 1736; it seems that there is until now no printed English translation, only one by Ian Bruce in the internet (German translation by Wolfers, Greifswald 1848)

EULER L., Theoria motus corporum solidorum seu rigidorum (Theory of the motion of rigid bodies), Rostock 1765; also here there is an English translation by Ian Bruce available in the internet (German translation by Wolfers, Greifswald 1853)

GALILEI, G., Dialogue Concerning the Two Chief World Systems: Ptolemaic and Copernican, translated by Stillman Drake, Modern Library, New York 2001

HERTZ H., The principles of mechanics: presented in a new form, translation by D.E. Jones and J.T. Walley, Dover Publications, New York 1956

LAGRANGE J.L., Mécanique analytique, Paris 1811; Engl. translation "Analytical Mechanics" by A. Boissonnade and V.N. Vagliente, Kluwer Academic Publishers, Dordrecht 1997

LAPLACE P.S., Mécanique Céleste, Paris 1798-1825; Engl. transl. of the first 4 volumes (of five) by N. Bowditch, Celestial Mechanics, Boston 1829-1839

NEWTON I., Philosophiae naturalis principia mathematica, London 1687; English translation by A. Motte: I. Newton, The Principia (1727), Prometheus Books, Amherst, New York, 1995.

POISSON S.D., Traité de Mécanique, Bachelier, Paris 1811; English translation: A treatise of mechanics, Univ. of Michigan Library, 2009

WHITTAKER E.T., A Treatise on the Analytical Dynamics of Particles and Rigid Bodies, Cambridge University Press, Cambridge 1904

Modern books

ARNOLD V.I., Mathematical methods of classical mechanics, [Graduate texts in mathematics. 60], Springer, New York 1978

BECKER R.A., Introduction to Theoretical Mechanics, McGraw-Hill, New York 1954

CORBEN H.C. and STEHLE P., Classical Mechanics, R.E. Krieger, Huntington 1974

FRENCH A.P., Newtonian Mechanics, W.W. Norton & Company, New York 1971

GOLDSTEIN H., Classical mechanics, Addison-Wesley, Reading, Mass. 1980

KIBBLE T.W.B., Classical Mechanics, Longman, London 1985

KNUDSEN J.M. and HJORTH P.G., Elements of Newtonian Mechanics Including Nonlinear Dynamics, Springer, Berlin 2000

LANDAU L.D. and LIFSHITZ E.M., Course of theoretical physics, Vol. 1: Mechanics, transl. from the Russian by J.B. Sykes and J.S. Bell, Pergamon P., Oxford 1976

MATZNER R.A. and SHEPLEY L.C., Classical Mechanics, Prentice Hall, New Jersey 1991

McCAULEY J.L., Classical Mechanics - Transformations, Flows, Integrable and Chaotic Dynamics, Cambridge University Press, Cambridge 1997

SALETAN E.J. and CROMER A.H., Theoretical Mechanics, J. Wiley, New York 1971

SCHECK F., Mechanics · From Newton's Laws to Deterministic Chaos, Springer, New York 1999

SOMMERFELD A., Vorlesungen über theoretische Physik, Band I: Mechanik, Harri Deutsch, Frankfurt 1977

Engl. transl.: Lectures on theoretical physics, vol 1: Mechanics, Academic Press, New York 1964

SPIEGEL M.R., Theory and Problems of Theoretical Mechanics, Schaum's Outline Series, McGraw-Hill, New York 1967

SPOSITO G., An introduction to classical mechanics, Wiley, New York 1976

THIIRRING W., A course in mathematical physics 1 and 2, Classical dynamical systems and classical field theory, Springer, Vienna 1992

Special topics

BALESCU R., Equilibrium and nonequilibrium statistical mechanics, J. Wiley, New York 1975

BERRY M.V., Regular and irregular motion, in: Topics in Nonlinear Dynamics, AIP Conference Proceedings **46**, ed. S. Jorna, American Institute of Physics, New York 1978

BORN M., The Mechanics of the Atom, G. Bell & Sons, London 1927; a valuable introduction to canonical transformations and Hamilton-Jacobi theory

CRABTREE H., Spinning Tops and Gyroscopic Motion. Longman, London 1909

DUMAS H.S., The KAM Story, World Scientific, Singapore 2014

KLEIN F. and SOMMERFELD A., Über die Theorie des Kreisels, reprint, Teubner, Stuttgart 1965

LICHTENBERG A.J. and LIEBERMAN M.A., Regular and Stochastic Motion, Springer, New York 1983

NAYFEH A.H.,Introduction to perturbation techniques, Wiley, New York 1981

NAYFEH A.H.,Perturbation methods, Wiley, New York 2000

POINCARÉ H., Les méthodes nouvelles de la mécanique céleste, 3 vols., Paris 1892-1899; Engl. transl. by D. L. Goroff, New Methods of Celestial Mechanics, Hist. Mod. Phys. Astron., Vol. 13, American Institute of Physics 1993

POINCARÉ H., Science and Method, reprint of the 1914 edition, Thoemmes Press, Bristol 1996

TABOR M., Chaos and integrability in nonlinear dynamics, J. Wiley, New York 1989

Books on historical aspects, Biographies

CAJORI F., A History of Mathematics (1st ed. 1893), Chelsea Publ. Company, New York 1991

CASPAR M., Kepler, Dover Publications, New York 1993

CHANDRASEKHAR S., Newton's Principia for the Common Reader, Clarendon Press, Oxford 1996

DRAKE Stillman, Galileo at work, The University of Chicago Press, Chicago 1978

DUHEM P., L'Évolution de la Mécanique, Joanin Paris 1903; English translation: The Evolution of Mechanics, Sijthoff & Noordhoff 1980

DUGAS R., A History of Mechanics, Dover Publications, New York 1988

FAUVEL J., FLOOD R., SHORTLAND M., and WILSON R. (Eds.), Let Newton be!, Oxford University Press, Oxford 1988

FÖLSING A., Galileo Galilei - Prozeß ohne Ende, Piper, München, 1989

GLEICK J., Chaos - Making A New Science, Viking, New York 1987

GUICCIARDINI N., Reading the Principia. The Debate on Newton's Mathematical Methods for Natural Philosophy from 1687 to 1736, Cambridge University Press, Cambrigde 2003

JAMMER M., Concepts of Space, Harvard University Press, Cambridge (Mass.) 1970

JAMMER M., Concepts of mass, in classical and modern physics, Harvard University Press, Cambridge, 1961

JAMMER M., Concepts of Force, Harper and Brothers, New York 1962

KOESTLER A., The sleepwalkers. a history of man's changing vision of the universe, Penguin, Harmondsworth 1986

MACH E., The Science of Mechanics, a critical account of its development, Open Court Publications, La Salle (Ill.) 1902

MANUEL F. E., A Portrait of Isaac Newton, Harvard Univ. Press 1968

NORTH J., The Fontana history of astronomy and cosmology, Fontana Press, London 1994

PEMBERTON H., A view of Sir Isaac Newton's philosophy, reprinted from the 1728 edition, Johnson Reprint Corporation, New York 1972

PETERSON I., Newton's Clock - Chaos in the Solar System, Freeman, New York 1993

THIELE R. Leonhard Euler, Teubner, Leipzig 1982 (German); I found no printed biographies in English

508   *BIBLIOGRAPHY*

VOLTAIRE, The Elements of Sir Isaac Newton's Philosophy, Transl.
    by J. Hanna, London 1738

WESTFALL R. S., Never at Rest: A Biography of Isaac Newton,
    Cambridge University Press, Cambridge 1983.

Mathematical literature

BOURNE D.E. and. KENDALL P.C., Vector Analysis, Oldbourne
    Book Co, London 1967

FORSYTH, A.R., Calculus of Variations, Dover, New York 1960

LA SALLE J. and LEFSCHETZ S., Stability by Liapunov's Direct
    Method with Applications, Academic Press, New York 1961

LICHNEROWICZ A., Éléments de calcul tensoriel, Colin, Paris 1950

LIGHTHILL M.J., Introduction to Fourier Analysis and Generalised
    Functions, Cambridge University Press, Cambridge 1958

SAGAN, H., Introduction to the Calculus of Variations, Dover, New
    York 1992

SPIEGEL M.R., Vector Analysis and an introduction to Tensor Analy-
    sis, Schaum's Outline Series, McGraw-Hill, New York 1959

Index

acceleration, 21
    Coriolis, 217
action integral, 249, 381
action variable, 441
adiabatic invariance, 473
angle variable, 441
angular momentum, 31
    of center of mass, 145, 269
    of relative motion, 145, 269
angular velocity, 104
anomaly
    eccentric, 157
    mean, 157
aphelion, 110
apocenter, 108
approximation
    harmonic, 41, 341
apsidal angle, 108
apsidal distance, 108
apsidal vector, 108
apsis, 108
Archimedes, 289
areal law, 104, 156
Aristotle, 18
asteroid belt, 470
attractor, 54
    strange, 96
autonomous system
    of equations, 24

Bernoulli, Daniel, 231

Bernoulli, Jacob, 231
Bernoulli, Johann, 231
Boltzmann, Ludwig, 6
boost, 206
brachistochrone problem, 231
Brahe, Tycho, 153

Cantor set, 90
Cavendish experiment, 168
center of mass
    coordinate, 143, 165
center of mass motion, 144
    law of the, 266
chaotic behavior, 89
collision
    elastic, 178
    inelastic, 179
compound pendulum, 333
configuration space, 23
conservation
    of angular momentum, 32,
        103
    of angular momentum of cen-
        ter of mass, 145
    of angular momentum of rel-
        ative motion, 145, 270
    of center of mass momentum,
        146, 178, 267
    of conjugate momentum, 251
    of energy, 29, 102

of energy of center of mass, 145, 273
of energy of relative motion, 145, 273
of total angular momentum, 142, 270
of total energy, 142, 272
conserved quantity, 26, 263, 378
constraint, 223
  force of, 224
  holonomic, 224
  non-holonomic, 224, 255
  rheonomic, 224
  scleronomic, 224
continuum mechanics, 363
coordinate
  cyclic, 251, 368
  elliptic, 412
  generalized, 276
  normal, 345
  relative, 143, 265
Copernicus, Nicolas, 151
Coriolis, Gaspard Gustave de, 217
Coulomb interaction, 186
couple of forces, 292
cycloid, 237
cyclotron frequency, 129

D'Alembert, Jean-Baptiste le Rond., 223
degree of freedom, 22, 261
derivative
  functional, 250
Descartes, René, 4
dimension
  fractal, 90
dipole moment, 165
Dirac delta function, 493
displacement
  virtual, 290
displacement field, 361
double pendulum, 333, 338
dynamical quantity, 377
dynamics, 12

eccentricity, 113
eigenfrequency, 345
eigenmode, 345
ellipsoid of inertia, 306
energy
  rotational, 299
  translational, 298
equilibrium
  conditions, 291
  displacement from, 287
  position, 41, 287, 342
Euclid (of Alexandria), 19
Euler's angles, 203
Euler's equations of motion, 310
Euler, Leonhard, 20
Euler-Lagrange equation
  see also Lagrange's equation, 235
Euler-Poisson equations, 321

fixed point, 58, 81
flow in phase space, 392
force, 21
  central, 32, 102, 171, 409
  centrifugal, 105, 217
  centripetal, 111, 229
  conservative, 28
  Coriolis, 217
  field, 22
  frictional, 49
  generalized, 342
  harmonic, 37
  interaction, 140

forces
    parallelogram law of, 17
    superposition principle of, 17
Foucault gyrocompass, 332
Foucault pendulum, 259
Foucault, (Jean Bernard) Léon,
    221
fractal structure, 90
frame
    accelerated, 210, 256
    body-fixed, 300
    center of mass, 144, 180, 265
    inertial, 198, 298
    laboratory, 180
    rotating, 257
frame of reference, 11, 179, 196
free fall, 43
frequency
    oscillation, 40
    renormalized, 66
    resonant, 463
friction
    Newtonian, 49
    static, 49
    Stokesian, 49
functional, 233

Galilean group
    proper, 206
Galilean transformation, 206
    special, 206
Galilei, Galileo, 8
gauge function, 281
generating function
    see also generator, 383
generator
    of an infinitesimal canonical
        transformation, 388
geodesic, 241

gravitation
    universal law of, 148
gravitational constant, 148
gravitational potential, 148
Green's function, 51, 497
gyroscope, 300

Hamilton' characteristic function,
    403
Hamilton's equations, 367
Hamilton's principle, 249
Hamilton, William Rowan, 249
Hamilton-Jacobi differential equa-
    tion, 404
Hamiltonian, 367
hard sphere, 171
Hausdorff dimension, 90
Heaviside step function
    see also theta function, 494
Hénon-Heiles system, 79
    modified, 464
Hero of Alexandria, 290
herpolhode, 307
Hertz, Heinrich, 6
homogeneity
    of space, 140, 205
    of time, 206
Huygens, Christiaan, 17

impact parameter, 174
inertia, 197
inertia tensor, 301
infinitesimal transformation
    of coordinates, 279
infinitesimal translation
    in time, 278
integrability, 398
integral
    complete, 404

first, 26
isolating, 89
nonisolating, 89
of the motion, 26
interaction region, 176
intersection points
sequence of, 74
invariable plane, 307
involution, 399
isotropy
of space, 140, 205

Jacobi's identity, 378
Jacobi, Carl Gustav Jacob, 277
Jacobian determinant, 387
Jacobian integral, 277, 368, 425
Jordanus de Nemore, 290

Kepler's equation, 157
Kepler's law
first, 155
second, 156
third, 156, 451
Kepler, Johannes, 151
kinematics, 12
kinetic energy, 28
of center of mass, 272
of relative motion, 273
Kolmogorov-Arnold-Moser theorem, 463

Lagrange multipliers, 241
Lagrange's equation, 247
Lagrange, Joseph-Louis, 20
Lagrangian, 275
Lagrangian density, 361
Lagrangian points, 427
Laplace, Pierre. Simon, 2
Larmor frequency, 218, 373

least action principle
see Hamilton's principle, 381
Legendre transformation, 366
Leibniz, Gottfried Wilhelm, 13
Leonardo da Vinci, 290
lever, 292
law of the, 293
Liapunov exponent, 92
libration, 441
linear chain, 357
Liouville's theorem
on integrability, 398
on the phase space volume,
395
Liouville, Joseph, 395
Lissajous's figure, 72

Mach, Ernst, 1
machine
simple, 292
map
area preserving, 94
iterative, 94
mass
active gravitational, 149
gravitational, 16, 149
inertial, 16, 149
passive gravitational, 149
reduced, 144
mass density, 160, 304
mass matrix, 343
Maupertuis, Pierre-Louis Moreau
de, 249
moment
of force, 31, 269
moment of inertia, 304
principal, 305
momentum, 16, 21
canonical, 247

center of mass, 178, 266
conjugate, 247
generalized, 251
of relative motion, 178
motion
bounded, 39
constant of the, 25, 263
externally bounded, 106
internally bounded, 106
orbital stability of, 159
periodic, 40
stable, 55
totally bounded, 106
unbounded, 41
multiply periodic functions, 453

Newton' pail experiment, 11
Newton's equation of motion, 21
Newton's law
first, 17
second, 17
third, 17, 140, 261
Newton, Isaac, 3
Noether's theorem, 280
Noether, Emmy, 280
nutation, 330

orbit, 23
closed, 108
equation of, 106, 172
rosette-shaped, 120
oscillation
damped, 51
overdamped, 52
oscillator
damped, 50
driven, 52
Duffing's, 64, 459

harmonic, one-dimensional,
46
kicked, 51
three-dimensional, isotropic,
124
two-dimensional, 71, 406
two-dimensional, isotropic, 76,
125

pendulum
plane, 43
spherical, 228, 254
pericenter, 108
perihelion, 110
perihelion precession rates of the
planets, 159
period, 40
phase space, 24, 392
Poincaré, Henri, 1
Poincaré map, 74, 91
Poincaré section
see Poincaré surface, 74
Poincaré surface, 74, 375
Poincaré's invariant, 387
Poinsot's construction, 307
Poinsot, Louis, 298
point mass, 22
point of reference, 195
point transformation, 384
Poisson bracket, 378
fundamental, 378
Poisson, Siméon Denis, 298
polhode, 307
potential, 29
effective, 105
interaction, 141
potential energy, 29
precession, 312, 330
principal axes

of the inertia tensor, 305
principle of virtual work, 290
product of inertia, 304
pseudo-force, 210, 217
Ptolemy, Claudius, 19
pulley, 292

quadrupole moment
    of a mass distribution, 165,
        301

reflection, 489
relative motion, 144
resonance catastrophe, 54
response, 51
return point, 74
rigid body, 288
    free, 314
    free symmetric, 311
    Hamiltonian of, 370
rings of Saturn, 470
Roberval, Gilles Personne de, 290
rotation, 202, 441
    infinitesimal, 204
    proper, 489
Runge-Lenz vector, 122, 174, 380,
        396, 451
Rutherford scattering formula, 191

scattering angle, 173
scattering cross section
    differential, 186
    total, 189
secular term, 66
selfsimilarity, 96
semilatus rectum, 112
separation of variables, 406
similarity
    mechanical, 120

stability analysis
    linear, 58, 315
statics, 12
    basic equations of, 291
    of the rigid body, 289
stationary solution, 58
Steiner's theorem, 319
Steiner, Jakob, 319
Stevin, Simon, 290
Stoke's theorem, 29
string
    vibrating, 361
summation convention, 198
symplectic, 393
system
    closed, 264
    completely separable, 439
    conservative, 30, 102, 142,
        264
    dissipative, 96
    integrable, 27, 399
    separable, 406

tensor, 490
    metric, Euclidean, 479
theta function, 494
three body problem
    restricted, 420
three-body problem
    Euler's, 411
    planar, 436
time
    reversal of, 24, 262
top, 300
    motion of, 320
    sleeping, 330
    symmetric, 324
torque, 31
torus, 401, 452

total energy, 29
track
    inclined, 42, 226
trajectory, 24
transformation
    active, 200
    canonical, 381
    infinitesimal, canonical, 388
    passive, 200
    velocity, 206
translation, 201
Trojans, 429
turning point, 39
two-particle interaction, 261

variable
    canonically conjugate, 377
    dynamical, 24
Varignon, Pierre, 20
velocity, 21
    areal, 104
    asymptotic, 174
vis viva, 28

wave equation, 361
work, 28

Printed in the United States
by Booksmasters.

Printed in the United States
By Bookmasters